Nazli Goharian · Nicola Tonellotto · Yulan He ·
Aldo Lipani · Graham McDonald ·
Craig Macdonald · Iadh Ounis
Editors

Advances in Information Retrieval

46th European Conference on Information Retrieval, ECIR 2024
Glasgow, UK, March 24–28, 2024
Proceedings, Part IV

Editors
Nazli Goharian
Georgetown University
Washington, WA, USA

Nicola Tonellotto
University of Pisa
Pisa, Italy

Yulan He
King's College London
London, UK

Aldo Lipani
University College London
London, UK

Graham McDonald
University of Glasgow
Glasgow, UK

Craig Macdonald
University of Glasgow
Glasgow, UK

Iadh Ounis
University of Glasgow
Glasgow, UK

ISSN 0302-9743 ISSN 1611-3349 (electronic)
Lecture Notes in Computer Science
ISBN 978-3-031-56065-1 ISBN 978-3-031-56066-8 (eBook)
https://doi.org/10.1007/978-3-031-56066-8

This Springer imprint is published by the registered company Springer Nature Switzerland AG
The registered company address is: Gewerbestrasse 11, 6330 Cham, Switzerland

Paper in this product is recyclable.

Preface

The 46th European Conference on Information Retrieval (ECIR 2024) was held in Glasgow, Scotland, UK, during March 24–28, 2024, and brought together hundreds of researchers from the UK, Europe and abroad. The conference was organised by the University of Glasgow, in cooperation with the British Computer Society's Information Retrieval Specialist Group (BCS IRSG) and with assistance from the Glasgow Convention Bureau.

These proceedings contain the papers related to the presentations, workshops, tutorials, doctoral consortium and other satellite tracks that took place during the conference. This year's ECIR program boasted a variety of novel work from contributors from all around the world. In addition, we introduced a number of novelties in this year's ECIR. First, ECIR 2024 included for the first time a new "Findings" track, which was offered to some full papers that were deemed to be solid, but which could not make the main conference track. Second, ECIR 2024 ran a new special IR4Good track that presented high-quality, high-impact, original IR-related research on societal issues (such as algorithmic bias and fairness, privacy, and transparency) at the interdisciplinary level (e.g., philosophy, law, sociology, civil society), which go beyond the purely technical perspective. Third, ECIR 2024 featured a new innovation called the "Collab-a-thon", intended to provide an opportunity for participants to foster new collaborations that could lead to exciting new research, and forge lasting relationships with like-minded researchers. Finally, ECIR 2024 introduced a new award to encourage and recognise researchers who have made significant contributions in using theory to develop the information retrieval field. The award was named after Professor Cornelis "Keith" van Rijsbergen (University of Glasgow), a pioneer in modern information retrieval, and a strong advocate of the development of models and theories in information retrieval.

The ECIR 2024 program featured a total of 578 papers from authors in 61 countries in its various tracks. The final program included 57 full papers (23% acceptance rate), an additional 18 finding papers, 36 short papers (24% acceptance rate), 26 IR4Good papers (41%), 18 demonstration papers (56% acceptance rate), 9 reproducibility papers (39% acceptance rate), 8 doctoral consortium papers (57% acceptance rate), and 15 invited CLEF papers. All submissions were peer-reviewed by at least three international Program Committee members to ensure that only submissions of the highest relevance and quality were included in the final ECIR 2024 program. The acceptance decisions were further informed by discussions among the reviewers for each submitted paper, led by a Senior Program Committee member. Each track had a final PC meeting where final recommendations were discussed and made, trying to reach a fair and equal outcome for all submissions.

The accepted papers cover the state-of-the-art in information retrieval and recommender systems: user aspects, system and foundational aspects, artificial intelligence & machine learning, applications, evaluation, new social and technical challenges, and other topics of direct or indirect relevance to search and recommendation. As in previous years, the ECIR 2024 program contained a high proportion of papers with students as first authors, as well as papers from a variety of universities, research institutes, and commercial organisations.

In addition to the papers, the program also included 4 keynotes, 7 tutorials, 10 workshops, a doctoral consortium, an IR4Good event, a Collab-a-thon and an industry day. Keynote talks were given by Charles L. A. Clarke (University of Waterloo), Josiane Mothe (Université de Toulouse), Carlos Castillo (Universitat Pompeu Fabra), and this year's Keith van Rijsbergen Award winner, Maarten de Rijke (University of Amsterdam). The tutorials covered a range of topics including explainable recommender systems, sequential recommendation, social good applications, quantum for IR, generative IR, query performance prediction and PhD advice. The workshops brought together participants to discuss narrative extraction (Text2Story), knowledge-enhanced retrieval (KEIR), online misinformation (ROMCIR), understudied users (IR4U2), graph-based IR (IRonGraphs), open web search (WOWS), technology-assisted review (ALTARS), geographic information extraction (GeoExT), bibliometrics (BIR) and search futures (SearchFutures).

The success of ECIR 2024 would not have been possible without all the help from the strong team of volunteers and reviewers. We wish to thank all the reviewers and meta-reviewers who helped to ensure the high quality of the program. We also wish to thank: the reproducibility track chairs Claudia Hauff and Hamed Zamani, the IR4Good track chairs Ludovico Boratto and Mirko Marras, the demo track chairs Giorgio Maria Di Nunzio and Chiara Renso, the industry day chairs Olivier Jeunen and Isabelle Moulinier, the doctoral consortium chairs Yashar Moshfeghi and Gabriella Pasi, the CLEF Labs chair Jake Lever, the workshop chairs Elisabeth Lex, Maria Maistro and Martin Potthast, the tutorial chairs Mohammad Aliannejadi and Johanne R. Trippas, the Collab-a-thon chair Sean MacAvaney, the best paper awards committee chair Raffaele Perego, the sponsorship chairs Dyaa Albakour and Eugene Kharitonov, the proceeding chairs Debasis Ganguly and Richard McCreadie, and the local organisation chairs Zaiqiao Meng and Hitarth Narvala. We would also like to thank all the student volunteers who worked hard to ensure an excellent and memorable experience for participants and attendees. ECIR 2024 was sponsored by a range of learned societies, research institutes and companies. We thank them all for their support. Finally, we wish to thank all of the authors and contributors to the conference.

March 2024

Nazli Goharian
Nicola Tonellotto
Yulan He
Aldo Lipani
Graham McDonald
Craig Macdonald
Iadh Ounis

Organization

General Chairs

Craig Macdonald	University of Glasgow, UK
Graham McDonald	University of Glasgow, UK
Iadh Ounis	University of Glasgow, UK

Program Chairs – Full Papers

Nazli Goharian	Georgetown University, USA
Nicola Tonellotto	University of Pisa, Italy

Program Chairs – Short Papers

Yulan He	King's College London, UK
Aldo Lipani	University College London, UK

Reproducibility Track Chairs

Claudia Hauff	Spotify & TU Delft, Netherlands
Hamed Zamani	University of Massachusetts Amherst, USA

IR4Good Chairs

Ludovico Boratto	University of Cagliari, Italy
Mirko Marras	University of Cagliari, Italy

Demo Chairs

Giorgio Maria Di Nunzio	Università degli Studi di Padova, Italy
Chiara Renso	ISTI - CNR, Italy

Industry Day Chairs

Olivier Jeunen — ShareChat, UK
Isabelle Moulinier — Thomson Reuters, USA

Doctoral Consortium Chairs

Yashar Moshfeghi — University of Strathclyde, UK
Gabriella Pasi — Università degli Studi di Milano Bicocca, Italy

CLEF Labs Chair

Jake Lever — University of Glasgow, UK

Workshop Chairs

Elisabeth Lex — Graz University of Technology, Austria
Maria Maistro — University of Copenhagen, Denmark
Martin Potthast — Leipzig University, Germany

Tutorial Chairs

Mohammad Aliannejadi — University of Amsterdam, Netherlands
Johanne R. Trippas — RMIT University, Australia

Collab-a-thon Chair

Sean MacAvaney — University of Glasgow, UK

Best Paper Awards Committee Chair

Raffaele Perego — ISTI-CNR, Italy

Sponsorship Chairs

Dyaa Albakour	Signal AI, UK
Eugene Kharitonov	Google, France

Proceeding Chairs

Debasis Ganguly	University of Glasgow, UK
Richard McCreadie	University of Glasgow, UK

Local Organisation Chairs

Zaiqiao Meng	University of Glasgow, UK
Hitarth Narvala	University of Glasgow, UK

Senior Program Committee

Mohammad Aliannejadi	University of Amsterdam, Netherlands
Omar Alonso	Amazon, USA
Giambattista Amati	Fondazione Ugo Bordoni, Italy
Ioannis Arapakis	Telefonica Research, Spain
Jaime Arguello	The University of North Carolina at Chapel Hill, USA
Javed Aslam	Northeastern University, USA
Krisztian Balog	University of Stavanger & Google Research, Norway
Patrice Bellot	Aix-Marseille Université CNRS (LSIS), France
Michael Bendersky	Google, USA
Mohand Boughanem	IRIT University Paul Sabatier Toulouse, France
Jamie Callan	Carnegie Mellon University, USA
Charles Clarke	University of Waterloo, Canada
Fabio Crestani	Università della Svizzera italiana (USI), Switzerland
Bruce Croft	University of Massachusetts Amherst, USA
Maarten de Rijke	University of Amsterdam, Netherlands
Arjen de Vries	Radboud University, Netherlands
Tommaso Di Noia	Politecnico di Bari, Italy
Carsten Eickhoff	University of Tübingen, Germany
Tamer Elsayed	Qatar University, Qatar

Liana Ermakova	HCTI/Université de Bretagne Occidentale, France
Hui Fang	University of Delaware, USA
Nicola Ferro	University of Padova, Italy
Norbert Fuhr	University of Duisburg-Essen, Germany
Debasis Ganguly	University of Glasgow, UK
Lorraine Goeuriot	Université Grenoble Alpes (CNRS), France
Marcos Goncalves	Federal University of Minas Gerais, Brazil
Julio Gonzalo	UNED, Spain
Jiafeng Guo	Institute of Computing Technology, China
Matthias Hagen	Friedrich-Schiller-Universität, Germany
Allan Hanbury	TU Wien, Austria
Donna Harman	NIST, USA
Claudia Hauff	Spotify, Netherlands
Jiyin He	Signal AI, UK
Ben He	University of Chinese Academy of Sciences, China
Dietmar Jannach	University of Klagenfurt, Germany
Adam Jatowt	University of Innsbruck, Austria
Gareth Jones	Dublin City University, Ireland
Joemon Jose	University of Glasgow, UK
Jaap Kamps	University of Amsterdam, Netherlands
Jussi Karlgren	SiloGen, Finland
Udo Kruschwitz	University of Regensburg, Germany
Jochen Leidner	Coburg University of Applied Sciences, Germany
Yiqun Liu	Tsinghua University, China
Sean MacAvaney	University of Glasgow, UK
Craig Macdonald	University of Glasgow, UK
Joao Magalhaes	Universidade NOVA de Lisboa, Portugal
Giorgio Maria Di Nunzio	University of Padua, Italy
Philipp Mayr	GESIS, Germany
Donald Metzler	Google, USA
Alistair Moffat	The University of Melbourne, Australia
Yashar Moshfeghi	University of Strathclyde, UK
Henning Müller	HES-SO, Switzerland
Julián Urbano	Delft University of Technology, Netherlands
Marc Najork	Google, USA
Jian-Yun Nie	Université de Montreal, Canada
Harrie Oosterhuis	Radboud University, Netherlands
Iadh Ounis	University of Glasgow, UK
Javier Parapar	University of A Coruña, Spain
Gabriella Pasi	University of Milano Bicocca, Italy
Raffaele Perego	ISTI-CNR, Italy

Benjamin Piwowarski	CNRS/ISIR/Sorbonne Université, France
Paolo Rosso	Universitat Politècnica de València, Spain
Mark Sanderson	RMIT University, Australia
Philipp Schaer	TH Köln (University of Applied Sciences), Germany
Ralf Schenkel	Trier University, Germany
Christin Seifert	University of Marburg, Germany
Gianmaria Silvello	University of Padua, Italy
Fabrizio Silvestri	University of Rome, Italy
Mark Smucker	University of Waterloo, Canada
Laure Soulier	Sorbonne Université-ISIR, France
Torsten Suel	New York University, USA
Hussein Suleman	University of Cape Town, South Africa
Paul Thomas	Microsoft, USA
Theodora Tsikrika	Information Technologies Institute/CERTH, Greece
Suzan Verberne	LIACS/Leiden University, Netherlands
Marcel Worring	University of Amsterdam, Netherlands
Andrew Yates	University of Amsterdam, Netherlands
Shuo Zhang	Bloomberg, UK
Min Zhang	Tsinghua University, China
Guido Zuccon	The University of Queensland, Australia

Program Committee

Amin Abolghasemi	Leiden University, Netherlands
Sharon Adar	Amazon, USA
Shilpi Agrawal	Linkedin, USA
Mohammad Aliannejadi	University of Amsterdam, Netherlands
Satya Almasian	Heidelberg University, Germany
Giuseppe Amato	ISTI-CNR, Italy
Linda Andersson	Artificial Researcher IT GmbH TU Wien, Austria
Negar Arabzadeh	University of Waterloo, Canada
Marcelo Armentano	ISISTAN (CONICET - UNCPBA), Argentina
Arian Askari	Leiden University, Netherlands
Maurizio Atzori	University of Cagliari, Italy
Sandeep Avula	Amazon, USA
Hosein Azarbonyad	Elsevier, Netherlands
Leif Azzopardi	University of Strathclyde, UK
Andrea Bacciu	Sapienza University of Rome, Italy
Mossaab Bagdouri	Walmart Global Tech, USA

Evgenia Christoforou	CYENS Centre of Excellence, Cyprus
Abu Nowshed Chy	University of Chittagong, Bangladesh
Charles Clarke	University of Waterloo, Canada
Stephane Clinchant	Naver Labs Europe, France
Fabio Crestani	Università della Svizzera Italiana (USI), Switzerland
Shane Culpepper	The University of Queensland, Australia
Hervé Déjean	Naver Labs Europe, France
Célia da Costa Pereira	Université Côte d'Azur, France
Maarten de Rijke	University of Amsterdam, Netherlands
Arjen De Vries	Radboud University, Netherlands
Amra Deli	University of Sarajevo, Bosnia and Herzegovina
Gianluca Demartini	The University of Queensland, Australia
Danilo Dess	Leibniz Institute for the Social Sciences, Germany
Emanuele Di Buccio	University of Padua, Italy
Gaël Dias	Normandie University, France
Vlastislav Dohnal	Masaryk University, Czechia
Gregor Donabauer	University of Regensburg, Germany
Zhicheng Dou	Renmin University of China, China
Carsten Eickhoff	University of Tübingen, Germany
Michael Ekstrand	Drexel University, USA
Dima El Zein	Université Côte d'Azur, France
David Elsweiler	University of Regensburg, Germany
Ralph Ewerth	Leibniz Universität Hannover, Germany
Michael Färber	Karlsruhe Institute of Technology, Germany
Guglielmo Faggioli	University of Padova, Italy
Fabrizio Falchi	ISTI-CNR, Italy
Zhen Fan	Carnegie Mellon University, USA
Anjie Fang	Amazon.com, USA
Hossein Fani	University of Windsor, UK
Henry Field	Endicott College, USA
Yue Feng	UCL, UK
Marcos Fernández Pichel	Universidade de Santiago de Compostela, Spain
Antonio Ferrara	Polytechnic University of Bari, Italy
Komal Florio	Università di Torino - Dipartimento di Informatica, Italy
Thibault Formal	Naver Labs Europe, France
Eduard Fosch Villaronga	Leiden University, Netherlands
Maik Fröbe	Friedrich-Schiller-Universität Jena, Germany
Giacomo Frisoni	University of Bologna, Italy
Xiao Fu	University College London, UK
Norbert Fuhr	University of Duisburg-Essen, Germany

Petra Galuščáková	University of Stavanger, Norway
Debasis Ganguly	University of Glasgow, UK
Eric Gaussier	LIG-UGA, France
Xuri Ge	University of Glasgow, UK
Thomas Gerald	Université Paris Saclay CNRS SATT LISN, France
Kripabandhu Ghosh	ISSER, India
Satanu Ghosh	University of New Hampshire, USA
Daniela Godoy	ISISTAN (CONICET - UNCPBA), Argentina
Carlos-Emiliano González-Gallardo	L3i, France
Michael Granitzer	University of Passau, Germany
Nina Grgic-Hlaca	Max Planck Institute for Software Systems, Germany
Adrien Guille	Université de Lyon, France
Chun Guo	Pandora Media LLC, USA
Shashank Gupta	University of Amsterdam, Netherlands
Matthias Hagen	Friedrich-Schiller-Universität Jena, Germany
Fatima Haouari	Qatar University, Qatar
Maram Hasanain	Qatar University, Qatar
Claudia Hauff	Spotify, Netherlands
Naieme Hazrati	Free University of Bozen-Bolzano, Italy
Daniel Hienert	Leibniz Institute for the Social Sciences, Germany
Frank Hopfgartner	Universität Koblenz, Germany
Gilles Hubert	IRIT, France
Oana Inel	University of Zurich, Switzerland
Bogdan Ionescu	Politehnica University of Bucharest, Romania
Thomas Jaenich	University of Glasgow, UK
Shoaib Jameel	University of Southampton, UK
Faizan Javed	Kaiser Permanente, USA
Olivier Jeunen	ShareChat, UK
Alipio Jorge	University of Porto, Portugal
Toshihiro Kamishima	AIST, Japan
Noriko Kando	National Institute of Informatics, Japan
Sarvnaz Karimi	CSIRO, Australia
Pranav Kasela	University of Milano-Bicocca, Italy
Sumanta Kashyapi	University of New Hampshire, USA
Christin Katharina Kreutz	Cologne University of Applied Sciences, Germany
Abhishek Kaushik	Dublin City University, Ireland
Mesut Kaya	Aalborg University Copenhagen, Denmark
Diane Kelly	University of Tennessee, USA

Jae Keol Choi — Seoul National University, South Korea
Roman Kern — Graz University of Technology, Austria
Pooya Khandel — University of Amsterdam, Netherlands
Johannes Kiesel — Bauhaus-Universität, Germany
Styliani Kleanthous — CYENS CoE & Open University of Cyprus, Cyprus
Anastasiia Klimashevskaia — University of Bergen, Italy
Ivica Kostric — University of Stavanger, Norway
Dominik Kowald — Know-Center & Graz University of Technology, Austria
Hermann Kroll — Technische Universität Braunschweig, Germany
Udo Kruschwitz — University of Regensburg, Germany
Hrishikesh Kulkarni — Georgetown University, USA
Wojciech Kusa — TU Wien, Austria
Mucahid Kutlu — TOBB University of Economics and Technology, Turkey
Saar Kuzi — Amazon, USA
Jochen L. Leidner — Coburg University of Applied Sciences, Germany
Kushal Lakhotia — Outreach, USA
Carlos Lassance — Naver Labs Europe, France
Aonghus Lawlor — University College Dublin, Ireland
Dawn Lawrie — Johns Hopkins University, USA
Chia-Jung Lee — Amazon, USA
Jurek Leonhardt — TU Delft, Germany
Monica Lestari Paramita — University of Sheffield, UK
Hang Li — The University of Queensland, Australia
Ming Li — University of Amsterdam, Netherlands
Qiuchi Li — University of Padua, Italy
Wei Li — University of Roehampton, UK
Minghan Li — University of Waterloo, Canada
Shangsong Liang — MBZUAI, UAE
Nut Limsopatham — Amazon, USA
Marina Litvak — Shamoon College of Engineering, Israel
Siwei Liu — MBZUAI, UAE
Haiming Liu — University of Southampton, UK
Yiqun Liu — Tsinghua University, China
Bulou Liu — Tsinghua University, China
Andreas Lommatzsch — TU Berlin, Germany
David Losada — University of Santiago de Compostela, Spain
Jesus Lovon-Melgarejo — Université Paul Sabatier IRIT, France
Alipio M. Jorge — University of Porto, Portugal
Weizhi Ma — Tsinghua University, China

Joel Mackenzie	The University of Queensland, Australia
Daniele Malitesta	Polytechnic University of Bari, Italy
Antonio Mallia	New York University, USA
Behrooz Mansouri	University of Southern Maine, USA
Masoud Mansoury	University of Amsterdam, Netherlands
Jiaxin Mao	Renmin University of China, China
Stefano Marchesin	University of Padova, Italy
Giorgio Maria Di Nunzio	University of Padua, Italy
Franco Maria Nardini	ISTI-CNR, Italy
Mirko Marras	University of Cagliari, Italy
Monica Marrero	Europeana Foundation, Netherlands
Bruno Martins	University of Lisbon, Portugal
Flavio Martins	Instituto Superior Técnico, Lisbon
David Massimo	Free University of Bolzano, Italy
Noemi Mauro	University of Turin, Italy
Richard McCreadie	University of Glasgow, UK
Graham McDonald	University of Glasgow, UK
Giacomo Medda	University of Cagliari, Italy
Parth Mehta	IRSI, India
Ida Mele	IASI-CNR, Italy
Chuan Meng	University of Amsterdam, Netherlands
Zaiqiao Meng	University of Glasgow, UK
Tristan Miller	University of Manitoba, Canada
Alistair Moffat	The University of Melbourne, Australia
Jose Moreno	IRIT/UPS, France
Gianluca Moro	University of Bologna, Italy
Josiane Mothe	Univ. Toulouse, France
Philippe Mulhem	LIG-CNRS, France
Cataldo Musto	University of Bari, Italy
Suraj Nair	University of Maryland, USA
Hitarth Narvala	University of Glasgow, UK
Julia Neidhardt	TU Wien, Austria
Wolfgang Nejdl	L3S and University of Hannover, Germany
Thong Nguyen	University of Amsterdam, Netherlands
Diana Nurbakova	INSA Lyon, France
Hiroaki Ohshima	Graduate School of Information Science, Japan
Harrie Oosterhuis	Radboud University, Netherlands
Salvatore Orlando	Università Ca' Foscari Venezia, Italy
Panagiotis Papadakos	Information Systems Laboratory - FORTH-ICS, Greece
Andrew Parry	University of Glasgow, UK
Pavel Pecina	Charles University, Czechia

Georgios Peikos	University of Milano-Bicocca, Italy
Gustavo Penha	Spotify Research, Netherlands
Marinella Petrocchi	IIT-CNR, Italy
Aleksandr Petrov	University of Glasgow, UK
Milo Phillips-Brown	University of Edinburgh, UK
Karen Pinel-Sauvagnat	IRIT, France
Florina Piroi	Vienna University of Technology, Austria
Alessandro Piscopo	BBC, UK
Marco Polignano	Università degli Studi di Bari Aldo Moro, Italy
Claudio Pomo	Polytechnic University of Bari, Italy
Lorenzo Porcaro	Joint Research Centre European Commission, Italy
Amey Porobo Dharwadker	Meta, USA
Martin Potthast	Leipzig University, Germany
Erasmo Purificato	Otto von Guericke University Magdeburg, Germany
Xin Qian	University of Maryland, USA
Yifan Qiao	University of California, USA
Georges Quénot	Laboratoire d'Informatique de Grenoble CNRS, Germany
Alessandro Raganato	University of Milano-Bicocca, Italy
Fiana Raiber	Yahoo Research, Israel
Amifa Raj	Boise State University, USA
Thilina Rajapakse	University of Amsterdam, Netherlands
Jerome Ramos	University College London, UK
David Rau	University of Amsterdam, Netherlands
Gábor Recski	TU Wien, Austria
Navid Rekabsaz	Johannes Kepler University Linz, Austria
Zhaochun Ren	Leiden University, Netherlands
Yongli Ren	RMIT University, Australia
Weilong Ren	Shenzhen Institute of Computing Sciences, China
Chiara Renso	ISTI-CNR, Italy
Kevin Roitero	University of Udine, Italy
Tanya Roosta	Amazon, USA
Cosimo Rulli	University of Pisa, Italy
Valeria Ruscio	Sapienza University of Rome, Italy
Yuta Saito	Cornell University, USA
Tetsuya Sakai	Waseda University, Japan
Shadi Saleh	Microsoft, USA
Eric Sanjuan	Avignon Université, France
Javier Sanz-Cruzado	University of Glasgow, UK
Fabio Saracco	Centro Ricerche Enrico Fermi, Italy

Harrisen Scells	Leipzig University, Germany
Philipp Schaer	TH Köln (University of Applied Sciences), Germany
Jörg Schlötterer	University of Marburg, Germany
Ferdinand Schlatt	Friedrich-Schiller-Universität Jena, Germany
Christin Seifert	University of Marburg, Germany
Giovanni Semeraro	University of Bari, Italy
Procheta Sen	University of Liverpool, UK
Ismail Sengor Altingovde	Bilkent University, Türkiye
Vinay Setty	University of Stavanger, Norway
Mahsa Shahshahani	Accenture, Netherlands
Zhengxiang Shi	University College London, UK
Federico Siciliano	Sapienza University of Rome, Italy
Gianmaria Silvello	University of Padua, Italy
Jaspreet Singh	Amazon, USA
Sneha Singhania	Max Planck Institute for Informatics, Germany
Manel Slokom	Delft University of Technology, Netherlands
Mark Smucker	University of Waterloo, Canada
Maria Sofia Bucarelli	Sapienza University of Rome, Italy
Maria Soledad Pera	TU Delft, Germany
Nasim Sonboli	Brown University, USA
Zhihui Song	University College London, UK
Arpit Sood	Meta Inc, USA
Sajad Sotudeh	Georgetown University, USA
Laure Soulier	Sorbonne Université-ISIR, France
Marc Spaniol	Université de Caen Normandie, France
Francesca Spezzano	Boise State University, USA
Damiano Spina	RMIT University, Australia
Benno Stein	Bauhaus-Universität, Germany
Nikolaos Stylianou	Information Technologies Institute, Greece
Aixin Sun	Nanyang Technological University, Singapore
Dhanasekar Sundararaman	Duke University, UK
Reem Suwaileh	Qatar University, Qatar
Lynda Tamine	IRIT, France
Nandan Thakur	University of Waterloo, Canada
Anna Tigunova	Max Planck Institute, Germany
Nava Tintarev	University of Maastricht, Germany
Marko Tkalcic	University of Primorska, Slovenia
Gabriele Tolomei	Sapienza University of Rome, Italy
Antonela Tommasel	Aarhus University, Denmark
Helma Torkamaan	Delft University of Technology, Netherlands
Salvatore Trani	ISTI-CNR, Italy

Giovanni Trappolini	Sapienza University, Italy
Jan Trienes	University of Duisburg-Essen, Germany
Andrew Trotman	University of Otago, New Zealand
Chun-Hua Tsai	University of Omaha, USA
Radu Tudor Ionescu	University of Bucharest, Romania
Yannis Tzitzikas	University of Crete and FORTH-ICS, Greece
Venktesh V	TU Delft, Germany
Alberto Veneri	Ca' Foscari University of Venice, Italy
Manisha Verma	Amazon, USA
Federica Vezzani	University of Padua, Italy
João Vinagre	Joint Research Centre - European Commission, Italy
Vishwa Vinay	Adobe Research, India
Marco Viviani	Università degli Studi di Milano-Bicocca, Italy
Sanne Vrijenhoek	Universiteit van Amsterdam, Netherlands
Vito Walter Anelli	Politecnico di Bari, Italy
Jiexin Wang	South China University of Technology, China
Zhihong Wang	Tsinghua University, China
Xi Wang	University College London, UK
Xiao Wang	University of Glasgow, UK
Yaxiong Wu	University of Glasgow, UK
Eugene Yang	Johns Hopkins University, USA
Hao-Ren Yao	National Institutes of Health, USA
Andrew Yates	University of Amsterdam, Netherlands
Fanghua Ye	University College London, UK
Zixuan Yi	University of Glasgow, UK
Elad Yom-Tov	Microsoft, USA
Eva Zangerle	University of Innsbruck, Austria
Markus Zanker	University of Klagenfurt, Germany
Fattane Zarrinkalam	University of Guelph, Canada
Rongting Zhang	Amazon, USA
Xinyu Zhang	University of Waterloo, USA
Yang Zhang	Kyoto University, Japan
Min Zhang	Tsinghua University, China
Tianyu Zhu	Beihang University, China
Jiongli Zhu	University of California San Diego, USA
Shengyao Zhuang	The University of Queensland, Australia
Md Zia Ullah	Edinburgh Napier University, UK
Steven Zimmerman	University of Essex, UK
Lixin Zou	Wuhan University, China
Guido Zuccon	The University of Queensland, Australia

Additional Reviewers

Pablo Castells
Ophir Frieder
Claudia Hauff
Yulan He
Craig Macdonald
Graham McDonald
Iadh Ounis
Maria Soledad Pera
Fabrizio Silvestri
Nicola Tonellotto
Min Zhang

Contents – Part IV

Short Papers

Reproducibility Papers

IR for Good Papers

Short Papers

ChatGPT Goes Shopping: LLMs Can Predict Relevance in eCommerce Search

Beatriz Soviero[1], Daniel Kuhn[2], Alexandre Salle[3], and Viviane Pereira Moreira[1](✉)

[1] Institute of Informatics, UFRGS, Porto Alegre, Brazil
{bfsoviero,viviane}@inf.ufrgs.br
[2] Institute of Education, Science and Technology of Rio Grande do Sul (IFRS), Ibirubá, Brazil
daniel.kuhn@ibiruba.ifrs.edu.br
[3] VTEX, Porto Alegre, Brazil
alexandre.salle@vtex.com

Abstract. The dependence on human relevance judgments limits the development of information retrieval test collections that are vital for evaluating these systems. Since their launch, large language models (LLMs) have been applied to automate several human tasks. Recently, LLMs started being used to provide relevance judgments for document search. In this work, our goal is to assess whether LLMs can replace human annotators in a different setting – product search in eCommerce. We conducted experiments on open and proprietary industrial datasets to measure LLM's ability to predict relevance judgments. Our results found that LLM-generated relevance assessments present a strong agreement (~82%) with human annotations indicating that LLMs have an innate ability to perform relevance judgments in an eCommerce setting. Then, we went further and tested whether LLMs can generate annotation guidelines. Our results found that relevance assessments obtained with LLM-generated guidelines are as accurate as the ones obtained from human instructions.[1](The source code for this work is available at https://github.com/danimtk/chatGPT-goes-shopping)

Keywords: relevance judgment prediction · LLM · eCommerce

1 Introduction

Test collections consisting of documents, queries, and relevance judgments are a crucial asset for the development of Information Retrieval (IR) tools and techniques. Obtaining human-generated relevance judgments has been the main bottleneck in the creation of IR test collections. Having humans evaluate relevance is costly in terms of time and money. Sanderson [14] reports that the 73K judgments for each year of the TREC ad-hoc tracks took over 600 h (considering a rate of two judgments per minute). Oliveira *et al.* [12] mention a much lower rate of about 20 query-document pairs per hour, which meant that 230 h of human volunteers were necessary to make relevance assessments for their small test collection.

In an eCommerce scenario, test collections are scarce as most datasets are proprietary. Although judging relevance for query-product pairs can be faster than for the

B. Soviero and D. Kuhn—Work conducted during an internship at VTEX.

N. Goharian et al. (Eds.): ECIR 2024, LNCS 14611, pp. 3–11, 2024.
https://doi.org/10.1007/978-3-031-56066-8_1

traditional query-document scenario, the creators of the WANDS dataset [4] reported that human assessors had a throughput of around 190 to 200 product-query pairs per hour. Considering the dataset has 233K query-product pairs that were assessed by three judges, we can estimate that over 3.5K h were spent to generate the relevance assessments.

The advent of Large Language Models (LLMs) has enabled the automation of a set of tasks that previously required direct human effort. This could be the case with the relevance judgment task – recent work has demonstrated that LLMs are promising for judging relevance in classical text collections, such as TREC [8, 18].

In this work, our goal is to assess whether LLMs can replace human relevance assessments in eCommerce test collections. Search in eCommerce differs from web search in general since the documents are typically very short, consisting of a product catalog. Queries are also short, emphasizing the use of keywords over long phrases in natural language. Attributes such as brands, measurements, and dosage are commonly used to describe the desired product [16]. We designed a set of experiments using GPT-3.5-turbo and GPT-4 performed on two datasets of query-product pairs: WANDS [4] and a dataset comprised of proprietary data sourced from the production environment of a large eCommerce technology provider. The results showed that LLM-generated judgments have an average overlap of around 82% with human judgments. This number is very high considering that human judges have shown much lower levels of agreement – *e.g.,* between 42 and 49% on TREC collections [19].

Our second contribution relates to the creation of annotation guidelines – a task that can be quite laborious. As pointed out by Faggioli *et al.* [8], Google search guidelines (geared towards human assessors) span over 170 pages. In addition, domain knowledge is important since the instructions may vary depending on the type of search. With that in mind and taking advantage of LLMs good summarization skills [13], we go a step further and prompt the LLM to generate the annotation guidelines. The generated guidelines are fed back into the LLM along with the query-product pairs to be annotated. Our results showed that the annotations obtained with LLM-generated guidelines are as accurate as the ones obtained from the human-generated guidelines.

2 Background and Related Work

Evaluation is paramount in IR and has been a constant focus throughout the years. The standard evaluation paradigm requires a set of documents, query topics, and human relevance assessments. Since the early days of the Cranfield experiments [5] in which relevance judgments were exhaustive (*i.e.,* for all query-document pairs) and made by human experts, the research community has constantly been trying to find means to create test collections in a more scalable way. The first step in this direction was the development of the pooling method [17] in which only a small subset of the documents are judged for each query topic. Then, several other strategies were devised, including choosing a small set of documents to judge [3] and relying on crowd workers [1], who are less expensive than experts.

Since LLMs were launched, the research community has been testing their abilities to automate a series of language tasks, including question answering, summarization, translation, and reading comprehension [2]. Recently, the use of LLMs for relevance prediction started being explored. Faggioli *et al.* [8] discuss the pros and cons of using LLMs for automatic relevance judgments. In an experimental evaluation, they found an agreement of $\kappa = .38$ between GPT-3.5-turbo ad human assessors on TREC-8 ad hoc [9]. The non-relevant documents were correctly predicted 90% of the time but, on the relevant documents, only half of the predictions were accurate. Thomas *et al.* [18] also applied LLMs (*i.e.*, GPT-3.5) for automatic relevance judgments. Using data from TREC Robust [20], they found an agreement of $\kappa = .64$ in their best run. But they also point out the sensitivity of results to prompt variations.

3 Method

This section describes the method we adopted to answer the following research questions:
RQ1 Can an LLM effectively perform relevance judgments for eCommerce search?
RQ2 Can an LLM effectively create a set of guidelines to instruct itself on making relevance judgments?

3.1 Datasets

Our experiments used two datasets from different domains and languages. In all cases, we used binary relevance judgments (*i.e.*, "Relevan" and "Not relevant"). Statistics of the datasets are in Table 1. The instances include a mix of easier and harder cases to test the capabilities of the LLM in different situations.

WANDS [4] is a publicly available eCommerce dataset consisting of 480 queries, 49K products, and 233K human relevance labels considering three classes ("relevant", "partially relevant", and "irrelevant"). The products in WANDS are household goods (furniture and decoration) described in English. Our experiments were performed on a random sample of query-product pairs covering 409 out of the 480 queries (85%). Our sample was equally balanced between the two classes, and half our relevant instances were mapped from the *partially relevant* instances in WANDS – these are our hard positives.

Pharma is a dataset based on production data from a large eCommerce solution provider consisting of 28K unique queries and 20K products. The data is in Portuguese and comes from logs with result-sets for user queries on an online pharmacy. The approach adopted for assigning relevance labels to query-product pairs relied on user clicks as an implicit relevance signal. While clicks are less reliable than explicit relevance judgments, they have been extensively used as a proxy for relevance. In a user study, Joachims *et al.* [10] found a reasonable level of agreement between clicks and explicit feedback for document relevance. Products in the catalog were ranked for each query in decreasing order of the number of clicks the product had received when retrieved for the query. Then, the result-set was divided into three bins where the first has the products with the most clicks, and the third bin has the fewest. The relevant products for each query were taken from the first bin. We considered as *hard positives* the products that were relevant to the query, and yet there were no words in common between the product name and the query. This sample was manually checked. The *easy positives*

Table 1. Statistics of Query-Product pairs annotated by the LLMs.

Dataset	Relevant		Not Relevant		Total
	Easy	Hard	Easy	Hard	
WANDS	700	700	1400	–	2800
Pharma	1260	140	700	700	2800

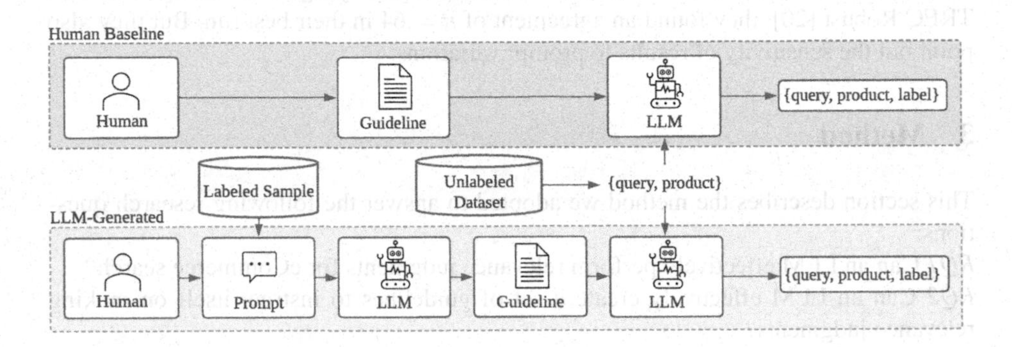

Fig. 1. Strategies for guideline generation and relevance annotation

were selected only considering the first bin. The *hard negatives* are instances that share terms between query and product name yet do not satisfy the user's intent. In contrast, *easy negative* instances share no common terms. Our experiments were performed with 2612 queries.

Because WANDS is a public dataset, we cannot attest that no data leakage occurred, as its contents may have been seen during the training of LLMs such as GPT. The Pharma dataset, on the other hand, is private. Thus, we can be sure that GPT models had no access to the queries and click-rates.

3.2 Prompting Strategies

In our prompting strategies, we varied how the guidelines were created and the number of examples provided. The process used to obtain the relevance assessments from the LLM is depicted in Fig. 1, and the details are as follows.

Annotation Guidelines.

Human Baseline – Initially, we created a baseline prompt with basic instructions asking the LLM to act as an expert for a relevance judgment task in the eCommerce context. This process is shaded in blue in Fig. 1. Basic instructions were provided with the levels of relevance ("Relevant" and "Not relevant"), the format in which the data would be fed, and brief definitions on when to assign the relevant and not relevant labels. In addition, we provided a query-product pair to be used as an example for each label. The guidelines also informed that the goal was to match the user's intent rather than focusing on the exact words in the query – the idea was to encourage the LLM not to act as a mere keyword-matching judge. Finally, we asked the model to judge a query-product pair based on the instructions and requested the response to be returned in the format *(query, product, label)*. Throughout this work, *product* is represented by the product's name in the catalog.

LLM-generated – We created a prompt asking the LLM to create a guideline for the relevance judgment task. Along with these instructions, we provided a set of 200 annotated query-product pairs for the model to use as a source for generating the guidelines. This process is shaded in green in Fig. 1. The goal was to enable the model to extract relevant patterns and, based on them, compose (hopefully) richer guidelines. There is no intersection between these query-product pairs and those used for annotation.

Examples. In addition to the guidelines, we also aim to assess whether adding annotation examples to the prompt can improve results. Thus, two settings were used: *+Zero-shot* (no further examples are given to the LLM), and *+Ten-shot* (ten annotated examples in the form *(query, product, label)* are given to the LLM).

Without Guideline. In order to evaluate the contribution guidelines, we created a prompt containing just the ten examples (as in *+Ten-shot*) and the request to judge a given tuple as relevant or not relevant.

4 Experimental Evaluation

4.1 Experimental Setup and Reproducibility

GPT models were accessed through the OpenAI API via the Chat Completion API endpoint. Each message sent to the LLM was composed of two objects: (*i*) the guidelines informed in *system* role and (*ii*) the query-product pair to be judged for relevance in the *user* role. The parameter settings were all default except for the *temperature*, which was set to zero.

The +zero-shot prompts had an average of 407 tokens, while the prompts in the +ten-shot scenario averaged 667 tokens. Completions averaged 26 tokens.

4.2 Evaluation Metrics

The results were evaluated according to two metrics of the agreement between LLM assessments and the human labels (*i.e.*, the ground truth): accuracy and Cohen's κ. While accuracy measures the overlap percentage of agreement, Cohen's κ also takes into account the possibility of the agreement occurring by chance. To verify statistical significance, we ran Friedman tests considering α =.05.

4.3 Results

To answer *RQ*1 (*Can an LLM effectively perform relevance judgments for eCommerce search?*), we evaluate the model's ability to judge relevance using human-generated prompts. The results in Table 2 are promising, with accuracy up to 85%. The agreement with the ground truth relevance labels was as high as κ = .7.

When we look at the accuracy at different levels of difficulty for the query, as expected, the scores are higher in the easy instances (~90% in both datasets) and

Table 2. Agreement with the ground truth labels – best scores in bold.

Dataset	Prompt	Examples	GPT-3.5-turbo		GPT-4	
			Accuracy	Kappa	Accuracy	Kappa
WANDS	human baseline	+ten-shot	.801	.602	**.829**	**.658**
	human baseline	+zero-shot	.804	.609	.797	.595
	LLM-generated	+ten-shot	.819	.638	.794	.587
	LLM-generated	+zero-shot	.790	.580	.751	.501
	without guideline	+ten-shot	.785	.570	.799	.598
Pharma	human baseline	+ten-shot	.834	.669	.841	.682
	human baseline	+zero-shot	.806	.613	.838	.676
	LLM-generated	+ten-shot	.837	.674	**.851**	**.701**
	LLM-generated	+zero-shot	.797	.594	.849	.697
	without guideline	+ten-shot	.838	.676	.847	.694

lower in the harder instances (∼52%). The worst results were for the hard positives in WANDS. We attribute the errors of judgment to the mismatch among the adjectives present in the query and product, *e.g.*, the pairs *('rug plum', 'ophir faux-fur pink area rug')*, *('card table', 'rian coffee table')* and *('bathroom lighting', 'chante 1 -bulb outdoor bulkhead light')* were labeled as relevant by human judges but not by the LLMs. Misjudged easy positives happened when the query was very generic and the product name was much more specific, *e.g.*, *('dinosaur', 'dinosaur ii holiday shaped ornament')* and *('flamingo', 'palm sprints flamingo graphic art')*.

+zero-shot *vs* **+ten-shot** – As expected, the scores were higher when annotated examples were provided (in 7 out of 8 possible comparisons). The differences were larger in Pharma, but the statistical test did not find them to be significant.

GPT-3.5-turbo *vs* **GPT-4** – The results show that GPT-4 achieved the best scores in both datasets. However, if we compare the individual configurations, we see that GPT-3.5-turbo wins in some cases (*e.g.*, in the LLM-generated prompts in WANDS. Yet, the only statistically significant difference was in the +zero-shot scenario in Pharma. Taking into consideration that GPT-3.5-turbo costs 20 times less than GPT-4, it may be the preferable model in many situations.

Human baseline *vs* **LLM-generated guidelines** – Agreement scores for the relevance judgments made in response to LLM-generated guidelines were comparable to the scores obtained with human prompts, sometimes even slightly better. This was the case of the winning configuration on the Pharma. Statistical tests did not find significant differences in the accuracy scores (p-values were always $\geq$0.05 when comparing human baseline and LLM-generated runs). We conclude that the answer to *RQ2* (*Can an LLM effectively create a set of guidelines to instruct itself on making relevance judgments?)* is yes.

Examples *vs* **guidelines** – The *+zero-shot* prompting strategies consist solely of the guidelines, whereas the prompts *without guidelines* are basically composed of annotation examples. By comparing the results of these two configurations we can compare the importance of providing guidelines versus examples to the LLM. In most cases, agreement scores in experiments conducted with examples only (without guideline) were similar to or higher than those where only the guideline was provided (*+zero-shot*), except for WANDS using GPT-3.5-turbo. However, both cases that achieved the best scores use guidelines (human-baseline or LLM-generated) and examples (*+ten-shot*) – this is more evident for GPT-4, which is better at following instructions.

Comparison with a Non-LLM Baseline – We establish a lower bound on the difficulty of our datasets by including a baseline based on BERT [7]; BERT-based models are strong baselines on sentence-pair tasks in GLUE [21]. We use the XML-RoBERTa-base variant [6], which employs the RoBERTa [11] pre-training scheme for BERT and is multilingual, which is a requirement since one of our datasets is in Portuguese. XLM-RoBERTa-base was fine-tuned for binary sentence-pair classification using the CLS token on the exact same 200 examples we use in our LLM-generated prompt strategy. For comparison, [22] report strong results in the paraphrase identification task from the FewCLUE dataset (a Chinese version of FewGLUE [15], which is a few-shot version of GLUE) fine-tuning RoBERTa using *only 32 examples*. We fine-tune for 20 epochs using AdamW, learning rate of 2e-5, and linear decay schedule with 10% warm-up steps. This resulted in an accuracy = .67 and κ = .35 for WANDS and accuracy = .78, κ = .57 for Pharma. These scores are significantly lower than the ones reported in Table 2 – the difference in terms of κ is as large as 30 p.p for WANDS and 13 p.p for Pharma.

Practical findings – In our early experiments with GPT-3.5-turbo, we found that two approaches (which we avoided for our final experiments) severely degraded κ to near-chance levels: (1) submitting multiple tuples for annotation *in the same prompt* instead of a single tuple (which by the nature of causal LMs conditions future annotations on previous ones), and (2) prompting only for the *label* in the model's annotated response instead of the tuple in the format *(product, name, label)*.

5 Conclusion

This work assessed the ability of LLMs to produce relevance judgments for product search. The results showed that LLMs can perform binary relevance judgments with a high overlap (~82%) with human judgments. These results come at a fraction of the time and cost taken by human judges – estimating 30 judgments per minute, LLMs are nine times faster than humans. We have also used the LLMs to generate annotation guidelines which yielded agreement scores that are not statistically different from the ones obtained with human-generated guidelines. Although more than 20 times cheaper than GPT-4, GPT-3.5-turbo achieved similar results. Our experiments also provided some practical findings in prompt engineering by highlighting the sensitivity of GPT to response formatting and the inability to deal with batch requests.

The focus of our work was to answer two research questions, and the experiments done here by no means exhaust this topic. For guideline generation, we relied on 200 annotated tuples and did not assess the impact that varying this number would have on the agreement scores.

Some limitations of our work are the usage of exclusively commercial LLMs and the lack of domains other than pharmacy and household goods. Nevertheless, we believe our findings can be useful for practitioners and contribute to the understanding of the potential of LLMs. In future work, we plan to experiment with open-source LLMs, use more datasets, and further investigate sensitivity to prompt variations.

Acknowledgments. The authors thank Shervin Malmasi for his helpful comments and suggestions. This work has been financed in part by VTEX BRASIL (EMBRAPII PCEE1911.0140), CAPES Finance Code 001, and CNPq/Brazil.

References

1. Blanco, R., et al.: Repeatable and reliable search system evaluation using crowdsourcing. In: Proceedings of the 34th International ACM SIGIR Conference on Research and Development in Information Retrieval, pp. 923–932 (2011)
2. Brown, T., et al.: Language models are few-shot learners. Adv. Neural. Inf. Process. Syst. **33**, 1877–1901 (2020)
3. Carterette, B., Allan, J., Sitaraman, R.: Minimal test collections for retrieval evaluation. In: Proceedings of the 29th Annual International ACM SIGIR Conference on Research and Development in Information Retrieval, pp. 268–275 (2006)
4. Chen, Y., Liu, S., Liu, Z., Sun, W., Baltrunas, L., Schroeder, B.: WANDS: dataset for product search relevance assessment. In: Hagen, M., et al. (eds.) ECIR 2022. LNCS, vol. 13185, pp. 128–141. Springer, Cham (2022). https://doi.org/10.1007/978-3-030-99736-6_9
5. Cleverdon, C.W.: The ASLIB cranfield research project on the comparative efficiency of indexing systems. In: ASLIB Proceedings, vol. 12, pp. 421–431. MCB UP Ltd. (1960)
6. Conneau, A., et al.: Unsupervised cross-lingual representation learning at scale. In: Proceedings of the 58th Annual Meeting of the Association for Computational Linguistics, pp. 8440–8451. Association for Computational Linguistics, Online (2020). https://doi.org/10.18653/v1/2020.acl-main.747, https://aclanthology.org/2020.acl-main.747
7. Devlin, J., Chang, M.W., Lee, K., Toutanova, K.: BERT: pre-training of deep bidirectional transformers for language understanding. In: Proceedings of the 2019 Conference of the North American Chapter of the Association for Computational Linguistics: Human Language Technologies, vol. 1 (Long and Short Papers), pp. 4171–4186. Association for Computational Linguistics, Minneapolis, Minnesota (2019). https://doi.org/10.18653/v1/N19-1423, https://aclanthology.org/N19-1423
8. Faggioli, G., et al.: Perspectives on large language models for relevance judgment. In: Proceedings of the 2023 ACM SIGIR International Conference on Theory of Information Retrieval, pp. 39–50 (2023)
9. Harman, D., Voorhees, E.: Overview of the eighth text retrieval conference (TREC-8). In: Proceedings of the Eight Text Retrieval Conference (TREC-8), pp. 1–19 (1999)
10. Joachims, T., Granka, L., Pan, B., Hembrooke, H., Gay, G.: Accurately interpreting clickthrough data as implicit feedback. In: ACM SIGIR Forum, vol. 51, pp. 4–11. ACM New York, NY, USA (2017)
11. Liu, Y., et al.: RoBERTa: a robustly optimized BERT pretraining approach (2020). https://openreview.net/forum?id=SyxS0T4tvS
12. Lima de Oliveira, L., Romeu, R.K., Moreira, V.P.: REGIS: a test collection for geoscientific documents in Portuguese. In: Proceedings of the 44th International ACM SIGIR Conference on Research and Development in Information Retrieval, pp. 2363–2368 (2021)

13. Ouyang, L., et al.: Training language models to follow instructions with human feedback. In: Koyejo, S., Mohamed, S., Agarwal, A., Belgrave, D., Cho, K., Oh, A. (eds.) Advances in Neural Information Processing Systems, vol. 35, pp. 27730–27744 (2022)
14. Sanderson, M., et al.: Test collection based evaluation of information retrieval systems. Found. Trends® Inf. Retrieval 4(4), 247–375 (2010)
15. Schick, T., Schütze, H.: It's not just size that matters: small language models are also few-shot learners. In: Proceedings of the 2021 Conference of the North American Chapter of the Association for Computational Linguistics: Human Language Technologies, pp. 2339–2352. Association for Computational Linguistics, Online (2021). https://doi.org/10.18653/v1/2021.naacl-main.185, https://aclanthology.org/2021.naacl-main.185
16. Sondhi, P., Sharma, M., Kolari, P., Zhai, C.: A taxonomy of queries for e-commerce search. In: The 41st International ACM SIGIR Conference on Research & Development in Information Retrieval, pp. 1245–1248 (2018)
17. Spark-Jones, K., van Rijsbergen, C.J.: Report on the need for and provision of an "ideal" information retrieval test collection. University of Cambridge, Computer Laboratory (1975)
18. Thomas, P., Spielman, S., Craswell, N., Mitra, B.: Large language models can accurately predict searcher preferences. arXiv preprint arXiv:2309.10621 (2023)
19. Voorhees, E.M.: Variations in relevance judgments and the measurement of retrieval effectiveness. Inform. Process. Manag. **36**(5), 697–716 (2000)
20. Voorhees, E.M., et al.: Overview of the TREC 2003 robust retrieval track. In: Proceedings of the Text Retrieval Conference, pp. 69–77 (2003)
21. Wang, A., Singh, A., Michael, J., Hill, F., Levy, O., Bowman, S.R.: GLUE: a multi-task benchmark and analysis platform for natural language understanding. In: International Conference on Learning Representations (2019). https://openreview.net/forum?id=rJ4km2R5t7
22. Xu, L., et al.: FewCLUE: a Chinese few-shot learning evaluation benchmark (2021)

Taxonomy of Mathematical Plagiarism

Ankit Satpute[1,2](✉), André Greiner-Petter[2], Noah Gießing[1], Isabel Beckenbach[1], Moritz Schubotz[1], Olaf Teschke[1], Akiko Aizawa[3], and Bela Gipp[2]

[1] FIZ Karlsruhe Leibniz Institute for Information Infrastructure, Berlin, Germany
{ankit.satpute,noah.gieBing,isabel.beckenbach,moritz.schubotz,olaf.teschke}@fiz-karlsruhe.de
[2] Georg August University of Göttingen, Göttingen, Germany
[3] National Institute of Informatics, Tokyo, Japan

Abstract. Plagiarism is a pressing concern, even more so with the availability of large language models. Existing plagiarism detection systems reliably find copied and moderately reworded text but fail for idea plagiarism, especially in mathematical science, which heavily uses formal mathematical notation. We make two contributions. First, we establish a taxonomy of mathematical content reuse by annotating potentially plagiarised 122 scientific document pairs. Second, we analyze the best-performing approaches to detect plagiarism and mathematical content similarity on the newly established taxonomy. We found that the best-performing methods for plagiarism and math content similarity achieve an overall detection score (PlagDet) of 0.06 and 0.16, respectively. The best-performing methods failed to detect most cases from all seven newly established math similarity types. Outlined contributions will benefit research in plagiarism detection systems, recommender systems, question-answering systems, and search engines. We make our experiment's code and annotated dataset available to the community: https://github.com/gipplab/Taxonomy-of-Mathematical-Plagiarism.

Keywords: Math Reuse · Plagiarism · Math Similarity Taxonomy

1 Introduction

Plagiarism is "the use of ideas, concepts, words, or structures without appropriately acknowledging the source in settings expecting originality" [7, p. 5]. Plagiarism is a pressing concern as it wastes peer reviewers' time, funding an unoriginal work and depriving original authors of the benefits [12,13]. Large Language Models (LLMs) make plagiarism easier due to produced unreferenced content [19]. Most plagiarism detection systems (PDS) have addressed identifying unoriginal text [2,15] and, to a lesser extent, unoriginal non-textual content such as formulae or images [8]. Research on plagiarized mathematical content is incipient, but prior studies have highlighted publications with questionable reuse of math content, leading to retractions [18,23].

N. Goharian et al. (Eds.): ECIR 2024, LNCS 14611, pp. 12–20, 2024.
https://doi.org/10.1007/978-3-031-56066-8_2

A common approach [11] to find similar mathematical content is determining whether or not two mathematical formulae are identical. Due to the complex nature of math and possible underlying assumptions not stated explicitly, it is difficult to judge semantic similarities between expressions since similarity in presentation does not imply a semantic similarity. The unavailability of an annotated dataset of similar mathematical content in a machine-processable format like LaTeX or MathML currently hinders progress in detecting human-modified similar mathematical content. For tasks like paraphrase identification (PI), established corpora like ETPC [14] provide a taxonomy of paraphrasing types. Establishing these types has been greatly helpful in shaping advanced PI methods [3,28] to detect various ways similar text can be represented. Thus far, a taxonomy of mathematical content similarity in math plagiarised instances hasn't been produced due to the subject's complicated nature [18], impeding methods to detect human-modified similar math. Establishing annotated corpora would benefit analysis of the performance of existing math similarity detection systems and the development of new systems.

In this work, we analyze and annotate 122 scientific document pairs from zbMATH Open that experts judged to be potential cases of plagiarism. We do not claim that the cases are plagiarized. The final decision of whether they are plagiarized is up to the field experts. We establish a taxonomy of mathematical content similarity by identifying a type of reuse and classifying it into a set of rules that describes the reuse. It has great potential to accelerate LLMs, enabling them to identify similar structural and semantic math contents.

2 Related Work

The datasets of the workshop series *Plagiarism Analysis, Authorship Identification, and Near-Duplicate Detection* (PAN) [24] are frequently utilized to develop and evaluate PDS. The PAN datasets consist of artificially created plagiarism, e.g., by randomly removing, inserting, or replacing words or phrases. The representativeness of simulated plagiarism in PAN datasets to real plagiarism is unclear [9], limiting the generalizability of evaluation results from these datasets.

For mathematical content, resources of a similar scale to text reuse are missing. Only two works analyzed mathematical content in scientific documents to identify plagiarism [17,18]. Both studied basic mathematical symbol occurrences and used a small evaluation dataset of 10 document pairs. Currently, No PDS considers semantic textual and non-textual content similarity [8,15]. While identifying mathematical plagiarism has barely been studied, most mathematics-related tasks focused on search engines and similarity analysis, such as NTCIR [26] and ARQMath [16]. In NTCIR-12, given a formulae query, relevant formulae were retrieved from the arXiv collection and Wikipedia articles. The ArqMath-3 [16] competition has released formula pairs from MathStackExchange question-answer pairs. Formulae from the questions and their answers are considered candidates, and the results submitted by participants are ranked on a scale of 0 (Irrelevant) to 3 (Highly relevant) by student annotators. Even though both NTCIR and ARQMath contain relevant formulae pairs, they were annotated by human post-detection, and none of the pairs have any information about how the content is modified.

Meuschke et al. [18] have formed categories of similar mathematical content by analyzing 10 document pairs retracted for plagiarism collected from computer science, biology, etc., research areas. However, most categories focus on presentation rather than semantic manipulation. The most comprehensive dataset of real-world plagiarised cases in mathematical academia was extracted from zbMATH Open [23]. They analyzed 10 document pairs out of the suspected 446, only visually pointing to similar math in a document under inspection and a potential source document.

3 Dataset

This work uses real-world plagiarised cases in mathematical documents from zbMATH Open [23]. zbMATH Open is the most comprehensive and longest-running (1868 - present) abstracting and reviewing service in mathematics. The majority of 4.5 million entries in zbMATH Open are in English, and even if the underlying article is not in English, the review, summary, or abstract is typically written in English. We use 446 entries from zbMATH Open identified for noticeable content reuse [23] and analyze each entry to annotate which exact parts of the documents are similar, i.e., reuse cases. Further, we categorize each reuse case into a type, eventually forming a taxonomy of reuse. We call these reuse types *Obfuscation Operators* since they present obstacles to automatic reuse detection. *Obfuscation Operators* does not imply malicious motivation to any author.

Annotation procedure: zbMATH Open does not contain full-text documents. We manually collected the full texts by accessing the publisher or repository hosting them. Unfortunately, we could only obtain full texts for 122 pairs (inspected document - potential source document) out of 446 due to inactive full-text hyperlinks present at zbMATH Open, removal of full texts by publishers due to retraction notices, etc. In most cases, we could find the document in PDF format. Hence, we used MathPiX [1] to convert PDF to LaTeX, eventually having access to all math formulae in machine-processable format LaTeX.

A group of 7 field experts with a postgraduate degree in Mathematics and 6 having Ph.D. annotated similar mathematical contents. They used the publicly available annotation tool TEIMMA [22] for annotating reuse. They regarded content from an inspected document as an instance of content reuse if it overlaps significantly with the content from the source document. They assigned *Obfuscation Operators* to each case, depicting the possible way in which content from the source document was modified. In most annotations, multiple Obfuscation Operators were found simultaneously. In the following, we describe each obfuscation operator formed by annotators.

Table 1. Overview of all Obfuscation Operators and count of annotated cases per obfuscation operator. Comb: cases occurring with other obfuscation. Uniq: cases occurring with only one particular obfuscation.

Obfuscation Operator	Abbreviation	Short Explanation	Cases count	
			Comb	Uniq
Paraphrasing	P	Text or Math rewording	1354	133
Insertions and Deletions	ID	Insert or delete text/math expressions	766	21
Substitutions	S	Substitute a term/expression or replace by cross reference	94	2
Text to Math/ Math to Text	TMMT	Math formulae explained with text words or vice versa	164	1
Different Presentation	DP	Objects with the same meaning but different presentations	733	20
Formula Manipulation	FM	Substituting formulae with a different repr. of semantic content through transformations	228	2
Variation of Subject	VS	Semantics different but argumentative structure similar	959	57

3.1 Proposed Taxonomy of Mathematical Content Similarity

Table 1 presents a summary of all obfuscations. The definitions and rules for each obfuscation operator were fixed gradually, with frequent discussions with experts with experience reviewing documents for zbMATH Open.

A taxonomy needs a precise vocabulary for categories. Terms like formulae and expressions are often used interchangeably [6, 25]. Thus, we initially set definitions for obfuscation operator terms. **Formula:** A combination of mathematical symbols formed in accordance with the rules of mathematical syntax. **Maximal Expressions:** A well-formed expression inside a mathematical formula that cannot be extended further without becoming a mathematical statement. For example, $a + 2$ and 0 in mathematical statement $a + 2 = 0$. Non-maximal expressions will be called subexpressions. For the above example, subexpressions would be a, 2, and 0. **Synonymous:** Two expressions of almost identical semantics are synonymous when representing the same mathematical object. For example, 1 and 1.0 are synonymous.

Obfuscation Operators: 1. Paraphrasing (P): Paraphrases convey the same meaning while using different words or sentence structures [5]. We extend the definition to mathematical content paraphrases by placing two restrictions. First, a mathematical content paraphrase requires that all maximal expressions from the source document must be preserved without changes. In addition, relators

(greater than, less than, etc.) may be changed according to necessity.
Example Case: { *Src* :The proof reduces to showing that $x \leq 1$. } { *Insp* : It is sufficient to prove that $1 \geq x$.}

2. Insertions and Deletions (ID): A pair of text passages are subject to Insertions if a similar passage adds redundant or insubstantial content. The definition for deletions is similar, only reversed: insubstantial content from the source passage is deleted.
Example Case: Notice the middle expression in *Insp* : { *Src* : $\langle x+y, x+y\rangle \leq \|x\|^2 + 2|\langle x,y\rangle| + \|y\|^2$ }
{ *Insp* : $\langle x+y, x+y\rangle = \|x\|^2 + \langle x,y\rangle + \langle y,x\rangle + \|y\|^2 \leq \|x\|^2 + 2|\langle x,y\rangle| + \|y\|^2$. }

3. Substitutions (S): We consider a reuse case as a substitution when either of the following occurs: 1. In formulae: replacement of expression e in the source by another expression f, which is related to e by a formula eRf for some relator R, or vice versa. 2. In-text: replacement of textual content in the source by cross-reference to an earlier passage with equivalent content, or vice versa.
Example Case: Content in *Insp* eventually leads to *Src* by substitutions. {*Src* : $x + 3 = y - 2$ } {*Insp* : $A(x) = B(y)$, where $A(x) = x + 3$ and $B(y) = y - 2$.}

4. Text to Math/Math to Text (TMMT): In the Text to Math obfuscation operator, mathematical content from the source document expressed in text is substituted by mathematical notations. Similarly, Math to Text describes the reverse obfuscation operation.
Example Case: { *Src* : The force F acting on a body equals the product of its mass m and acceleration a. } { *Insp* : $F = ma$.}

5. Different Presentation (DP): This operator allows two notational changes. First, an expression is replaced by a synonymous expression. Second, if two expressions are single objects, all root and child objects in one expression are matched to synonymous operators and objects in another.
Example Case: A simple change of variable names. {*Src* :$f(x)$ } {*Insp* :$g(x)$ }

6. Formula Manipulation (FM): Formula Manipulation refers to substituting formulae with different representations of their semantic content obtained through transforming expressions using algebraic identities or statements using logical equivalences or implications. In contrast to the *substitutions* operator, the equivalence of transformed expressions is not announced elsewhere in the document and has to be logically deduced. Examples of methods to transform expressions could be expanding a single term into multiple terms using rules such as distributivity, associativity, etc.
Example Case: The following example involves applying coordinate changes, e.g., expressing a complex number from *Src* in polar coordinates in *Insp*. { *Src* : $z = a + bi$ } { *Insp* : $z = re^{i\phi}$ }

7. Variation of Subject (VS): In the Variation of the Subject obfuscation operator, the semantic content of a passage after the application is strictly different from the original passage, but the argumentative structure remains similar.
Example Case: This is a non-mathematical (Included due to space limitation.

Please refer to the repository[1] for a real example) example that illustrates how content might be modified in several places to accommodate the new setting. { *Src* : I went to [Paris] for a few days and visited the (Eiffel Tower).} { *Insp* : I went to [Rome] for a few days and visited the (Colosseum).}

Annotation Verification by an Expert: Since similar mathematical content annotations were done in a single instance by field experts, verifying if their annotations are consistent is essential. We provided 10% document pairs to an independent expert with a Ph.D. in mathematics to do similar annotations. Creating manual annotations is time-consuming, and it is hard to find experts in multiple disciplines to produce adequate annotations. We provided obfuscation categories defined above to assign to the selected spans. The independent expert annotated the document pairs using TEIMMA [22]. We use Jaccard similarity [4] to find the overlap between the original annotations and the independent expert's annotations, i.e., for each document pair. On average, there was 80.01% token overlap (both text and math). Similarly, for exact *Case Type* (Whether reuse contains text, math, or both) and exact *Obfuscation Type* matches, there was 95.21% and 74.23% overlap, respectively. For obfuscation types, we further calculate inter-annotator agreement using the kappa value and achieve a score of 0.39, indicating a fair agreement. We understand that agreeing upon the same obfuscations is difficult as it is subjective.

4 Evaluation

Table 2. Evaluation of plagiarism detection and Math Content Similarity methods on the new dataset. Values in parentheses indicate the percentage of cases of all cases in which a particular obfuscation is present. (Lower G and higher F1 & PD represents better detection)

Obfusc.→	All			P(78.55%)			ID(44.86%)			S(5.39%)		
Method ↓	F1	G	PD	F1	G	PD	F1	G	PD	F1	G	PD
LCIS	0.00	**1.00**	0.00	0.00	**1.00**	0.00	0.00	**1.00**	0.00	0.00	**1.00**	0.00
GIT	0.05	1.28	0.04	0.05	1.35	0.04	0.02	1.38	0.02	0.00	**1.00**	0.00
AdaPlag	0.08	1.34	0.06	0.08	1.39	0.06	**0.09**	1.29	**0.07**	**0.10**	1.36	**0.08**
MABOWDOR	**0.17**	1.10	**0.16**	**0.13**	1.27	**0.11**	0.04	1.02	0.04	0.00	**1.00**	0.00
Obfusc.→	TMMT(9.50%)			DP(45.89%)			FM(12.92%)			VS(55.30%)		
Method↓	F1	G	PD	F1	G	PD	F1	G	PD	F1	G	PD
LCIS	0.00	**1.00**	0.00	0.00	**1.00**	0.00	0.00	**1.00**	0.00	0.00	**1.00**	0.00
GIT	0.00	2.00	0.00	0.02	2.00	0.01	0.01	2.00	0.01	0.02	1.42	0.02
AdaPlag	**0.07**	1.38	**0.05**	**0.08**	1.30	**0.07**	**0.09**	1.33	**0.07**	0.07	1.17	0.06
MABOWDOR	0.00	**1.00**	0.00	0.07	1.24	0.06	0.02	1.09	0.02	**0.09**	1.09	**0.09**

[1] https://github.com/gipplab/Taxonomy-of-Mathematical-Plagiarism.

We use four detection approaches. The first two are the Longest Common Subsequence of Identifier (LCIS) and Greedy Identifier Tiles (GIT) presented in the only prior work on math plagiarism detection using mathematical content by Meuschke et al. [18]. Third, AdaPlag, the winning approach of the PAN plagiarism competition [21]. Fourth, a Math-Aware Best of-Worlds Domain Optimized Retriever (MABOWDOR) [27], which is the best-performing math content similarity approach on the ARQMath dataset [16].

We create a test collection to represent a real-world retrieval scenario. zbMATH Open does not contain full texts, but around 464K entries correspond to documents from the arXiv. We obtained LaTeX sources of these 464K entries from the arXMLiv dataset [10]. We include the corresponding source documents of all annotated 122 inspected documents in the test collection and remove inspected documents from the test collection to avoid exact matches. For each of the four methods, we query our test collection of 464K documents by each of the 122 inspected documents. We combine two steps (source retrieval and detailed comparison) of typical PDS into one, i.e., direct comparison without any hyperparameters tuning. For evaluation, we use the F1 score, i.e., the harmonic mean of precision and recall), Granularity (G), which determines whether a reuse case was detected as a whole or in several pieces, and PlagDet (PD), which combines recall, precision, and G to allow for ranking, are standard evaluation metrics for plagiarism detection [20]. For each of the 122 inspected documents, we take the top 10 most similar documents out of the test collection and calculate evaluation scores. Table 2 shows the evaluation of approaches on the newly constructed dataset. Results indicate that most of the cases from all Obfuscation Operators remain undetected by existing methods. Improvements can likely be made by hyperparameter tuning.

5 Conclusion

In this work, we established a novel taxonomy of math content similarity. We formed 7 math content similarity types by annotating 122 document pairs identified for potential plagiarism. We analyzed the performance of the best-performing plagiarism detection methods and math content similarity in detecting cases from newly established taxonomy. It was found that the current best-performing methods do not detect most human modifications on math content (Overall PlagDet score of 0.06 and 0.16 for the best-performing plagiarism and math content similarity methods, respectively). The dataset curated in this work and the code of the experiments are publicly available to aid future research on math content similarity detection. A resource and experiments presented in this paper set up a base to develop advanced detection techniques capable of identifying modified math content reuse. This work will accelerate research identifying concealed plagiarism instances in academia and research involving mathematical content similarity.

Acknowledgements. This work was funded by the Deutsche Forschungsgemeinschaft (DFG, German Research Foundation) - 437179652, the Deutscher Akademischer Austauschdienst (DAAD, German Academic Exchange Service - 57515245), and the Lower Saxony Ministry of Science and Culture and the VW Foundation.

References

1. Mathpix: AI-powered document automation. – mathpix.com. https://mathpix.com/. Accessed 15 Oct 2023
2. Alvi, F., Stevenson, M., Clough, P.: Plagiarism detection in texts obfuscated with homoglyphs. In: Jose, J.M., et al. (eds.) ECIR 2017. LNCS, vol. 10193, pp. 669–675. Springer, Cham (2017). https://doi.org/10.1007/978-3-319-56608-5_64
3. Babakov, N., Dale, D., Logacheva, V., Panchenko, A.: A large-scale computational study of content preservation measures for text style transfer and paraphrase generation. In: Proceedings of the 60th Annual Meeting of the Association for Computational Linguistics: Student Research Workshop, pp. 300–321. Association for Computational Linguistics, Dublin, Ireland (2022). https://doi.org/10.18653/v1/2022.acl-srw.23
4. Bank, J., Cole, B.: Calculating the jaccard similarity coefficient with map reduce for entity pairs in Wikipedia. Wikipedia Similarity Team **1**, 94 (2008)
5. Bhagat, R., Hovy, E.: What Is a Paraphrase? Computational linguistics **39**(3), 463–472 (2013). https://doi.org/10.1162/COLI_a_00166
6. Edwards, B.S., Ward, M.B.: Surprises from mathematics education research: student (Mis)use of mathematical definitions. Am. Math. Mon. **111**(5), 411–424 (2004). https://doi.org/10.1080/00029890.2004.11920092
7. Fishman, T.: "we know it when we see it" is not good enough: toward a standard definition of plagiarism that transcends theft, fraud, and copyright (2009)
8. Foltynek, T., Meuschke, N., Gipp, B.: Academic Plagiarism Detection: a Systematic literature review. ACM Comput. Surv. **52**(6), 112:1–112:42 (2019). https://doi.org/10.1145/3345317
9. Foltỳnek, T., Meuschke, N., Gipp, B.: Academic plagiarism detection: a systematic literature review. ACM Comput. Surv. (CSUR) **52**(6), 1–42 (2019)
10. Ginev, D.: arxmliv:2020 dataset, an html5 conversion of arxiv.org. https://sigmathling.kwarc.info/resources/arxmliv-dataset-2020/ (2020), sIGMathLing - Special Interest Group on Math Linguistics
11. Guidi, F., Sacerdoti Coen, C.: A survey on retrieval of mathematical knowledge. Math. Comput. Sci. **10**(4), 409–427 (2016)
12. Hall, S.E.: Is it happening? How to avoid the deleterious effects of plagiarism and cheating in your courses. Bus. Commun. Q. **74**, 179–182 (2011)
13. Hart, K., Mano, C., Edwards, J.: Plagiarism deterrence in cs1 through keystroke data. In: Proceedings of the 54th ACM Technical Symposium on Computer Science Education, vol. 1, pp. 493–499. SIGCSE 2023, Association for Computing Machinery, New York, NY, USA (2023). https://doi.org/10.1145/3545945.3569805
14. Kovatchev, V., Martí, M.A., Salamó, M.: ETPC-a paraphrase identification corpus annotated with extended paraphrase typology and negation. In: Proceedings of the Eleventh International Conference on Language Resources and Evaluation (LREC 2018) (2018)

15. Lovepreet, V.G., Kumar, R.: Survey on plagiarism detection systems and their comparison. In: Behera, H., Nayak, J., Naik, B., Pelusi, D., (eds.) Computational Intelligence in Data Mining: Proceedings of the International Conference on ICCIDM 2018, vol. 990, p. 27. Springer, Singapore (2019). https://doi.org/10.1007/978-981-13-8676-3_3
16. Mansouri, B., Agarwal, A., Oard, D.W., Zanibbi, R.: Advancing Math-Aware Search: the ARQMath-3 Lab at CLEF 2022. In: Hagen, M., et al. (eds.) ECIR 2022. LNCS, vol. 13186, pp. 408–415. Springer, Cham (2022). https://doi.org/10.1007/978-3-030-99739-7_51
17. Meuschke, N., Schubotz, M., Hamborg, F., Skopal, T., Gipp, B.: Analyzing mathematical content to detect academic plagiarism. In: Proceedings of the International Conference on Information and Knowledge Management (CIKM) (2017). https://doi.org/10.1145/3132847.3133144
18. Meuschke, N., Stange, V., Schubotz, M., Kramer, M., Gipp, B.: Improving academic plagiarism detection for stem documents by analyzing mathematical content and citations. In: Proceedings of the ACM/IEEE Joint Conference on Digital Libraries (JCDL) (2019). https://doi.org/10.1109/JCDL.2019.00026
19. Meyer, J.G., et al.: ChatGPT and large language models in academia: opportunities and challenges. BioData Mining **16**(1), 20 (2023)
20. Potthast, M., Stein, B., Barrón-Cedeño, A., Rosso, P.: An evaluation framework for plagiarism detection. In: Coling 2010: Posters, pp. 997–1005. Coling 2010 Organizing Committee, Beijing, China (2010). https://aclanthology.org/C10-2115
21. Sanchez-Perez, M.A., Gelbukh, A., Sidorov, G.: Adaptive algorithm for plagiarism detection: the best-performing approach at PAN 2014 text alignment competition. In: Mothe, J., et al. (eds.) CLEF 2015. LNCS, vol. 9283, pp. 402–413. Springer, Cham (2015). https://doi.org/10.1007/978-3-319-24027-5_42
22. Satpute, A., et al.: TEIMMA: the first content reuse annotator for text, images, and math. In: 2023 ACM/IEEE Joint Conference on Digital Libraries (JCDL) (2023). https://doi.org/10.1109/JCDL57899.2023.00056
23. Schubotz, M., Teschke, O., Stange, V., Meuschke, N., Gipp, B.: Forms of plagiarism in digital mathematical libraries. In: Intelligent Computer Mathematics - 12th International Conference, CICM 2019, Prague, Czech Republic, 8–12 July 2019, Proceedings (2019). https://doi.org/10.1007/978-3-030-23250-4_18
24. Stein, B., Koppel, M., Stamatatos, E.: Plagiarism analysis, authorship identification, and near-duplicate detection PAN 2007. ACM SIGIR Forum **41**(2), 68–71 (2007). https://doi.org/10.1145/1328964.1328976
25. Torkildsen, H.A., Forbregd, T.A., Kaspersen, E., Solstad, T.: Toward a unified account of definitions in mathematics education research: a systematic literature review. Int. J. Math. Educ. Sci. Technol., 1–28 (2023). https://doi.org/10.1080/0020739X.2023.2180678. Taylor & Francis
26. Zanibbi, R., Aizawa, A., Kohlhase, M., Ounis, I., Topic, G., Davila, K.: NTCIR-12 mathIR task overview. In: NTCIR (2016)
27. Zhong, W., Lin, S.C., Yang, J.H., Lin, J.: One blade for one purpose: Advancing math information retrieval using hybrid search. In: Proceedings of the 46th International ACM SIGIR Conference on Research and Development in Information Retrieval, p. 141–151. SIGIR 2023, Association for Computing Machinery, New York, NY, USA (2023). https://doi.org/10.1145/3539618.3591746
28. Zhou, C., Qiu, C., Acuna, D.E.: Paraphrase identification with deep learning: a review of datasets and methods. arXiv preprint arXiv:2212.06933 (2022)

Unraveling Disagreement Constituents in Hateful Speech

Giulia Rizzi[1,2(✉)], Alessandro Astorino[1], Paolo Rosso[2,3], and Elisabetta Fersini[1]

[1] University of Milano-Bicocca, Milan, Italy
{a.astorino2,g.rizzi10}@campus.unimib.it, elisabetta.fersini@unimib.it
[2] Universitat Politècnica de València, Valencia, Spain
prosso@dsic.upv.es
[3] ValgrAI - Valencian Graduate School and Research Network of Artificial Intelligence, Valencia, Spain

Abstract. This paper presents a probabilistic semantic approach to identifying disagreement-related textual constituents in hateful content. Several methodologies to exploit the selected constituents to determine if a message could lead to disagreement have been defined. The proposed approach is evaluated on 4 datasets made available for the SemEval 2023 Task 11 shared task, highlighting that a few constituents can be used as a proxy to identify if a sentence could be perceived differently by multiple readers. The source code of our approaches is publicly available (https://github.com/MIND-Lab/Unrevealing-Disagreement-Constituents-in-Hateful-Speech).

Keywords: Disagreement · Hate Speech · Perspectivism

1 Introduction

Detecting hate speech and toxic language, especially in the online environment, has become an increasingly important task, as online platforms seek to mitigate the harmful effects of such language. However, the detection of hate speech and toxicity is a complex task that is often subject to disagreement among human annotators. The subjectivism of hate detection refers to the fact that individuals' perceptions of what constitutes hate speech can vary widely [14]. Identifying a potential disagreement in hateful content, and in particular ascertaining those constituents that are more controversial, is of paramount importance for several reasons. For those contents that could lead to disagreement, specific annotation policies could be adopted or, specific highlights could be provided to the annotators to focus more on specific constituents that could be perceived differently. Identifying disagreement-related constituents can be also beneficial in Information Retrieval, for instance, to address ambiguities emerging from the difficulties of annotating document relevance. Identifying specific components linked with disagreements, such as controversial opinions or ambiguous claims, enables a more nuanced understanding of document content. This not only helps to refine

N. Goharian et al. (Eds.): ECIR 2024, LNCS 14611, pp. 21–29, 2024.
https://doi.org/10.1007/978-3-031-56066-8_3

Table 1. Datasets characteristics.

Dataset	Language	N. items	Task	Annotators	Pool Ann.	% of items with full agr.
HS-Brexit [1]	En	1,120	Hate Speech	6	6	69%
ArMis [2]	Ar	943	Misogyny and sexism detection	3	3	86%
ConvAbuse [5]	En	4,050	Abusive Language detection	2–7	7	65%
MD-Agreement [10]	En	10,753	Offensiveness detection	5	>800	42%

annotations but also mitigates disagreements that may arise from differing perspectives, thereby improving the precision and accuracy of Information Retrieval systems. In this work, we propose a probabilistic semantic approach for detecting disagreement in hateful content and identifying those elements in a sentence that could potentially convey such a lack of agreement among different readers. The proposed approach is aimed at identifying disagreement-related elements by exploring several textual constituents (i.e., words, emoji, and hashtags) based on the assumption that such constituents may appear more frequently in ambiguous or subjective samples and therefore in samples characterized by a lower confidence level related to the hard (gold) labels. To evaluate our approach, we conducted a few experiments on the hate speech datasets made available at the SemEval 2023 - Task 11 about Learning With Disagreements (Le-Wi-Di) [11], summarized in Table 1. These benchmarks cover a wide range of characteristics, including textual types, language, goals, and annotation methods.

2 Related Work

Several natural language tasks have been proven to be ambiguous or subjective [15], and this subjectivity is reflected in the datasets by the presence of different annotations or by a confidence level attached to the labels. A growing area of research is, in fact, abandoning the assumption that a unique perception and interpretation exists for each instance, in favor of multiple labels that could represent the multiple perceptions of the annotators with different points of view and understanding [15]. The information that represents annotators' disagreement has primarily been used in three ways: (i) to improve the quality of the dataset by removing instances characterized by annotators' disagreement [4], (ii) to weight the instances during the training phase in order to give more importance to the ones with a higher level of confidence [7], or (iii) to directly train a machine learning model from disagreement without taking into account any aggregated label [8,15]. Despite several techniques have been proposed to deal with disagreement in the dataset creation phase or to exploit disagreement information in the realization of machine learning classification models [12], less attention has been paid to explaining and recognizing disagreement. As demonstrated in [9] different annotators adopt different work strategies to complete a given task, for instance developing shortcut patterns or focusing their attention on specific elements. [13,14] demonstrated that age, personality, cultural background, and beliefs of an annotator might have a significant impact on their perception of offensive or hateful content and therefore recognition of hate speech. The authors of [3] use integrated gradients to effectively discern both disagreement and hate speech, also introducing a *filtering strategy* of textual constituents that contributes remarkably to explain hateful messages. In contrast, our approach diverges by specifically targeting constituents related to disagreement and leveraging them in the detection of disagreement itself. Our approach verifies the hypothesis that disagreement could be also grasped by specific textual constituents.

3 Proposed Approach

3.1 Identification of Disagreement-Related Constituents

The first phase aims to evaluate the relationship between textual constituents and annotators' disagreement. Preliminary preprocessing operations have been performed before identifying disagreement-related constituents, i.e., tokenization, lemmatization, lower casing and stop word removal.

The preprocessing steps allowed us to extract a list of constituents from each tweet in the dataset. Since the disagreement among annotators is strictly related to the specific task's objectives and context, a given constituent can be considered as a source of disagreement in some cases and neutral in others. For instance, the term *immigrant* can convey disagreement in the Brexit dataset focused on politics, but not in Armis that is centered on misogyny.

Therefore, the datasets exploited in this study have not been joined for the evaluation of a unique disagreement score per constituent.

Given a constituent t, a corresponding Constituent Disagreement Score (*CDS(t)*) has been computed according to the following equation:

$$CDS(t) = P(Agree|t) - P(\neg Agree|t) \quad (1)$$

where $P(Agree|t)$ represents the conditional probability that there is agreement on a sentence given that the sample contains a constituent t. Analogously, $P(\neg Agree|t)$ denotes the conditional probability that there is no agreement on a sample given that that sample contains a constituent t. Since CDS represents a difference between two complementary probabilities, it is bounded within the range $[-1; +1]$. Higher positive scores indicate stronger agreement, while lower negative scores suggest annotators' disagreement. The score can be estimated on the training data[1] and exploited to identify additional disagreement-related constituent (and sentences) on unseen text. A similar score can be estimated to qualify the presence of hate and used, together with the CDS, as coordinates in a Cartesian plane bounded between -1 and $+1$ for each axe.

3.2 Disagreement Identification

Once the Constituent Disagreement Scores have been estimated for each constituent in the training dataset, it can be exploited to qualify the level of disagreement on unseen messages. The overall hypothesis is that the identified

[1] The selected datasets contain information about the disagreement among annotators in the form of soft labels. Such agreement values have been transformed into boolean values to represent complete agreement and disagreement among the annotators.

constituents present a different distribution in samples with and without an agreement, and therefore they can be exploited for identifying the disagreement at the sentence level. For each message in the test set, the corresponding constituents have been qualified according to the Constituent Disagreement Score estimated on the training data: after applying preprocessing operations, resulting in a constituents vector, a corresponding vector containing the constituent's score has been created. In particular, for each constituent in a given sample, its score has been extracted, when available. Several approaches have been investigated to aggregate the scores of the constituents in a sentence. The Sentence Disagreement Score (SDS) has been estimated according to the following strategies: **Sum**, **Mean**, **Median**, and **Minimum**. The threshold τ denotes a hyper-parameter that has been estimated according to a grid-search approach for each strategy.

The CDS score has been evaluated from a qualitative point of view, while the SDS has been validated by computing both the F1-score for the two considered classes (+ denotes agreement, - denotes disagreement) and globally (by estimating the average F1-score).

3.3 Generalization Towards Unseen Constituents

CDS estimation (and consequently SDS) is strongly based on what is observed in the training data. When making predictions on unseen messages, a few constituents could not be qualified in terms of disagreement because they do not belong to the training lexicon. To overcome this limitation and therefore generalize to unseen constituents, we approximate the corresponding CDS by exploiting their latent representation in a contextualized embedding space. In particular, given an unseen constituent $\hat{t}$ we approximate its CDS score as follows:

- **Embeddings of the training lexicon:** the embedding representation of each constituent t in the training lexicon is obtained by determining their contextualized representation using mBert [6]. Since t can have multiple embedding representations according to the context where it occurs, an average embedding vector representation $\vec{\mathbf{x}}_t$ is computed. In particular, given an element t and N sentences containing it, we aggregated the embeddings in a vector representation by a simple average $\vec{\mathbf{x}}_t = \sum_{i=1}^{N} \vec{\mathbf{v}}_i/N$, where $\vec{\mathbf{x}}_t$ is the final average embeddings representing t, $\vec{\mathbf{v}}_i$ is the constituent contextualized embedding vector related to the ith occurrence of t and obtained through mBert. The final average embeddings not only represent the contexts in which the constituent appears but also capture the frequency in which a constituent is related to a given context.
- **Embeddings of unseen constituents:** given an unseen constituent $\hat{t}$ within a given sentence, its contextualized embedding representation $\vec{\mathbf{v}}_{\hat{t}}$ is obtained by means of the mBert model.
- **Most similar constituent:** given an unseen constituent $\hat{t}$ with the corresponding embedding $\vec{\mathbf{v}}_{\hat{t}}$ and the average embedding of a training element t,

the set D of most similar constituents to $\hat{t}$ is determined according to:

$$D = \bigcup_{t} \{t | cos(\vec{\mathbf{x}}_t, \vec{\mathbf{v}}_{\hat{t}}) \leq \psi\} \tag{2}$$

where $cos(\vec{\mathbf{x}}_t, \vec{\mathbf{v}}_{\hat{t}})$ is the cosine similarity between the average contextualized embedding representation of element t and $\hat{t}$, while ψ is a threshold that has been estimated via a grid search approach.

- **Unseen constituent score:** the CDS score for an unseen constituent $\hat{t}$ is computed as the weighted average of the most similar constituents t of the training lexicon:

$$CDS(\hat{t}) = \frac{\sum_{t \in D} \left[cos(t, \hat{t}) \cdot CDS(t)\right]}{\sum_{t \in D} cos(t, \hat{t})} \tag{3}$$

- **Sentence Disagreement Score with unseen constituents:** SDS are extended in order to account also constituents that do not belong to the training lexicon. In particular, given a sentence s, the aggregation functions presented in Sect. 3.2 will now consider the CDS values of both seen and unseen constituents. For the constituents t belonging to the training lexicon, $CDS(t)$ will be taken. For those constituents that do not belong to the training lexicon $CDS(\hat{t})$ will be used. We will denote such generalized aggregation functions using the prefix $G-$.

4 Results

The proposed approach has been used to compute the CDS for each constituent and the SDS for each sentence. Examples of the computed CDS for the Brexit and for the MD dataset are shown respectively in Tables 2.a and 2.b. In particular, both tables report the top-5 highest positive and negative scores.

We can easily notice that agreement/disagreement appears to be related to different types of constituents, for instance representing specific persons (e.g. Merkel) or hashtags (e.g. #trump2016). Figure 1 reports a visual representation

Table 2. Constituents with the highest positive and the lowest negative scores.

Disagreement		Agreement	
Constituent	Score	Constituent	Score
illegal	-0.8181	economic	0.8666
invasion	-0.7647	crisis	0.8750
merkel	-0.4800	foreign	0.8909
islam	-0.4800	issue	1
#trump2016	-0.2727	minister	1

(a) HS-Brexit dataset

Disagreement		Agreement	
Constituent	Score	Constituent	Score
drink	-1	*crying emoji*	0.5000
encourage	-1	article	0.6250
swamp	-1	council	0.6363
bs	-0.8888	beautiful	0.6363
cult	-0.8888	update	0.7142

(b) MD-Agreement dataset

Hope so. Obama has 'arranged' so illegals can vote over Internet. This will be completely fraudulent. <url>

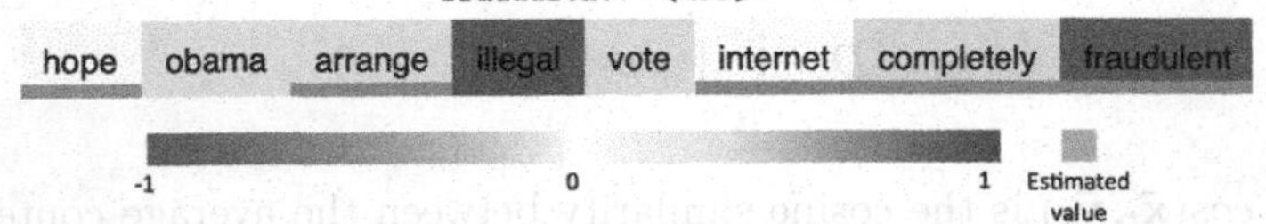

Fig. 1. Visual representation of disagreement scores on a sentence from the Brexit dataset. Positive and negative values are represented with green and pink respectively. The grey bar denotes unseen constituents for which the CDS has been estimated, while the white color is used for constituents with CDS values equal to zero. (Color figure online)

Table 3. Comparison of the different approaches on the test set for disagreement detection. In particular, the agreement label is set equal to (+) when there is a 100% agreement among the annotators, regardless of the value of the hard label, while equal to (−) in all the other cases. **Bold** denotes the best approach according to the F1-Score, while underline represents the best approach according to the disagreement label. (*) denotes that model outperforms mBERT and obtains results that are statistically different according to a McNemar Test ($p < 0.05$).

Approach	HS-Brexit			ConvAbuse			MD-Agreement			ArMIS		
	F1$^+$	*F1*$^-$	*F1-score*	*F1*$^+$	*F1*$^-$	*F1-score*	*F1*$^+$	*F1*$^-$	*F1-score*	*F1*$^+$	*F1*$^-$	*F1-score*
Sum	0.83	0.47	0.65	0.83	0.15	0.49*	0.50	0.67	0.58*	0.69	0.36	**0.53***
Mean	0.78	0.63	0.71	0.88	0.23	0.56*	0.53	0.58	0.55	0.64	0.36	0.50*
Median	0.76	0.44	0.60	0.74	0.30	0.52*	0.52	0.60	0.56*	0.65	0.37	0.51*
Minimum	0.84	0.67	**0.76***	0.86	0.28	**0.57***	0.55	0.63	**0.59***	0.55	0.45	0.50*
mBERT	0.85	0.51	0.68	0.93	0.05	0.49	0.38	0.63	0.50	0.37	0.43	0.40
G-Sum	0.81	0.46	0.64	0.79	0.19	0.49*	0.47	0.69	0.58*	0.64	0.20	0.42*
G-Mean	0.72	0.41	0.70	0.89	0.22	0.55*	0.54	0.57	0.56	0.47	0.54	0.51*
G-Median	0.73	0.42	0.57	0.76	0.32	0.54*	0.53	0.54	0.54	0.71	0.25	0.48*
G-Minimum	0.84	0.69	**0.77***	0.85	0.33	**0.59***	0.54	0.64	**0.59***	0.59	0.48*	**0.54***

of CDS computed for a Tweet from the Brexit dataset. It's easy to notice the presence of constituents associated with disagreement (pink color) and to notice how the estimated scores related to unseen constituents (grey color), correlate to the sentence labels and therefore could help in the prediction of disagreement.

All the proposed aggregation strategies have been evaluated and summarized in Table 3. The results obtained by the proposed approaches have been reported distinguishing between the prediction based on the training lexicon only (i.e., *Sum*, *Mean*, *Median*, and *Minimum*) and the ones based also on the unseen constituents (i.e., *G-Sum*, *G-Mean*, *G-Median* and *G-Minimum*). Moreover, in order to provide a comparison with a baseline, mBERT [6] has been fine-tuned to predict agreement/disagreement on each sentence. Overall, the proposed approach performs better on the majority class, which in general is related to the complete agreement. By analyzing the results in terms of F1-Score, we can conclude that the strategies based on the *Minimum* are the most performing ones. In particular, *G-Minimum* achieves the highest F1-Score on all the considered datasets.

In order to understand whether the proposed approaches obtain significant results compared with mBert, a McNemar Test has been performed (confidence level equal to 0.95). If a given model outperforms mBert and its error distribution is different compared to mBert, then the corresponding F1-Score is marked with a wildcard symbol (*) in Table 3. The test confirms that the strategies based on the *Minimum* are not only the best-performing ones and statistically significant.

However, the performance of the proposed approaches can be distinguished into two main categories. For what concerns the HS-Brexit dataset, the F1-Score can be considered almost satisfactory (77%), while for the other datasets, the performances are inferior (54–59%). HS-Brexit is less challenging because (i) it is composed of tweets that cover a single macro-topic, (ii) it is based on a well-represented language and (iii) the text does not involve any interaction between users. On the other hand, the other datasets are more difficult due to multiple domains of discussion (MD-Agreement), the user-bot interplay (ConvAbuse), and the extreme variability of the lexicon (ArMIS). More specifically, MD-Agreement is focused on three main macro-topic of discussion, ConvAbuse contains multiple empty messages either from the conversational agent or from the user, and finally, ArMIS is characterized by a reduced-size training lexicon that leads to approximate a large number of CSC values related to unseen constituents.

5 Conclusions and Future Work

This paper presents a probabilistic semantic approach to identify textual constituents for disagreement representation in hate speech detection problems. The achieved results highlight that the identified constituents could be used as a proxy to identify if a sentence could be perceived differently by the annotators and, therefore, lead to a disagreement about the final gold labels. Highlighting potentially disagreement-related constituents for human annotators can be beneficial in the annotation phase by inviting annotators to exercise extra caution: this not only helps to distinguish different meanings of the given sentence but also enhances awareness of potential biases that may unintentionally influence their evaluation. Regarding future work, two main research issues will be addressed. The first one relates to the quantification of the level of disagreement in a sentence: models able to predict the level of disagreement starting from the constituent disagreement scores will be trained. Finally, an additional research direction regards the exploration of syntactic properties: the impact of part-of-speech classes and dependency structures could be considered to improve the estimation of disagreement and hatefulness scores.

Acknowledgments. We acknowledge the support of the PNRR ICSC National Research Centre for High Performance Computing, Big Data and Quantum Computing (CN00000013), under the NRRP MUR program funded by the NextGenerationEU. The work of Paolo Rosso was in the framework of the FairTransNLP-Stereotypes research project (PID2021-124361OB-C31) funded by MCIN/AEI/10.13039/501100011033 and by ERDF, EU A way of making Europe.

References

1. Akhtar, S., Basile, V., Patti, V.: Whose opinions matter? Perspective-aware models to identify opinions of hate speech victims in abusive language detection. arXiv preprint arXiv:2106.15896 (2021)
2. Almanea, D., Poesio, M.: ArMIS - the Arabic misogyny and sexism corpus with annotator subjective disagreements. In: Proceedings of the Thirteenth Language Resources and Evaluation Conference, pp. 2282–2291. European Language Resources Association, Marseille, France (2022). https://aclanthology.org/2022.lrec-1.244
3. Astorino, A., Rizzi, G., Fersini, E.: Integrated gradients as proxy of disagreement in hateful content. In: CEUR WORKSHOP PROCEEDINGS, vol. 3596. CEUR-WS.org (2023)
4. Beigman Klebanov, B., Beigman, E.: From annotator agreement to noise models. Comput. Linguist. **35**(4), 495–503 (2009)
5. Cercas Curry, A., Abercrombie, G., Rieser, V.: ConvAbuse: data, analysis, and benchmarks for nuanced abuse detection in conversational AI. In: Proceedings of the 2021 Conference on Empirical Methods in Natural Language Processing, pp. 7388–7403. Association for Computational Linguistics, Online and Punta Cana, Dominican Republic (2021). https://doi.org/10.18653/v1/2021.emnlp-main.587, https://aclanthology.org/2021.emnlp-main.587
6. Devlin, J., Chang, M.W., Lee, K., Toutanova, K.: BERT: pre-training of deep bidirectional transformers for language understanding. In: Proceedings of NAACL-HLT, pp. 4171–4186 (2019)
7. Dumitrache, A., Mediagroep, F., Aroyo, L., Welty, C.: A crowdsourced frame disambiguation corpus with ambiguity. In: Proceedings of NAACL-HLT, pp. 2164–2170 (2019)
8. Fornaciari, T., Uma, A., Paun, S., Plank, B., Hovy, D., Poesio, M., et al.: Beyond black & white: Leveraging annotator disagreement via soft-label multi-task learning. In: Proceedings of the 2021 Conference of the North American Chapter of the Association for Computational Linguistics: Human Language Technologies. Association for Computational Linguistics (2021)
9. Han, L., et al.: Crowd worker strategies in relevance judgment tasks. In: Proceedings of the 13th International Conference on Web Search and Data Mining, pp. 241–249 (2020)
10. Leonardelli, E., Menini, S., Palmero Aprosio, A., Guerini, M., Tonelli, S.: Agreeing to disagree: Annotating offensive language datasets with annotators' disagreement. In: Proceedings of the 2021 Conference on Empirical Methods in Natural Language Processing, pp. 10528–10539. Association for Computational Linguistics, Online and Punta Cana, Dominican Republic (2021). https://doi.org/10.18653/v1/2021.emnlp-main.822, https://aclanthology.org/2021.emnlp-main.822
11. Leonardelli, E., et al.: SemEval-2023 task 11: Learning with disagreements (LeWiDi) (2023)
12. Rizzi, G., Astorino, A., Scalena, D., Rosso, P., Fersini, E.: Mind at SemEval-2023 task 11: From uncertain predictions to subjective disagreement. In: Proceedings of the The 17th International Workshop on Semantic Evaluation (SemEval-2023), pp. 556–564 (2023)

13. Sandri, M., Leonardelli, E., Tonelli, S., Jezek, E.: Why don't you do it right? analysing annotators' disagreement in subjective tasks. In: Vlachos, A., Augenstein, I. (eds.) Proceedings of the 17th Conference of the European Chapter of the Association for Computational Linguistics, pp. 2428–2441. Association for Computational Linguistics, Dubrovnik, Croatia (2023). https://doi.org/10.18653/v1/2023.eacl-main.178, https://aclanthology.org/2023.eacl-main.178
14. Sang, Y., Stanton, J.: The origin and value of disagreement among data labelers: A case study of individual differences in hate speech annotation. In: Smits, M. (eds.) Information for a Better World: Shaping the Global Future: 17th International Conference, iConference 2022, Virtual Event, February 28-March 4, 2022, Proceedings, Part I, pp. 425–444. Springer, Cham (2022). https://doi.org/10.1007/978-3-030-96957-8_36
15. Uma, A.N., Fornaciari, T., Hovy, D., Paun, S., Plank, B., Poesio, M.: Learning from disagreement: a survey. J. Artif. Intell. Res. **72**, 1385–1470 (2021)

Context-Aware Query Term Difficulty Estimation for Performance Prediction

Abbas Saleminezhad[1](✉), Negar Arabzadeh[2], Soosan Beheshti[1], and Ebrahim Bagheri[1]

[1] Toronto Metropolitan University, Toronto, ON, Canada
{abbas.saleminezhad,soosan,bagheri}@torontomu.ca
[2] University of Waterloo, Toronto, ON, Canada
narabzad@uwaterloo.ca

Abstract. Research has already found that many retrieval methods are sensitive to the choice and order of terms that appear in a query, which can significantly impact retrieval effectiveness. We capitalize on this finding in order to predict the performance of a query. More specifically, we propose to learn query term difficulty weights specifically within the context of each query, which could then be used as indicators of whether each query term has the likelihood of making the query more effective or not. We show how such difficulty weights can be learnt through the finetuning of a language model. In addition, we propose an approach to integrate the learnt weights into a cross-encoder architecture to predict query performance. We show that our proposed approach shows a consistently strong performance prediction on the MSMARCO collection and its associated widely used Trec Deep Learning tracks query sets. Our findings demonstrate that our method is able to show consistently strong performance prediction over different query sets (MSMARCO Dev, TREC DL'19, '20, Hard) and a range of evaluation metrics (Kendall, Spearman, sMARE).

1 Introduction

With the diverse range of user queries, a single Information Retrieval (IR) method faces challenges in effectively addressing all query types. Certain retrieval methods excel for specific queries but may fall short for others [1,6,7,12,39]. To assess how well a retrieval method can meet a query's needs, researchers have delved into the realm of *Query Performance Prediction (QPP)*. The primary aim of QPP is to predict the potential retrieval effectiveness of a method for a given query [3,4,11,22–24,27,33,36,41]. Numerous QPP methods exist in the literature, broadly categorized as *pre-retrieval* and *post-retrieval* methods. Post-retrieval methods, despite their superior performance, incur additional overhead by necessitating the complete retrieval of the query for estimating effectiveness [11]. On the other hand, pre-retrieval methods are lightweight, concentrating solely on query and document collection characteristics [20]. Most recent pre-retrieval methods focus on injecting external knowledge into performance estimation by benefiting from contextual language models. For instance, the work by

N. Goharian et al. (Eds.): ECIR 2024, LNCS 14611, pp. 30–39, 2024.
https://doi.org/10.1007/978-3-031-56066-8_4

Arabzadeh et al. [8,9] advocates for the use of neural-embedding representation of the query to determine query *specificity*, as an indicator of query performance. Roy et al. [35] also promote the idea of using contextual embeddings to measure the *ambiguity* of a query by estimating the number of senses each query term is associated with.

Similar to the works by Arabzadeh et al. and Roy et al. [5,9,10,35], we also benefit from contextual embeddings; however, in contrast rather than estimating *query specificity* or *ambiguity*, we are interested in using contextual embeddings to learn the impact of query terms on the overall query performance. Our proposed approach builds on the foundational premise that certain terms in the query and document spaces have a higher impact on retrieval effectiveness [15,16]. These terms are often more discriminative and can hence more effectively discern between relevant and irrelevant documents to a given query. For this reason and during the retrieval process, such terms would need to play a more important role and have a higher weight. On the same basis, there are terms that negatively impact the performance of a query and would hence need to receive a lower weight. We build on this premise and propose to learn weights for terms in the query space so as to understand which terms have the potential to contribute positively or negatively to query performance. A query with a large number of terms that can positively contribute to retrieval effectiveness is more likely to be an easier query, whereas conversely, a query with many terms with a negative prospect would be harder queries [2,37]. We have extensively evaluated our proposed approach based on four widely-used MSMARCO query sets, namely Dev set [28], TREC DL 2019 [14], DL 2020 [13], and DL Hard [26]. We show that our approach has strong performance compared to the baselines on various evaluation metrics and query sets. For reproducibility purposes, we made our code and models publicly available at https://github.com/Saleminezhad/context-aware-qpp.git.

2 Proposed Approach

Objective. The goal of our work in this paper is to propose a *pre-retrieval query performance predictor*, denoted by $\mu(q, C)$, where q and C represent a query, and a collection of documents, respectively. This predictor would need to estimate the performance of query q with respect to a specific IR evaluation metric M, resulting in an estimated performance score represented as $\widehat{M_q}$ [11].

Approach Overview. We are interested in estimating which query terms are likely to impact the performance of a query positively or negatively. Terms that positively impact the performance of a query can be seen as softer terms, while terms with a negative impact would be harder. To distinguish between soft and hard terms, we can compare pairs of queries expressing the same information need but differently. In essence, such two queries could be the same from an information seeking objective perspective, but in practice, their retrieval effectiveness could (potentially vastly) differ from each other. Given two such queries and their retrieval effectiveness, it would be possible to determine which query is softer and which is harder. Consequently, depending on the terms in each query and how overlapping they are between the two queries, one could also make inferences about whether and to what extent query terms can impact query performance and hence be considered soft or hard. Our work in this paper offers a systematic approach to identify a collection of comparable query pairs based on which the likelihood of soft or hard query terms are learnt. Once the likelihood of a query term contributing to query performance is learnt, we incorporate this information to predict the performance of the query on a pre-retrieval basis.

Methodology. Our proposed work consists of three main steps, namely: (1) developing a collection of comparable pairs of queries that address the same information need but have varying retrieval effectiveness, (2) given the pairs of queries, learning the likelihood of query terms contributing to the softness or hardness of the query, and (3) adopting the learnt term likelihood information to estimate query performance. We provide the details of each step in the following.

Developing Comparable Query Pairs. Here, the objective is to develop a collection of comparable query pairs that address the same information need but the performance of the queries in each pair is not the same. To achieve this goal, we are inspired by the DocT5Query [30,31], which has suggested that a translation function can be learned based on the T5 transformer to map documents to queries [32]. The idea is simple yet intuitive: given a collection of relevance judgements, one can finetune a T5 transformer to learn to generate the query from its relevant document. Once the transformer is finetuned, one could then use it to generate queries from any given document. In the context of our work, we adopt a similar strategy where we finetune a T5 transformer architecture based on a large relevance judgment collection. Using the finetuned T5 transformer and for the documents in the MSMARCO collection, we generate multiple queries per document. The queries generated for each document would be addressing the same information need as they have been generated for the same document but their degree of retrieval effectiveness is not the same. We create pairs of such queries where queries in each pair have differing effectiveness.

More formally, let $C = \{d_1, d_2, ..., d_m\}$ be a collection of m documents. Let us assume that a T5 transformer architecture can be finetuned, as outlined in [29,30], to serve as translation function $\mathcal{T} : \mathcal{D} \rightarrow \mathcal{Q}$, to facilitate the transition from the document space to the query space. With $\mathcal{T}$, we can generate queries for documents in $\mathcal{D}$ where each query q_d^q is seeking information from d. Succinctly, given a document d and the translation function $\mathcal{T}$, we generate q_d^g as $\mathcal{T}(d) = q_d^g$. Now, for any query q and its relevant judged document d_q, it is possible to generate alternative variations for q through $\mathcal{T}(d_q)$. Based on $\mathcal{T}(d_q)$, we develop query pairs (q, q') where $q' \in \mathcal{T}(d_q)$ and $M'_q \neq M_q$.

Learning Query Term Weights. Using the created pairs of queries, we aim to learn query term weights that signify the likelihood of terms influencing the query's *softness* or *hardness*. We utilize contextualized word embeddings for q and q', facilitating the prediction of term difficulty weights via linear regression. To discern a term's difficulty within a given query, we employ an attention mechanism, allowing the term to gradually incorporate contextual information from its interactions with other terms in the same query. Let us define $TD(q_t)$ to denote the term difficulty weight of query term q_t as follows when $M'_q > M_q$:

$$TD(q_t) = \begin{cases} -1 & \text{if } q_t \in q \text{ and } q_t \notin q' \\ 1 & \text{if } q_t \notin q \text{ and } q_t \in q' \\ 0 & \text{if } q_t \in q \text{ and } q_t \in q' \end{cases} \quad (1)$$

Equation 1 illustrates that terms present in more challenging queries have a higher likelihood of lowering query performance, whereas terms contributing to improved query performance could be considered to be easier query terms. We utilize contextualized word embeddings representing query terms and their associated weights to train a linear regression model. This model predicts $TD(q_t)$ for each query term q_t in query q. We train a model via per-token regression, aiming to minimize the Mean Squared Error (MSE) between predicted weights $\widehat{TD}(q_t)$ and target weights $TD(q_t)$ derived from Eq. 1. In other words, the model's goal is to minimize loss MSE $= \sum_{q_t \in q} \left(TD(q_t) - \widehat{TD}(q_t)\right)^2$. The regression model will be able to predict $\widehat{TD}(q_t)$ for any term within a query.
Pre-retrieval Predictor. The objective of the final step is to incorporate term difficulty weights for performance prediction. To do so, we develop two *term sets* based on the term difficulty weights predicted for each term in the query. The two term sets represent soft, $\phi^+(q)$, and hard, $\phi^-(q)$, terms, respectively:

$$\phi^+(q) = \{\Omega(q_t) \mid \hat{TD}(q_t) > 0\}, \qquad \phi^-(q) = \{\Omega(q_t) \mid \hat{TD}(q_t) < 0\} \quad (2)$$

where $\Omega()$ is a weighting function based on term difficulty weight. We adopt the weighting function in [15] to implement Ω. Given the two sets $\phi^+(q)$ and $\phi^-(q)$ for each query q, we utilize a cross-encoder architecture to estimate the performance of a query directly. To achieve this, we feed both the weighted concatenated representations of easy query terms ($\phi^+(q)$) and hard query terms ($\phi^-(q)$) into a cross-encoder network and train this network for the efficient development of μ. The goal of this network is to learn a continuous difficulty score M_q by examining the relationship between the weighted representation of the predicted easy query terms $\phi^+(q)$ and hard query terms $\phi^-(q)$. This is achieved by concatenating the query terms w.r.t their weights as suggested in [15,16]. Subsequently, we apply a linear layer to the initial vector generated by the transformer, resulting in a scalar value as $\mu(\phi^+(q), \phi^-(q))$. To further refine the network's performance, we employ a sigmoid layer σ in conjunction with a one-class Binary cross-entropy loss function l. More formally, the loss function for our cross-encoder network can be defined as follows:

$$l\left(\mu(\phi^+(q), \phi^-(q)), M(q)\right) = -\left[M(q) \cdot \log\ \sigma(\mu(\phi^+(q), \phi^-(q))) \right. \quad (3)$$

$$\left. +(1 - M(q)) \cdot \log\ (1 - \sigma(\mu(\phi^+(q), \phi^-(q)))\right] \quad (4)$$

where μ is our proposed pre-retrieval query performance predictor.

Table 1. Performance comparison between our proposed approach and other pre-retrieval QPP baselines over MS MARCO Dev on MRR@10 and TREC DL 2019,Trec DL 2020 and TREC DL HARD on ndcg@10 in terms of sMARE, Kendall τ and Spearman ρ. The highest value in each column is in bold.

QPP Method	MS MARCO Dev			TREC DL 2019			TREC DL 2020			DL Hard		
	$K-\tau$	$S-\rho$	sMARE	$K-\tau$	$S-\rho$	sMARE	$K-\tau$	$S-\rho$	sMARE	$K-\tau$	$S-\rho$	sMARE
IDF	0.116	0.154	0.330	0.158	0.245	0.321	0.245	0.353	0.374	0.116	0.152	0.342
VAR	0.062	0.083	0.333	0.107	0.152	0.290	0.059	0.077	0.318	0.016	0.035	0.349
PMI	0.017	0.023	0.323	0.009	0.017	0.341	0.040	0.056	0.344	0.022	0.031	0.349
SCS	0.037	0.049	0.333	**0.194**	**0.287**	0.316	0.272	0.397	0.333	0.106	0.140	0.326
SCQ	0.011	0.014	0.334	0.116	162	0.387	0.076	0.132	0.365	0.127	0.179	0.369
ICTF	0.114	0.152	0.330	0.153	0.240	0.360	0.345	0.330	0.330	0.107	0.115	0.314
DC	0.107	0.144	0.333	0.095	0.053	0.293	0.091	0.035	0.327	0.123	0.165	0.335
CC	0.065	0.085	0.333	0.099	0.055	0.319	0.106	0.026	0.327	0.103	0.141	0.310
IEF	0.094	0.104	0.330	0.187	0.166	0.387	0.064	0.081	0.334	0.140	0.191	0.377
our model	**0.303**	**0.401**	**0.321**	0.158	0.224	0.320	**0.290**	**0.423**	**0.314**	**0.355**	**0.492**	**0.284**
ours - $\phi^{-}(q)$	0.297	0.402	0.335	0.114	0.176	0.334	0.189	0.356	0.327	0.305	0.424	0.305
ours - $\phi^{+}(q)$	0.288	0.400	0.336	0.105	0.124	0.352	0.204	0.350	0.383	0.300	0.427	0.310

3 Experiments

3.1 Data

We employ the MSMARCO passage collection dataset [28], featuring 8.8 million passages and over 500,000 queries, each associated with at least one relevance-judged document. Following [30], we fine-tune the T5 transformer with default settings to create the translation function $\mathcal{T}$ [32] from the MSMARCO collection. Using $\mathcal{T}$, we generate queries for passages with conditions $M'_q > M_q$ and $M'_q = 1$, where q' is a generated query for q based on the relevant document. This results in 188,398 query pairs, used to train the regression model for predicting $TD()$. We evaluate our approach on four widely used query sets: MSMARCO Development set (Dev set, 6,980 queries), TREC DL 2019 [14] (43 queries), TREC DL 2020 [13] (53 queries), and DL-Hard [26] (50 queries).

3.2 Evaluation Metrics

The QPP evaluation involves correlating predicted and actual query performance [11,22] using Kendall's τ and Spearman ρ coefficients, and the scaled Mean Absolute Relative Error (sMARE) [18]. Higher Spearman and Kendall and lower sMARE values indicate better prediction accuracy. We assess based on predicting BM25 performance using official metrics MRR@10 for MS MARCO dev set and nDCG@10 for the other datasets [38].

3.3 Baselines

For the sake of comparative analysis with the state of the art, we adopt the following *pre-retrieval QPP* baselines including term-frequency baselines which utilize index statistics, including such as IDF [25] and ICTF [25]. The Simplified Clarity Score (SCS) metric [21] measures query specificity through Kullback-Leibler divergence, while SCQ [40] introduces vector-space-based query and collection similarity. Pointwise Mutual Information (PMI) analyzes term co-occurrence [19], and VAR assesses coherency based on term weight distributions [40]. From the neural-embedding based baselines, we include CC, DC, and IEF [8] that operate on term specificity. We note that for metrics that require an aggregation function, the best aggregator (average, maximum or minimum) was chosen based on the best performance of another set (DL 2019 for DL 2020, Dev for DL Hard, and vice versa).

3.4 Experimental Setup

We adopted the BERT-base-uncased [17] and got it fine-tuned for the weight prediction as the regression task for 10 epochs with a learning rate of 2e-5 and the maximum input length for queries was set to 9 (covering the maximum query length of 90% of queries in MSMARCO). For the Cross-encoder training, we employed the SentenceTransformer library [34]. This architecture underwent one epoch of training on the query pairs generated based on the finetuned T5 transformer ($\mathcal{T}$), with a batch size of 8.

3.5 Findings

We make the following observations based on the reported results in Table 1 : **(1)** we find that our approach shows the best performance on all three metrics over three of the query sets, namely MSMARCO Dev, DL 2020 and DL Hard. On the DL 2019 set, while competitive, our method does not show the best performance. However, we note that on DL 2019, there is no single baseline that shows the best performance on all metrics. Specifically, `VAR` shows the best performance on sMARE while `SCS` exhibits a stronger performance on Kendall and Spearman correlations. **(2)** In contrast to the baseline methods, our proposed approach consistently maintains robust performance across all three query sets. Notably, stronger baselines like `SCS` exhibit strong performance in specific query sets, such as TREC DL 2019 and 2020, but fall short in competitiveness on MARCO dev and DL HARD. A similar trend is observed for `IDF`, which excels in TREC DL 2020 but lacks competitiveness in the other three query sets. Despite `SCS` and `IDF` outperforming other baselines, our proposed method surpasses them by a significant margin on MS MARCO Dev set and TREC DL HARD. For instance, on the MS MARCO Dev set, `IDF` and our proposed method show Kendall τ correlations of 0.116 and 0.303, respectively. On TRECDL 2020, `SCS` achieves a Spearman of 0.397, while our method achieves a higher correlation of 0.423. **(3)** On the MS MARCO dev set, all baseline methods report Kendall τ correlations below 0.12, which is negligible compared to our model's correlation of 0.303. Our method consistently demonstrates superior consistency and performance, indicating its robustness across diverse query subsets and evaluation strategies. **(4)** Neural-based baselines exhibit less impressive results across the four query

Table 2. Sample queries color-coded to show term difficulty weights. Darker blue color indicate softer terms, and darker red colors show harder terms. Terms with no background denote terms that are neither hard or soft.

product level activity define	define a multichannel radio
definition of capias issued on a background	how far back do employment background checks
what is the gas called that they give you at the dentist	calculate the mass in grams of 2.74 l of co gas

sets, potentially because they consider the embedding representation of each query term without fine-tuning contextual representations for this specific task. In contrast, our approach predicts a term difficulty weight for each query term, learned through a fine-tuning process over the BERT language model. Therefore, the representations used by our proposed approach may be better tailored for this purpose. **(5)** Considering the impact of $\phi^{+}(q)$ and $\phi^{-}(q)$ on our proposed method's overall performance (last two rows of both tables), we observe significant performance improvement when both sets are taken into account, particularly on DL 2019, 2020, and Hard. On MSMARCO Dev, our method's performance improves on Kendall τ correlation and sMARE metric with both sets, while the performance remains consistent on Spearman correlation.

Finally, to illustrate the learned query term difficulty weights, we color-coded six sample queries in Table 2. Each row includes two queries with at least one overlapping term. In the first row, the common term between the queries is 'define'. In the first row, both queries share the term 'define.' The user seeks definitions for two phrases. Our model recognizes that the phrases enhance retrieval, but the term 'define' diminishes effectiveness. This may be due to BM25, which seeks relevant documents based on query terms, yet documents with phrase definitions may lack the term 'define.' The second example consists of two queries where the common term between them is 'background'. In this case, our model has determined that this term does not have much impact on the first query but improves retrieval effectiveness if included along with 'employment' and 'checks'. In the third row, the shared term is 'gas' and we would like to show that our model considers context to decide on term weights. As seen in this example, 'gas' was considered a hard term on the right and soft on the left. This helps us understand how our model adapts term difficulty to different situations.

4 Concluding Remarks

In this paper, we have proposed to learn contextualized query term difficulty weights that can inform the process of query performance prediction. We have shown that query term weights can be learnt, through finetuning a contextual language model, that estimate how each term can possibly impact the difficulty of a query. Through extensive experiments on five widely used query sets, we have shown that our proposed approach is both effective and consistent for predicting the performance of a range of queries.

References

1. Arabzadeh, N., Bigdeli, A., Hamidi Rad, R., Bagheri, E.: Quantifying ranker coverage of different query subspaces. In: Proceedings of the 46th International ACM SIGIR Conference on Research and Development in Information Retrieval, pp. 2298–2302 (2023)
2. Arabzadeh, N., Bigdeli, A., Seyedsalehi, S., Zihayat, M., Bagheri, E.: Matches made in heaven: toolkit and large-scale datasets for supervised query reformulation. In: Proceedings of the 30th ACM International Conference on Information and Knowledge Management, pp. 4417–4425 (2021)
3. Arabzadeh, N., Bigdeli, A., Zihayat, M., Bagheri, E.: Query performance prediction through retrieval coherency. In: Hiemstra, D., et al. (eds.) ECIR 2021. LNCS, vol. 12657, pp. 193–200. Springer, Cham (2021). https://doi.org/10.1007/978-3-030-72240-1_15
4. Arabzadeh, N., Hamidi Rad, R., Khodabakhsh, M., Bagheri, E.: Noisy perturbations for estimating query difficulty in dense retrievers. In: CIKM (2023)
5. Arabzadeh, N., Khodabakhsh, M., Bagheri, E.: Bert-qpp: contextualized pre-trained transformers for query performance prediction. In: CIKM (2021)
6. Arabzadeh, N., Mitra, B., Bagheri, E.: MS marco chameleons: challenging the MS marco leaderboard with extremely obstinate queries. In: Proceedings of the 30th ACM International Conference on Information and Knowledge Management, pp. 4426–4435 (2021)
7. Arabzadeh, N., Yan, X., Clarke, C.L.A.: Predicting efficiency/effectiveness trade-offs for dense vs. sparse retrieval strategy selection. arXiv preprint arXiv:2109.10739 (2021)
8. Arabzadeh, N., Zarrinkalam, F., Jovanovic, J., Al-Obeidat, F., Bagheri, E.: Neural embedding-based specificity metrics for pre-retrieval query performance prediction. Inf. Process. Manag. **57**(4), 102248 (2020)
9. Arabzadeh, N., Zarrinkalam, F., Jovanovic, J., Bagheri, E.: Neural embedding-based metrics for pre-retrieval query performance prediction. In: Jose, J.M., et al. (eds.) ECIR 2020. LNCS, vol. 12036, pp. 78–85. Springer, Cham (2020). https://doi.org/10.1007/978-3-030-45442-5_10
10. Arabzadeh, N., Zarrinkalam, F., Jovanovic, J., Bagheri, E.: Geometric estimation of specificity within embedding spaces. In: Proceedings of the 28th ACM International Conference on Information and Knowledge Management, pp. 2109–2112 (2019)
11. Carmel, D., Yom-Tov, E.: Estimating the query difficulty for information retrieval. Synth. Lect. Inf. Concept. Retriev. Serv. **2**(1), 1–89 (2010)
12. Carmel, D., Yom-Tov, E., Darlow, A., Pelleg, D.: What makes a query difficult? In: Proceedings of the 29th Annual International ACM SIGIR Conference on Research and Development in Information Retrieval, pp. 390–397 (2006)
13. Craswell, N., Mitra, B., Yilmaz, E., Campos, D.: Overview of the TREC 2020 deep learning track. arXiv preprint arXiv:2102.07662 (2021)
14. Craswell, N., Mitra, B., Yilmaz, E., Campos, D., Voorhees, E.M.: Overview of the trec 2019 deep learning track. arXiv preprint arXiv:2003.07820 (2020)
15. Dai, Z., Callan, J.: Context-aware sentence/passage term importance estimation for first stage retrieval. arXiv preprint arXiv:1910.10687 (2019)
16. Dai, Z., Callan, J.: Context-aware term weighting for first stage passage retrieval. In: Proceedings of the 43rd International ACM SIGIR Conference on Research and Development in Information Retrieval, pp. 1533–1536 (2020)

17. Devlin, J., Chang, M.W., Lee, K., Toutanova, K.: Bert: pre-training of deep bidirectional transformers for language understanding. arXiv preprint arXiv:1810.04805 (2018)
18. Faggioli, G., Zendel, O., Culpepper, J.S., Ferro, N., Scholer, F.: Smare: a new paradigm to evaluate and understand query performance prediction methods. Inf. Retriev. J. **25**(2), 94–122 (2022)
19. Hauff, C.: Predicting the effectiveness of queries and retrieval systems. In: SIGIR Forum, vol. 44, p. 88 (2010)
20. Hauff, C., Hiemstra, D., de Jong, F.: A survey of pre-retrieval query performance predictors. In: Proceedings of the 17th ACM Conference on Information and Knowledge Management, CIKM 2008, Napa Valley, California, 26–30 October 2008, pp. 1419–1420 (2008). https://doi.org/10.1145/1458082.1458311
21. He, B., Ounis, I.: Inferring query performance using pre-retrieval predictors. In: Apostolico, A., Melucci, M. (eds.) String Processing and Information Retrieval. LNCS, vol. 3246, pp. 43–54. Springer, Heidelberg (2004). https://doi.org/10.1007/978-3-540-30213-1_5
22. He, B., Ounis, I.: Query performance prediction. Inf. Syst. **31**(7), 585–594 (2006)
23. Khodabakhsh, M., Bagheri, E.: Semantics-enabled query performance prediction for ad hoc table retrieval. Inf. Process. Manag. **58**(1), 102399 (2021)
24. Khodabakhsh, M., Bagheri, E.: Learning to rank and predict: multi-task learning for ad hoc retrieval and query performance prediction. Inf. Sci. **639**, 119015 (2023)
25. Kwok, K.L.: A new method of weighting query terms for ad-hoc retrieval. In: Proceedings of the 19th Annual International ACM SIGIR Conference on Research and Development in Information Retrieval (SIGIR 1996), 18–22 August 1996, Zurich (Special Issue of the SIGIR Forum), pp. 187–195 (1996). https://doi.org/10.1145/243199.243266
26. Mackie, I., Dalton, J., Yates, A.: How deep is your learning: the dl-hard annotated deep learning dataset. In: Proceedings of the 44th International ACM SIGIR Conference on Research and Development in Information Retrieval (2021)
27. Meng, C., Arabzadeh, N., Aliannejadi, M., de Rijke, M.: Query performance prediction: from ad-hoc to conversational search. arXiv preprint arXiv:2305.10923 (2023)
28. Nguyen, T., et al.: MS marco: a human generated machine reading comprehension dataset. In: CoCo@ NIPS (2016)
29. Nogueira, R., Cho, K.: Passage re-ranking with bert. arXiv preprint arXiv:1901.04085 (2019)
30. Nogueira, R., Lin, J., Epistemic, A.: From doc2query to docttttquery. Online preprint **6**, 2 (2019)
31. Nogueira, R., Yang, W., Lin, J., Cho, K.: Document expansion by query prediction. arXiv preprint arXiv:1904.08375 (2019)
32. Raffel, C., et al.: Exploring the limits of transfer learning with a unified text-to-text transformer. J. Mach. Learn. Res. **21**(140), 1–67 (2020). http://jmlr.org/papers/v21/20-074.html
33. Raiber, F., Kurland, O.: Query-performance prediction: setting the expectations straight. In: Proceedings of the 37th International ACM SIGIR Conference on Research and Development in Information Retrieval, pp. 13–22 (2014)
34. Reimers, N., Gurevych, I.: Sentence-bert: sentence embeddings using siamese bert-networks. arXiv preprint arXiv:1908.10084 (2019)
35. Roy, D., Ganguly, D., Mitra, M., Jones, G.J.F.: Estimating Gaussian mixture models in the local neighbourhood of embedded word vectors for query performance

prediction. Inf. Process. Manag. **56**(3), 1026–1045 (2019). https://doi.org/10.1016/j.ipm.2018.10.009
36. Salamat, S., Arabzadeh, N., Seyedsalehi, S., Bigdeli, A., Zihayat, M., Bagheri, E.: Neural disentanglement of query difficulty and semantics. In: CIKM, pp. 4264–4268 (2023)
37. Tamannaee, M., Fani, H., Zarrinkalam, F., Samouh, J., Paydar, S., Bagheri, E.: Reque: a configurable workflow and dataset collection for query refinement. In: Proceedings of the 29th ACM International Conference on Information and Knowledge Management, pp. 3165–3172 (2020)
38. Yang, P., Fang, H., Lin, J.: Anserini: enabling the use of lucene for information retrieval research. In: Proceedings of the 40th International ACM SIGIR Conference on Research and Development in Information Retrieval, pp. 1253–1256 (2017)
39. Yom-Tov, E., Fine, S., Carmel, D., Darlow, A.: Learning to estimate query difficulty: including applications to missing content detection and distributed information retrieval. In: Proceedings of the 28th Annual International ACM SIGIR Conference on Research and Development in Information Retrieval, pp. 512–519 (2005)
40. Zhao, Y., Scholer, F., Tsegay, Y.: Effective pre-retrieval query performance prediction using similarity and variability evidence. In: Advances in Information Retrieval, 30th European Conference on IR Research, ECIR 2008, Glasgow, 30 March–3 April 2008. Proceedings. pp. 52–64 (2008). https://doi.org/10.1007/978-3-540-78646-7_8
41. Zhou, Y., Croft, W.B.: Query performance prediction in web search environments. In: Proceedings of the 30th Annual International ACM SIGIR Conference on Research and Development in Information Retrieval, pp. 543–550 (2007)

Learning to Jointly Transform and Rank Difficult Queries

Amin Bigdeli[1(✉)], Negar Arabzadeh[1], and Ebrahim Bagheri[2]

[1] University of Waterloo, Waterloo, Canada
{abigdeli,narabzad}@uwaterloo.ca

[2] Toronto Metropolitan University, Toronto, Canada
bagheri@torontomu.ca

Abstract. Recent empirical studies have shown that while neural rankers exhibit increasingly higher retrieval effectiveness on tasks such as ad hoc retrieval, these improved performances are not experienced uniformly across the range of all queries. There are typically a large subset of queries that are not satisfied by neural rankers. These queries are often referred to as *difficult queries*. Given the fact that neural rankers operate based on the similarity between the embedding representations of queries and their relevant documents, the poor performance of difficult queries can be due to the sub-optimal representations learnt for difficult queries. As such, the objective of our work in this paper is to learn to rank documents and also transform query representations in tandem such that the representation of queries are transformed into one that shows higher resemblance to their relevant document. This way, our method will provide the opportunity to satisfy a large number of difficult queries that would otherwise not be addressed. In order to learn to jointly rank documents and transform queries, we propose to integrate two forms of triplet loss functions into neural rankers such that they ensure that each query is moved along the embedding space, through the transformation of its embedding representation, in order to be placed close to its relevant document(s). We perform experiments based on the MS MARCO passage ranking task and show that our proposed method has been able to show noticeable performance improvement for queries that were extremely difficult for existing neural rankers. On average, our approach has been able to satisfy 277 queries with an MRR@10 of 0.21 for queries that had a reciprocal rank of zero on the initial neural ranker.

1 Introduction

The Information Retrieval (IR) community has immensely benefited from Large Language Models (LLMs), such as BERT [9] for improving the performance of ad hoc retrieval [21,27]. There has been a significant boost in the performance of ad hoc retrieval task, which is primarily due to how representations are learnt for terms, queries and documents within a dense embedding space [16]. However, while the overall effectiveness of neural rankers have increased by at least two folds, e.g., based on the MS MARCO passage ranking dataset [18], these improvements have not been uniformly observed over different query subsets [3].

N. Goharian et al. (Eds.): ECIR 2024, LNCS 14611, pp. 40–48, 2024.
https://doi.org/10.1007/978-3-031-56066-8_5

Researchers have traditionally understood that not all queries are satisfied to the same extent by various retrievers [11]; however, given the rather high performance of neural rankers, the number of queries that are not satisfied at all by neural rankers can be surprising [1,4]. A recent study on the effectiveness of neural rankers on the MS MARCO passage dataset showed that out of the 6,980 queries in its development set, there are at least 2,500 queries that are not addressed by any state-of-the-art neural ranker (queries with a reciprocal rank of zero). This indicates that neural rankers improve overall average retrieval effectiveness by focusing on a particular subset of queries at the expense of another subset [3,5].

In addition, there have works that have reported that the performance of neural rankers can be sensitive to the input query and the retrieval effectiveness is very dependent on how the input query is formulated [2,26,28]. Various researchers have already extensively explored how methods such as query expansion [6,8,30] and query reformulation [15,17,22] can be used to improve the effectiveness of rankers. However, most, if not all, of such methods are designed specifically for reformulating the textual surface form of the query by adding, replacing or removing terms from the original query such that the retrieval effectiveness of the reformulated query is stronger than that of the original query. While such methods have shown a strong impact on sparse retrievers [24], recent research has shown that they are not effective for improving the performance of neural rankers [2] and can even lead to degraded overall performance.

Based on observations, neural rankers (1) are quite effective in retrieving highly relevant documents for a subset of queries. According to [3], these queries are common among various neural rankers and are often referred to as easy queries; and, (2) fall short of effectively, or even minimally, addressing another large subset of queries, which we refer to as *difficult queries*. A typical neural ranker would not require any assistance in addressing easy queries; however, they would require additional mechanisms to help them effectively address difficult queries. To this end, we hypothesize that difficult queries are those whose embedding representations are not placed well within the embedding space, i.e., they are placed closer to irrelevant documents and further away from relevant ones. A possible solution to this problem would be to transform the representation of queries such that difficult queries would be placed closer to their relevant documents; hence, leading to their improved retrieval effectiveness.

The goal in this paper is to transform the representation of queries in order to improve the performance of difficult queries. To achieve this, we propose to perform (1) query representation transformation, and (2) passage ranking tasks in tandem where the transformation of the query does not happen on its surface form but rather happens through the translation of the query representation into a position within the embedding space that places it closer to its relevant passage and moves it away from other irrelevant passages. We perform our experiments on the MS MARCO passage ranking dataset and over a range of state-of-the-art neural ranking methods. Based on our experimental results, we show that our strategy to learn to transform queries and rank documents in tandem results in higher retrieval effectiveness specially on difficult queries.

The key contributions of our work can be enumerated as: (1) we propose an approach to learn to rank documents and transform queries at the same time; (2) we show that our approach is able to consistently improve the performance of the most difficult queries for various neural rankers. Furthermore, the advantages of our proposed training strategy include (a) it is widely applicable to a range of neural rankers as it does not require any changes to the architecture of the neural ranking methods; and (b) it does not require any additional overhead for learning to transform queries as this happens at the same time as when the ranking method is being trained.

2 Proposed Approach

Typically neural rankers are only trained based on relevance triplets where the network is finetuned to transform the query representation whereby the transformed representation is closer to the relevant document. Our proposed approach in this paper is based on the idea of learning to rank documents and transform queries in tandem. As such, and in order to train a neural-based retrieval method R, we need to consider two different types of triplets, namely (1) relevance triplets $< q, d_q^+, d_q^- >$ where q is the original query from a set of training queries Q, d_q^+ is the relevant document for query q, d_q^- is a negative sample document for query q; and, (2) transformation triplets $< q, q_q^+, q_q^- >$ where q_q^+ is an ideal transformation of q and q_q^- is an irrelevant (negative) query to q. We propose that the inclusion of transformation triplets in the training process of neural rankers will enable neural rankers to transform the representation of queries such that they are placed closer to the representation of their ideal alternative query.

To operationalize the training of neural rankers based on these two types of triplets, marginal ranking loss [21] is used for both the relevance triplets and transformation triplets to calculate the total loss of the neural network as follows:

$$L_{Relevance} = \frac{1}{|Q|} \sum_{i=1}^{N^+} \sum_{j=1}^{N^-} \max(0, m - ||q - d_i^+|| + ||q - d_j^-||) \tag{1}$$

$$L_{QueryTransformation} = \frac{1}{|Q|} \sum_{i=1}^{N^+} \sum_{j=1}^{N^-} \max(0, m - ||q - q_i^+|| + ||q - q_j^-||) \tag{2}$$

We define the total loss for the neural ranker as the sum of the loss of the relevance triplets and query transformation triplets expressed as $L = L_{Relevance} + L_{QueryTransformation}$. As a result, during the training procedure of the model, the goal of the loss function is to tune the network in a way that $sim(q, d_q^+) > sim(q, d_q^-)$ and $sim(q, q_q^+) > sim(q, q_q^-)$ conditions are maximized. In other words, the network is fine-tuned to transform the query representation in a way that the transformed representation is closer to the relevant document. On this basis, the model not only learns relevance but also transforms the representation of the original query to be better suited for retrieving the relevant document.

The training of a neural ranker based on the proposed loss function will require access to data samples to form the two types of triplets. It is not difficult to obtain relevance triplets based on relevance judgment collections. Let us define $D^k_{q_i} = \{d^1_{q_i}, d^2_{q_i}, ..., d^k_{q_i}\}$ as a list of k retrieved documents by a first-stage retrieval method M for query q_i from Q. Also let $D^+_{q_i} = \{d^1_{q_i}, d^2_{q_i}, ..., d^n_{q_i}\}$ and $D^-_{q_i} = \{d^1_{q_i}, d^2_{q_i}, ..., d^l_{q_i}\}$ be the list of n judged documents and a set of l irrelevant documents selected randomly from $D^k_{q_i}$ for query q_i, respectively. Relevance triplets can be randomly sampled from $D^+_{q_i}$ and $D^-_{q_i}$. The challenge; however, is to produce transformation triplets as they require access to ideal query representation for the initial input query. For this purpose, let us assume that we have access to a transformation function $\mathcal{T}$ capable of generating synthetic queries for a judged document $d^r_{q_i}$ randomly selected from $D^+_{q_i}$ of query q_i.

Based on $\mathcal{T}$, it would be possible to generate a set of queries for $d^r_{q_i}$ that would maximize retrieval effectiveness by retrieving $d^r_{q_i}$ in response to that query. For instance, given a document $d^r_{q_i}$, and the transformation function $\mathcal{T}$, we can generate $Q'_{d^r_{q_i}}$ by applying $\mathcal{T}$, expressed as $Q'_{d^r_{q_i}} = \mathcal{T}(d^r_{q_i})$.

By using $\mathcal{T}$, we can define $Q'^+_{q_i} = \{q'^1_{q_i}, q'^2_{q_i}, ..., q'^p_{q_i}\}$ as a list of p queries generated based on the judged documents $D^+_{q_i}$ associated with query q_i. Having $Q'^+_{q_i}$, we can calculate the retrieval effectiveness of each of the generated queries using a first-stage retrieval method M and then select the best performing query $q'^+_{q_i}$ as the ideal query representation of query q_i. We note that earlier research, such as [19,20], have shown that $\mathcal{T}$ can be learnt using a T5 transformer architecture and trained on a large relevance judgment collection.

Furthermore and in order to pair query q_i with a negative query that is irrelevant to q_i, we randomly select a query from the training query set Q and consider it as $q'^-_{q_i}$ as suggested in the literature [12,31]. Representing $q'^+_{q_i}$ and $q'^-_{q_i}$ as the relevant and irrelevant pairs of q_i would help the model to transform the representation of q_i closer to $q'^+_{q_i}$ that is capable of gaining higher retrieval effectiveness for retrieving the relevant document of q_i and also place it far away from the negative query $q'^-_{q_i}$ that is irrelevant.

3 Experiments

3.1 Experimental Setup

Code. The artifacts of our work are released on Github[1] for general use.
Document and Query Collection. We conduct our experiments on the widely used MSMARCO collection that consists of over 8.8M passages and over 500k queries. For the query set, we use the small dev set of the MS MARCO collection, which is frequently used for evaluation purposes. This dataset consists of 6,980 queries along with their relevance judged documents.

[1] https://github.com/aminbigdeli/query_transformation.

Table 1. Comparison between the performance (MRR@10) of different neural ranking models and ours.

Architecture	LLM	Original	Ours
Sentence Transformer	`BERT`	0.334	0.337
	`MiniLM`	0.319	0.325†
	`DistilRoBERTa`	0.305	0.308
`ColBERT`	`BERT`	0.338	0.342†
`RepBERT`	`BERT`	0.287	0.290†

Neural Rankers. To show the effectiveness of our proposed training strategy, we use different types of neural-based retrieval methods and compare the retrieval effectiveness of those models when they are trained on the original training dataset and when they are trained based on our proposed training strategy. The neural models used in our experiments include (1) `S-BERT` (Sentence Transformer) [21] with different pre-trained language models including `BERT-base-uncased` [10], `DistilRoBERTa-base` [23], and `MiniLM` [25]; (2) `ColBERT` [13], and (3) `RepBERT` [29].

Model Details and Hyperparameters. To train `S-BERT` with different pre-trained natural language models, we used the bi-encoder architecture of Sentence Transformer library with batch size being set to 64, number of epochs to 5, and the maximum sequence length is set to 350. To train the `RepBERT` model, batch size is set to 26, the number of epochs is 1, the maximum document length is set to 256, and the maximum query length is set to 20. The `ColBERT` model is also trained with the default parameters with a batch size of 32, number of epochs to 1, the maximum document length of 180, and a maximum query length of 32. In order to implement the transformation function $\mathcal{T}$ introduced in Sect. 2, as suggested by [19], we employed the T5 Transformer trained based on the 500k queries in the MS MARCO collection and their relevant documents and generate queries from relevant documents that are semantically similar to the training query and might achieve a higher retrieval effectiveness when used for retrieval. Having a set of generated queries, we select the one with the highest retrieval effectiveness and consider it as the positive pair for each of the training queries as released in [2]. As for the negative pair of each training query, we randomly select a query from the training set and consider it as the negative query. Finally, having the relevant and irrelevant documents as well as relevant and irrelevant queries for each of the training set queries, we can train the aforementioned networks by passing the representation of the query and each of the other sequences (relevant/irrelevant document and relevant/irrelevant query) to each of the networks for training.

Table 2. Comparison between the MRR@10 of the different neural ranking models trained based on the original dataset and our proposed one on difficult queries among five buckets each consisting of 698 queries.

Architecture	LLM	Training	0–10%	10–20%	20–30%	30–40%	40–50%
Sentence Transformer	BERT	Original	0	0	0	0.002	0.133
		Ours	0.024†	0.02†	0.022†	0.026†	0.148†
	MiniLM	Original	0	0	0	0	0.096
		Ours	0.022†	0.017†	0.015†	0.023†	0.115†
	DistilRoBERTa	Original	0	0	0	0	0.062
		Ours	0.021†	0.023†	0.022†	0.019†	0.078†
ColBERT	BERT	Original	0	0	0	0.017	0.142
		Ours	0.015†	0.013†	0.021†	0.034†	0.166†
RepBERT	BERT	Original	0	0	0	0	0.04
		Ours	0.011†	0.016†	0.014†	0.018†	0.05†

3.2 Findings

The main purpose of performing query transformation is to help *difficult queries* achieve better retrieval effectiveness. As such, we first report the performance of our proposed fused approach [7] compared to the performance of the various state-of-the-art neural baselines [13,21,29] over all queries in Table 1. We note that statistical significance is measured based on paired t-test with p-value of 0.05 and denoted with superscript † in the tables. As seen in Table 1, our approach exhibits a slightly superior performance compared to the baselines. Therefore, across all queries, our proposed approach that learns to rank and transform queries maintains slightly better levels of retrieval effectiveness. We additionally show, as also already reported in the relevant literature [14], that **Pseudo-Relevant Feedback (PRF) methods** do not necessarily lead to improvement on these baselines. For instance when applying PRF on the baselines, namely SBERT-PRF and ColBERT-PRF, we obtain an MRR@10 of 0.3035 and 0.2816, respectively. These are consistently lower than the performance of our model. Given our focus is on *difficult queries*, we evaluate the impact of our approach on the retrieval effectiveness of such queries. We define difficult queries for a ranking method to be those that have a poor performance when retrieved by that ranking method. Ensan et al., suggested [11] that one can consider the lowest-performing queries to be the set of difficult queries for a ranking method. More specifically, we define the most difficult queries for a ranker to be those that fall in the lower half of retrieval effectiveness compared to other queries. To identify such queries, we rank-order queries based on their MRR@10 values and choose the bottom 50% of queries to represent difficult queries. Given there are 6,980 queries in the small dev set, the bottom 50% of the queries include 3,490 queries in total. We further split these queries into five finer-grained difficulty buckets and report the

Table 3. The number of MS MARCO dev small queries helped by our method from the bottom 50% of the most difficult queries for each ranker.

Architecture	LLM	Number of Queries	MRR@10
Sentence Transformer	BERT	270 out of 2,776	0.23
	MiniLM	275 out of 2,942	0.20
	DistilRoBERTa	308 out of 3,106	0.22
ColBERT	BERT	258 out of 2,671	0.17
RepBERT	BERT	291 out of 3,233	0.17

performance of the baseline rankers as well as the performance of our proposed method in each of these difficulty buckets. Table 2 reports the results of method performance across different difficulty buckets. As seen in the table, the bottom 3 buckets (bottom 30% of queries) in all neural rankers have a reciprocal rank value of zero i.e., on these queries ($\sim 2,000$), none of the neural rankers are able to retrieve a relevant document in its top-10 ranked documents. In contrast, our approach has been able to effectively address a subset of the hard queries in each difficulty bucket. While the MRR@10 scores obtained by our method are relatively small, this should be interpreted in the context of the fact that no neural baseline included in our experiments was able to address any of the queries in this set and the absolute MRR@10 score for all of them is zero.

We further qualify the findings in Table 2 by exploring the number of queries that have been helped by our approach in each baseline. In particular, we report the number of queries that had an original reciprocal rank score of zero on the original neural ranker but later received a higher reciprocal rank score by our proposal approach. The number of such queries along with their average MRR@10 values are reported in Table 3. As seen in this table, on average, our proposed approach has been able to improve the performance of 277 queries per neural ranker. This means that there are on average 277 queries that were originally completely unsatisfied by the base ranker and then addressed by the proposed approach. The average improvement on MRR@10 over these queries is 0.21. This shows that when the representation of the original query is transformed in the embedding space, a noticeable number of difficult queries are then effectively, or at least partially, addressed by the neural ranker.

4 Concluding Remarks

In this paper, we have proposed a strategy to learn to rank documents while at the same time learning an embedding transformation from the original query to one that maximizes retrieval effectiveness. The idea of our approach is that extremely difficult queries can be addressed if their embedding representations are moved closer to an ideal query representation that has a high retrieval effectiveness. Based on our experiments on MS the MARCO passage collection development set, we have shown that the representation of the transformed queries from our proposed approach can lead to increased performance on difficult queries.

References

1. Arabzadeh, N., Bigdeli, A., Hamidi Rad, R., Bagheri, E.: Quantifying ranker coverage of different query subspaces. In: Proceedings of the 46th International ACM SIGIR Conference on Research and Development in Information Retrieval, pp. 2298–2302 (2023)
2. Arabzadeh, N., Bigdeli, A., Seyedsalehi, S., Zihayat, M., Bagheri, E.: Matches made in heaven: toolkit and large-scale datasets for supervised query reformulation. In: Proceedings of the 30th ACM International Conference on Information and Knowledge Management, pp. 4417–4425 (2021)
3. Arabzadeh, N., Mitra, B., Bagheri, E.: MS marco chameleons: challenging the MS marco leaderboard with extremely obstinate queries. In: Proceedings of the 30th ACM International Conference on Information and Knowledge Management, pp. 4426–4435 (2021)
4. Arabzadeh, N., Vtyurina, A., Yan, X., Clarke, C.L.A.: Shallow pooling for sparse labels. arXiv preprint arXiv:2109.00062 (2021)
5. Arabzadeh, N., Yan, X., Clarke, C.L.A.: Predicting efficiency/effectiveness trade-offs for dense vs. sparse retrieval strategy selection. arXiv preprint arXiv:2109.10739 (2021)
6. Azad, H.K., Deepak, A.: Query expansion techniques for information retrieval: a survey. Inf. Process. Manag. **56**(5), 1698–1735 (2019)
7. Bassani, E., Romelli, L.: ranx.fuse: a python library for metasearch. In: Hasan, M.A., Xiong, L. (eds.) Proceedings of the 31st ACM International Conference on Information and Knowledge Management, Atlanta, 17–21 October 2022, pp. 4808–4812. ACM (2022). DOI: https://doi.org/10.1145/3511808.3557207
8. Carpineto, C., Romano, G.: A survey of automatic query expansion in information retrieval. ACM Comput. Surv. **44**(1), 1–50 (2012)
9. Devlin, J., Chang, M.W., Lee, K., Toutanova, K.: Bert: pre-training of deep bidirectional transformers for language understanding. arXiv preprint arXiv:1810.04805 (2018)
10. Devlin, J., Chang, M.W., Lee, K., Toutanova, K.: Bert: pre-training of deep bidirectional transformers for language understanding. arXiv preprint arXiv:1810.04805 (2018)
11. Ensan, F., Bagheri, E.: Document retrieval model through semantic linking. In: Proceedings of the Tenth ACM International Conference on Web Search and Data Mining, pp. 181–190 (2017)
12. Karpukhin, V., et al.: Dense passage retrieval for open-domain question answering. arXiv preprint arXiv:2004.04906 (2020)
13. Khattab, O., Zaharia, M.: Colbert: efficient and effective passage search via contextualized late interaction over bert. In: Proceedings of the 43rd International ACM SIGIR Conference on Research and Development in Information Retrieval, pp. 39–48 (2020)
14. Li, H., Mourad, A., Zhuang, S., Koopman, B., Zuccon, G.: Pseudo relevance feedback with deep language models and dense retrievers: successes and pitfalls. ACM Trans. Inf. Syst. **41**(3), 1–40 (2023)

15. Li, X., et al.: A cooperative neural information retrieval pipeline with knowledge enhanced automatic query reformulation. In: Proceedings of the Fifteenth ACM International Conference on Web Search and Data Mining, pp. 553–561 (2022)
16. Lin, J., Nogueira, R., Yates, A.: Pretrained transformers for text ranking: BERT and beyond. arXiv preprint arXiv:2010.06467 (2020)
17. Lioma, C., Ounis, I.: A syntactically-based query reformulation technique for information retrieval. Inf.Process. Manag. **44**(1), 143–162 (2008)
18. Nguyen, T., et al.: MS marco: a human generated machine reading comprehension dataset. Choice **2640**, 660 (2016)
19. Nogueira, R., Lin, J., Epistemic, A.: From doc2query to docttttquery. Online Preprint **6** (2019)
20. Nogueira, R.F., Yang, W., Lin, J., Cho, K.: Document expansion by query prediction. arXiv preprint arXiv:1904.08375 (2019)
21. Reimers, N., Gurevych, I.: Sentence-bert: sentence embeddings using siamese bert-networks. arXiv preprint arXiv:1908.10084 (2019)
22. Rieh, S.Y., et al.: Analysis of multiple query reformulations on the web: the interactive information retrieval context. Inf. Process. Manag. **42**(3), 751–768 (2006)
23. Sanh, V., Debut, L., Chaumond, J., Wolf, T.: Distilbert, a distilled version of bert: smaller, faster, cheaper and lighter. arXiv preprint arXiv:1910.01108 (2019)
24. Tamannaee, M., Fani, H., Zarrinkalam, F., Samouh, J., Paydar, S., Bagheri, E.: Reque: a configurable workflow and dataset collection for query refinement. In: Proceedings of the 29th ACM International Conference on Information and Knowledge Management, pp. 3165–3172 (2020)
25. Wang, W., Wei, F., Dong, L., Bao, H., Yang, N., Zhou, M.: Minilm: deep self-attention distillation for task-agnostic compression of pre-trained transformers. Adv. Neural Inf. Process. Syst. **33**, 5776–5788 (2020)
26. Wang, X., Macdonald, C., Ounis, I.: Deep reinforced query reformulation for information retrieval. arXiv preprint arXiv:2007.07987 (2020)
27. Xiong, L., et al.: Approximate nearest neighbor negative contrastive learning for dense text retrieval. arXiv preprint arXiv:2007.00808 (2020)
28. Zerveas, G., Zhang, R., Kim, L., Eickhoff, C.: Brown university at trec deep learning 2019. arXiv preprint arXiv:2009.04016 (2020)
29. Zhan, J., Mao, J., Liu, Y., Zhang, M., Ma, S.: Repbert: contextualized text embeddings for first-stage retrieval. arXiv preprint arXiv:2006.15498 (2020)
30. Zheng, Z., Hui, K., He, B., Han, X., Sun, L., Yates, A.: Bert-qe: contextualized query expansion for document re-ranking. arXiv preprint arXiv:2009.07258 (2020)
31. Zhou, K., et al.: Simans: simple ambiguous negatives sampling for dense text retrieval. arXiv preprint arXiv:2210.11773 (2022)

Estimating Query Performance Through Rich Contextualized Query Representations

Sajad Ebrahimi[1](✉), Maryam Khodabakhsh[2], Negar Arabzadeh[3], and Ebrahim Bagheri[4]

[1] University of Guelph, Guelph, ON, Canada
sebrah05@uoguelph.ca
[2] Shahrood University of Technology, Shahrood, Iran
m_khodabakhsh@shahroodut.ac.ir
[3] University of Waterloo, Waterloo, ON, Canada
narabzad@uwaterloo.ca
[4] Toronto Metropolitan University, Toronto, ON, Canada
bagheri@torontomu.ca

Abstract. The state-of-the-art query performance prediction methods rely on the fine-tuning of contextual language models to estimate retrieval effectiveness on a per-query basis. Our work in this paper builds on this strong foundation and proposes to learn rich query representations by learning the interactions between the query and two important contextual information, namely (1) the set of documents retrieved by that query, and (2) the set of similar historical queries with known retrieval effectiveness. We propose that such contextualized query representations can be more accurate estimators of query performance as they embed the performance of past similar queries and the semantics of the documents retrieved by the query. We perform extensive experiments on the MSMARCO collection and its accompanying query sets including MSMARCO Dev set and TREC Deep Learning tracks of 2019, 2020, 2021, and DL-Hard. Our experiments reveal that our proposed method shows robust and effective performance compared to state-of-the-art baselines.

1 Introduction

Information Retrieval (IR) researchers have been concerned with both the *effectiveness* and *robustness* of retrieval methods [5,12,31]. A successful IR method would be one that simultaneously shows both effective and robust performance, i.e., shows strong and consistent performance over a large range of queries. While not ideal, but in practice, IR methods are often only effective on a subset of queries and less effective on others. By identifying challenging queries for an IR method, it would be possible to adopt alternative strategies to satisfy these queries such as query routing [37], query reformulation [34], and asking users to clarify their intents [1,6]. The task of *Query Performance Prediction (QPP)* speaks directly to this need and focuses on estimating the effectiveness of retrieval methods on input queries.

N. Goharian et al. (Eds.): ECIR 2024, LNCS 14611, pp. 49–58, 2024.
https://doi.org/10.1007/978-3-031-56066-8_6

Broadly speaking, QPP methods have been classified into the categories of *pre-retrieval* and *post-retrieval* methods [10]. The former is concerned with predicting the performance of a query prior to retrieval, whereas the latter counterparts consider additional information accessible after an initial retrieval stage, such as retrieval scores of retrieved documents [2,7–9,36,39,42], among others. Given access to a wider range of information, post-retrieval methods often exhibit stronger performance and can themselves be seen as unsupervised [3,15,28,39,42,44] and supervised [4,16,18,23,43] variations. Most recently and thanks to large-scale relevance judgment collections such as MS MARCO [30], researchers have more extensively explored supervised post-retrieval QPP methods [4,16,18,43]. For example, NeuralQPP [43] was among the first neural frameworks that used unsupervised QPP methods as weak signals to learn a more effective supervised method. There has also been interest in employing contextualized Large Language Models (LLM) [11] in the QPP task where the performance of a query is estimated through the finetuning of an LLM [4,16,24]. Most existing supervised post-retrieval QPP methods focus on estimating the performance of a query by finetuning an LLM for this purpose [4,16]. The underlying assumption of these methods is that the semantic finetuned representation of the query obtained from an LLM may be correlated with the performance of the query.

Our work in this paper aligns closely with earlier works [4,16] and provides a more generalizable framework to learn rich and contextualized query representations that can more effectively estimate the performance of the query. We propose that a rich query representation suitable for query performance prediction would be one that is informed by the interaction between the query and (1) the documents retrieved by that query, and (2) the set of similar historical queries with known effectiveness. The underlying premise of our work is that contextual language models capture meaningful geometric relations [17,21,29]; therefore, contents that are placed closer to each other in the embedding space carry similar semantics and hence would exhibit similar characteristics, such as comparable retrieval effectiveness, when used in applications like retrieval. On this basis, learning rich contextualized representations for queries that are influenced by *relevant content from the document space* as well as *relevant content from the query space* can provide insight into the potential effectiveness of the query. Our work is guided by the hypothesis that queries that are embedded in close proximity to effective historical queries and semantically relevant documents are more likely to be effective queries themselves. In contrast, queries that are embedded in close proximity to ineffective queries and whose set of retrieved documents do not have a semantic resemblance to the query are more likely to be ineffective.

For this reason, we propose to learn rich contextualized query representations based on the interaction between the query, its retrieved documents, and past similar historical queries in order to predict the query's potential effectiveness. In order to learn such rich representations, we propose to finetune a contextual language model to capture these interactions through a cross-encoder architecture. To show the effectiveness of our approach, we have performed extensive experiments on five widely used MS MARCO datasets, namely the Dev, TREC DL 2019, 2020, 2021 and DL-Hard sets [11,13,26,30]. Our experiments show that our method enjoys significantly higher effectiveness on the QPP task compared to other state-of-the-art approaches. For reproducibility purposes, we made our code and model publicly available at GitHub[1].

2 Proposed Approach

Problem Definition. Let C, q, R, D_q, be the collection of documents, input query, a retrieval method, and a ranked list of documents retrieved by R in response to query q, respectively. The task of QPP is concerned with developing a predictor μ to estimate the performance of R on q based on a given retrieval metric M, e.g., average precision or reciprocal rank, without accessing the relevance judgments. This can be expressed as: $\widehat{M_q} = \mu(q, D_q, C)$ where $\widehat{M_q}$ is an estimated value of M for query q.

Hypothesis. The underlying premise of our work is that a rich contextualized representation for a query, which can effectively predict the performance of the query should not only consider the representation of the query itself but also capture (1) *The association between the retrieved documents by the query and the query*: Earlier research has shown that the qualities of the list of documents retrieved for a query, such as coherence [14], can be indicators of the possible effectiveness of that query. As such, a rich query representation that can encode and embody the characteristics of the retrieved set of documents for that query is more likely to effectively estimate query performance; and (2) *The relation with past similar queries with known effectiveness*: A rich query representation that can effectively identify past similar queries with comparable retrieval effectiveness will have a higher likelihood of estimating the effectiveness of the query based on its association with queries with analogous performance.

Proposed Formulation. To encode the above three characteristics in our rich query representation, we contextualize the query as follows: (1) We capture the individual characteristics of the query through its contextualized representation using a pre-trained language model. This ensures that queries that are both semantically and syntactically similar to each other receive comparable representations; (2) The properties of the retrieved documents are also considered by representing them through their contextualized representations, and (3) The association between each query and its most similar historical queries is also computed through their geometric distance in the embedding space. Two queries would be considered to be more similar if they have smaller distances from each other. Similar queries can be identified through a *nearest neighbor* scheme. We systematically incorporate this contextual information into our query representation through a cross-encoder architecture that finetunes a language model to estimate query performance. In particular, for query q, we let the cross-encoder architecture estimate the performance of q, through regression, based on the contextualized representation of q, D_q, and a set of most similar queries to q with

[1] https://github.com/sadjadeb/nearest-neighbour-qpp.

Table 1. The performance of our proposed approach and baselines on Dev set with MRR@10 and DL-Hard with NDCG@10. The correlations are statistically significant with 95% confidence interval.

	MS MARCO Dev			DL-Hard		
	$p-\rho$	$k-\tau$	$s-\rho$	$p-\rho$	$k-\tau$	$s-\rho$
Clarity	0.149	0.258	0.345	0.149	0.099	0.126
WIG	0.154	0.170	0.227	0.331	0.260	0.348
QF	0.170	0.210	0.264	0.210	0.164	0.217
NeuralQPP	0.193	0.171	0.227	0.173	0.111	0.134
n($\sigma_{\%}$)	0.221	0.217	0.284	0.195	0.120	0.147
RSD	0.310	0.337	0.447	0.362	0.322	0.469
SMV	0.311	0.271	0.357	0.375	0.269	0.408
NQC	0.315	0.272	0.358	0.384	0.288	0.417
UEF_{NQC}	0.316	0.303	0.398	0.359	0.319	0.463
NQA-QPP	0.451	0.364	0.475	0.386	0.297	0.418
BERT-QPP	0.517	0.400	0.520	0.404	0.345	0.472
qpp-BERT-PL	0.520	0.413	0.522	0.330	0.266	0.390
qpp-PRP	0.302	0.311	0.412	0.090	0.061	0.063
Ours	**0.555**	**0.421**	**0.544**	**0.434**	**0.412**	**0.508**

known effectiveness $\hat{Q}_q$. We encode q, D_q, and $\hat{Q}_q$ and their retrieval effectiveness by concatenating them using a special separator token and then apply a linear layer on the first vector produced to estimate a scalar value of $\widehat{M}$ as the difficulty of the query. We leverage a sigmoid layer and a one-class Binary cross-entropy loss function. Given $M(q, D_q)$ as the desired ranking metric, such as average precision, we adopt the following loss function to train for query performance:

$$\ell(\widehat{M_q}, M(q, D_q)) = -w[M(q, D_q).log(\sigma(\widehat{M_q})) + (1 - M(q, D_q).log(1 - \sigma(\widehat{M_q})))] \quad (1)$$

Nearest Neighbor Queries. In addition to the query itself, and the set of retrieved documents retrieved by R for q, our approach also requires access to Q_q. In order to identify the set of most similar queries to q, we first collect a set of historical queries with known effectiveness, namely $\chi = \{(q_1, M_1), (q_2, M_2), ..., (q_n, M_n)\}$. Given a representation function f that maps each q_i to a vector in the embedding space, we define a query store QS consisting of a set of key-value pairs where each key is the embedding representation of a previously seen query and a corresponding value that denotes the performance of that query. The query store QS can be formulated as:

$$QS \stackrel{\text{def}}{=} \{(\kappa, v)\} = \{(f(q_i), M(q_i))\} \forall i \in \{1, ..., n\} \quad (2)$$

The query store QS can be indexed using an approximate nearest neighborhood indexing mechanism [22], allowing for the efficient retrieval of $\hat{Q}_q$ for query q. Given a distance function $\Psi(.,.)$, we first generate a contextualized representation of query $f(q)$ and then find the nearest neighbors of the query from QS.

Table 2. The performance of our proposed approach and baselines on Trec Deep Learning Track 2019, 2020, and 2021. The correlations are statistically significant on NDCG@10 with 95% confidence interval.

	2019			2020			2021		
	$p-\rho$	$k-\tau$	$s-\rho$	$p-\rho$	$k-\tau$	$s-\rho$	$p-\rho$	$k-\tau$	$s-\rho$
Clarity	0.271	0.229	0.332	0.360	0.215	0.296	0.111	0.070	0.094
WIG	0.310	0.158	0.226	0.204	0.117	0.166	0.197	0.195	0.270
QF	0.295	0.240	0.340	0.358	0.266	0.366	0.132	0.101	0.142
NeuralQPP	0.289	0.159	0.224	0.248	0.129	0.179	0.134	0.221	0.188
n($\sigma_{\%}$)	0.371	0.256	0.377	0.480	0.329	0.478	0.269	0.169	0.256
RSD	0.460	0.262	0.394	0.426	0.364	0.508	0.256	0.224	0.340
SMV	0.495	0.289	0.440	0.450	**0.391**	**0.539**	0.252	0.192	0.278
NQC	0.466	0.267	0.399	0.464	0.294	0.423	0.271	0.201	0.292
UEF$_{NQC}$	0.507	0.293	0.432	**0.511**	0.347	0.476	0.272	0.223	0.327
NQA-QPP	0.348	0.164	0.255	0.507	0.347	0.496	0.258	0.185	0.265
BERT-QPP	0.491	0.289	0.412	0.467	0.364	0.448	0.262	0.237	0.34
qpp-BERT-PL	0.432	0.258	0.361	0.427	0.280	0.392	0.247	0.172	0.292
qpp-PRP	0.321	0.181	0.229	0.189	0.157	0.229	0.027	0.004	0.015
Ours	**0.519**	**0.318**	**0.459**	0.462	0.318	0.448	**0.322**	**0.266**	**0.359**

3 Experiments

Training Set. For our experiments, we adopt the well-known MS MARCO passage retrieval dataset [30], which consists of 8.8M passages. The training set includes over 500k search queries that correspond with at least one relevance-judged passage. We utilize this set of queries to build the query store QS and also to train our model. For each query q in this dataset, we obtain the retrieved documents D_q, as well as nearest neighbor queries Q_q, which are then used for training the cross-encoder architecture and estimating M_q.

Test Set. We evaluate our model on five query sets: (1) The MS MARCO Development set, also referred to as the Dev set, which consists of 6,980 queries. (2) The TREC Deep Learning Track 2019, [13], (3) The TREC Deep Learning Track 2020 [11], (4) The TREC Deep Learning Track 2021 [41], and (5) The TREC Deep Learning Hard set (DL-Hard) [27]. These query sets consist of 43, 54, 47, and 50 queries, respectively, and differ from the MS MARCO Dev set in that they provide multiple relevant judged passages for each query whereas the Dev set consists of, on average, one judged passage per query.

Evaluation Metrics. For evaluation purposes, we follow the well-known strategy of computing the correlation between the set of queries that are ranked based on their predicted performance against their actual performance based on the standard performance metric for each dataset, i.e., MRR@10 on the Dev set and NDCG@10 on the others. To this end, we compute Pearson ρ for linear correlation, Kendall τ, and Spearman ρ for rank correlation. A higher correlation value shows more accurate query performance prediction. For the retrieval method, we estimate BM25 implemented by Pyserini [25].

Experimental Setup. We use the pre-trained language model, DeBERTa [20], which has shown huge success in different downstream IR and Natural Language processing tasks [19], to create vector representations and conduct nearest-neighbor sampling using Faiss [22]. In order to implement $\Psi(.,.)$, we adapt the widely-used cosine similarity distance. In order to train our model, we adopt the implementation offered for the cross-encoder architecture in the SentenceTransformers package [33]. Without loss of generality and as suggested in [4], we set $|D_q| = 1$ as well as $|\hat{Q}_q| = 1$ and we used Map@20 as for retrieval effectiveness labels of our model as well as the effectiveness of $\hat{Q}_q$. The model was trained on a 24 GB NVIDIA GeForce RTX 3090 GPU with a batch size of 16 for one epoch, and the training process took an hour and a half.

Baselines. We compare our model against the state-of-the-art post-retrieval QPP baselines. These include the following methods: `Clarity` which works based on the KL divergence between language models induced from retrieved documents and the corpus. `WIG` [44], `NQC` [39], $n(\sigma\%)$ [15], RSD [35] and `SMV` [42] which are all score-based methods that predict query performance by computing different statistics of the retrieval scores of the top-ranked documents. Unlike the above predictors that are unsupervised, NeuralQPP [43] is the first supervised QPP method that uses existing unsupervised QPP methods as signals to perform weakly-supervised learning. The Utility Estimation Framework (UEF) [38] is designed to function alongside highly effective QPP baseline methods such as NQC. NQA-QPP [18], is another supervised method that uses a BERT model to learn the representations of queries and documents. BERT-QPP [4] also which fine-tunes BERT to directly predict the retrieval score of the query. QppBERT-PL [16] is one of the most recent BERT-based methods that uses point-wise training on individual queries, and list-wise training over top-ranked pseudo-relevant documents. Additionally, we consider QPP-PRP [40], which was designed to evaluate the performance of neural rankers by assessing the level of agreement between a pairwise neural reranker, exemplified by DuoT5 [32], and the ranked list generated by the neural ranker for a given query.

Findings. The results of our experiments compared to the state-of-the-art baselines are shown in Tables 1 and 2. We can highlight the following findings based on the results in these two tables:

(1) Our proposed method has shown better performance compared to all baseline models on four of the five test sets. This includes the MS MARCO Dev set, which consists of the largest number of queries, and also the DL-Hard set, which includes challenging queries.

(2) On the DL 2020 dataset where our proposed approach does not show the best performance, we find that there is no single baseline that shows the best performance on all three metrics. In fact `NQA-QPP` shows the best performance on Pearson correlation while `SMV` shows better performance on Kendall and Spearman correlations. Therefore, there are no robust baselines for this dataset in the state of the art.

(3) We note that on the DL 2020 dataset where we do not outperform the baselines, our proposed approach has a similar performance to supervised

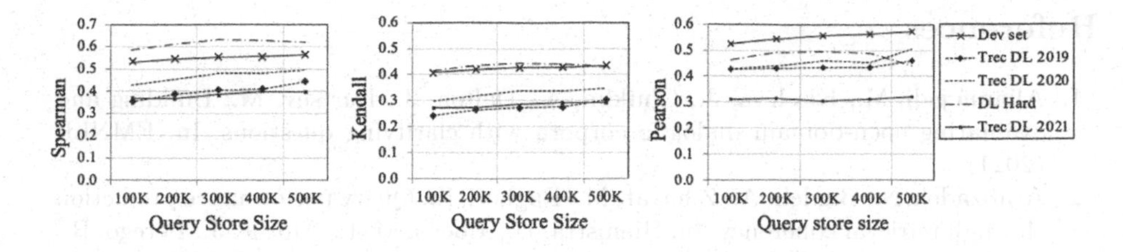

Fig. 1. The impact of the query store size on the performance of our approach.

baseline methods that fine-tune a contextual embedding model, i.e., `BERT-QPP`, `qpp-BERT-PL`, and `qpp-PRP`. In fact on this dataset, unsupervised baselines that do not use contextual embeddings such as `SMV` and `RSD` show better performance on rank correlation metrics, i.e., Spearman and Kendall correlations.

(4) In terms of robustness, our proposed approach shows the most consistent performance compared to all the baselines. For instance, `SMV` that shows a high rank correlation on DL 2020, does not show competitive performance on DL 2021 or MS MARCO Dev. Similarly, `qpp-BERT-PL`, which offers the best performance among the baselines on MS MARCO Dev, is not competitive on DL-Hard. However, our approach shows a consistent behavior across all datasets and metrics.

Finally, we explore the impact of the query store size on the performance of our proposed approach. Given we integrate most similar queries from the query store with known retrieval effectiveness, the size of the query store can have an impact on the performance of our model. We empirically study to what extent having a smaller query store size could have a negative impact on the performance of our method. To assess this, we randomly down-sampled the query store by only including 100k, 200k, 300k, 400k, and 500k from the MS MARCO training set. We report the impact of this down-sampling on all five query sets and using three correlation measures in Fig. 1. As seen in the figure, while larger query stores lead to improved overall performance, smaller query stores still show competitive performance. In fact, we find that the differences between the smallest query store size and the largest are not statistically significant.

4 Concluding Remarks

In this paper, we have shown that a rich contextualized query representation that encodes the semantics of the query itself, as well as the interactions of the query with its set of retrieved documents, along with its most similar historical queries, can be quite effective for predicting the performance of the query. It means the model can perform better when it knows the performance of similar data to the given input. Our experiments performed on five widely used datasets show that our proposed approach offers strong and robust performance on a range of QPP metrics.

References

1. Aliannejadi, M., Kiseleva, J., Chuklin, A., Dalton, J., Burtsev, M.: Building and evaluating open-domain dialogue corpora with clarifying questions. In: EMNLP (2021)
2. Arabzadeh, N., Bigdeli, A., Zihayat, M., Bagheri, E.: Query performance prediction through retrieval coherency. In: Hiemstra, D., Moens, M.-F., Mothe, J., Perego, R., Potthast, M., Sebastiani, F. (eds.) ECIR 2021, Part II. LNCS, vol. 12657, pp. 193–200. Springer, Cham (2021). https://doi.org/10.1007/978-3-030-72240-1_15
3. Arabzadeh, N., Hamidi Rad, R., Khodabakhsh, M., Bagheri, E.: Noisy perturbations for estimating query difficulty in dense retrievers. In: Proceedings of the 32nd ACM International Conference on Information and Knowledge Management, pp. 3722–3727 (2023)
4. Arabzadeh, N., Khodabakhsh, M., Bagheri, E.: BERT-QPP: contextualized pre-trained transformers for query performance prediction. In: Proceedings of the 30th ACM International Conference on Information & Knowledge Management, pp. 2857–2861 (2021)
5. Arabzadeh, N., Mitra, B., Bagheri, E.: MS Marco Chameleons: challenging the MS Marco leaderboard with extremely obstinate queries. In: Proceedings of the 30th ACM International Conference on Information & Knowledge Management, pp. 4426–4435 (2021)
6. Arabzadeh, N., Seifikar, M., Clarke, C.L.: Unsupervised question clarity prediction through retrieved item coherency. In: Proceedings of the 31st ACM International Conference on Information & Knowledge Management, pp. 3811–3816 (2022)
7. Arabzadeh, N., Zarrinkalam, F., Jovanovic, J., Al-Obeidat, F., Bagheri, E.: Neural embedding-based specificity metrics for pre-retrieval query performance prediction. Inf. Process. Manage. **57**(4), 102248 (2020)
8. Arabzadeh, N., Zarrinkalam, F., Jovanovic, J., Bagheri, E.: Neural embedding-based metrics for pre-retrieval query performance prediction. In: Jose, J.M., et al. (eds.) ECIR 2020, Part II 42. LNCS, vol. 12036, pp. 78–85. Springer, Cham (2020). https://doi.org/10.1007/978-3-030-45442-5_10
9. Arabzadeh, N., Zarrinkalam, F., Jovanovic, J., Bagheri, E.: Geometric estimation of specificity within embedding spaces. In: Proceedings of the 28th ACM International Conference on Information and Knowledge Management, pp. 2109–2112 (2019)
10. Carmel, D., Yom-Tov, E.: Estimating the Query Difficulty for Information Retrieval. Synthesis Lectures on Information Concepts, Retrieval, and Services, pp. 1–89 (2010). https://doi.org/10.1007/978-3-031-02272-2
11. Craswell, N., Mitra, B., Yilmaz, E., Campos, D.: Overview of the TREC 2020 deep learning track. CoRR abs/2102.07662 (2021). https://arxiv.org/abs/2102.07662
12. Craswell, N., Mitra, B., Yilmaz, E., Campos, D., Lin, J.: MS Marco: benchmarking ranking models in the large-data regime. In: Proceedings of the 44th International ACM SIGIR Conference on Research and Development in Information Retrieval, pp. 1566–1576 (2021)
13. Craswell, N., Mitra, B., Yilmaz, E., Campos, D., Voorhees, E.M.: Overview of the TREC 2019 deep learning track. In: Text REtrieval Conference (TREC) (2020)
14. Cronen-Townsend, S., Zhou, Y., Croft, W.B.: Predicting query performance. In: Proceedings of the 25th Annual International ACM SIGIR Conference on Research and Development in Information Retrieval, pp. 299–306 (2002)
15. Cummins, R., Jose, J., O'Riordan, C.: Improved query performance prediction using standard deviation. In: Proceedings of the 34th International ACM SIGIR

Conference on Research and Development in Information Retrieval, pp. 1089–1090 (2011)
16. Datta, S., MacAvaney, S., Ganguly, D., Greene, D.: A 'pointwise-query, listwise-document' based QPP approach. In: Proceedings of the 45th International ACM SIGIR Conference on Research and Development in Information Retrieval (2022). https://doi.org/10.1145/3477495.3531821
17. Devlin, J., Chang, M., Lee, K., Toutanova, K.: BERT: pre-training of deep bidirectional transformers for language understanding. In: Proceedings of the 2019 Conference of the North American Chapter of the Association for Computational Linguistics: Human Language Technologies, NAACL-HLT 2019, Minneapolis, MN, USA, 2–7 June 2019, Volume 1 (Long and Short Papers), pp. 4171–4186. Association for Computational Linguistics (2019). https://doi.org/10.18653/v1/n19-1423
18. Hashemi, H., Zamani, H., Croft, W.B.: Performance prediction for non-factoid question answering. In: Proceedings of the 2019 ACM SIGIR International Conference on Theory of Information Retrieval, pp. 55–58 (2019)
19. He, P., Gao, J., Chen, W.: DeBERTaV33: improving DeBERTa using ELECTRA-style pre-training with gradient-disentangled embedding sharing. arXiv preprint arXiv:2111.09543 (2021)
20. He, P., Liu, X., Gao, J., Chen, W.: DeBERTa: decoding-enhanced BERT with disentangled attention. In: International Conference on Learning Representations (2021). https://openreview.net/forum?id=XPZIaotutsD
21. Hofmann, V., Pierrehumbert, J.B., Schütze, H.: Dynamic contextualized word embeddings. arXiv preprint arXiv:2010.12684 (2020)
22. Johnson, J., Douze, M., Jégou, H.: Billion-scale similarity search with GPUs. IEEE Trans. Big Data **7**(3), 535–547 (2019)
23. Khodabakhsh, M., Bagheri, E.: Semantics-enabled query performance prediction for ad hoc table retrieval. Inf. Process. Manag. **58**(1), 102399 (2021). https://doi.org/10.1016/J.IPM.2020.102399
24. Khodabakhsh, M., Bagheri, E.: Learning to rank and predict: multi-task learning for ad hoc retrieval and query performance prediction. Inf. Sci. **639**, 119015 (2023)
25. Lin, J., Ma, X., Lin, S.C., Yang, J.H., Pradeep, R., Nogueira, R.: Pyserini: a Python toolkit for reproducible information retrieval research with sparse and dense representations. In: Proceedings of the 44th International ACM SIGIR Conference on Research and Development in Information Retrieval, pp. 2356–2362 (2021)
26. Mackie, I., Dalton, J., Yates, A.: How deep is your learning: the DL-HARD annotated deep learning dataset. In: Proceedings of the 44th International ACM SIGIR Conference on Research and Development in Information Retrieval (2021)
27. Mackie, I., Dalton, J., Yates, A.: How deep is your learning: the DL-HARD annotated deep learning dataset. In: Proceedings of the 44th International ACM SIGIR Conference on Research and Development in Information Retrieval, pp. 2335–2341 (2021)
28. Meng, C., Arabzadeh, N., Aliannejadi, M., de Rijke, M.: Query performance prediction: from ad-hoc to conversational search. arXiv preprint arXiv:2305.10923 (2023)
29. Mikolov, T., Chen, K., Corrado, G., Dean, J.: Efficient estimation of word representations in vector space. arXiv preprint arXiv:1301.3781 (2013)
30. Nguyen, T., et al.: MS Marco: a human generated machine reading comprehension dataset. In: CoCo@ NIPs (2016)
31. Penha, G., Câmara, A., Hauff, C.: Evaluating the robustness of retrieval pipelines with query variation generators. In: Hagen, M., et al. (eds.) ECIR 2022. LNCS,

vol. 13185, pp. 397–412. Springer, Cham (2022). https://doi.org/10.1007/978-3-030-99736-6_27
32. Pradeep, R., Nogueira, R., Lin, J.: The Expando-Mono-Duo design pattern for text ranking with pretrained sequence-to-sequence models (2021)
33. Reimers, N., Gurevych, I.: Sentence-BERT: sentence embeddings using Siamese BERT-networks. In: Proceedings of the 2019 Conference on Empirical Methods in Natural Language Processing. Association for Computational Linguistics (2019)
34. Roitman, H., Erera, S., Feigenblat, G.: A study of query performance prediction for answer quality determination. In: Proceedings of the 2019 ACM SIGIR International Conference on Theory of Information Retrieval, pp. 43–46 (2019)
35. Roitman, H., Erera, S., Weiner, B.: Robust standard deviation estimation for query performance prediction. In: Kamps, J., Kanoulas, E., de Rijke, M., Fang, H., Yilmaz, E. (eds.) Proceedings of the ACM SIGIR International Conference on Theory of Information Retrieval, ICTIR 2017, Amsterdam, The Netherlands, 1–4 October 2017, pp. 245–248. ACM (2017). https://doi.org/10.1145/3121050.3121087
36. Salamat, S., Arabzadeh, N., Seyedsalehi, S., Bigdeli, A., Zihayat, M., Bagheri, E.: Neural disentanglement of query difficulty and semantics. In: Proceedings of the 32nd ACM International Conference on Information and Knowledge Management, pp. 4264–4268 (2023)
37. Sarnikar, S., Zhang, Z., Zhao, J.L.: Query-performance prediction for effective query routing in domain-specific repositories. J. Am. Soc. Inf. Sci. **65**(8), 1597–1614 (2014)
38. Shtok, A., Kurland, O., Carmel, D.: Using statistical decision theory and relevance models for query-performance prediction. In: Proceedings of the 33rd International ACM SIGIR Conference on Research and Development in Information Retrieval, pp. 259–266 (2010)
39. Shtok, A., Kurland, O., Carmel, D., Raiber, F., Markovits, G.: Predicting query performance by query-drift estimation. ACM Trans. Inf. Syst. (TOIS) **30**(2), 1–35 (2012)
40. Singh, A., Ganguly, D., Datta, S., McDonald, C.: Unsupervised query performance prediction for neural models with pairwise rank preferences. In: Proceedings of the 46th International ACM SIGIR Conference on Research and Development in Information Retrieval, pp. 2486–2490 (2023)
41. Soboroff, I.: Overview of TREC 2021. In: 30th Text REtrieval Conference, Gaithersburg, Maryland (2021)
42. Tao, Y., Wu, S.: Query performance prediction by considering score magnitude and variance together. In: Proceedings of the 23rd ACM International Conference on Conference on Information and Knowledge Management, pp. 1891–1894 (2014)
43. Zamani, H., Croft, W.B., Culpepper, J.S.: Neural query performance prediction using weak supervision from multiple signals. In: The 41st International ACM SIGIR Conference on Research & Development in Information Retrieval, pp. 105–114 (2018)
44. Zhou, Y., Croft, W.B.: Query performance prediction in web search environments. In: Proceedings of the 30th Annual International ACM SIGIR Conference on Research and Development in Information Retrieval, pp. 543–550 (2007)

Instant Answering in E-Commerce Buyer-Seller Messaging Using Message-to-Question Reformulation

Besnik Fetahu(✉), Tejas Mehta, Qun Song, Nikhita Vedula, Oleg Rokhlenko, and Shervin Malmasi

Amazon.com, Inc., Seattle, WA, USA
{besnikf,mehtejas,qunsong,veduln,olegro,malmasi}@amazon.com

Abstract. E-commerce customers frequently seek detailed product information for purchase decisions, commonly contacting sellers directly with extended queries. This manual response requirement imposes additional costs and disrupts buyer's shopping experience with response time fluctuations ranging from hours to days. We seek to automate buyer inquiries to sellers in a leading e-commerce store using a domain-specific federated Question Answering (QA) system. The main challenge is adapting current QA systems, designed for single questions, to address detailed customer queries. We address this with a low-latency, sequence-to-sequence approach, MESSAGE-TO-QUESTION (M2Q). It reformulates buyer messages into *succinct* questions by identifying and extracting the most salient information from a message. Evaluation against baselines shows that M2Q yields relative increases of 757% in question understanding, and 1,746% in answering rate from the federated QA system. Live deployment shows that automatic answering saves sellers from manually responding to millions of messages per year, and also accelerates customer purchase decisions by eliminating the need for buyers to wait for a reply.

1 Introduction

The rapid growth of e-commerce, with hundreds of millions of global users, hinges on customer satisfaction tied to easy, instant access to comprehensive product details. However, the available product descriptions, customer reviews, and QAs on e-commerce sites may not provide enough information for users to make informed purchasing decisions.

E-commerce platforms use a *Seller Messaging* feature for facilitating communication between buyers and sellers through an online messaging system about product details, shipping, or post-purchase concerns. However, large platforms which handle over 100K messages daily experience high wait times ranging from *1–2 h* to *several days* due to the time and monetary costs of manual responses, thus delaying purchase decisions.

We address these gaps by leveraging a federated, multiple backend QA system for instantly answering customer questions using knowledge sources such as reviews, product catalogs, manuals, and community QA. However, *style mismatch* is a key challenge:

B. Fetahu and T. Mehta—Contributed equally to this work.

N. Goharian et al. (Eds.): ECIR 2024, LNCS 14611, pp. 59–67, 2024.
https://doi.org/10.1007/978-3-031-56066-8_7

Table 1. Examples of buyer messages reformulated by M2Q with the buyer's main intent in red.

Buyer's Message to Seller	Ground Truth	Reformulated Message
I'm trying to look into Forids trash bags, and have found that the website and facebook on the box dont exist :(***I'm curious about their sourcing and material is used*** !!	what is the sourcing material of this product?	*what is the sourcing material used for this product?*
Hello! I'm curious whether ***this Re:Zero REM figure you're selling is from the authentic Taito Coreful brand?*** Thank you	is this product from the taito coreful brand?	*is this product from the authentic taito coreful brand?*
My family surname is not a known surname. Are you ***able to create a family crest with all the emblems and mottos?*** Looking forward to hear from you	can this product be customized with all the emblems and mottos?	*can this product be created with all the emblems and mottos?*

customers write lengthy email-style messages, while traditional QA systems operate on simple direct questions. We tackle this via an end-to-end approach, MESSAGE-TO-QUESTION (M2Q), using sequence-to-sequence models to reformulate lengthy buyer messages into short standalone questions. M2Q is optimized to distill relevant details from buyer's messages into *answerable* questions and use a state-of-the-art QA model to generate precise answers. This allows the Seller Messaging feature to provide instant QA by leveraging existing resources without incurring additional time or cost overheads (see Table 1). During message reformulation, M2Q considers the buyer's primary needs, combines multiple needs if necessary into a concise request for the QA system, and ensures customer privacy by omitting personal information.

2 Background

Buyer-Seller Interactions & QA: Prior research has highlighted the financial and commercial importance of studying interactions between buyers and sellers on e-commerce stores, as well as answering buyer inquiries in a fast and effective manner [1,13,18]. We propose to improve buyers' access to instantaneous answers, and reduce product sellers' burden and expenses on e-commerce stores by integrating product question answering (QA) systems [7,12] into the Seller Messaging feature. Prior work on automatically answering customer queries within customer service applications, focuses on retrieving answers from a knowledge base [6,14,15,23] as well as generating answers [3,20].

Text Reformulation: Buyer messages can be verbose with long descriptions or irrelevant personal details (see Table 1). This distracts QA models and makes it difficult for them to provide accurate answers [2,16,24]. M2Q ensures that buyer message reformulations can *adapt* to and maximize the understanding and answering rate of existing QA systems. Several Large Language Models (LLMs) have been fine-tuned to summarize and extract salient information from dialogues and email threads [10,22]; identify the appropriate context to be input to a model to answer questions effectively [17,19];

reformulate questions for easy answering [9,11,25]; as well as to select the most relevant conversation history as context in case of conversational QA [8,27].

3 M2Q: Message-to-Question

M2Q reformulate buyer's messages into succinct questions that are instantly answered using a federated QA system. If the question cannot be answered, or buyers are dissatisfied with the automatic response, they can forward their message to the seller for a manual response. Figure 1 shows an overview of the approach. Messages sent to sellers are reformulated into questions, which is then used to retrieve an instant answer from the QA system. Next, we describe the components from the figure.

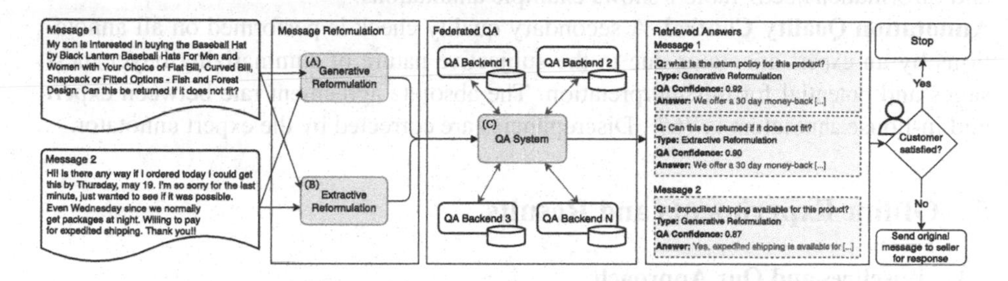

Fig. 1. M2Q approach overview. For each buyer message that may contain multiple questions/intents, M2Q reformulates them (using (B) and (C)) into a *succinct question* with the most salient intent (or conjunction of intents). These are sent to a federated QA system (C) for instant answering. If customers are unsatisfied with the response or receive none, the original message is forwarded to the seller for manual reply.

Generative Reformulation (A): The main objective of this step is to convert buyers' lengthy messages into a shorter, concise form that correctly captures their primary intent, filtering out any irrelevant details or personal information. Table 1 shows examples of such reformulations. For the generative reformulation, we fine-tune different sequence-to-sequence models to perform the rewriting automatically, using a parallel dataset of buyer messages and their human generated reformulations (c.f. §4).

Extractive Reformulation (B): A simpler reformulation method involves splitting the buyer message into sentences, and running them through a question classifier provided by the QA system. The sentence with the highest confidence is chosen to represent the entire message. This method works well for simple messages containing a direct question, but fails on more complex messages, and may not identify any questions at all.

Question Answering System (C): The reformulated messages from the previous steps are input to a federated QA system, which passes them to numerous in-house answer retrieval systems. This system returns answers and a *QA confidence score*. The QA system is a black box in this work; we do not modify or tune it. We show in §5.3 that shorter messages reformulated by M2Q maximize the performance of downstream QA systems, leading to an increase in the automatic question answering rates.

4 Dataset Construction

We created a dataset of ~6k pairs of buyer messages and their target reformulations, divided into 5k/600/450 for train/dev/test sets. Most messages are between 25–75 words.

Annotation Guidelines. We engaged in-house annotators to transform messages into answerable questions based on certain guidelines, ensuring the quality of the parallel dataset[1]. First, they checked for English language and then identified the number of intents in each message. With multi-intent messages, they focused on the primary intent considering the context, while the remaining were most likely follow-up questions. The message was transformed into a precise question that pinpointed the buyer's main intent and information need. Table 1 shows example annotations.

Annotation Quality Control. A secondary quality check is performed on all annotations by an expert annotator due to the subjective nature of summarizing buyer messages and potential for misinterpretation. The absolute agreement rate between expert and in-house annotators is 76%. Discrepancies are corrected by the expert annotator.

5 Offline Experiments and Results

5.1 Baselines and Our Approach

We compare several reformulation methods[2] against the same QA system.

Extractive Reformulation: The sentence with the highest question classifier confidence score from the buyer message is selected. This works best with short messages that contain a clear, single question.

Flan-T5-XXL: We assess FLAN-T5's reformulations in zero- and few-shot (3 exemplars) settings.

Vicuna-13B: We assess the Vicuna LLM [4], which reportedly obtains 90% of GPT-4's performance, in the same zero-shot and few-shot settings as FLAN-T5.

Approach: We evaluate our M2Q approach with two underlying base models. M2Q-T5 uses the smaller 220M parameter `T5-base` model [21], while M2Q-FT5 uses FLAN-T5-XXL [5] with 11B parameters. We consider two configurations:

- **M2Q:** The buyer's original message is converted into a standalone question via generative reformulation only, and sent to the QA system.
- **M2Q-HYBRID**: We combine generative and extractive reformulation methods. The extractive approach responds when it obtains an answer, else generative reformulation is employed. M2Q-HYBRID thus allows direct responses to shorter messages, decreasing error probability from reformulations.

[1] Our human annotators are expert in-house annotators that provide their relevance judgements based on a pre-determined annotation protocol, which was designed specifically for this task.

[2] Due to privacy regulations, we cannot use external API-based LLMs like ChatGPT.

5.2 Evaluation Strategies

We measure reformulation quality in several ways:

- **Text Generation Performance**: We use BLEU and ROUGE to measure how closely the revised messages align with their originals.
- **Reformulation Accuracy:** Human annotators evaluate reformulated messages based on binary score relevance, verifying the buyer's true intent.
- **Understand Rate:** Measures the QA system's capability to understand buyer messages or the corresponding reformulations.
- **Answer Rate:** Measures the answerability of buyer messages by our QA system.
- **Answer Relevance:** We manually rate each message or its revision as either "*helpful*" or "*unhelpful*" based on the answer's correctness or if it's unanswered.

5.3 Results

Table 2. We report reformulation performance (left side of the table) measured in terms of reformulation accuracy, BLEU and ROUGE, and QA performance.

	Reformulation Accuracy	Generation Performance							QA Performance		
		BLEU1	BLEU2	BLEU3	BLEU4	ROUGE1	ROUGE2	ROUGEL	Understand Rate	Answer Rate	Answer Relevance
Extractive Baseline	41%	0.143	0.060	0.036	0.025	0.228	0.099	0.214	-	-	-
VICUNA-ZERO-SHOT	-	0.139	0.058	0.037	0.027	0.330	0.121	0.292	+235%	+827%	+320%
VICUNA-FEW-SHOT	31%	0.335	0.206	0.145	0.098	0.504	0.280	0.484	+578%	+1,182%	+1,280%
FT5-ZERO-SHOT	-	0.356	0.182	0.136	0.108	0.489	0.231	0.450	+616%	+1,246%	+1,240%
FT5-FEW-SHOT	58%	0.390	0.210	0.153	0.117	0.520	0.265	0.489	+651%	+1,282%	+1,300%
M2Q-T5	79%	**0.547**	0.369	0.295	0.243	**0.606**	0.384	**0.586**	+746%	+1,478%	+1,820%
M2Q-FLAN-T5	82%	0.546	**0.394**	**0.319**	**0.273**	0.599	**0.406**	**0.586**	+755%	+1,727%	**+2,220%**
M2Q-HYBRID-T5	-	-	-	-	-	-	-	-	+749%	+1,500%	+1,860%
M2Q-HYBRID-FLAN-T5	-	-	-	-	-	-	-	-	**+757%**	**+1,746%**	**+2,220%**

Generation Performance: Table 2 shows the BLEU and ROUGE metric results. Both FT5 and VICUNA, achieve lower performance than M2Q for both zero-shot and few-shot scenarios. For FT5 in the few shot setting the exemplars help to obtain better generation performance with an improvement of 3.4 BLEU-1 points. While VICUNA underperforms FT5, few-shot exemplars allow it to significantly improve performance, with an increase of 19 BLEU-1 points. Both LLMs are significantly outperformed by M2Q, with an increase of 9 ROUGE-L points.

The M2Q results highlight two factors: (1) generative models are better for this task, as the extractive baseline achieves the lowest performance on all metrics. (2) reformulating buyer messages is complex, and accurate performance requires fine-tuning.

Reformulation Accuracy: Due to limitations of automated metrics [26], we manually assess reformulation accuracy, computed for: VICUNA–FS, FT5–FS, M2Q-T5 and M2Q-FT5. M2Q obtains the highest reformulation accuracy with 79% for M2Q-T5 and 82% for M2Q-FT5. Note that the size of the base model is important: M2Q-FT5 has 3% higher accuracy than M2Q-T5. Without fine-tuning, FT5 only obtains a 58%

accuracy, a drop of ▼24% compared to M2Q-FT5. Similarly, VICUNA obtains a reformulation accuracy of 31%, a ▼51% drop compared to M2Q-FT5.

Question Answering Performance: Table 2 shows the QA results in terms of the understanding confidence scores and answer rates.[3] For reasons of confidentiality, the QA results are reported as relative improvements over the extractive baseline. On question understanding, M2Q-T5 and M2Q-FT5 obtain relative improvements over HB with 746% and 755%, respectively. Similarly, on answering rate, M2Q-FT5 obtains the highest improvement with 1,727%, while for M2Q-T5 the improvement is 1,478%. This result shows that the buyer's messages in their original form are unsuitable for QA.

M2Q-HYBRID achieves the highest relative improvement across all metrics. This validates our intuition that for shorter messages already in question form, reformulation is not necessary, and in such cases an extractive method provides accurate answers.

Finally, in terms of answer relevance, we see a relative improvement of 2,220% over the extractive baseline, and see no difference between M2Q-FT5 and M2Q-HYBRID-FT5. This shows that not only do we increase answer rates, but also answer precision, as the relative increases in relevance from M2Q are much higher compared to other baselines.

6 Online Deployment and Evaluation

We deployed the more cost-effective M2Q-HYBRID-T5 model, exhibiting a performance equal to M2Q-HYBRID-FT5 according to offline results.

We assess user satisfaction and purchase metrics from millions of e-commerce customers. We split users into two cohorts: a control group (C) whose messages are manually answered by sellers, and a treatment group (T), whose questions are answered by M2Q-HYBRID-T5. We also consider T_{pos}, a subset of T, who provide explicit positive feedback on M2Q answers. We consider the following online evaluation metrics:

- **Purchase Rate – PR**: The ratio of *unique users* who ask a question about a product and buy it within a week, to the total number of users asking a question.
- **Successful Answer Rate – SAR**: The proportion of messages that received an instant answer where buyers were satisfied and did not send it to the seller.[4]

6.1 Results

Due to confidentiality, the results are reported as relative improvements over the control cohort (C). Table 3 shows the results for the PR and SAR rate metrics.

Purchase Rate – PR: Both treatment groups have significant increases in PR. T_{pos} obtains the highest purchase rate. This is intuitive given that the instant answers are explicitly marked as helpful. This result demonstrates that providing instant answers accelerates customer purchase decisions.

Successful Answer Rate – SAR: On T cohort, users submit significantly fewer messages to sellers (50% relative increase of SAR). For T_{pos}, SAR increases by 276%.

[3] We assess the proportion of questions answered when the QA confidence surpasses a threshold.
[4] If unsatisfied with an instant answer, users can forward their question to the seller.

Table 3. Online evaluation results with real customers. The reported metrics represent relative improvement over the control cohort (C).

	Control	T	T_{pos}
PR	0.0	+28.57%	+50.88%
SAR	0.0	+57.14%	+276.73%

7 Conclusion

We proposed M2Q, an approach for automatically answering messages that are sent from buyers to sellers. Offline experiments validated our approach, and live deployment demonstrated that it improves the shopping experience for both buyers and sellers.

Our method efficiently reformulates messages into concise, salient questions optimal for understanding and response by a federated QA system, providing instant answers to buyers. The instant answers feature significantly influences both buyers and sellers, evidenced by a reduction of up to 276% in buyer-to-seller messages. This decrease likely reflects users' satisfaction with the instant responses from M2Q, contributing to enhanced buyer experiences and decreased seller overhead. An empirical online study involving real e-commerce users demonstrated a substantial relative increase in purchase rates by 57.14% when compared to a control group not utilizing M2Q instant answers.

References

1. Ahearne, M., Atefi, Y., Lam, S.K., Pourmasoudi, M.: The future of buyer-seller interactions: a conceptual framework and research agenda. J. Acad. Mark. Sci. **50**, 22–45 (2022). https://doi.org/10.1007/s11747-021-00803-0
2. Cao, Y., et al.: TASA: deceiving question answering models by twin answer sentences attack. arXiv preprint arXiv:2210.15221 (2022)
3. Chen, M., et al.: The JDDC corpus: a large-scale multi-turn Chinese dialogue dataset for e-commerce customer service. In: Proceedings of the 12th Language Resources and Evaluation Conference, pp. 459–466 (2020)
4. Chiang, W.L., et al.: Vicuna: an open-source chatbot impressing GPT-4 with 90%* ChatGPT quality, March 2023. https://lmsys.org/blog/2023-03-30-vicuna/
5. Chung, H.W., et al.: Scaling instruction-finetuned language models. CoRR abs/2210.11416 (2022). https://doi.org/10.48550/arXiv.2210.11416
6. Cui, L., Huang, S., Wei, F., Tan, C., Duan, C., Zhou, M.: SuperAgent: a customer service Chatbot for e-commerce websites. In: Proceedings of ACL 2017, System Demonstrations, pp. 97–102 (2017)
7. Deng, Y., Zhang, W., Yu, Q., Lam, W.: Product question answering in e-commerce: a survey. arXiv preprint arXiv:2302.08092 (2023)
8. Do, X.L., Zou, B., Pan, L., Chen, N., Joty, S., Aw, A.: CoHS-CQG: context and history selection for conversational question generation. In: Proceedings of the 29th International Conference on Computational Linguistics, pp. 580–591 (2022)

9. Faustini, P., Chen, Z., Fetahu, B., Rokhlenko, O., Malmasi, S.: Answering unanswered questions through semantic reformulations in spoken QA. In: Sitaram, S., Klebanov, B.B., Williams, J.D. (eds.) Proceedings of The 61st Annual Meeting of the Association for Computational Linguistics: Industry Track, ACL 2023, Toronto, Canada, 9–14 July 2023, pp. 729–743. Association for Computational Linguistics (2023). https://aclanthology.org/2023.acl-industry.70
10. Feng, X., Feng, X., Qin, B.: A survey on dialogue summarization: recent advances and new frontiers. arXiv preprint arXiv:2107.03175 (2021)
11. Ferguson, N., Guillou, L., Nuamah, K., Bundy, A.: Investigating the use of paraphrase generation for question reformulation in the FRANK QA system. arXiv preprint arXiv:2206.02737 (2022)
12. Gao, S., Chen, X., Ren, Z., Zhao, D., Yan, R.: Meaningful answer generation of e-commerce question-answering. ACM Trans. Inf. Syst. (TOIS) **39**(2), 1–26 (2021)
13. Kumar, G., Henderson, M., Chan, S., Nguyen, H., Ngoo, L.: Question-answer selection in user to user marketplace conversations. In: D'Haro, L.F., Banchs, R.E., Li, H. (eds.) 9th International Workshop on Spoken Dialogue System Technology. LNEE, vol. 579, pp. 397–403. Springer, Singapore (2019). https://doi.org/10.1007/978-981-13-9443-0_35
14. Li, Y., et al.: Question answering for technical customer support. In: Zhang, M., Ng, V., Zhao, D., Li, S., Zan, H. (eds.) NLPCC 2018, Part I 7. LNCS (LNAI), vol. 11108, pp. 3–15. Springer, Cham (2018). https://doi.org/10.1007/978-3-319-99495-6_1
15. Liao, L.Y., Fares, T.: A practical 2-step approach to assist enterprise question-answering live chat. In: Proceedings of the 22nd Annual Meeting of the Special Interest Group on Discourse and Dialogue, pp. 457–468 (2021)
16. Lyu, Q., Zhang, H., Sulem, E., Roth, D.: Zero-shot event extraction via transfer learning: challenges and insights. In: Proceedings of the 59th Annual Meeting of the Association for Computational Linguistics and the 11th International Joint Conference on Natural Language Processing (Volume 2: Short Papers), pp. 322–332 (2021)
17. Mao, Y., et al.: Generation-augmented retrieval for open-domain question answering. arXiv preprint arXiv:2009.08553 (2020)
18. Masterov, D.V., Mayer, U.F., Tadelis, S.: Canary in the e-commerce coal mine: detecting and predicting poor experiences using buyer-to-seller messages. In: Proceedings of the Sixteenth ACM Conference on Economics and Computation, pp. 81–93 (2015)
19. McDonald, T., et al.: Detect, retrieve, comprehend: a flexible framework for zero-shot document-level question answering. arXiv preprint arXiv:2210.01959 (2022)
20. Peng, B., et al.: Check your facts and try again: improving large language models with external knowledge and automated feedback. arXiv preprint arXiv:2302.12813 (2023)
21. Raffel, C., et al.: Exploring the limits of transfer learning with a unified text-to-text transformer. J. Mach. Learn. Res. **21**(1), 5485–5551 (2020)
22. Rennard, V., Shang, G., Hunter, J., Vazirgiannis, M.: Abstractive meeting summarization: a survey. arXiv preprint arXiv:2208.04163 (2022)
23. Samarakoon, L., Kumarawadu, S., Pulasinghe, K.: Automated question answering for customer helpdesk applications. In: 2011 6th International Conference on Industrial and Information Systems, pp. 328–333. IEEE (2011)
24. Shi, F., et al.: Large language models can be easily distracted by irrelevant context. arXiv preprint arXiv:2302.00093 (2023)
25. Vakulenko, S., Longpre, S., Tu, Z., Anantha, R.: Question rewriting for conversational question answering. In: Proceedings of the 14th ACM International Conference on Web Search and Data Mining, pp. 355–363 (2021)

26. Yang, A., Liu, K., Liu, J., Lyu, Y., Li, S.: Adaptations of ROUGE and BLEU to better evaluate machine reading comprehension task. In: Choi, E., Seo, M., Chen, D., Jia, R., Berant, J. (eds.) Proceedings of the Workshop on Machine Reading for Question Answering, ACL 2018, Melbourne, Australia, 19 July 2018, pp. 98–104. Association for Computational Linguistics (2018). https://doi.org/10.18653/v1/W18-2611. https://aclanthology.org/W18-2611/
27. Zaib, M., Zhang, W.E., Sheng, Q.Z., Mahmood, A., Zhang, Y.: Conversational question answering: a survey. Knowl. Inf. Syst. **64**(12), 3151–3195 (2022)

SoftQE: Learned Representations of Queries Expanded by LLMs

Varad Pimpalkhute[1(✉)], John Heyer[2], Xusen Yin[2], and Sameer Gupta[2]

[1] University of Massachusetts Amherst, Amherst, USA
pimpalkhutevarad@gmail.com
[2] Alexa AI, Amazon, Seattle, USA
{heyjohn,yxusen,gupsam}@amazon.com

Abstract. We investigate the integration of Large Language Models (LLMs) into query encoders to improve dense retrieval without increasing latency and cost, by circumventing the dependency on LLMs at inference time. SoftQE incorporates knowledge from LLMs by mapping embeddings of input queries to those of the LLM-expanded queries. While improvements over various strong baselines on in-domain MS-MARCO metrics are marginal, SoftQE improves performance by 2.83 absolute percentage points on average on five out-of-domain BEIR tasks.

1 Introduction

Query expansion [15,22] methods aim to expand search queries with additional terms to improve downstream information retrieval (IR) performance. Expansion terms can come directly from highly ranked documents, as in pseudo relevance feedback based methods like RM3 [15,17], or from generative models as in methods like GAR [19]. While query expansion can mitigate the *token mismatch* problem that plagues sparse retrieval methods like BM25 [21], which depend on token overlap between queries and documents, dense retrieval methods [11,14] offer a natural solution by embedding queries and documents in a shared feature space wherein queries and documents with strong *semantic* overlap are close.

Recent methods [9,13,30] prompt Large Language Models (LLMs) [1,4,26] to expand queries with relevant terms or "pseudo-documents" that resemble real passages from the corpus. Perhaps surprisingly, *query2doc* (Q2D) [30] demonstrates improved performance of *dense* retrievers, indicating that LLM-based query expansion can facilitate learning the semantic overlap between underspecified queries and document corpora. However, adding an LLM to a real-time IR pipeline is often prohibitively expensive in terms of both cost and latency. Motivated by both the promise of LLM-based query expansion for dense retrieval *and* its impracticality, we propose ***Soft Query*** **E***xpansion* (SoftQE), wherein we learn to estimate the representations of LLM expansions *offline* during training, thus circumventing the dependency on LLMs at runtime as shown in Fig. 1.

V. Pimpalkhute—Work done as an intern at Amazon.

N. Goharian et al. (Eds.): ECIR 2024, LNCS 14611, pp. 68–77, 2024.
https://doi.org/10.1007/978-3-031-56066-8_8

SoftQE performs at least as well as baseline dense retrievers such as DPR [14], and stronger alternatives combining large-scale pretraining and cross-encoder distillation such as SimLM [28] and E5 [29], on in-domain MS-MARCO [2], TREC DL 2019 and 2020 datasets [5,6]. Further, SoftQE significantly improves upon these baselines for a majority of out-of-domain BEIR [25] tasks. Our findings corroborate those of Q2D, specifically that the increase in retrieval performance diminishes when combined with stronger encoders. However, we observe measurable improvements in the zero-shot setting, suggesting that information learned through the SoftQE objective is complementary to other forms of distillation, such as distillation from a cross-encoder.

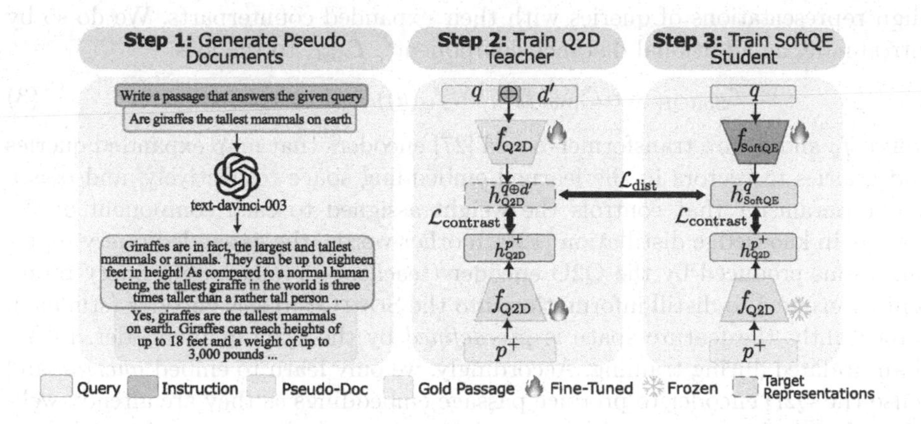

Fig. 1. Overview of the SoftQE approach. **Step 1:** Given a query, prompt an LLM to generate a pseudo-document d', as in Q2D [30]. **Step 2:** Train teacher encoder using the Q2D method and expanded queries from Step 1 ($q \oplus d'$). **Step 3:** Train SoftQE encoder to align query representations with the expanded query representations from Step 2, in addition to the standard contrastive objective. h_y^x denotes the *representation* (e.g., the last hidden state of the `CLS` token) given an input x and encoder y.

2 Method

Expanded Queries. An expanded query q^+ is formed by appending a pseudo-document d' to the original query, q:

$$q^+ = q \oplus g_\phi(\mathcal{I}, q), \tag{1}$$

where, g_ϕ is an LLM that generates pseudo document (d') with prompt $\mathcal{I}$, employing techniques such few-shot, chain of thought [31], etc. We use the pseudo-documents released[1] with Q2D [30], which were generated by *text-davinci-003* [1] using an instruction and examples of positive query/document pairs from MS MARCO [2]. An example pseudo document is shown in Fig. 1.

[1] Pseudo-documents generated using text-davinci-003 for MS MARCO queries are released by [30] here: https://huggingface.co/datasets/intfloat/query2doc_msmarco.

Dual-Encoder Training. Dual encoders are typically trained by optimizing a contrastive objective [14]:

$$\mathcal{L}_{\text{cont}} = -\log\left(\frac{e^{h_q \cdot h_{p^+}}}{e^{h_q \cdot h_{p^+}} + \sum_{i=1}^{N} e^{h_q \cdot h_{p_i^-}}}\right), \tag{2}$$

where h_q and h_p represent query and passage embeddings, respectively, and N is the number of negative passages. In Q2D, embeddings of *expanded* query inputs (h_{q^+}) are learned, and BM25 hard negatives are used.

SoftQE Objective. Driven by the superior performance of Q2D, we seek to align representations of queries with their expanded counterparts. We do so by introducing an additional distance component[2], $\mathcal{L}_{\text{dist}}$, to the loss:

$$\mathcal{L}_{\text{SoftQE}} = \alpha \mathcal{L}_{\text{dist}}(f_\theta(q^+), f_\psi(q)) + (1-\alpha)\mathcal{L}_{\text{cont}}, \tag{3}$$

where f_θ and f_ψ are transformer-based [27] encoders that map expanded queries and queries to vectors in the learned embedding space respectively, and α is a hyper parameter that controls the weight assigned to each component of the loss, as in knowledge distillation [12]. In other words, the expanded query representations produced by the Q2D encoder (teacher) serve as target query representations used to distill information into the SoftQE query encoder (student). Importantly, the feature space is *pre-defined* by the Q2D dual-encoder, rather than updated during training. Accordingly, we only learn to embed *queries*, and reuse the Q2D encoder to produce passage embeddings as they are already well-aligned with the target query representations.

We additionally experiment with state-of-the-art dense retrievers [28,29] that are trained using KL divergence from cross-encoder scores [20]. We apply SoftQE to distilled retrievers by simply combining the 3 objective terms with an additional weight controlled by β:

$$\mathcal{L}_{\text{SQE+KD}} = \alpha \mathcal{L}_{\text{dist}}(f_\theta(q^+), f_\psi(q)) + (1-\alpha)\left[\beta \text{KL}(f_\theta, f_{\text{CE}}) + (1-\beta)\mathcal{L}_{\text{cont}}\right], \tag{4}$$

as we find the information distilled through cross-encoder scores and expanded query representations to be complementary.

3 Experiments

Datasets, Metrics, and Baselines. For in-domain evaluation, we use the MS MARCO Passage Ranking [2], TREC DL 2019 [5] and TREC DL 2020 [6] datasets. Following Q2D [30], we evaluate zero-shot performance on five low-resource tasks from the BEIR benchmark [25], namely: SciFact, NFCorpus, Trec-Covid, DBPedia and Touche-2020. Evaluation metrics include MRR@10, R@50, R@1k, and nDCG@10. We benchmark SoftQE against a DPR [14] dense retrieval baseline, and two state-of-the-art dense retrievers: SimLM [28], and E5 [29].

[2] In practice, we find no significant difference between distance metrics, so we simply use mean squared error (MSE).

Hyperparameters. We follow the hyperparameter settings used in [30], with a few distinctions. We initialize our DPR models from $\text{BERT}_{\text{base}}$ [7], and our SoftQE variants of SimLM [28], and E5 [29] from their corresponding public checkpoints. When fine-tuning with cross-encoder distillation, β is set to 0.2, following SimLM [28]. We set α to 1.0 for 3 epochs in order to establish an initial alignment with the target expanded query embeddings, then relax α to 0.2 as well. This choice is further discussed in Sect. 4.

Table 1. Results on in-domain MS MARCO and TREC DL datasets, grouped by retrievers trained with and without distillation from cross-encoders. Underline: best result including Q2D, which requires an LLM at inference time; **Bold**: highest result among non-Q2D solutions; *: our reproduction; †: denotes statistical significance with a p-value less than 0.05 using a paired T-test.

Method	MS MARCO Dev Set			TREC DL 19	TREC DL 20
	MRR@10	R@50	R@1k	nDCG@10	nDCG@10
Dual-encoder without distillation					
DPR*	33.74	80.90	96.18	64.04	62.81
+SoftQE	**33.87**	**81.24**†	**96.25**†	**65.22**†	**63.80**†
+ Q2D*	<u>35.26</u>	<u>82.78</u>	<u>97.21</u>	<u>70.54</u>	<u>66.68</u>
Dual-encoders distilled from cross-encoders					
SimLM*	41.13	87.78	**98.69**	71.40	69.68
+SoftQE	**41.15**	**87.93**†	98.61†	70.50†	70.10†
+ Q2D*	<u>41.45</u>	<u>88.43</u>	<u>98.82</u>	74.59	71.37
E5*	40.70	87.13	98.50	72.52	71.38
+SoftQE	40.30†	87.22	98.50	**72.82**†	**71.73**†
+ Q2D*	40.93	87.95	98.76	<u>75.03</u>	<u>73.27</u>

Results. We first evaluate the performance on in-domain datasets (Table 1). SoftQE consistently improves upon DPR across all metrics on MS MARCO, TREC DL 19 and TREC DL 20 datasets. When evaluating the performance against dual-encoders distilled from cross-encoders, we notice that SoftQE and SimLM perform closely with SoftQE slightly underperforming in R@1k on MS MARCO and nDCG@10 on TREC DL2019. Similarly, SoftQE results in marginal improvements over E5. This finding corroborates the claim in [30] that improvements diminish when encoders are distilled from strong cross-encoders.

Table 2 highlights the zero-shot evaluation results on out-of-domain datasets from BEIR. SoftQE considerably outperforms DPR and SimLM, by 2.94 and 2.72 absolute percentage points, respectively, averaged across tasks. SoftQE yields marginal differences in performance on tasks where Q2D results in regressions (SciFact and NFCorpus), but substantial improvements on the remaining

Table 2. Results on out-of-domain BEIR benchmark datasets by nDCG@10. Underline: best result including Q2D, which requires an LLM at inference time; **Bold**: highest result among non-Q2D solutions; *: our reproduction[5]; †: denotes statistical significance.

Method	SciFact	NFCorpus	Trec-Covid	DBPedia	Touche-2020	Average
DPR*	**51.85**	25.72	44.81	30.99	19.91	34.65
+SoftQE	49.81†	**25.73**	60.02†	**31.82**†	**20.59**	**37.59**
SimLM*	61.42	<u>**32.38**</u>	52.90	35.06	19.21	40.19
+SoftQE	<u>**61.72**</u>	32.34	<u>**61.78**</u>	**36.75**	**21.94**	**42.91**
+ Q2D [30]	59.50	32.10	59.90	<u>38.30</u>	<u>25.60</u>	<u>43.08</u>

tasks when applied to either DPR or SimLM, indicating that SoftQE is complementary to cross-encoder distillation.

4 Discussion

Is Fine-Tuning on Expanded Queries Necessary? Traditional query expansion methods applied to lexical systems do not require modifications to the retrieval algorithm. Q2D [30], however, requires fine-tuning the dense retriever on expanded queries, as demonstrated by the difference between the first 2 rows in Table 3. Simply passing expanded queries to an off-the-shelf DPR model actually deteriorates performance, which is somewhat surprising given the model's ability to effectively embed queries and passages *independently*.

Table 3. MS Marco MRR@10 of DPR and Q2D with query (q) and expanded query (q^+) inputs. DPR has not been trained with expanded query inputs, while Q2D has.

Method	TREC DL19	TREC DL20
DPR(q^+)	61.65	59.45
+ Q2D(q^+)	**70.54**	**66.68**
DPR(q)	64.04	62.81
+ Q2D(q)	57.78	57.12

Table 4. TREC nDCG@10 across four variations of α in the training objective: $(\alpha\mathcal{L}_{\text{cont}}+(1-\alpha)\mathcal{L}_{\text{dist}})$. In "Warm up", we set alpha to 1 for the first 3 epochs, then to 0.2 for the remaining 3.

Method	α	TREC DL19	TREC DL20
$\mathcal{L}_{\text{dist}}$ Only	1	61.62	62.81
$\mathcal{L}_{\text{cont}}$ Only	0	64.66	63.42
Combined	0.2	63.33	63.03
Warm up	$1 \rightarrow 0.2$	**65.23**	**63.79**

Combining L_{dist} and L_{cont}. In Table 4, we explore four variations of the training objective in Eq. 3 to determine how to balance supervision from labeled passages vs. target representations. A perfect mapping ($L_{\text{dist}} = 0$) between query and expanded query representations would yield Q2D performance, but is not realistic, as evident by the subpar performance of "L_{dist} Only". Combining the two losses by setting α to 0.2 results in query embeddings that are no closer to the target embeddings produced when using only a contrastive loss, as shown by the MSE Loss plot in Fig. 2 (right). To remedy this, we propose a step-wise "warm up" method, in which we set α to 1 (L_{dist} only) for 3 epochs to establish a strong alignment with the target representations, then relax α to 0.2 for the remaining 3 epochs. Figure 2 demonstrates that this reduces L_{dist} while negligibly impacting L_{cont}, resulting in the best performance in Table 4.

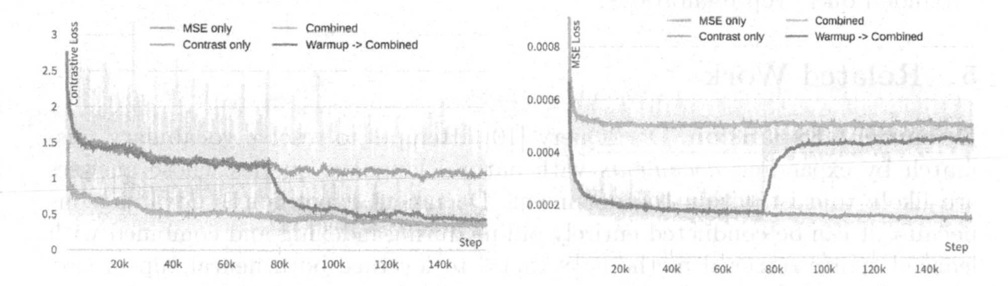

Fig. 2. Training curves of four settings of α shown in Table 4. **Left**: Contrastive Loss - *does it reject negative documents?* `MSE-only` performs the worst in terms of contrastive loss, while `Warmup` → `Combined` converges to the same loss as *Combined*. **Right**: MSE Loss - *is it close to the teacher?* `Contrast-only` has the highest MSE loss, while `Warmup` → `Combined` MSE loss increases after the warmup, but converges to a value noticeably lower than `Combined`.

Should we also Fine-Tune the Passage Encoder? We decided not to fine-tune the passage encoder during SoftQE training, because it allowed us to re-use the Q2D passage-encoder. Intuitively, this means that the space in which passages and queries are embedded is the same as in Q2D. In Table 5 we show that, on average, fine-tuning the passage encoder results in reduced performance. This is assuring – if the passage representations were to change, our Q2D representation targets would be unfounded, as they would no longer be optimally aligned with the passages.

Table 5. Unfreezing the passage encoder during training results in a degradation of performance on TREC nDCG@10.

Freeze Encoder	Method	DL19	DL20
✗	SoftQE	**65.59**	62.06
✓	SoftQE	65.22	**63.80**

Table 6. Comparing our method to traditional knowledge distillation (using only model predictions) on TREC nDCG@10.

Method	DL19	DL20
DPR	64.04	62.81
+ Traditional KD	61.12	62.14
+ SoftQE	**65.22**	**63.79**

SoftQE vs. Traditional Knowledge Distillation. Our method distills the high-dimensional *representation* of the teacher model, as opposed to teacher's *predictions*, as in traditional knowledge distillation. In Table 6, we compare our method to a variant in which we compute MSE *only over the scalar-valued scores* produced by the Q2D teacher as our distillation loss. This results in reduced performance, as the score-only distillation model underperforms the DPR baseline, indicating that the teacher's predictions alone do not provide sufficient supervision for estimating the nuanced information contained in the high-dimensional expanded query representations.

5 Related Work

Document Expansion. Doc2Query [10] attempts to resolve vocabulary mismatch by expanding *documents* with natural language queries whose answers are likely to exist within the document. Document expansion is advantageous because it can be conducted entirely offline during indexing and combined with learned sparse retrieval methods [8,16,18] to leverage both neural supervision and efficient inverted index algorithms. However, document expansion techniques can significantly increase the size of the index, and must be applied to the entire corpus each time the expansion method is changed, which might be too costly for corpora containing billions of documents.

Knowledge Distillation. Knowledge distillation [12] methods use the predictions of large *teacher* models to improve the performance of smaller, more practical *student* models. Knowledge distillation is ubiquitous, and has been used to improve dense dual-encoder retrievers via distillation from a cross-encoder [20]. SoftQE is a form of *indirect* knowledge distillation, wherein a student encoder targets the continuous representations of an architecturally-equivalent teacher model whose discrete, natural language *inputs* have been augmented by an LLM. Recent methods such as Alpaca [24] and Vicuna [3] use generations of superior LLMs to improve the performance of smaller, instruction-following models, but these methods imitate "teacher" LLMs by using their outputs as training data directly.

6 Conclusion

We present SoftQE, a technique to align query-encoder representations with the representations of queries expanded by LLMs. Empirical evaluations demonstrate improvements across several retrieval benchmarks and models, and suggest that SoftQE improves generalization to new domains, as made evident by zero-shot evaluations on BEIR datasets. Importantly, this improvement comes without increasing the cost or latency of dense retrieval at runtime compared to other single vector dual-encoder methods, because an LLM is not required at time of inference. Future work might consider improved prompting strategies, or applying LLM-based supervision to higher-capacity retrieval methods like ColBERTv2 [23]. To the best of our knowledge, SoftQE is the first attempt to distill strong representations through natural language generation, and we hope that this will inspire efficient solutions to new tasks in the future.

A Additional BEIR Results

See Table 7.

Table 7. nDCG@10 for the remaining BEIR tasks; *: our reproduction (not tested for significance). Q2D was not evaluated on these datasets.

Method	Signal 1m	Trec-News	Quora	NQ	Fiqa	Arguana
DPR*	**24.15**	35.21	84.05	43.72	24.47	28.80
+SoftQE	22.18	**36.43**	**84.32**	**44.19**	**25.07**	**31.11**

Method	Scidocs	BioASQ	HotpotQA	Climate Fever	Fever	Avg.
DPR*	**11.79**	25.09	**49.58**	17.20	65.36	37.22
+SoftQE	11.35	**27.05**	48.36	**18.40**	**67.38**	37.80

References

1. Brown, T.B., et al.: Language models are few-shot learners. arXiv arXiv:2005.14165 (2020). https://api.semanticscholar.org/CorpusID:218971783
2. Campos, D.F., et al.: MS Marco: a human generated machine reading comprehension dataset. arXiv arXiv:1611.09268 (2016). https://api.semanticscholar.org/CorpusID:1289517
3. Chiang, W.L., et al.: Vicuna: an open-source Chatbot impressing GPT-4 with 90%* ChatGPT quality, March 2023. https://lmsys.org/blog/2023-03-30-vicuna/
4. Chowdhery, A., Narang, S., Devlin, J., et al.: Palm: scaling language modeling with pathways. arXiv arXiv:2204.02311 (2022). https://api.semanticscholar.org/CorpusID:247951931
5. Craswell, N., Mitra, B., Yilmaz, E., Campos, D.F., Voorhees, E.M.: Overview of the TREC 2019 deep learning track. arXiv arXiv:2003.07820 (2020). https://api.semanticscholar.org/CorpusID:253234683
6. Craswell, N., Mitra, B., Yilmaz, E., Campos, D.F., Voorhees, E.M.: Overview of the TREC 2020 deep learning track. arXiv arXiv:2102.07662 (2021). https://api.semanticscholar.org/CorpusID:212737158
7. Devlin, J., Chang, M.W., Lee, K., Toutanova, K.: BERT: pre-training of deep bidirectional transformers for language understanding. In: Proceedings of the 2019 Conference of the North American Chapter of the Association for Computational Linguistics: Human Language Technologies, Volume 1 (Long and Short Papers), Minneapolis, Minnesota, June 2019, pp. 4171–4186. Association for Computational Linguistics (2019). https://doi.org/10.18653/v1/N19-1423. https://aclanthology.org/N19-1423

8. Formal, T., Piwowarski, B., Clinchant, S.: SPLADE: sparse lexical and expansion model for first stage ranking. In: Proceedings of the 44th International ACM SIGIR Conference on Research and Development in Information Retrieval, SIGIR 2021, pp. 2288–2292. Association for Computing Machinery, New York (2021). https://doi.org/10.1145/3404835.3463098
9. Gao, L., Ma, X., Lin, J., Callan, J.: Precise zero-shot dense retrieval without relevance labels. arXiv arXiv:2212.10496 (2022). https://api.semanticscholar.org/CorpusID:254877046
10. Gospodinov, M., MacAvaney, S., Macdonald, C.: Doc2Query–: when less is more. In: Kamps, J., et al. (eds.) Advances in Information Retrieval, ECIR 2023. LNCS, vol. 13981, pp. 414–422. Springer, Cham (2023). https://doi.org/10.1007/978-3-031-28238-6_31
11. Guu, K., Lee, K., Tung, Z., Pasupat, P., Chang, M.W.: REALM: retrieval-augmented language model pre-training. arXiv arXiv:2002.08909 (2020). https://api.semanticscholar.org/CorpusID:211204736
12. Hinton, G.E., Vinyals, O., Dean, J.: Distilling the knowledge in a neural network. arXiv arXiv:1503.02531 (2015). https://api.semanticscholar.org/CorpusID:7200347
13. Jagerman, R., Zhuang, H., Qin, Z., Wang, X., Bendersky, M.: Query expansion by prompting large language models. arXiv arXiv:2305.03653 (2023). https://arxiv.org/pdf/2305.03653.pdf
14. Karpukhin, V., et al.: Dense passage retrieval for open-domain question answering. In: Conference on Empirical Methods in Natural Language Processing (2020). https://api.semanticscholar.org/CorpusID:215737187
15. Lavrenko, V., Croft, W.B.: Relevance based language models. In: Proceedings of the 24th Annual International ACM SIGIR Conference on Research and Development in Information Retrieval, SIGIR 2001, pp. 120–127. Association for Computing Machinery, New York (2001). https://doi.org/10.1145/383952.383972
16. Lin, J., Ma, X.: A few brief notes on Deepimpact, coil, and a conceptual framework for information retrieval techniques (2021)
17. Lv, Y., Zhai, C.: Positional language models for information retrieval. In: Proceedings of the 32nd International ACM SIGIR Conference on Research and Development in Information Retrieval, SIGIR 2009, pp. 299–306. Association for Computing Machinery, New York (2009). https://doi.org/10.1145/1571941.1571994
18. Mallia, A., Khattab, O., Suel, T., Tonellotto, N.: Learning passage impacts for inverted indexes. In: Proceedings of the 44th International ACM SIGIR Conference on Research and Development in Information Retrieval, SIGIR 2021, pp. 1723–1727. Association for Computing Machinery, New York (2021). https://doi.org/10.1145/3404835.3463030
19. Mao, Y., et al.: Generation-augmented retrieval for open-domain question answering. In: Proceedings of the 59th Annual Meeting of the Association for Computational Linguistics and the 11th International Joint Conference on Natural Language Processing (Volume 1: Long Papers), August 2021, pp. 4089–4100. Association for Computational Linguistics (2021). https://doi.org/10.18653/v1/2021.acl-long.316. https://aclanthology.org/2021.acl-long.316
20. Qu, Y., et al.: RocketQA: an optimized training approach to dense passage retrieval for open-domain question answering. In: Proceedings of the 2021 Conference of the North American Chapter of the Association for Computational Linguistics: Human Language Technologies, June 2021, pp. 5835–5847. Association for Computational Linguistics (2021). https://doi.org/10.18653/v1/2021.naacl-main.466. https://aclanthology.org/2021.naacl-main.466

21. Robertson, S., Zaragoza, H.: The probabilistic relevance framework: BM25 and beyond. Found. Trends Inf. Retr. **3**(4), 333–389 (2009). https://doi.org/10.1561/1500000019
22. Rocchio, J.J.: Relevance Feedback in Information Retrieval, p. 1. Prentice Hall, Englewood, Cliffs, New Jersey (1971). http://www.is.informatik.uni-duisburg.de/bib/docs/Rocchio_71.html
23. Santhanam, K., Khattab, O., Saad-Falcon, J., Potts, C., Zaharia, M.: ColBERTv2: effective and efficient retrieval via lightweight late interaction. CoRR abs/2112.01488 (2021). https://arxiv.org/abs/2112.01488
24. Taori, R., et al.: Stanford alpaca: an instruction-following llama model (2023). https://github.com/tatsu-lab/stanford_alpaca
25. Thakur, N., Reimers, N., Ruckl'e, A., Srivastava, A., Gurevych, I.: BEIR: a heterogenous benchmark for zero-shot evaluation of information retrieval models. arXiv arXiv:2104.08663 (2021). https://api.semanticscholar.org/CorpusID:233296016
26. Touvron, H., et al.: LLaMA: open and efficient foundation language models. arXiv arXiv:2302.13971 (2023). https://api.semanticscholar.org/CorpusID:257219404
27. Vaswani, A., et al.: Attention is all you need. In: Proceedings of the 31st International Conference on Neural Information Processing Systems, NIPS 2017, pp. 6000–6010. Curran Associates Inc., Red Hook (2017)
28. Wang, L., et al.: SimLM: pre-training with representation bottleneck for dense passage retrieval. arXiv arXiv:2207.02578 (2022). https://api.semanticscholar.org/CorpusID:250311114
29. Wang, L., et al.: Text embeddings by weakly-supervised contrastive pre-training. arXiv arXiv:2212.03533 (2022). https://api.semanticscholar.org/CorpusID:254366618
30. Wang, L., Yang, N., Wei, F.: Query2doc: query expansion with large language models. arXiv arXiv:2303.07678 (2023). https://api.semanticscholar.org/CorpusID:257505063
31. Wei, J., et al.: Chain of thought prompting elicits reasoning in large language models. arXiv arXiv:2201.11903 (2022). https://api.semanticscholar.org/CorpusID:246411621

Towards Automated End-to-End Health Misinformation Free Search with a Large Language Model

Ronak Pradeep(✉) and Jimmy Lin

David R. Cheriton School of Computer Science, University of Waterloo, Waterloo, ON, Canada
{rpradeep,jimmylin}@uwaterloo.ca

Abstract. In the information age, health misinformation remains a notable challenge to public welfare. Integral to addressing this issue is the development of search systems adept at identifying and filtering out misleading content. This paper presents the automation of Vera, a state-of-the-art consumer health search system. While Vera can discern articles containing misinformation, it requires expert ground truth answers and rule-based reformulations. We introduce an answer prediction module that integrates GPT_x with Vera and a GPT-based query reformulator to yield high-quality stance reformulations and boost downstream retrieval effectiveness. Further, we find that chain-of-thought reasoning is paramount to higher effectiveness. When assessed in the TREC Health Misinformation Track of 2022, our systems surpassed all competitors, including human-in-the-loop configurations, underscoring their pivotal role in the evolution towards a health misinformation-free search landscape. We provide all code necessary to reproduce our results at https://github.com/castorini/pygaggle.

Keywords: Neural Retrieval · Large Language Models · Misinformation

1 Introduction

Individuals often resort to web search engines for acquiring health-related information, motivated either by curiosity or the need for self-diagnosis. However, the pervasive presence of false or misleading content complicates the distinction between accurate and inaccurate information or credible and non-credible sources. Misinformation can lead to reliance on ineffective and potentially harmful treatments, exacerbating health risks.

In recent years, the integration of information retrieval and deep learning has been instrumental in enhancing the accuracy and reliability of search results, highlighted by the advent of pretrained models like BERT [4]. These models have eclipsed traditional IR methods such as BM25 [9] in effectiveness. A multi-stage ranking approach has emerged, balancing model complexity and search latency by progressively narrowing candidate sets, facilitating the application of powerful yet slower rerankers [5].

N. Goharian et al. (Eds.): ECIR 2024, LNCS 14611, pp. 78–86, 2024.
https://doi.org/10.1007/978-3-031-56066-8_9

Pradeep et al. [7] pinpointed the pitfalls of training multi-stage ranking models on datasets primarily consisting of credible information, leading to an unintended emphasis on harmful misinformation. In response, they introduced Vera, a stance prediction strategy effective at discerning useful from harmful content. Additionally, they advocated for the manual alignment of search questions to the appropriate stances to enhance result quality. This approach, however, is not without its flaws. The dependency on human-labeled stances and handcrafted rules for question reformulation restricts its adaptability and comprehensiveness, prompting a need for automation and generalization to accommodate the dynamic and diverse nature of information queries.

Through this paper, we attempt to tackle some drawbacks of Vera by automating the labeling of answers by organizers and generating arbitrary reformulations using large language models from the GPT class of models in combination with Vera [7].

More specifically, we build an answer prediction module using GPT_x in a zero-shot, in-context manner. We explore enhancing its capabilities by integrating it with evidence-aware Vera repurposed for the answer prediction task. Evaluating various prompting methods for GPT_x, including chain-of-thought reasoning, we develop a system that rivals human experts without requiring additional fine-tuning. Finally, we introduce a GPT_x reformulator that, taking into account these predicted answers, rephrases the question into a more naturally structured sentence, yielding improved retrieval effectiveness. On evaluating our systems in the TREC 2022 Health Misinformation Track, we find that our best systems outperform competing systems involving humans in the loop.

2 Datasets

2.1 MS MARCO Passage Ranking

We leverage relevance ranking models trained on the MS MARCO passage ranking dataset [1], which provides a corpus of 8.8M passages gathered from Bing search engine results. The collection has a training set of 500K (query, relevant passage) pairs, which helps finetune large language models, known to require a lot of training data to avoid overfitting the training data distribution.

2.2 TREC Health Misinformation

The TREC Health Misinformation Track is a concerted effort to advance retrieval methods that elevate accurate and credible health information while mitigating the spread of misinformation. In our study, we meticulously assess our systems utilizing the TREC 2022 Health Misinformation Track, drawing insights from questions and their associated medical consensus-based stances from the 2021 iteration to enrich our answer prediction module. Documents in this track are judged and categorized as "helpful" or "harmful", contingent on the relevance grade assigned, leading to the compilation of two distinct judgment sets.

In evaluating retrieval effectiveness, organizers employed three primary metrics: helpful compatibility measure ($\text{COMP}_{\text{HELP}}$), harmful compatibility measure ($\text{COMP}_{\text{HARM}}$), and the difference between these measures (COMP_{Δ}) [2,3]. The objective is to amplify helpful content and demote harmful documents, benchmarked by the COMP_{Δ} metric.

The TREC 2022 Health Misinformation Track inaugurated the "Answer Prediction" task that requires teams to submit predicted answers for all topics. Unlike previous versions where the organizers supplied the medical consensus answer, this new requirement augments the complexity of the tasks and adds a layer to system evaluations. However, this addition is critical, given that end-to-end systems capable of automatically determining the medical consensus and leveraging them to improve retrieval effectiveness are crucial to dealing with the evolving nature of health information.

3 Multi-stage Relevance Ranking

3.1 First-Stage Retrieval

The first stage receives as input the user query, q, and produces top-k_0 candidates R_0 from the corpus. The intent is to curate a refined candidate set to be scored by a sophisticated neural reranker.

The query is treated as a "bag of words" for ranking documents from the corpus. We used a standard inverted index based on BM25 in the Anserini IR toolkit [12,13], built on the popular open-source Lucene search engine with default hyperparameters. In all experiments, we set $k_0 = 1000$.

3.2 Neural Rerankers

In this stage, documents retrieved by first stage retrieval, R_0, are reranked by a pointwise reranker called monoT5 [6]. The model estimates a score s_i, quantifying how relevant a candidate $d_i \in R_0$ is to a query q.

monoT5 uses T5 [8], a popular pretrained sequence-to-sequence transformer model. During training, the model takes in query–document pairs from MS MARCO passage [1] and produces the words "true" or "false" depending on whether the document is relevant to the query. Nogueira et al. [6] finetuned monoT5-3B (around 2.8B model parameters) with a constant learning rate of 10^{-3} for 10k iterations with a batch size of 128. Following [6], we reranked the documents according to the probabilities assigned to the "true" token.

We further refined our candidate set with duoT5-3B, which aims to predict $p_{i,j}$, the probability d_i is more relevant than d_j to the query. To do so efficiently, we used the representative segment of both d_i and d_j based on the highest monoT5 score. To account for the longer input length resulting from pairs of document segments, we increased the maximum number of input tokens from the default of 512 to 1024. At inference time, we aggregated the pairwise scores $p_{i,j}$ so that each document received a single score s_i.

Following Pradeep et al. [7], we evaluated pointwise and pairwise variants that used the topic description as the query (monoT5$_{base}$ and duoT5$_{base}$) and also those that rephrased the query based on the predicted answer into a natural language sentence form (monoT5$_{NL}$ and duoT5$_{NL}$).

```
We are a committee of leading scientific experts and medical doctors
reviewing the latest and highest quality of research from PubMED. For
each question, we have chosen an answer, either 'yes' or 'no', based on
our best understanding of current medical practice and literature.
Q: Will wearing an ankle brace help achilles tendonitis? E: There is
little scientific evidence to suggest that orthotics alone will be
effective in healing it. A: no
Q: Can vitamin b12 and sun exposure together help treat vitiligo? E: The
spread of vitiligo stopped in the majority of the patients after this
treatment. A: yes
...
Q: {query} E:
```

Fig. 1. The GPT$_x$ Prompt for the Answer Prediction module.

4 Stance and Answer Prediction Modules

In this section, *stance prediction* refers to the task where the model is given a question and a relevant text snippet and returns a label corresponding to the snippet's stance to the question. The *answer prediction* task is more global, where we aim to predict a single stance to a question that represents the medical consensus.

4.1 Vera—Stance Prediction

Pradeep et al. [7] addressed the problem of discerning misinformation by leveraging a stance prediction module called Vera. Given the topic q and a document d_i, the model is tasked to predict a label $\hat{y}(q, d_i) \in \{\text{true}, \text{weak}, \text{false}\}$. They leveraged the same input template as monoT5. To train Vera, they utilized effectiveness judgments from the TREC 2019 Decision (Medical Misinformation) Track and finetuned the Vera-3B model using a constant learning rate of 10^{-3} for 500 iterations with batch sizes of 128.

During inference, for a particular document d_i, given t_i and f_i are the probabilities assigned to the "true" and "false" tokens, respectively, they used the scoring scheme:

$$s_i^{\text{final}} = \lambda \cdot s_i^z + (1 - \lambda) \begin{cases} t_i - f_i, \text{answer field is "yes"} \\ f_i - t_i, \text{answer field is "no"} \end{cases} \quad (1)$$

which they denoted as Vera (λ, z), where $z \in \{\text{mono}, \text{duo}\}$ is referred to as the "relevance setting" and λ is the linear combination constant. The "weak" labels do not factor into inference as we are only concerned with how "true" or "'false" the model believes the stance is.

4.2 GPT$_x$—Answer Prediction

The success of Vera as a stance detection model relies on a single established medical consensus stance. These are absent (by choice) before judging in the 2022 edition. To this end, we formulate two ways to deal with this issue, the first of which leverages OpenAI's large language models, GPT$_x$.

We experimented with one completion model, GPT$_3$ (`text-davinci-002`) and two chat models, GPT$_{3.5}$ (`gpt-3.5-turbo`) and GPT$_4$ (`gpt-4`) using the prompt seen in Fig. 1. The prompt begins with a preamble of what we ideally strive for when we have access to labor and resources, a consensus among experts.

Providing such information in the prompt helps in grounding the model. Then, we included eight examples of questions from the TREC 2020 Health Misinformation Track, followed by an explanation leading to the answer inspired by chain-of-thought (CoT) reasoning [11].reasoning [11]. We handcrafted a short and simple explanation based on a quick skim of the "PubMED" article the decision makers cite. Finally, we added to the prompt a query from the TREC 2022 Health Misinformation Track and appended with the "E:" tag that signifies what comes next is the explanation. Note that this method does not include information on *any* documents in the prompt. Finally, we generated a single token, took the scores corresponding to the "yes" and "no" tokens, and normalized over these two scores when both were available.

We experimented with self-consistency checks [10] with multiple (5) target sequences but found the results are always consistent, especially with larger models. We crafted the prompt before submissions and stuck with it for post-hoc analysis and ablations, albeit introducing and removing components.

The costs of querying the GPT$_3$, GPT$_{3.5}$, and GPT$_4$ API are at most 0.002 USD, 0.001 USD, and 0.03 USD, respectively. Computationally, we can get answer predictions in less than a minute for the entire test query set, with GPT$_{3.5}$ being considerably faster.

4.3 Vera—Answer Prediction

To extend Vera [7] to the task of answer prediction, we first calculated the probabilities Vera assigns to the true and false tokens for the top 50 monoT5 documents. With these probabilities, we calculated the means of "true" and "false" scores over all documents and predicted the answer "yes" if that of the "true" token is higher and "no" otherwise. This approach essentially forms a consensus based on the top-retrieved documents; employing more effective retrieval systems could enhance the precision of answer predictions. Given the corpus-aware consensus from Vera and corpus-free prediction from GPT$_x$, we also evaluated the effectiveness of the mean prediction from Vera and GPT$_x$.

4.4 GPT_x Reformulation

Pradeep et al. [7] found that reformulating health questions to a natural language sentence based on the predicted answer results in better relevance ranking results. However, they handcrafted rules to solve this task, which does not generalize to arbitrary questions. To avoid this complication, we leveraged GPT_3, which can easily handle this natural language reformulation task. We used the following prompt: "Rephrase the questions to sentence-long answers based on the stance. Question: Will wearing an ankle brace help heal achilles tendonitis? Stance: no Answer: Wearing ankle brace does not help heal achilles tendonitis ...", with eight in-context examples followed by the question and stance from the topic set. Variants with and without the reformulator are denoted by $*_{NL}$ and $*_{base}$, respectively.

Table 1. Model Effectiveness on the TREC 2022 Health Misinformation Answer Prediction task.

Model	AUC	Accuracy	TPR	FPR
(a) Median	70.7	64.0	80.0	48.0
(b) Humans	94.0	94.0	88.0	0.0
(c) GPT_3	95.2	86.0	76.0	4.0
(d) Vera	82.1	68.0	84.0	48.0
(e) Hyb(GPT_3, Vera)	93.4	88.0	80.0	4.0
(f) $GPT_{3.5}$	86.0	86.0	80.0	8.0
(g) + CoT	94.0	94.0	88.0	0.0
(h) GPT_4 + CoT	94.0	94.0	96.0	8.0

5 Results

Table 1 reports the results on the Answer Prediction task from the TREC 2022 Health Misinformation Track. The reported metrics are Area Under the Curve (AUC), Accuracy, True Positive Rate (TPR), and False Positive Rate (FPR). For reference, row (a) provides the median TREC evaluation score, and row (b) provides the score from a human-in-the-loop submission. Rows (c)–(e) present our official submissions and (f)–(h) our post-hoc experiments.

Firstly, GPT_3 and Vera answer prediction models, rows (c)–(d), are more effective than the median. Among the submissions, GPT_3, row (c), has the highest AUC that demonstrates the effectiveness of these large language models even in a zero-shot setting. However, combining retrieval-based methods such as Vera with GPT_3, row (e), seems to improve the accuracy, the most critical for the retrieval task.

Moving from GPT_3 to the chat variant $GPT_{3.5}$ shows a similar accuracy but better TPR, row (f) vs. (c). Models post GPT_3 do not provide token probabilities, forcing us to set probabilities of selected tokens as a 1. Hence, we do not include hybrids with Vera, as they do not make sense anymore.

Finally, we find that chain-of-though prompting results in a considerable effectiveness boost, rows (g) vs. (f). Switching the chat variant to GPT_4, rows (h) vs. (g), does not seem to have a considerable effect.

Table 2 looks at the effectiveness of the retrieval task of the TREC 2022 Health Misinformation Track. For reference, row (a) provides the median score across all submissions in the track. Row (b) presents the second top-scoring submitted run, a manual submission. Rows (c)–(j) represent all our submitted runs, and rows (k)–(l) represent our post-hoc runs. Rows (j)–(l) consider progressively better answer prediction modules ending with an Oracle system.

With little surprise, pointwise reranking improves the helpful compatibility scores over that of BM25, rows (d) and (f) vs. (c). While pairwise rerankers generally improve over pointwise results, in this case, looking at rows (d) vs. (e) and (f) vs. (g), we see a surprising drop in effectiveness.

Table 2. Compatibility scores on the TREC 2022 Health Misinformation Ad Hoc Retrieval Task.

Retrieval Model	Answer Model	$COMP_{HELP}$	$COMP_{HARM}$	$COMP_{\Delta}$
(a) Median	-	0.2455	0.1465	0.0990
(b) WatS	Humans	0.287	0.140	0.147
(c) BM25	-	0.1928	0.1487	0.0441
(d) + $monoT5_{base}$	-	0.2838	0.1942	0.0896
(e) + $duoT5_{base}$	-	0.2780	0.1894	0.0886
(f) + $monoT5_{NL}$	GPT_3	0.3276	0.1264	0.2012
(g) + $duoT5_{NL}$	GPT_3	0.3216	0.1467	0.1749
(h) Vera ($\lambda = 0.0$)	GPT_3	0.2836	0.0971	0.1865
(i) Vera ($0.95, mono$)	GPT_3	0.3386	0.1168	0.2218
(j) Vera ($0.95, mono$)	Hyb(GPT_3, Vera)	0.3460	0.0894	0.2566
(k) Vera ($0.95, mono$)	GPT_4 + CoT	0.3528	0.0892	0.2636
(l) Vera ($0.95, mono$)	Oracle	0.3602	0.0797	0.2805

When introducing the answer prediction model module and the query reformulator, comparing rows (f) to (d) and (g) to (e), we notice it brings a considerable increase in helpful compatibility scores and a similar drop in harmful compatibility scores, as desired.

The introduction of Vera in the $\lambda = 0$ setting, i.e., label prediction alone, results in a model with worse helpful compatibility but better harmful compatibility score compared to $monoT5_{NL}$, row (h) vs. (f). Linear combinations with the neural relevance ranking system, as seen from row (i) onwards, bring an effectiveness boost, finding a spot with better helpful compatibility but slightly worse harmful compatibility scores.

Improving the answer prediction model (based on accuracy) results in progressively better results, as evidenced in rows (i)–(l). Row (k) illustrates our most effective automatic system, noting a 79% relative improvement over competing systems by other teams based on the primary metric, COMP_{Δ}. Compared to a system with an oracle answer prediction module, i.e., with ground truth answer predictions, row (l), this system demonstrates comparable effectiveness showcasing its strength.

6 Conclusion

In this paper, we focus on automating the consumer health search pipeline—building an end-to-end system capable of discerning helpful from harmful health search results with experimentation in the TREC 2022 Health Misinformation Track. We build an effective answer prediction module using GPT_{x} in a zero-shot in-context fashion and augment it with the evidence-aware Vera. We explore various ways of prompting GPT_{x}, incorporating chain-of-though reasoning to build a system on par with human experts without finetuning. We incorporate a reformulator that takes in these predicted answers and rephrases the question in a natural language sentence, resulting in improved results. Coupled with a state-of-the-art retrieval misinformation-free consumer health search pipeline, our models outperform runs from other teams by over 79% based on COMP_{Δ}.

Acknowledgements. This research was supported in part by the Natural Sciences and Engineering Research Council (NSERC) of Canada.

References

1. Bajaj, P., et al.: MS MARCO: a human generated MAchine Reading COmprehension dataset. arXiv arXiv:1611.09268 (2016)
2. Clarke, C.L.A., Smucker, M.D., Vtyurina, A.: Offline evaluation by maximum similarity to an ideal ranking. In: Proceedings of the 29th ACM International Conference on Information & Knowledge Management, CIKM 2020, pp. 225–234 (2020)
3. Clarke, C.L.A., Vtyurina, A., Smucker, M.D.: Offline evaluation without gain. In: Proceedings of the 2020 ACM SIGIR International Conference on Theory of Information Retrieval, ICTIR 2020, pp. 185–192 (2020)
4. Devlin, J., Chang, M.W., Lee, K., Toutanova, K.: BERT: pre-training of deep bidirectional transformers for language understanding. In: Proceedings of the 2019 Conference of the North American Chapter of the Association for Computational Linguistics: Human Language Technologies, Volume 1 (Long and Short Papers), NAACL 2019, pp. 4171–4186 (2019)
5. Nogueira, R., Cho, K.: Passage re-ranking with BERT. arXiv arXiv:1901.04085 (2019)
6. Nogueira, R., Jiang, Z., Pradeep, R., Lin, J.: Document ranking with a pretrained sequence-to-sequence model. In: Findings of the Association for Computational Linguistics, EMNLP 2020, pp. 708–718 (2020)

7. Pradeep, R., Ma, X., Nogueira, R., Lin, J.: Vera: prediction techniques for reducing harmful misinformation in consumer health search. In: Proceedings of the 44th International ACM SIGIR Conference on Research and Development in Information Retrieval, SIGIR 2021, pp. 2066–2070 (2021)
8. Raffel, C., et al.: Exploring the limits of transfer learning with a unified text-to-text transformer. J. Mach. Learn. Res. **21**, 1–67 (2020)
9. Robertson, S.E., Walker, S., Jones, S., Hancock-Beaulieu, M., Gatford, M.: Okapi at TREC-3. In: Proceedings of the 3rd Text REtrieval Conference (TREC-3), pp. 109–126 (1994)
10. Wang, X., et al.: Self-consistency improves chain of thought reasoning in language models. In: The Eleventh International Conference on Learning Representations, ICLR 2023 (2023)
11. Wei, J., et al.: Chain of thought prompting elicits reasoning in large language models. In: Advances in Neural Information Processing Systems, NeurIPS 2022 (2022)
12. Yang, P., Fang, H., Lin, J.: Anserini: enabling the use of Lucene for information retrieval research. In: Proceedings of the 40th Annual International ACM SIGIR Conference on Research and Development in Information Retrieval, SIGIR 2017, pp. 1253–1256 (2017)
13. Yang, P., Fang, H., Lin, J.: Anserini: reproducible ranking baselines using Lucene. J. Data Inf. Qual. **10**(4), 16:1–16:20 (2018)

Weighted AUReC: Handling Skew in Shard Map Quality Estimation for Selective Search

Gijs Hendriksen(✉), Djoerd Hiemstra, and Arjen P. de Vries

Radboud University, Nijmegen, The Netherlands
{gijs.hendriksen,djoerd.hiemstra,arjen.devries}@ru.nl

Abstract. In selective search, a document collection is partitioned into a collection of topical index shards. To efficiently estimate the topical coherence (or quality) of a shard map, the AUReC measure was introduced. AUReC makes the assumption that shards are of similar sizes, one that is violated in practice, even for unsupervised approaches. The problem might be amplified if supervised labelling approaches with skewed class distributions are used. To estimate the quality of such unbalanced shard maps, we introduce a weighted adaptation of the AUReC measure, and empirically evaluate its effectiveness using the ClueWeb09B and Gov2 datasets. We show that it closely matches the evaluations of the original AUReC when shards are similar in size, but captures better the differences in performance when shard sizes are skewed.

Keywords: selective search · clustering · evaluation · cluster-based retrieval

1 Introduction

With increasingly complex and expensive retrieval pipelines, efficient retrieval over large document collections can be quite difficult to achieve. Distributed systems may alleviate part of this problem by partitioning the index into shards and searching these in parallel. Another approach, *selective search* [19], avoids the exhaustive search over the full collection, reducing the total number of documents processed per query by partitioning the collection into topically coherent shards. This assumes the Cluster Hypothesis [25] to be true, that a query's relevant documents are allocated to the same (subset of) shards. The *resource selection algorithm* should select the relevant shards for each incoming query.

Previous work in selective search has investigated end-to-end effectiveness [19], runtime efficiency [16,17], partitioning strategies [12,18], resource selection algorithms [1,11,13,20,22,24], and robustness [10]. The AUReC (Area Under Recall Curve) measure [15] estimates *shard map quality*, considered high if, for any query, its relevant documents are indeed clustered in a few shards only. The AUReC measure circumvents the need for manual relevance assessments by marking the top documents retrieved by a strong ranker as pseudo-relevant.

Unfortunately, AUReC makes the assumption that shards have similar size, such that the top k shards can be returned. However, existing clustering algorithms cannot guarantee that the resulting shards have similar sizes; in fact, the

N. Goharian et al. (Eds.): ECIR 2024, LNCS 14611, pp. 87–96, 2024.
https://doi.org/10.1007/978-3-031-56066-8_10

shard maps on which AUReC was evaluated originally, generated by Dai et al. [12], also contain shards that are orders of magnitude larger than the smallest ones – even after applying size-bounded clustering [19]. In her dissertation, Kim [14, Sect. 8.2] already observed that AUReC could be biased towards unbalanced shard maps.

The situation can be expected to become much worse when these shard maps distribute Web documents by language, top level domain or category. Especially category assignments by a supervised classifier like in the ClueWeb22 corpus [23] are likely to follow a Zipfian distribution. In this case, splitting and merging shards to balance the shard sizes may not be desirable, as it might decrease the coherence and interpretability of affected shards. To mitigate this bias, we introduce a weighted variant of the AUReC that takes shard size into account. We show empirically how this *weighted AUReC* is a better measure of shard map quality when these shard maps exhibit skew.

We also consider a budget-constrained situation, where the system processes a fixed number of *documents* per query instead of a fixed number of shards. In this case, it may be a good strategy to select many small shards, instead of the few large shards that AUReC is biased towards. We show that weighted AUReC measures shard map quality also more accurately in this budget-constrained setting. The code used for our experiments is published to GitLab.[1]

2 Weighted AUReC

The AUReC measure [15] builds on the underlying goal of a selective search system: retrieving the same documents as an exhaustive search system, but more efficiently. As such, a strong ranker can be used to exhaustively retrieve the top k documents D_q for a given query q (usually, $k = 1000$). These documents are then marked as pseudo-relevant for the calculation of the AUReC.

Assume a shard map p with n_p shards. For each query q, let $count(D_q, s_i^p)$ be the number of documents from D_q that appear in shard s_i^p. Define a relevance-based ranking (RBR) order of shards, such that each shard s_i^p contains more pseudo-relevant documents from D_q than the next one. Formally:

$$count(D_q, s_i^p) \geq count(D_q, s_{i+1}^p) \qquad \text{for all } i \in \{1 \dots n_p - 1\}$$

Given this ordering, a recall-like measure $R_q(p, k)$ is defined to measure the percentage of pseudo-relevant documents that appear in the first k shards of shard map p[2]:

$$R_q(p, k) = \frac{1}{|D_q|} \sum_{i=1}^{k} count(D_q, s_i^p) \qquad \text{for } k \in \{0 \dots n_p\}$$

[1] https://gitlab.science.ru.nl/informagus/weighted-aurec.

[2] We use a simplified version of the formula from Kim and Callan [15], in which we assume that D_q is never empty. Other than that, the formulas are equivalent.

Using this definition, the recall curve is formed by the points $\langle k/n_p,\ R_q(p,k)\rangle$ for all $k \in \{0 \ldots n_p\}$. The AUReC for query q is the area under this curve:

$$AUReC_q(p) = \frac{1}{2n_p} \sum_{k=0}^{n_p-1} \left(R_q(p,k) + R_q(p,k+1)\right)$$

To obtain a final quality measurement, we average the $AUReC_q$ over all queries in the query set. AUReC scores range from 0.5 (relevant documents are uniformly distributed) to 1.0 (relevant documents are clustered together).

We will now adjust this measure to handle skewed shard maps, in two steps. First, instead of ordering the shards by the number of documents from D_q they contain, we simply divide that number by the size of each shard. Formally:

$$count(D_q, s_i^p)/|s_i^p| \geq count(D_q, s_{i+1}^p)/|s_{i+1}^p| \qquad \text{for all } i \in \{1 \ldots n_p - 1\}$$

This modification promotes smaller shards with a relatively large proportion of pseudo-relevant documents while pushing back large shards with a higher proportion of irrelevant documents.

Note that this has no impact on the definitions of $count(D_q, s_i^p)$ and $R_q(p,k)$; they are applied the same way, only for a different ordering.

Second, we scale each segment of the recall curve by the size of the corresponding shard. In other words, if D is the full collection, we define the recall curve as the points $\langle \sum_{i=1}^{k} |s_i^p| \ /\ |D|,\ R_q(p,k)\rangle$ for all $k \in \{0 \ldots n_p\}$.

Weighted AUReC (wAUReC) follows as the area beneath this adjusted curve:

$$wAUReC_q(p) = \sum_{k=0}^{n_p-1} \frac{|s_{k+1}^p|}{2\sum_{i=1}^{n_p} |s_i^p|} \cdot \left(R_q(p,k) + R_q(p,k+1)\right)$$

When documents are distributed evenly across shards (i.e., every shard has the same size), the value of wAUReC equals the normal AUReC. Therefore, wAUReC is not only applicable in the case of a skewed shard map; it can be used as a full substitute for the normal AUReC.

3 Experimental Setup

3.1 Documents, Queries and Runs

To empirically evaluate the effectiveness of the wAUReC and compare it to the normal AUReC, we ran a set of experiments with a similar setup to Kim and Callan [15]. We used the same document collections: Gov2[3] and ClueWeb09B[4]. We used the topics and relevance assessments provided by the TREC Terabyte Track from 2004 until 2006 [3,4,9] for evaluation on Gov2, and the TREC Web Track from 2009 until 2012 [5–8] for evaluation on ClueWeb09B. Finally, we also

[3] http://ir.dcs.gla.ac.uk/test_collections/access_to_data.html.
[4] https://lemurproject.org/clueweb09/.

used SlideFuse-MAP [2,21] to fuse the top 10 runs submitted each year to obtain the results of a 'strong ranker'. This fusion run was used both for gathering D_q for the AUReC measures and for the evaluation of an end-to-end selective search system using different resource selection algorithms.

Since we use topics, relevance judgments and submitted runs from TREC, we do not have to index the corpus ourselves and run a retrieval system against it. This makes our experimental setting easy to setup and replicate.

3.2 Shard Maps and Shard Selection

We reuse the 6 shard maps (3 per dataset) generated by Dai et al. [12] for a basic comparison between the AUReC and wAUReC.[5]

Because these shard maps are fairly balanced in terms of shard size, the differences between the AUReC and its weighted variant might not become fully apparent. To make the problem more pronounced, we therefore generate extra shard maps, using different size distributions: uniform, linear and quadratic. For each collection and distribution type, we generate 10 different shard maps (60 in total). Per query, we distribute the relevant documents randomly over up to 10 shards. We repeat this procedure 50 times per shard map (resulting in 3000 shard maps), which allows us to perform a robust comparison between the two AUReC variants on unbalanced shard maps.

Evidently, these simulated random shard maps are unlikely to be used in practice, and lose the topical coherence that real-world clustering algorithms provide. However, we evaluate the end-to-end selective search performance using oracle resource selection algorithms only, meaning we can still evaluate whether relevant documents are clustered together – both for the end-to-end performance and for the AUReC and wAUReC. As such, the random shard maps are still useful to demonstrate the advantages of wAUReC, even if experiments with more realistic distribution approaches are warranted in future research.

Relevance-Based Shard Ranking. Kim and Callan [15] evaluated AUReC on selective search systems using different resource selection algorithms: Rank-S [20], Taily [1], and relevance-based ranking (RBR). RBR is the oracle that provides the theoretically maximum performance, selecting the shards with the highest number of relevant documents at static cutoff k. We only use RBR for our comparison between AUReC and wAUReC (for cutoffs $k \in \{1, 3, 5\}$).

Budget-Based Shard Ranking. Like AUReC, RBR assumes that shard sizes are balanced, as it orders the shards based on the absolute number of relevant documents they contain. As a result, the RBR oracle method might not showcase the limitations of the AUReC when it comes to unbalanced shard maps: they follow the same shard ordering. To illustrate the difference between AUReC and wAUReC more clearly, we also consider an alternate setting.

[5] Downloaded from https://boston.lti.cs.cmu.edu/appendices/CIKM2016-Dai/.

When shard maps are skewed, selecting a static number of shards may not suffice. Consider a system with a maximum number of documents to process, e.g. in order to keep latency below a certain threshold or limit the allocated resources per query. We call this number the *budget* of the system.

With a budget of 1000 documents, one can either search one shard with 1000 documents, or 10 shards with 100 documents. The larger shard may have a larger absolute number of relevant documents, but the smaller shards combined may contain even more relevant documents. Relevance-based ranking will always return the largest shard first, even if this would result in sub-optimal results.

We therefore introduce *budget-based ranking* (BBR), an alternative for RBR. Unfortunately, finding the optimal set of shards given a budget b is an instance of the knapsack problem, infeasible to solve in practice. Instead, we approximate the optimal selection by ordering the documents according to the fraction of relevant documents they contain, and selecting shards greedily, such that the total number of processed documents stays below b. Since the ordering used for the BBR is similar to that for computing wAUReC, the measure relates to BBR shard selection as normal AUReC relates to RBR.

3.3 Metrics

We follow Kim and Callan's [15] example to evaluate how wAUReC relates to the performance of an end-to-end system. We compute the correlation (Pearson's r) between each of the AUReC measures and the selective search system's end-to-end performance, with either shard selection algorithm. Like Kim and Callan [15], we evaluate the end-to-end system using P@1000, a deep, recall-focused measure that assumes selective search is used as a first-stage retrieval system.

Table 1. Correlation (Pearson's r) between AUReC variants and the P@1000 of end-to-end systems using relevance-based ranking for resource selection.

End-to-end	Dai et al. [12]		Random	
	AUReC	*wAUReC*	*AUReC*	*wAUReC*
RBR ($k = 1$)	0.932	**0.934**	0.258	0.891
RBR ($k = 3$)	0.929	**0.930**	0.261	0.897
RBR ($k = 5$)	0.925	**0.927**	0.261	0.899

4 Experimental Results

4.1 Relevance-Based Ranking

The left-hand side of Table 1 shows the correlation between an RBR-based end-to-end system and both AUReC variants, across all six shard maps from Dai et al. [12]. We clearly see that the AUReC and wAUReC are more or less equivalent in this setting. In fact, the correlation between AUReC and wAUReC over all topics and datasets is 0.990, indicating the similar outcomes of the two variants.

For the randomly generated shard maps (right-hand side of Table 1), the wAUReC seems to be more correlated with the system's end-to-end performance, showing the added benefit of using shard size in the evaluation of shard maps. Figure 1 additionally shows the correlation as a function of the standard deviation of shard sizes in a shard map: the lower the standard deviation, the more balanced a shard map is. For Gov2, the correlation stays roughly the same, but for ClueWeb09B we clearly see a drop in performance for the AUReC when shard maps become more unbalanced, while the wAUReC remains more or less consistent. We observed similar outcomes for different RBR cutoff values k.

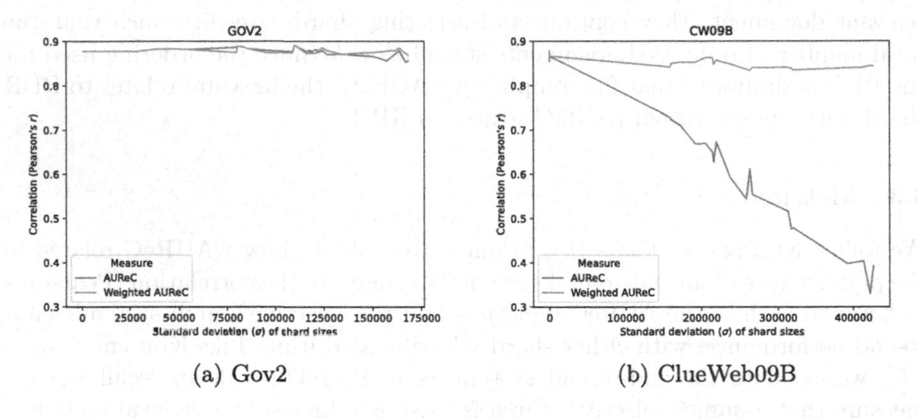

(a) Gov2 (b) ClueWeb09B

Fig. 1. Correlation (Pearson's r) between AUReC variants and P@1000 of end-to-end systems (with RBR and $k = 3$) for varying degrees of shard size skew.

4.2 Budget-Based Ranking

Figure 2 shows the correlation between the AUReC measures and end-to-end systems using BBR, for a wide range of budgets b. There is a large difference between the performance of the regular AUReC and the wAUReC for both datasets. As expected, AUReC performs sub-optimally in the setting where a system is limited in the number of documents it can process, rather than the number of shards. The wAUReC is better able to capture this budget-constrained environment, though its correlation also still leaves room for improvement (especially for small values b). A possible explanation for this outcome is that the artificial nature of the generated shard maps makes selective search more difficult in general. Alternatively, our greedy heuristic for determining the optimal BBR could result in suboptimal performance of the end-to-end system. Too small values for b might even make effective retrieval impossible altogether.

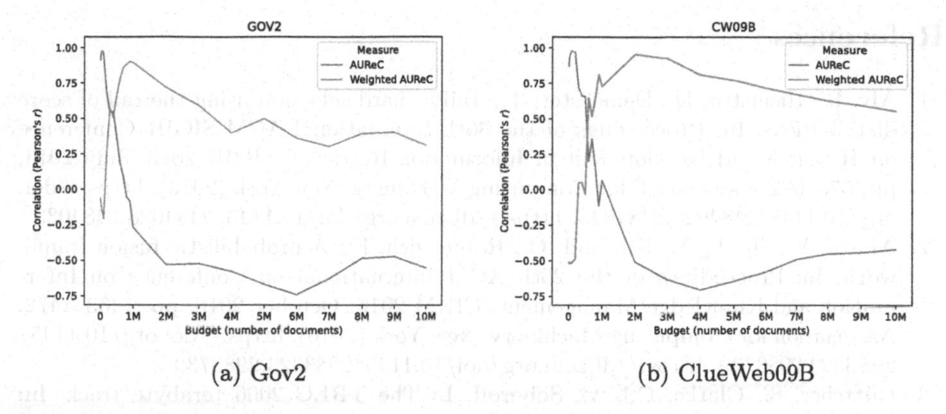

(a) Gov2 (b) ClueWeb09B

Fig. 2. Correlation (Pearson's r) between AUReC variants and the P@1000 of end-to-end systems using budget-based ranking for resource selection.

5 Conclusion

This paper introduced a weighted variant of the AUReC measure (wAUReC), which can be used to evaluate shard maps for use in a selective search system when the shards are skewed in size. First, we have shown that the wAUReC performs similarly to the normal AUReC when shard maps are balanced. Then, we showed that the AUReC performance degrades when shard sizes are skewed, and that its weighted counterpart can handle such shard maps better.

We also studied a setup in which a system does not select k shards but instead has a fixed budget b of documents that it can process given limited time or resources. In this case, it might be more worthwhile to select smaller shards with a higher relative number of relevant documents first, to not fill up the budget with a few large shards. In this setting, AUReC was unable to accurately measure the quality of a shard map. The wAUReC achieved a much higher correlation with the end-to-end system, for a wide range of budgets b.

We aim to continue this work and apply the wAUReC on datasets and shard maps with inherent size skew, to evaluate its performance in more realistic scenarios and ensure it can be applied in practice.

Unlike Kim and Callan did for AUReC [15], we have not yet investigated whether wAUReC can be used in significance testing. However, because of the strong similarities between the measures, we hypothesise that those findings also translate to the wAUReC. This hypothesis can be verified in future work.

Acknowledgments. This work has received funding from the European Union's Horizon Europe research and innovation programme under grant agreement No. 101070014 (OpenWebSearch.EU, https://doi.org/10.3030/101070014). We also thank Yubin Kim, who kindly helped us with our experimental setup by making her code available.

References

1. Aly, R., Hiemstra, D., Demeester, T.: Taily: shard selection using the tail of score distributions. In: Proceedings of the 36th International ACM SIGIR Conference on Research and Development in Information Retrieval, SIGIR 2013, July 2013, pp. 673–682. Association for Computing Machinery, New York (2013). https://doi.org/10.1145/2484028.2484033. https://dl.acm.org/doi/10.1145/2484028.2484033
2. Anava, Y., Shtok, A., Kurland, O., Rabinovich, E.: A probabilistic fusion framework. In: Proceedings of the 25th ACM International on Conference on Information and Knowledge Management, CIKM 2016, October 2016, pp. 1463–1472. Association for Computing Machinery, New York (2016). https://doi.org/10.1145/2983323.2983739. https://dl.acm.org/doi/10.1145/2983323.2983739
3. Büttcher, S., Clarke, C.L.A., Soboroff, I.: The TREC 2006 terabyte track. In: Voorhees, E.M., Buckland, L.P. (eds.) Proceedings of the Fifteenth Text REtrieval Conference, TREC 2006, Gaithersburg, Maryland, USA, 14–17 November 2006. NIST Special Publication 500-272. National Institute of Standards and Technology (NIST) (2006). https://trec.nist.gov/pubs/trec15/papers/TERA06.OVERVIEW.pdf
4. Clarke, C.L.A., Craswell, N., Soboroff, I.: Overview of the TREC 2004 terabyte track. In: Voorhees, E.M., Buckland, L.P. (eds.) Proceedings of the Thirteenth Text REtrieval Conference, TREC 2004, Gaithersburg, Maryland, USA, 16–19 November 2004. NIST Special Publication 500-261. National Institute of Standards and Technology (NIST) (2004). https://trec.nist.gov/pubs/trec13/papers/TERA.OVERVIEW.pdf
5. Clarke, C.L.A., Craswell, N., Soboroff, I.: Overview of the TREC 2009 web track. In: Voorhees, E.M., Buckland, L.P. (eds.) Proceedings of the Eighteenth Text REtrieval Conference, TREC 2009, Gaithersburg, Maryland, USA, 17–20 November 2009. NIST Special Publication 500-278. National Institute of Standards and Technology (NIST) (2009). https://trec.nist.gov/pubs/trec18/papers/WEB09.OVERVIEW.pdf
6. Clarke, C.L.A., Craswell, N., Soboroff, I., Cormack, G.V.: Overview of the TREC 2010 web track. In: Voorhees, E.M., Buckland, L.P. (eds.) Proceedings of the Nineteenth Text REtrieval Conference, TREC 2010, Gaithersburg, Maryland, USA, 16–19 November 2010. NIST Special Publication 500-294. National Institute of Standards and Technology (NIST) (2010). https://trec.nist.gov/pubs/trec19/papers/WEB.OVERVIEW.pdf
7. Clarke, C.L.A., Craswell, N., Soboroff, I., Voorhees, E.M.: Overview of the TREC 2011 web track. In: Voorhees, E.M., Buckland, L.P. (eds.) Proceedings of the Twentieth Text REtrieval Conference, TREC 2011, Gaithersburg, Maryland, USA, 15–18 November 2011. NIST Special Publication 500-296. National Institute of Standards and Technology (NIST) (2011). https://trec.nist.gov/pubs/trec20/papers/WEB.OVERVIEW.pdf
8. Clarke, C.L.A., Craswell, N., Voorhees, E.M.: Overview of the TREC 2012 web track. In: Voorhees, E.M., Buckland, L.P. (eds.) Proceedings of the Twenty-First Text REtrieval Conference, TREC 2012, Gaithersburg, Maryland, USA, 6–9 November 2012. NIST Special Publication 500-298. National Institute of Standards and Technology (NIST) (2012). https://trec.nist.gov/pubs/trec21/papers/WEB12.overview.pdf

9. Clarke, C.L.A., Scholer, F., Soboroff, I.: The TREC 2005 terabyte track. In: Voorhees, E.M., Buckland, L.P. (eds.) Proceedings of the Fourteenth Text REtrieval Conference, TREC 2005, Gaithersburg, Maryland, USA, 15–18 November 2005. NIST Special Publication 500-266. National Institute of Standards and Technology (NIST) (2005). https://trec.nist.gov/pubs/trec14/papers/TERABYTE.OVERVIEW.pdf
10. Dai, Z., Kim, Y., Callan, J.: How random decisions affect selective distributed search. In: Proceedings of the 38th International ACM SIGIR Conference on Research and Development in Information Retrieval, SIGIR 2015, August 2015, pp. 771–774. Association for Computing Machinery, New York (2015). https://doi.org/10.1145/2766462.2767796. https://dl.acm.org/doi/10.1145/2766462.2767796
11. Dai, Z., Kim, Y., Callan, J.: Learning to rank resources. In: Proceedings of the 40th International ACM SIGIR Conference on Research and Development in Information Retrieval, SIGIR 2017, August 2017, pp. 837–840. Association for Computing Machinery, New York (2017). https://doi.org/10.1145/3077136.3080657. https://dl.acm.org/doi/10.1145/3077136.3080657
12. Dai, Z., Xiong, C., Callan, J.: Query-biased partitioning for selective search. In: Proceedings of the 25th ACM International on Conference on Information and Knowledge Management, CIKM 2016, October 2016, pp. 1119–1128. Association for Computing Machinery, New York (2016). https://doi.org/10.1145/2983323.2983706. https://dl.acm.org/doi/10.1145/2983323.2983706
13. Ergashev, U., Dragut, E., Meng, W.: Learning to rank resources with GNN. In: Proceedings of the ACM Web Conference 2023, WWW 2023, April 2023, pp. 3247–3256. Association for Computing Machinery, New York (2023). https://doi.org/10.1145/3543507.3583360. https://doi.org/10.1145/3543507.3583360
14. Kim, Y.: Robust Selective Search. Ph.D. thesis, Carnegie Mellon University (2019)
15. Kim, Y., Callan, J.: Measuring the effectiveness of selective search index partitions without supervision. In: Proceedings of the 2018 ACM SIGIR International Conference on Theory of Information Retrieval, ICTIR 2018, September 2018, pp. 91–98. Association for Computing Machinery, New York (2018). https://doi.org/10.1145/3234944.3234952. https://dl.acm.org/doi/10.1145/3234944.3234952
16. Kim, Y., Callan, J., Culpepper, J.S., Moffat, A.: Does selective search benefit from WAND optimization? In: Ferro, N., et al. (eds.) ECIR 2016. LNCS, vol. 9626, pp. 145–158. Springer, Cham (2016). https://doi.org/10.1007/978-3-319-30671-1_11
17. Kim, Y., Callan, J., Culpepper, J.S., Moffat, A.: Load-balancing in distributed selective search. In: Proceedings of the 39th International ACM SIGIR conference on Research and Development in Information Retrieval, SIGIR 2016, July 2016, pp. 905–908. Association for Computing Machinery, New York (2016). https://doi.org/10.1145/2911451.2914689. https://dl.acm.org/doi/10.1145/2911451.2914689
18. Kulkarni, A., Callan, J.: Document allocation policies for selective searching of distributed indexes. In: Proceedings of the 19th ACM international conference on Information and knowledge management, CIKM 2010, October 2010, pp. 449–458. Association for Computing Machinery, New York (2010). https://doi.org/10.1145/1871437.1871497. https://dl.acm.org/doi/10.1145/1871437.1871497
19. Kulkarni, A., Callan, J.: Selective search: efficient and effective search of large textual collections. ACM Trans. Inf. Syst. **33**(4), 17:1–17:33 (2015). https://doi.org/10.1145/2738035. https://dl.acm.org/doi/10.1145/2738035
20. Kulkarni, A., Tigelaar, A.S., Hiemstra, D., Callan, J.: Shard ranking and cutoff estimation for topically partitioned collections. In: Proceedings of the 21st ACM International Conference on Information and Knowledge Management, Maui, Hawaii,

USA, October 2012, pp. 555–564. ACM (2012). https://doi.org/10.1145/2396761.2396833. https://dl.acm.org/doi/10.1145/2396761.2396833
21. Lillis, D., Toolan, F., Collier, R., Dunnion, J.: Extending probabilistic data fusion using sliding windows. In: Macdonald, C., Ounis, I., Plachouras, V., Ruthven, I., White, R.W. (eds.) ECIR 2008. LNCS, vol. 4956, pp. 358–369. Springer, Heidelberg (2008). https://doi.org/10.1007/978-3-540-78646-7_33
22. Mohammad, H.R., Xu, K., Callan, J., Culpepper, J.S.: Dynamic shard cutoff prediction for selective search. In: The 41st International ACM SIGIR Conference on Research & Development in Information Retrieval, SIGIR 2018, June 2018, pp. 85–94. Association for Computing Machinery, New York (2018). https://doi.org/10.1145/3209978.3210005. https://dl.acm.org/doi/10.1145/3209978.3210005
23. Overwijk, A., Xiong, C., Callan, J.: ClueWeb22: 10 billion web documents with rich information. In: Proceedings of the 45th International ACM SIGIR Conference on Research and Development in Information Retrieval, SIGIR 2022, July 2022, pp. 3360–3362. Association for Computing Machinery, New York (2022). https://doi.org/10.1145/3477495.3536321. https://dl.acm.org/doi/10.1145/3477495.3536321
24. Si, L., Callan, J.: Relevant document distribution estimation method for resource selection. In: Proceedings of the 26th Annual International ACM SIGIR Conference on Research and Development in Informaion Retrieval, SIGIR 2003, Jul 2003, pp. 298–305. Association for Computing Machinery, New York (2003). https://doi.org/10.1145/860435.860490. https://dl.acm.org/doi/10.1145/860435.860490
25. Van Rijsbergen, C.J.: Information Retrieval. Butterworths (1979)

Reproducibility Papers

A Second Look on BASS – Boosting Abstractive Summarization with Unified Semantic Graphs

A Replication Study

Osman Alperen Koraş[1,2(✉)], Jörg Schlötterer[3,4], and Christin Seifert[4]

[1] Institute for AI in Medicine (IKIM), University Medicine Essen, Essen, Germany
[2] University of Duisburg-Essen, Essen, Germany
osman.koras@uni-due.de
[3] University of Mannheim, Mannheim, Germany
joerg.schloetterer@uni-marburg.de
[4] University of Marburg, Marburg, Germany
christin.seifert@uni-marburg.de

Abstract. We present a detailed replication study of the BASS framework, an abstractive summarization system based on the notion of Unified Semantic Graphs. Our investigation includes challenges in replicating key components and an ablation study to systematically isolate error sources rooted in replicating novel components. Our findings reveal discrepancies in performance compared to the original work. We highlight the significance of paying careful attention even to reasonably omitted details for replicating advanced frameworks like BASS, and emphasize key practices for writing replicable papers.

Keywords: Replication · Abstractive Summarization · Graph-Enhanced Transformer

1 Introduction

The goal of automatic text summarization is to generate fluent, concise, informative, and faithful summaries of source documents [10]. Extractive summarization methods select salient phrases from the source document and concatenate them to form the summary. In contrast, abstractive summarization methods freely generate text conditioned on an intermediate representation of the source document [10]. Consequently, the capabilities of abstractive summarization systems depend on the richness of this intermediate representation. Many state-of-the-art abstractive summarization systems are based on Pre-trained Language Models (PLM), such as BERT [4], PEGASUS [34], or T5 [25]. The success of transformers [27] across many domains shows that they are capable of generating rich representations for a wide range of signals, including vision [6], audio [5] and graphs [33]. The Graphormer [33] is one of many Attentive Graph Neural Networks [3,28,33], which have been successfully adapted for transformers to leverage graphs in abstractive summarization systems [7,11,15–17,20,24,32,36],

N. Goharian et al. (Eds.): ECIR 2024, LNCS 14611, pp. 99–114, 2024.
https://doi.org/10.1007/978-3-031-56066-8_11

with the aim to complement or guide the rich representation of transformers with explicitly structured data to improve accuracy and faithfulness of the generated summaries.

One of the graph-enhanced transformer models is BASS [30], which is of specific interest because i) it introduces a compressed dependency graph structure based on the idea of semantic units and ii) the authors report competitive performance in abstractive summarization while being only half the size (201M parameters for BASS vs. 406M parameters for BART [19] and PEGASUS). Because the original paper [30] is not accompanied by source code, we conduct a replication study of the BASS framework. Our results contribute to the broader discourse surrounding reproducibility concerns of Machine Learning [12,14] and in particular NLP research, sometimes even referred to as the "reproducibility crisis" [1,2]. Belz et al. report that fewer than 15% scores of their study were reproducible, and that "worryingly small differences in code have been found to result in big differences in performance" [1]. Even for performance scores reproduced under the same conditions, they discovered that almost 60% of reproduced scores were worse than the original score. Consequently, results from different works have to be compared with caution, even if similar components are employed, drawing attention to the importance of generating or replicating own baselines for meaningful comparisons and drawing conclusions. Concretely, the contributions of this paper are the following:

1. We conduct a replication study of the BASS framework and publish our implementation[1], including source code for the graph construction component provided by the authors of the original paper.
2. We conduct an ablation study to examine BASS' architectural adaptations to incorporate the graph information into transformers and find we can not replicate the performance improvement on the summarization task.
3. We detail our replication for each framework component, and summarize the specific and general challenges we faced during replication.

2 Replication Methodology

Our initial goal was to implement the BASS framework (cf. Fig. 1) in Python one component at a time, solely from information available to the community, i.e., the paper. We started by implementing the pre-processing ① and graph construction ②, but quickly identified missing information and uncertainties (see Sect. 6).

When key information was missing on a component, we inquired the authors via email for missing details and source code. When met with uncertainties, we contacted the authors only when we were not certain to have faithfully replicated the component. We exchanged multiple emails with over 20 questions out of which roughly three quarters have been answered. On average, the authors responded within 10 days. In the end, the authors provided a Java implementation for the graph construction component ② and snippets for pre-processing ①.

[1] https://github.com/osmalpkoras/bass-replication.

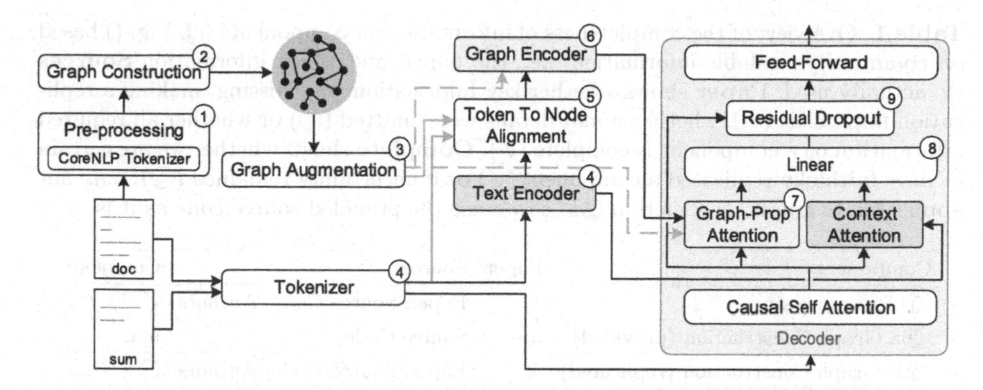

Fig. 1. Illustration of the BASS framework. The pre-processing and graph construction is done on the input document (left). The resulting graph information is used for token-to-node alignment ⑤, the graph encoder ⑥ and the respective cross-attention module ⑦ in the decoder.

However, we did not receive information about specific training details, e.g., the batch size, despite multiple inquiries.

Given the additional information, we identified inconsistencies between author information, paper details and source code (see Sect. 3 and Appendix A). We let details provided in the paper take precedence over implementation details of the Java source code, and let the source code take precedence over our correspondence with the authors. For details that still remained unclear, we made a best guess.

A summary of our replication is shown in Table 1, where we indicate per component which information sources we used, and whether uncertainties remained. We implemented the pre-processing ① in Java to use the authors' source code for graph construction ②a, but additionally replicated the graph construction ②b in Python. All other components ③–⑨ are replicated in Python only. To switch between programming languages, we save the pre-processing and graph construction output as needed for subsequent computation in Python.

3 Replicating the BASS Framework

BASS is an abstractive summarization framework (cf. Fig. 1), which uses i) dependency parse trees to generate *Unified Semantic Graphs* (USG) for documents to compress and relate information across the input document, and ii) a model architecture, which incorporates the graph information. As this work focuses on the replication of the framework, we refer to the original work [30] for details and only elaborate on necessary complementary information in the following.

Pre-processing ①. An input document is passed to a linguistic parser for POS tagging, co-reference resolution and dependency parsing. We used the latest CoreNLP library [22] (v4.5.2) and had the authors confirm a configuration we found in their source code, as no details were given in the paper, that is:

Table 1. Overview of the completeness of information on components (cf. Fig. 1) based on community-available information, i.e., the paper, and which information **Sources** we actually used. **Paper** shows whether key information was missing, making a replication impossible (×), whether minor details were omitted (◯) or whether all required information on a component is complete (✓). **Complete** shows whether we are certain to have faithfully replicated a component (✓) or if uncertainty remained (◯). n.a.: not applicable to graph construction ②a, as we use the provided source code as it is.

Component	Paper	Sources	Complete
① Pre-processing	×	Paper, Source Code, Authors	✓
②a Graph Construction (provided)	n.a.	Source Code	n.a.
②b Graph Construction (replicated)	×	Paper, Source Code, Authors	◯
③ Graph Augmentation	✓	Paper	✓
④ Tokenizer & Text Encoder	◯	Paper, Authors	✓
⑤ Token To Node Alignment	◯	Paper	✓
⑥–⑨ Model Architecture	◯	Paper, Authors	◯

```
"annotators": "tokenize, ssplit, pos, lemma, ner, parse, coref",
"coref.algorithm": "neural", "depparse.extradependencies": "MAXIMAL"
```

We were initially unable to pre-process many documents (over 20%) due to endless runtimes or out-of-memory errors of the CoreNLP parser. Upon inquiry, the authors confirmed to have used the pre-processing strategy indicated in their source code, so we chunk source documents into blocks of sentences with approx. 500 words and pre-process them separately. The resulting graphs per chunk are concatenated into a single document-level graph consisting of multiple disjoint sub-graphs. In particular, nodes across different sub-graphs (e.g., those referring to the same entity) are neither merged nor connected. We additionally set a maximum runtime (10 h) and RAM consumption (90 GB) for pre-processing a single chunk. The output is one dependency parse tree per sentence, co-reference chains across each chunk and a POS tag for each word.

Graph Construction ②. To construct USGs, we treat the dependency trees of all sentences as directed graphs with one node representing a single word and its POS tag and an edge representing the respective dependency relation. Nodes are merged in multiple steps: i) the nodes of entity mentions from co-reference resolution are merged into a single entity node, ii) nodes are removed or merged based on linguistic rules reasoning over the dependency and POS annotations, iii) entity nodes of mentions in the same co-reference chain are merged, iv) non-entity nodes representing the same phrase (exact lexical match) are merged.

We noticed that strictly complying with the paper [30] yields a different algorithm than the one provided with the Java source code (cf. Appendix A). Therefore, we consider two USG variants in this work: i) USG_{src} generated by original Java implementation ②a and ii) USG_{ppr} generated by our paper-compliant replication of the graph construction ②b in Python. Since the linguistic rules in step iii) are omitted from the paper, we adopted these from the Java code.

Graph Augmentation ③. Following the paper, the USG is augmented by adding self-loops and reverse edges. All nodes are connected with their two-hop neighbours and a supernode connecting to all nodes is added. The output of this step is two matrices, the graph construction matrix C (solid orange line) and the adjacency matrix A (dashed orange line). See Appendix B for more details.

Text Encoder and Tokenizer ④. Following our correspondence with the authors, we use the pre-trained RoBERTa-base model [37] for the text encoder with the tokenizer `RobertaTokenizerFast` from Hugging Face's transformer library ([29], v4.26.1). The context length is extended to $N = 1024$ by randomly initializing the extended part.

Aligning Text to Graph Embeddings ⑤. We align text embeddings with graph embeddings by multiplying the text encoder output with the graph construction matrix from step ③. We refer to Appendix B for more details.

Model Architecture ⑥–⑨. The model architecture builds upon a standard transformer encoder-decoder architecture for abstractive summarization, complemented with three additional components: i) a graph encoder ⑥, which is a standard two-layered transformer encoder using the adjacency matrix of the graph as attention mask, ii) a corresponding multi-head attention module ⑦ in a six-layered decoder attending over the graph encoder output and iii) a fully connected linear layer ⑧ to fuse graph and text information, followed by a Residual Dropout [27] layer ⑨.

The Causal Self Attention, i.e., masked self attention attending to left context only, and Feed-Forward [27] layers are also followed by a Residual Dropout layer, not indicated in the figure. In contrast to regular cross-attention modules, the attention weights in ⑦ are propagated using PageRank and the augmented adjacency matrix to leverage the graph structure. We refer to Appendix C for more details on our implementation. Our final model has approx. 205M trainable parameters, which is around 2% larger than reported in the BASS paper [30] (201M parameters), implying architectural differences we could not entirely resolve.

4 Evaluation

We conduct a replication study by training one model each for the author-provided graphs USG_{src} and the paper-compliant graphs USG_{ppr} (cf. Sect. 3) to better isolate potential errors rooted in replicating the model architecture from potential errors rooted in replicating the graph construction. Since replication studies are known to be challenging due to an overwhelming number of error sources, we follow up with an ablation study where we generate our own baselines (not using graphs) to measure the impact of architectural adaptations.

Datasets. We follow [30] and select the BigPatent [26] dataset for Single Document Summarization. However, we were only able to pre-process 99.79% documents, resulting in 1,204,631 documents for training and 66,962 documents each for validation and test. Using both graph construction methods (cf. Sect. 3) we generate two sets of graphs USG_{src} and USG_{ppr} and obtain two respective training datasets. Running the pre-processing and graph construction ②a on the Big Patent dataset took about four days on our cluster with 900 CPU cores (Intel(R) Xeon(R) Silver 4216 CPU) with an aggregated runtime of 3,360 days ± 1 day.

Models. We use the following baselines for our studies (cf. [30, Tab. 3] for original results): **TransS2S**, which is a standard text-only transformer encoder-decoder model [27], **RoBERTaS2S**, which differs from TransS2S by using RoBERTa-base for the encoder and decoder, and **BASS**, which is the original model we replicate.

For our replication study, we train the following models: $\mathbf{BASS_{ours/src}}$, which is our replicated BASS model trained with USG_{src}, and $\mathbf{BASS_{ours/ppr}}$, which is a full replication of the original paper differing from $BASS_{ours/src}$ only by it's use of USG_{ppr} graphs.

For the ablation study, we additionally train two transformer based encoder-decoder architectures on the BigPatent dataset without any graphs: **RTS2S**, which is a text-only model consisting of the text encoder ④ and a standard 6-layered decoder without any graph components (i.e., omitting ③,⑤–⑧). And **exRTS2S**, which extends RTS2S by the graph encoder and the decoder components ⑥–⑧ without informative graph structure (③,⑤), and by replacing Graph-prop Attention ⑦ with normal Context Attention.

RTS2S is most similar to TransS2S and RoBERTaS2S, using the encoder of RoBERTaS2S and the decoder of TransS2S. We therefore expect this model to perform somewhere in-between. exRTS2S is most similar to $BASS_{ours}$ and differs only in the lack of graph structure: every token is considered a graph node connected to every other node. Hence layer ⑤ is skipped and no graph structure is injected into the attention mechanism in the graph-encoder ⑥ and no attention is graph-propagated in the cross-attention module ⑦.

We want to measure the impact of all architectural adaptations proposed for BASS with RTS2S as our baseline. With exRTS2S as a baseline, we further isolate the impact of the USG_{src} structure from the impact of increasing model size. We choose USG_{src} assuming the graph construction method ②a reproduces the graphs from the original work.

Training Details. We use the same training and hyper-parameter setup as the original work [30, §5.2] for all models, which uses the learning rate schedule of Liu and Lapata [21]. Each model is trained once for 300,000 steps. Since we were not able to find out the batch size used in the original work, we use a batch size of 48 per step, which is the largest possible value on our hardware (6 RTX A6000 GPUs with a total RAM of 288 GB – the original authors reported the use of 8xV100, of which the latest version totals 256 GB). Training the models for 10,000 steps took about 4 h ± 30 min at average on our machines (cf. Table 2).

Table 2. Evaluation of our models on the BigPatent dataset. The baselines are all taken from prior work [30] and best scores are in bold. All scores are pairwise significantly different from each other, except those indicated by †.

	R-1	R-2	R-L	R-L$_{sum}$	BS	params	time
BASELINES AND PAPER							
TransS2S [30]	34.93	9.86	29.92[a]		9.42	n.a.	n.a.
RoBERTaS2S [30]	43.62	18.62	37.86[a]		18.18	n.a.	n.a.
BASS [30]	**45.04**	**20.32**	**39.21**[a]		**20.13**	201M	n.a.
REPLICATION STUDY							
BASS$_{ours/src}$	40.52	14.97	26.64	34.67	16.15	204.6M	5 d 15 h
BASS$_{ours/src\ 600k}$	**41.23**	**15.62**	**27.18**	**35.37**	**17.03**	204.6M	11 d 11 h
BASS$_{ours/ppr}$	39.30	14.65	26.31	33.74	16.51	204.6M	5 d 16 h
ABLATION STUDY							
RTS2S	**39.75**	**14.84**†	26.48	**34.04**	16.34	172.4M	4 d 9 h
exRTS2S	39.62	**14.86**†	**26.56**	33.91	**16.63**	204.6M	5 d 7 h

[a]While the BASS paper reports sentence-level R-L scores, they systematically match better with our summary-level R-L$_{sum}$ scores, which may indicate that prior results are actually R-L$_{sum}$ scores. Hence we place the scores reported by BASS between columns and compare them with our R-L$_{sum}$ scores.

Evaluation. We evaluate the models on the test set and apply beam search [31] with trigram blocking [23] for decoding using a beam size of 5 and a length penalty of 0.9. We enforce a maximum decoding length of 1024 and report ROUGE scores [13] R-1, R-2, sentence-level R-L, summary-level R-L$_{sum}$ and F_1 BERTScore [35][2] BS. We exclude summaries, for which no `eos` token has been generated during decoding[3], from evaluation and use paired bootstrap resampling [9] with $p = 0.05$ for significance testing.

4.1 Experimental Results

Replication. Comparing our replication with the original (**BASS**), we score more than 4 points lower than reported originally[4], performing even worse than the RoBERTaS2S baseline. Since the training batch size of the original work remains unknown, we investigated if our models might be undertrained by **extending the training** of BASS$_{ours/src}$ for another 300,000 steps[5] (cf. Table 2) and are indeed able to raise the scores but only by about 0.7 ± 0.2 points.

Seeing how the graph construction implementation ②a provided to us differs algorithmically from our paper-compliant implementation ②b, we also compare USG$_{src}$ and USG$_{ppr}$ **graphs** (cf. Table 3). The replicated graphs are slightly

[2] `lang="en"` `model_type="roberta-large"` `rescale_with_baseline=True` .

[3] We excluded less than 0.015% of the documents, having no effect on the final score.

[4] While the BASS paper reports sentence-level R-L scores, they systematically match better with our summary-level R-L$_{sum}$ scores, which may indicate that prior results are actually R-L$_{sum}$ scores. Hence we place the scores reported by BASS between columns and compare them with our R-L$_{sum}$ scores.

[5] We did not train further, because we already doubled the computational budget used for the original paper.

Table 3. Comparison of USG_{src} and USG_{ppr} structures for the BigPatent dataset $D = \{d_i\}_{i \in I}$, with d_i denoting a tokenized input document. We consider the subsets $D_T = \{d_i \mid |T - t(d_i)| \leq 20\}$ for token count $t(d_i)$ and $T \in \{400, 600, 800, 1000\}$. Let $\bar{t}_{D_T}$ denote the average token count of documents $d \in D_T$ and $|D_T|$ the cardinality of D_T. We report the average node count $\bar{n}$, average edge count $\bar{e}$ as well as the average count of tokens $\bar{t}_c$ covered by the graphs generated for $d \in D_T$. The bottom row shows the increase in respective quantities for USG_{ppr} w.r.t USG_{src}

$\bar{t}_{D_T}$	400			600			800			1000		
$\lvert D_T \rvert$	152			966			3045			6087		
	$\bar{n}$	$\bar{e}$	$\bar{t}_c$	$\bar{n}$	$\bar{e}$	$\bar{t}_c$	$\bar{n}$	$\bar{e}$	$\bar{t}_c$	$\bar{n}$	$\bar{e}$	$\bar{t}_c$
USG_{src}	117	142	232	168	211	340	227	281	463	278	349	569
USG_{ppr}	129	156	301	185	233	453	240	308	606	293	385	759
Increase	10%	10%	30%	10%	10%	33%	6%	10%	31%	5%	10%	33%

larger in size. They also cover at least 30% more tokens and while we expected $BASS_{ours/ppr}$ to perform better for this reason, we actually observe mixed results: a small increase in the BERTscore, and a decrease in ROUGE scores.

Ablation. We observe slight but mostly significant differences in model performances (cf. Table 2, ABLATION STUDY), with no clear winner. However, our ablation models perform in-between TransS2S and RoBERTaS2S, as expected. The introduction of graph components (exRTS2S vs. RTS2S) mostly improves the BERTScore with mixed results in terms of ROUGE. Further comparing exRTS2S with our replicated models shows the impact of using USGs: using USG_{ppr} slightly hurts performance overall, while using USG_{src} slightly improves ROUGE scores while hurting BERTScore.

4.2 Discussion

Replication. Our replicated **BASS** models substantially fall short in performance, even below baselines of the original work (cf. RoBERTaS2S in Table 2). Since this is true even for our model $BASS_{ours/src}$ which is trained on USG_{src} graphs, we mainly attribute this to the model architecture, assuming that the provided graph construction ②a is the same one used in the original work. By analzying the loss curve of the **extended training**, we find that the training has mostly saturated.[6] For this reason, we rule out undertraining as a potential reason for performance discrepancies between the original method and our replication.

Since the discrepancy between the replicated and the original performance of BASS can be attributed to the model architecture, the impact of using USG_{ppr} over USG_{src} **graphs** on downstream performance being minor does not surprise, despite substantial qualitative differences in graph structures. However,

[6] The loss curve is decreasing by 0.003 points every 10,000 steps in the range of 300,000 to 450,000 steps, and by 0.002 points every 10,000 steps in the range of 450,000 to 600,000 steps at average.

the authors' source code not complying with the paper, possibly having undergone some changes (see Sect. 3 & 6), sheds doubts on whether the provided graph construction ②a replicates the graphs of the original work and whether the mismatch in graph structures indicates a failed replication of the original work.

Ablation. Contrary to [30] we did not find substantial performance gains, neither in model adaptations (increased model size, cf. Table 2, RTS2S vs. exRTS2S), nor in USGs (additional structured information, cf. Table 2, exRTS2S vs. $\text{BASS}_{\text{ours/src}}$ and $\text{BASS}_{\text{ours/ppr}}$). This is shown by the minor impact the model adaptations and USGs have on our baselines. Our replicated USG_{ppr} graphs even consistently hurt the performance overall. Nonetheless the comparison with previous baselines (TransS2S and RoBERTaS2S) shows that RTS2S performs reasonably well. We therefore ascribe the lack of substantial gains in our replicated models solely to BASS' model adaptations and graph information not being as effective as expected.

5 Limitations

This study aimed to faithfully replicate the original work, adhering closely to the experimental setup and software artifacts described in the original paper, spot-checking the replicability of papers from high-level conferences such as ACL'21. Consequently, we are constricted in addressing performance discrepancies between our replication and the original method, and we leave the investigation of the effect of algorithmic and methodological deviations on model performance for future work. Furthermore, we do not investigate the effect of randomness on our results, partly because no information on random seeds were given, and partly because we infer that the impact of randomness is negligible, if anything. Since our models, each trained with a different random seed, perform relatively similar across all configurations, this shows that performance discrepancies between our replication and the original method are unlikely to be attributed to randomness.

6 Replication Challenges and Recommendations

In this section, we reflect on the replication process and highlight the main challenges we encountered, hoping to sensitize readers to the underlying issues that compromise the replicability of research papers. We conclude by recommending key practices for writing replicable papers, that would have significantly helped us with the replication.

Self-Explanatory Details. Some details are omitted from papers usually for good reasons: being straight-forward, well-known or trivial. However, our experience showed that the leeway in implementation choices for omitted details (e.g. in every step of the graph construction, or for aligning the tokenizations of

CoreNLP's tokenizer ① and RoBERTa's tokenizer ④, but also for the pre-processing strategy) entails ambiguities and consequently an avalanche of potential error sources and mitigation strategies to resolve them. Hence, these omitted details can make the difference between an accurate replication and an endless errand to fix errors. Although we got many details confirmed by the authors, studied the source code and strictly followed the paper, some uncertainties remained (cf. Table 1). In the end, we were neither able to pre-process the entire dataset considered in this work (see Sect. 4) nor achieve the same model size as the original authors (see Sect. 3).

Missing Third Party Information. One problem was missing version information and configuration of third-party components, i.e., of the CoreNLP pipeline, the RoBERTa model and the tokenizers. We were able to resolve most of these issues through correspondence with the authors for this paper.

Missing Key Information. Overall, we encountered many details that required additional information or clarification beyond the paper. However, not all missing details fundamentally obstructed our replication: our first attempt to replicate the paper (see Sect. 2) failed primarily due to the omission of the linguistic rules reasoning over the dependency trees for creating USGs. Nevertheless, following our correspondence with the authors and access to source code, we identified and resolved many misunderstandings.

Algorithmic Complexity and Error-Proneness. A thorough analysis of the original source code was necessary to fix (or not to fix) our replicated graph construction algorithm due to the many errors we encountered during runtime, often rooted in erroneous annotation results, such as wrong POS tag annotation, co-reference resolution, or even dependency graphs being rooted in punctuation tokens or sentences mistakenly being split at decimal points. To our surprise, the provided graph construction slightly differs from the description in the paper. This is likely because the source code has been used in other projects, as noted by the authors, and consequently might have undergone some changes before or after the paper's publication, emphasizing the importance of version control systems. The complexity of algorithms, whereas, can be lessened using tools during development to analyze and reduce cognitive complexity in software.

Recommendations. We found the mathematical and algorithmic descriptions (notation, equations, pseudo-code) most helpful along the way, allowing us to consolidate many misconceptions. Therefore we emphasize the importance of i) providing a clear and complete technical context, ii) a clear and (given the context) complete notation, iii) technical and mathematical precision particularly for describing how different components (novel or not) interleave, and iv) commented pseudo-code. We feel the latter can often replace a detailed description of an algorithm, while being shorter and less ambiguous.

We also strongly encourage to use technical terms coined by prior work wherever applicable, such as "Residual Dropout" [27] for layer ⑨, instead of short descriptions of well-known components. The latter can be inaccurate and leave readers questioning potential differences and misunderstandings in case of failed replications, while undermining the development of a well-defined and well-known terminology of a research domain.

7 Conclusion

We started implementing the BASS framework based on the paper, but found that most components were not sufficiently described (cf. Table 1). Some uncertainties persisted even after our correspondence with the authors, and the examination of the provided source code (cf. Sect. 2), partly because some inquiries (e.g. for the training batch size) had been left pending. On one hand, the provided graph construction ②a did not align with the paper (cf. Sect. 3). On the other hand, our model's parameter size was larger by approx 3.6M parameters than reported in the original work (cf. Sect. 3). Therefore, it is unsurprising we could neither replicate prior results of BASS on the BigPatent dataset, nor clear performance improvements on the summarization task as a result of the novel adaptations proposed for BASS (cf. Sect. 4.2). Assuming the graph construction method ②a provided by the original authors' reproduces the same USGs as in the original work, our results indicate the poor performance can be ascribed to the model architecture. However, some doubts remain on whether the provided graph construction method even reproduces the original USGs.

Moreover, we found the pre-processing ① and in extension the graph construction ② to be very error-prone and time consuming. Parsing one document of the BigPatent dataset with approx. 1,000 tokens took us about 3.5 min, not accounting for the 2,811 documents (approx. 0.2%) we had to exclude for not being parseable in less than 10 h. Additionally, erroneous dependency annotations make it difficult to construct USGs, leading to fractured graphs, isolated nodes or deletion of salient information. Based on our experiences, we suggest investigating the use of simpler semantic dependency parsing methods [8] which reportedly are more accurate, or to move away from systems that construct graphs from semantic annotations based on manually hand-crafted rules.

Overall, the replication was complicated by missing third-party information, the ambiguity of self-explanatory details, and omission of some key information (the latter requiring us to contact the authors for a faithful replication), despite the fact that the original paper is very detailed and comprehensive, representative of the high quality of the venue it was published on (ACL'21). However, our experience shows that the way information is detailed is just as vital as being comprehensive. Furthermore, as this lesson is learned only after attempting a replication, it may lead reviewers, who lack similar experiences, to overrated reproducibility assessments. We have therefore emphasized key issues and practices for replicable papers and recommend supplementing reproducibility as well as reviewer checklists with a corresponding section to address these problems.

Acknowledgements. We thank the authors [30] for their correspondence, their source code and for giving us their consent to share it. This work was partly funded by the German State Ministry of Culture and Science NRW for research under the Cancer Research Center Cologne Essen (CCCE) foundation. The funding was not provided specifically for this project.

A Discrepancies Between USG_{src} and USG_{ppr}

In the following, we point out algorithmic differences between the two graph constructions methods ②a and ②b (cf. Table 1). These differences arise from strictly complying with the paper for the replicated method ②b. For the sake of simplicity, we align our comparison with the pseudo-code in the appendix of the original work [30].

`REMOVE_PUNCTUATION` While ②b removes all tokens whose dependency relation equals `punct` or whose POS tag is an element of P ={ `.`, `,`, `:`, `!`, `?`, `(`, `)` }, ②a only removes tokens whose POS tag is an element of P. In ②a, punctuation is removed recursively as part of the `MERGE_NODES` routine.

`MERGE_COREF_PHRASE` ②b merges all tokens of a co-reference mention into a single node, while ②a i) uses merging rules not mentioned in the paper and also ii) immediately merges the resulting nodes in the same co-reference chain into a single node. ②b, on the other hand, merges nodes in the same co-reference chain only in the `MERGE_PHRASES` step.

`MERGE_NODES` While ②a and ②b use exactly the same rules to merge nodes, ②a traverses the dependency trees pre-order depth-first. We intuitively chose to traverse post-order depth-first for ②b without looking at the provided source code, as working through the tree bottom up from leaf to root nodes generally complies better with the intention to merge nodes (which includes rules to delete nodes). For example, if ②a deletes a child, all descendants are detached from the tree and never visited by the algorithm.

`MERGE_PHRASES` ②b merges the nodes of mentions in the same co-reference chain into a single node and later merges all nodes, whose phrases are equal (exact lexical match). ②a, on the other hand, only does the latter.

B Aligning Graph and Text Embeddings

The Unified Semantic Graph imposes a graph structure on the token embeddings returned by the text encoder. The text encoder output t must therefore be mapped to the graph encoder input g, before passing it to the graph encoder. For this, we match tokens with nodes based on text characters from left to right, as the CoreNLP tokenizer ① is different from RoBERTa's tokenizer ④.

Let $G := (V, E)$ be the augmented Unified Semantic Graph ③ with nodes $V := \{v_i\}_{i=0}^{N_V}$ and edges $E := \{e_{ij}\} \subset V \times V$ and let $S = \{s_{v_i}\}_{i=0}^{N_V}$ with $s_{v_j} = \{c_i\}_{i \in I_{v_j}}$ be the set of characters of the input document $D = \{c_i\}_{i=0}^{N_D}$ being represented by node v_j. As a result of merging nodes across a document, s_{v_j} may consist of multiple disconnected character sequences.

We map nodes v_j to tokens $t_{v_j} \subset T$, where t_{v_j} is the subset of input tokens $T = \{t_i\}_{i=0}^{N_T}$ associated with at least one character $c_i \in s_{v_j}$. This gives us the graph construction matrix $C = (c_{ij}) \in \mathbb{R}^{N_V \times N_T}$ and the adjacency matrix $A = (a_{ij}) \in \mathbb{R}^{N_V \times N_V}$ with

$$c_{ij} = \begin{cases} 1, & \text{if } t_j \in t_{v_i} \\ 0, & \text{otherwise} \end{cases} \qquad a_{ij} = \begin{cases} 1, & \text{if } e_{ij} \in E \\ 0, & \text{otherwise.} \end{cases}$$

Let d_{model} be the dimension of token embeddings and $t \in \mathbb{R}^{N_T \times d_{model}}$ be the output of the text encoder. The graph encoder input g is then given by $g := C't$ where C' is the degree normalized graph construction matrix C. Multiplicating t with C' is equal to calculating the representation of node v_j by averaging over the tokens t_{v_j}. The matrix g is then passed to the graph encoder alongside the node padding and the adjacency matrix A as attention mask.

C Graph Propagation

The paper suggests propagating attention weights in the cross attention module for the graph encoder using PageRank [18] and the adjacency matrix A given by the augmented Unified Semantic Graph. For this, we compute the graph propagation matrix $P = \omega^p \hat{A} + (1 - \omega)(\sum_{i=0}^{p-1} \omega^i \hat{A}^i)$, where p is the number of aggregation steps, ω is the teleport probability and $\hat{A}$ is the degree normalized adjacency matrix A, including self loops and reverse edges, supernode edges and shortcut edges. The graph propagated attention weights are then computed as $\alpha' = \alpha P^T$, where $\alpha = (\alpha_{ij}) \in \mathbb{R}^{N_V \times N_V}$ are the attention weights of a single head in the multi-headed cross-attention module for the graph encoder given by $\alpha_{ij} = (y_i W_Q)(v_j W_K)^T / \sqrt{d_k}$, with query and key projection weights W_Q, W_K, the i-th token representation as query y_i and the j-th node representation as key v_j. Here, d_k denotes the query and key dimensions.

References

1. Belz, A., Agarwal, S., Shimorina, A., Reiter, E.: A systematic review of reproducibility research in natural language processing. In: Proceedings of the 16th Conference of the European Chapter of the Association for Computational Linguistics: Main Volume, pp. 381–393. Association for Computational Linguistics, Online (2021). https://doi.org/10.18653/v1/2021.eacl-main.29
2. Belz, A., Thomson, C., Reiter, E., Mille, S.: Non-repeatable experiments and non-reproducible results: the reproducibility crisis in human evaluation in NLP. In: Findings of the Association for Computational Linguistics: ACL 2023, Toronto, Canada, pp. 3676–3687. Association for Computational Linguistics (2023). https://doi.org/10.18653/v1/2023.findings-acl.226. https://aclanthology.org/2023.findings-acl.226
3. Brody, S., Alon, U., Yahav, E.: How attentive are graph attention networks? In: International Conference on Learning Representations (2022). https://openreview.net/forum?id=F72ximsx7C1

4. Devlin, J., Chang, M.W., Lee, K., Toutanova, K.: BERT: pre-training of deep bidirectional transformers for language understanding. In: Proceedings of the 2019 Conference of the North American Chapter of the Association for Computational Linguistics: Human Language Technologies, Minneapolis, Minnesota (Volume 1: Long and Short Papers), pp. 4171–4186. Association for Computational Linguistics (2019). https://doi.org/10.18653/v1/N19-1423
5. Dong, L., Xu, S., Xu, B.: Speech-transformer: a no-recurrence sequence-to-sequence model for speech recognition. In: 2018 IEEE International Conference on Acoustics, Speech and Signal Processing (ICASSP), pp. 5884–5888 (2018). https://doi.org/10.1109/ICASSP.2018.8462506
6. Dosovitskiy, A., et al.: An image is worth 16 × 16 words: transformers for image recognition at scale. arXiv abs/2010.11929 (2020)
7. Dou, Z.Y., Liu, P., Hayashi, H., Jiang, Z., Neubig, G.: GSum: a general framework for guided neural abstractive summarization. In: Proceedings of the 2021 Conference of the North American Chapter of the Association for Computational Linguistics: Human Language Technologies, pp. 4830–4842. Association for Computational Linguistics, Online (2021). https://doi.org/10.18653/v1/2021.naacl-main.384
8. Dozat, T., Manning, C.D.: Simpler but more accurate semantic dependency parsing. In: Proceedings of the 56th Annual Meeting of the Association for Computational Linguistics, Melbourne, Australia (Volume 2: Short Papers), pp. 484–490. Association for Computational Linguistics (2018). https://doi.org/10.18653/v1/P18-2077
9. Dror, R., Baumer, G., Shlomov, S., Reichart, R.: The hitchhiker's guide to testing statistical significance in natural language processing. In: Proceedings of the 56th Annual Meeting of the Association for Computational Linguistics, Melbourne, Australia (Volume 1: Long Papers), pp. 1383–1392. Association for Computational Linguistics (2018). https://doi.org/10.18653/v1/P18-1128
10. El-Kassas, W.S., Salama, C.R., Rafea, A.A., Mohamed, H.K.: Automatic text summarization: a comprehensive survey. Expert Syst. Appl. **165**, 113679 (2021). https://doi.org/10.1016/j.eswa.2020.113679
11. Fan, A., Gardent, C., Braud, C., Bordes, A.: Using local knowledge graph construction to scale Seq2Seq models to multi-document inputs. In: Proceedings of the 2019 Conference on Empirical Methods in Natural Language Processing and the 9th International Joint Conference on Natural Language Processing (EMNLP-IJCNLP), Hong Kong, China, pp. 4186–4196. Association for Computational Linguistics (2019). https://doi.org/10.18653/v1/D19-1428
12. Gibney, E.: Could machine learning fuel a reproducibility crisis in science? Nature **608**, 250–251 (2022). https://api.semanticscholar.org/CorpusID:251102207
13. Google LLC: Rouge-score. https://pypi.org/project/rouge-score
14. Gundersen, O.E., Coakley, K., Kirkpatrick, C.R.: Sources of irreproducibility in machine learning: a review. arXiv abs/2204.07610 (2022). https://api.semanticscholar.org/CorpusID:248227686
15. Hu, J., et al.: Word graph guided summarization for radiology findings. In: Findings of the Association for Computational Linguistics: ACL-IJCNLP 2021, pp. 4980–4990. Association for Computational Linguistics, Online (2021). https://doi.org/10.18653/v1/2021.findings-acl.441
16. Huang, L., Wu, L., Wang, L.: Knowledge graph-augmented abstractive summarization with semantic-driven cloze reward. In: Proceedings of the 58th Annual Meeting of the Association for Computational Linguistics, pp. 5094–5107. Association for Computational Linguistics, Online (2020). https://doi.org/10.18653/v1/2020.acl-main.457

17. Jin, H., Wang, T., Wan, X.: Semsum: semantic dependency guided neural abstractive summarization. In: Proceedings of the AAAI Conference on Artificial Intelligence, vol. 34, no. 05, pp. 8026–8033 (2020). https://doi.org/10.1609/aaai.v34i05.6312. https://ojs.aaai.org/index.php/AAAI/article/view/6312
18. Klicpera, J., Bojchevski, A., Günnemann, S.: Predict then propagate: graph neural networks meet personalized pagerank. In: International Conference on Learning Representations (2018)
19. Lewis, M., et al.: BART: Denoising sequence-to-sequence pre-training for natural language generation, translation, and comprehension. In: Proceedings of the 58th Annual Meeting of the Association for Computational Linguistics, pp. 7871–7880. Association for Computational Linguistics, Online (2020). https://doi.org/10.18653/v1/2020.acl-main.703
20. Li, H., Peng, Q., Mou, X., Wang, Y., Zeng, Z., Bashir, M.F.: Abstractive financial news summarization via transformer-bilstm encoder and graph attention-based decoder. IEEE/ACM Trans. Audio Speech Lang. Process. **31**, 3190–3205 (2023). https://doi.org/10.1109/TASLP.2023.3304473
21. Liu, Y., Lapata, M.: Text summarization with pretrained encoders. In: Proceedings of the 2019 Conference on Empirical Methods in Natural Language Processing and the 9th International Joint Conference on Natural Language Processing, Hong Kong, China (EMNLP-IJCNLP), pp. 3730–3740. Association for Computational Linguistics (2019). https://doi.org/10.18653/v1/D19-1387
22. Manning, C.D., Surdeanu, M., Bauer, J., Finkel, J., Bethard, S.J., McClosky, D.: The Stanford CoreNLP natural language processing toolkit. In: Association for Computational Linguistics (ACL) System Demonstrations, pp. 55–60 (2014). https://www.aclweb.org/anthology/P/P14/P14-5010
23. Paulus, R., Xiong, C., Socher, R.: A deep reinforced model for abstractive summarization (2017)
24. Qi, P., Huang, Z., Sun, Y., Luo, H.: A knowledge graph-based abstractive model integrating semantic and structural information for summarizing Chinese meetings. In: 2022 IEEE 25th International Conference on Computer Supported Cooperative Work in Design (CSCWD), pp. 746–751 (2022). https://doi.org/10.1109/CSCWD54268.2022.9776298
25. Raffel, C., et al.: Exploring the limits of transfer learning with a unified text-to-text transformer. J. Mach. Learn. Res. **21**(1) (2020)
26. Sharma, E., Li, C., Wang, L.: BIGPATENT: a large-scale dataset for abstractive and coherent summarization. In: Proceedings of the 57th Annual Meeting of the Association for Computational Linguistics, Florence, Italy, pp. 2204–2213. Association for Computational Linguistics (2019). https://doi.org/10.18653/v1/P19-1212
27. Vaswani, A., et al.: Attention is all you need. In: Proceedings of the 31st International Conference on Neural Information Processing Systems, NIPS 2017, pp. 6000–6010. Curran Associates Inc., Red Hook (2017)
28. Velickovic, P., Cucurull, G., Casanova, A., Romero, A., Lio', P., Bengio, Y.: Graph attention networks. arXiv abs/1710.10903 (2017)
29. Wolf, T., et al.: Transformers: state-of-the-art natural language processing. In: Proceedings of the 2020 Conference on Empirical Methods in Natural Language Processing: System Demonstrations, pp. 38–45. Association for Computational Linguistics, Online (2020). https://doi.org/10.18653/v1/2020.emnlp-demos.6
30. Wu, W., et al.: BASS: boosting abstractive summarization with unified semantic graph. In: Proceedings of the 59th Annual Meeting of the Association for Computational Linguistics and the 11th International Joint Conference on Natural

Language Processing (Volume 1: Long Papers), pp. 6052–6067. Association for Computational Linguistics, Online (2021). https://doi.org/10.18653/v1/2021.acl-long.472
31. Wu, Y., et al.: Google's neural machine translation system: bridging the gap between human and machine translation. CoRR abs/1609.08144 (2016). https://arxiv.org/abs/1609.08144
32. Xu, J., Gan, Z., Cheng, Y., Liu, J.: Discourse-aware neural extractive text summarization. In: Proceedings of the 58th Annual Meeting of the Association for Computational Linguistics, pp. 5021–5031. Association for Computational Linguistics, Online (2020). https://doi.org/10.18653/v1/2020.acl-main.451
33. Ying, C., et al.: Do transformers really perform bad for graph representation? In: Neural Information Processing Systems (2021)
34. Zhang, J., Zhao, Y., Saleh, M., Liu, P.J.: Pegasus: pre-training with extracted gap-sentences for abstractive summarization. In: Proceedings of the 37th International Conference on Machine Learning, ICML 2020. JMLR.org (2020)
35. Zhang, T., Kishore, V., Wu, F., Weinberger, K.Q., Artzi, Y.: BERTScore: evaluating text generation with BERT. In: International Conference on Learning Representations (2020). https://openreview.net/forum?id=SkeHuCVFDr
36. Zhu, C., et al.: Enhancing factual consistency of abstractive summarization. In: Proceedings of the 2021 Conference of the North American Chapter of the Association for Computational Linguistics: Human Language Technologies, pp. 718–733. Association for Computational Linguistics, Online (2021). https://doi.org/10.18653/v1/2021.naacl-main.58
37. Zhuang, L., Wayne, L., Ya, S., Jun, Z.: A robustly optimized BERT pre-training approach with post-training. In: Proceedings of the 20th Chinese National Conference on Computational Linguistics, Huhhot, China, pp. 1218–1227. Chinese Information Processing Society of China (2021)

Performance Comparison of Session-Based Recommendation Algorithms Based on GNNs

Faisal Shehzad(✉) and Dietmar Jannach(✉)

University of Klagenfurt, Klagenfurt, Austria
{faisal.shehzad,dietmar.jannach}@aau.at

Abstract. In session-based recommendation settings, a recommender system has to base its suggestions on the user interactions that are observed in an ongoing session. Since such sessions can consist of only a small set of interactions, various approaches based on Graph Neural Networks (GNN) were recently proposed, as they allow us to integrate various types of side information about the items in a natural way. Unfortunately, a variety of evaluation settings are used in the literature, e.g., in terms of protocols, metrics and baselines, making it difficult to assess what represents the state of the art. In this work, we present the results of an evaluation of eight recent GNN-based approaches that were published in high-quality outlets. For a fair comparison, all models are systematically tuned and tested under identical conditions using three common datasets. We furthermore include k-nearest-neighbor and sequential rules-based models as baselines, as such models have previously exhibited competitive performance results for similar settings. To our surprise, the evaluation showed that the simple models outperform *all* recent GNN models in terms of the Mean Reciprocal Rank, which we used as an optimization criterion, and were only outperformed in three cases in terms of the Hit Rate. Additional analyses furthermore reveal that several other factors that are often not deeply discussed in papers, e.g., random seeds, can markedly impact the performance of GNN-based models. Our results therefore *(a)* point to continuing issues in the community in terms of research methodology and *(b)* indicate that there is ample room for improvement in session-based recommendation.

Keywords: Recommender systems · Evaluation · Methodology

1 Introduction

Recommender systems play a critical role in modern platforms by recommending personalized content to users according to their past preferences. Conventional recommender systems e.g., in particular ones based on collaborative filtering, rely heavily on rich user profiles and long-term historical interactions. Such systems may however perform poorly in real-world applications, where users interact with

N. Goharian et al. (Eds.): ECIR 2024, LNCS 14611, pp. 115–131, 2024.
https://doi.org/10.1007/978-3-031-56066-8_12

the service without logging in [22,32]. Consequently, session-based recommenders (SBRS), which recommend a next item solely based on an active session, received immense attention from industry and academia in recent years [13].

From a technical point of view, most recent session-based algorithms are based on neural methods. A landmark work in this area is GRU4Rec [9], which is often used as a baseline to benchmark newly developed models. One main characteristic of GRU4Rec and many subsequent models is that they are designed to operate solely on user–item interaction data. In the more recent literature, Graph Neural Networks (GNN) however received increased attention in the context of SBRS, because they allow us to consider various types of heterogeneous information in the learning process in a natural way. Examples of such recent works published in top-tier conferences in the last three years include [15,33,37–39,42,43].

Quite surprisingly, for the category of neural SBRS models that operate solely on user–item interactions like GRU4Rec, various studies have shown that simpler methods, e.g., based on nearest-neighbor techniques, can lead to competitive accuracy results and in many cases even outperform the more complex neural models [7,12,14,20]. Similar observations were made for traditional *top-n* recommendation tasks [1,5,6,24,25], where the latest neural models were found to be outperformed by longer-existing approaches, e.g., based on matrix factorization.[1] Various factors that can contribute to this phenomenon of 'phantom progress' were discussed in the literature [3]. Besides the issue that the baseline algorithms in many cases may not have been properly tuned in the reported experiments [26], a central issue lies in the selection of the baseline models themselves. In many published research works, only very recent neural models are considered, and a comparison with well-tuned longer-existing models is often missing.

In this present work, we examine to what extent such phenomena can be found in the most recent literature on SBRS that are built on Graph Neural Networks. For this purpose, we benchmarked eight GNN models—all were recently published at top-tier venues—with simple approaches. All algorithms are reproduced under identical settings by incorporating the original code published by the authors[2] into the *session-rec* SBRS evaluation framework [19]. The results of our study are again surprising. They show that under our independent evaluation, all of the analyzed GNN methods are outperformed by simple techniques, which do not even use side information, in terms of the Mean Reciprocal Rank, which was also our hyperparameter optimization criterion. Only in some situations, GNN-based models were favorable in terms of the Hit Rate. Overall, the results indicate that the problem of the inclusion of too weak baselines still exists in the recent SBRS literature.

While examining the research literature, we encountered a number of additional bad practices that may contribute to the apparently somewhat limited

[1] Such phenomena were also found outside the area of recommender systems, e.g., in information retrieval or time series forecasting [16,21].

[2] This is important, as third-party implementations may be unreliable [10].

progress in this area. First, we find that researchers often use the same embedding size for all compared models for 'fair comparison'. Since embedding sizes are however a hyperparameter to tune, such a comparison may instead be rather unfair [2,36]. Furthermore, the analyses in [29] showed that in a substantial fraction of today's research, the proposed models are tuned on test data instead of using a held-out validation dataset, potentially leading to data leakage and overfitting issues that produce overly optimistic results [3]. Finally, as reported in [23], even the choice of random seeds can have a non-negligible impact on the observed results.

In order to understand to what extent such bad practices may impact the performance results obtained by GNN models, we conducted a series of additional analyses using a subset of the models and datasets that were used in our main experiments. Our results show that the described factors can have a marked impact on the performance of the GNN-based models, thus potentially further contributing to a limited reliability of the reported progress in the literature.

2 Research Methodology

As discussed in earlier works [5], researchers often rely on a variety of experiment setups, using different protocols, metrics, datasets, pre-processing steps, and baselines. The primary objective of the study in this present paper is to provide an independent evaluation of recent GNN-based algorithms under identical conditions, i.e., same protocol, dataset, and metrics.

2.1 Algorithms

Compared GNN-Based Models: To ensure that our analysis is not focused on a few hand-picked examples, we followed a semi-systematic approach for the selection of algorithms to compare, using the following criteria:

- Publication outlets: We manually browsed important outlets for recommender system research (e.g., conferences such as RecSys or SIGIR and journals with a high impact factor), to identify recent works that propose GNN-based models for session-based recommendation.
- Reproducibility: We then only considered works for which the code was shared. We analyzed the shared repositories and contacted the authors if any part of the code was missing.[3] Ultimately, we were able to identify eight GNN models that could be trained and evaluated using the provided code and which we integrated into the *session-rec* framework.[4] The eight models are briefly described in Table 1.

Baselines: We included four non-neural baseline models that were also used in previous performance comparisons in [14] and [19], where it turned out that

[3] We considered articles to be non-reproducible if the authors did not reply after sending reminders.

[4] https://github.com/rn5l/session-rec.

Table 1. Description of compared GNN models

Model	Description
SR-GNN (AAAI '19)	Maybe the first work that uses a GNN model for SBRS. It constructs session graphs and uses a soft attention mechanism to aggregate information among the items [35]
TAGNN (SIGIR '20)	Uses an attention module attached to a GNN to adaptively learn the users' interests [39]
GCE-GNN (SIGIR '20)	Relies on a two-level item embedding integrated with a position-aware attention module to recursively learn the position of items in a session [33]
COTREC (CIKM '21)	Based on self-supervised learning with co-training to handle data sparsity and to generate self-supervised signals for SBRS [37]
GNRWW (ICDM '21)	Combines two kinds of embedding techniques to retrieve local and global patterns from the sessions [43]
FLCSP (J. Inf. Sci. '21)	Combines latent category abstractions with sequential data to obtain accurate recommendations [42]
CM-HGNN (KBS '22)	Differently to FLCSP, CM-HGNN leverages actual category information for accurate recommendations [38]
MGS (SIGIR '22)	Another model [15] which leverages category information. Uses a dual refinement mechanism to move the information between item and category graphs

Table 2. Description of baseline algorithms

Algorithm	Description
SR [18]	Learns sequential rules of order two by considering the co-occurrence of items in a session
STAN [7]	Builds on SKNN [18] and it uses three decay factors to compute the similarity of sessions and the relevance of the items
VSTAN [20]	Combines the particularities of VSKNN [18] and STAN and uses a sequence-aware item scoring procedure and an Inverse Document Frequency approach to promote less popular items
SFSKNN [18]	This variant of SKNN [18] also focuses on the recency of items by considering only those items for recommendations that appear in the neighbor sessions at least once after the last item of the current session

such more simple models can sometimes be quite difficult to beat. The baselines are briefly described in Table 2.

2.2 Datasets and Preprocessing

Datasets: We inspected the relevant literature from high-quality outlets to identify the most frequently used datasets. While a variety of datasets is used, the most prominent ones include RSC15 (RecSys Challenge 2015), DIGI (Diginetica), and RETAIL (Retailrocket).[5] Another important reason for choosing these

[5] https://www.kaggle.com/datasets/chadgostopp/recsys-challenge-2015, https://competitions.codalab.org/competitions/11161 https://www.kaggle.com/datasets/retailrocket/ecommerce-dataset.

datasets is that they contain category information, which is used by some GNN-based models. Table 3 provides summary statistics for the selected datasets.

Table 3. Dataset statistics

Dataset	Clicks	Items	# Categories	Sessions	Avg. sess. length
RSC15 (1/12)	574,157	20,979	58	145,202	3.95
RSC15 (1/64)	107,242	13,133	31	26,464	4.05
DIGI	1,004,598	43,100	995	216,134	4.65
RETAIL	1,045,413	44,540	944	304,902	3.42

Data Filtering and Splitting: We adopt common data preprocessing practices discussed from the literature [17,38,41]. Sessions of length one, items that appear less than five times in the dataset, and items that occur in test data but are not present in the training dataset are filtered out. From a training perspective, RSC15 consists of a large number of sessions, making it difficult to systematically tune all GNN models for large dataset without having access to massive GPU resources. Since previous works [18,27] show that using a recent fraction of the data leads to competitive or even superior results, we use the commonly used 1/12 and 1/64 fractions of the RSC15 dataset. For data splitting, we follow the common practice as mentioned in [31,32]. For DIGI, the sessions of the last seven days are put aside as test data and the remaining sessions are used for hyperparameter tuning and training. For RSC15 and RETAIL, the last-day sessions are used as test data.

Discussion. While our choices for data preprocessing align with what is common in the literature, our analysis of the relevant papers revealed that there is no agreed standard. Some works, for example, only consider items that appear in at least 30 sessions, others filter out sessions that contain more than 20 items. In yet another set of papers, data preprocessing is not mentioned at all. All of this further contributes to the difficulty of determining the state of the art. In terms of the used datasets, we focused on three widely used ones. We note that each of the 8 considered GNN-based models listed in Table 1 used at least one of these three datasets. Further experiments with other datasets to are left for future work.

2.3 Evaluation Metrics and Tuning

Metrics and Evaluation Protocol: We rely on Mean Reciprocal Rank and Hit Rate (MRR@K, HR@K) as measures that are widely used[6] in the relevant literature [40]. Furthermore, we report two beyond-accuracy metrics, coverage and

[6] Accuracy metrics such as Precision, Recall, or NDCG are used as well in session-based recommendation, but these metrics are often highly correlated [1,30].

popularity. We found that different evaluation protocols are used in the original papers that proposed the eight GNN-based models, e.g., leaving out the last item of each test session. In our evaluation, we relied on the commonly used procedure of incrementally "revealing" the items in the session. This entails that all items but the first one in a session are part of the ground truth in the evaluation at some stage, see also [9,18].

Hyperparameter Optimization: We tune the hyperparameters of all models using the training dataset and validate them using a subset of the training dataset. As it is commonly known, the tuning process of GNN models can be time-consuming, in particular when a larger hyperparameters space is searched. Therefore, we relied on a random optimization process where the number of training rounds is 25 to 60, depending on the time complexity of the selected models [19]. All selected models for the three datasets are optimized using MRR@20 as the target metric. The ranges and tuned values of the hyperparameters can be found in an online repository, where we also share all data and the code used for pre-processing, tuning, and evaluation for reproducibility.[7]

3 Results

In this section we first report the main results of our performance comparison in Sect. 3.1 and then provide additional analyses in Sect. 3.2.

3.1 Performance Measures

In this section, we first report the results for the accuracy and beyond-accuracy quality measures and then briefly discuss aspects of computational complexity.

Accuracy. Table 4 shows the accuracy results for the different datasets at the common list lengths of 10 and 20. We sort the results by MRR@20, as this was our optimization target. In terms of the MRR, we find that for all datasets one of the simple methods leads to the highest accuracy values. In particular STAN and VSTAN show consistently good results across several datasets. There is however no single 'winner' across datasets. For example, SR works particularly well for the larger RSC15 (1/12) dataset.[8] Generally, we observe that the differences between the winning simple method and the best GNN-based model are sometimes quite small. Nonetheless, it is surprising that several years after the first publication on the effectiveness of nearest-neighbor techniques in 2017 [12], the most recent published models are not consistently outperforming these baselines.

Looking only at the GNN models, we find there is also no model that is consistently better than the others. The ranking of the GNN-based models varies across datasets and depending on the chosen metric. While this is not surprising,

[7] https://faisalse.github.io/SessionRecGraphFusion/.
[8] The effectiveness of SR on some datasets in terms of MRR was observed also in [18].

this observation stands in contrast to almost all published works on SBRS, where any newly proposed model is usually reported to outperform all other baselines on all datasets and metrics. Moreover, we find that some of the recent GNN-based models, even though reported to outperform the state of the art in the original papers, actually perform very poorly in our comparison, leading to MRR values that are sometimes more than 50% lower than the best models.

In terms of the Hit Rate, we observe that VSTAN is the best model for the RETAIL dataset, and it has highly competitive performance for the RSC15 (1/12) dataset. For the DIGI and RSC15 (1/64) datasets, in contrast, several GNN-based models are better than the simpler approaches. In particular for the DIGI dataset, the margin between the best GNN model and the best simple model is quite large. This suggests that GNN models in fact can be effective and help us to achieve progress over existing models. The observed improvements at least in our experiment are limited to certain configurations and to the Hit Rate. We however recall that the Hit Rate was not the optimization goal during hyperparameter tuning.

Considering only the rankings of the GNN-based models, again no clear winner can be found. In some cases, models that use category information (in particular MGS) work quite well. In other cases, the reliance on category information does not seem to be too helpful, and other category-agnostic GNN models lead to higher accuracy.

Table 4. Accuracy results, sorted by MRR@20. Black circles in the table indicate simple baselines and empty circles indicate GNN-based models. The highest scores for each metric are printed in bold font while the second-best scores are underlined.

Metrics	MRR@10	MRR@20	HR@10	HR@20
RETAIL				
• VSTAN	**0.631**	**0.631**	**0.971**	**0.980**
∘ SR-GNN	<u>0.629</u>	**0.631**	0.886	0.931
• SFSKNN	0.599	<u>0.603</u>	<u>0.939</u>	**0.980**
∘ CM-HGNN	0.562	0.568	0.812	0.890
• SR	0.553	0.560	0.865	<u>0.959</u>
∘ GNRRW	0.553	0.558	0.804	0.869
∘ MGS	0.553	0.558	0.820	0.878
∘ COTREC	0.551	0.556	0.792	0.861
• STAN	0.544	0.548	0.873	0.931
∘ TAGNN	0.540	0.545	0.804	0.882
∘ FLCSP	0.451	0.455	0.800	0.865
∘ GCE-GNN	0.423	0.429	0.596	0.678

(*continued*)

Table 4. (*continued*)

RSC15 (1/64)				
• STAN	**0.290**	**0.296**	0.538	0.613
• VSTAN	0.286	0.289	0.546	0.595
∘ GCE-GNN	0.278	0.285	0.538	0.633
∘ CM-HGNN	0.278	0.284	**0.575**	**0.650**
• SFSKNN	0.264	0.264	0.422	0.428
∘ COTREC	0.274	0.279	0.543	0.616
∘ GNRRW	0.269	0.276	0.526	0.618
∘ MGS	0.264	0.270	0.543	0.630
• SR	0.263	0.266	0.462	0.506
∘ SR-GNN	0.245	0.251	0.497	0.581
∘ TAGNN	0.195	0.201	0.425	0.509
∘ FLCSP	0.176	0.183	0.393	0.497
RSC15 (1/12)				
• SR	**0.338**	**0.344**	0.557	0.627
∘ GNRRW	0.335	0.342	0.581	**0.682**
∘ SR-GNN	0.328	0.332	0.563	0.621
∘ MGS	0.326	0.331	0.566	0.642
• STAN	0.325	0.330	0.590	0.661
• VSTAN	0.325	0.330	**0.599**	0.673
• SFSKNN	0.321	0.325	0.550	0.609
∘ COTREC	0.307	0.312	0.560	0.639
∘ CM-HGNN	0.293	0.299	0.532	0.618
∘ FLCSP	0.284	0.290	0.480	0.560
∘ TAGNN	0.281	0.288	0.520	0.621
∘ GCE-GNN	0.220	0.225	0.391	0.453
DIGI				
• SFSKNN	**0.348**	**0.351**	0.559	0.604
• STAN	0.347	**0.351**	0.529	0.600
• VSTAN	0.342	0.346	0.520	0.581
• SR	0.333	0.337	0.568	0.617
∘ COTREC	0.330	0.335	0.555	0.637
∘ MGS	0.322	0.329	0.559	0.656
∘ SR-GNN	0.321	0.327	**0.591**	**0.688**
∘ GNRRW	0.318	0.324	0.578	0.667
∘ CM-HGNN	0.310	0.316	0.561	0.649
∘ TAGNN	0.279	0.287	0.544	0.645
∘ GCE-GNN	0.227	0.236	0.419	0.553
∘ FLCSP	0.175	0.184	0.398	0.525

An ANOVA analysis revealed that there are significant differences ($p < 0.05$) for the obtained MRR@20 values for the DIGI and RETAIL datasets across all examined models. However, the differences between the best-performing models for these datasets is too small to reach statistical significance according to a Tukey post-hoc test. Nonetheless, given these results, we can conclude that the best simple models are performing at least equally well as the neural ones.

Additional Quality Measures. In Table 5 we report additional beyond-accuracy metrics that are relevant in practice. Cov@20 refers to the percentage of items, which appear at least once in the top-20 recommendation lists for all test sessions. Pop@20 of an item is measured in terms of the number of times it appears in the training sessions. The metric reported here is the average (normalized) popularity of the recommended items in the top-20 recommendation lists for all test sessions. The results show that the simple models lead to high coverage values, indicating that these models consider a broader range of items in their recommendations at an aggregate level. In terms of popularity, we find that the simple methods are at about the same level as some of the GNN-based approaches. However, two GNN-based models, namely TAGNN, and GCE-GNN seem to have a stronger and usually undesired tendency to recommend more popular items.

Time Complexity. To illustrate the complexity of the problem, we report an example of the training time (T-Time) and prediction time (P-Time) for the different models on the DIGI dataset using our hardware[9] in Table 5. On this dataset, GCE-GNN is the slowest model which takes approximately 19 h to be trained once, and we recall that many training rounds are needed during hyperparameter tuning for each dataset and for each model. We also note that the training time may vary strongly between the GNN models across datasets, depending, e.g., on the number of items, the number of the sessions, and the length of the sessions. The kNN-based models, in contrast, do not have a training phase. The time needed for predictions for these models lies about in the range of the faster GNN-based models. However, we notice that the COTREC model is particularly slow and needs approximately 2 s to generate one single recommendation list, which is prohibitive in practice.

---

[9] A machine with an AMD EPYC 7H12 64-Core Processor 2600 Mhz 16 Core(s) and an NVIDIA RTX A4000 WDDM graphics card.

Table 5. Results for the DIGI dataset, sorted by Cov@20

Metrics	Cov@20	Pop@20	T-Time (m)	P-Time (ms)
• STAN	**0.087**	0.091	**0.038**	14.697
• VSTAN	<u>0.086</u>	0.084	<u>0.045</u>	13.911
• SR	<u>0.086</u>	0.075	0.121	**3.575**
◦ SR-GNN	0.079	0.078	55.119	27.468
◦ COTREC	0.078	0.079	480.709	1992.498
◦ MGS	0.076	0.083	390.894	14.635
◦ TAGNN	0.073	**0.102**	119.823	6.991
◦ GNRRW	0.073	0.078	154.657	173.524
◦ FLCSP	0.073	0.075	40.825	10.393
◦ CM-HGNN	0.072	0.083	944.725	12.839
• SFSKNN	0.072	0.071	0.081	<u>5.505</u>
◦ GCE-GNN	0.065	<u>0.097</u>	1,153.861	503.492

3.2 Additional Robustness Analyses for GNN-Based Models

As mentioned earlier, a number of factors can endanger the robustness and reliability of accuracy results reported in research papers, including the choice of embedding sizes and random seeds, or the practice of tuning on test data. When we were screening the literature for recent GNN-based models, we identified 34 relevant articles in top-level outlets.[10] An analysis of these papers leads to a number of interesting observations.

Embedding Size. We found that the authors frequently fix the embedding size and only tune the other hyperparameters such as learning rate or dropout rate. In several cases, we also observed that authors chose the embedding size of the baseline models from the original papers which may have been optimal for different datasets. However, they tune the embedding size of their own proposed models. Such a comparison, in which only one of the models is extensively tuned, certainly cannot lead to any reliable insights, as the embedding size in fact is a crucial hyperparameter [2,8,36].

To analyze the potential extent of the problem, we conducted experiments in which we varied the embedding size of the models while keeping the other hyperparameters constant (at their optimal values, as determined in the previous experiment). Since some GNN-based models are computationally expensive, we focused on five training-efficient models and on two datasets.

As a representative of a smaller dataset, we chose the RSC15 (1/64) dataset and conducted the experiments by using 15 different values for embedding sizes ranging from 16 to 500. We used the DIGI dataset as an example for a larger

[10] The list of considered papers can be found in the online material.

dataset, and the experiments are conducted using 10 different values for embedding sizes ranging from 16 to 250. In this case, the maximum value of the embedding size is 250, as we experienced memory problems with larger values.

The outcomes of these experiments are provided in Table 6.[11] We can observe that some models, depending on the embedding size, can lead to largely different results in terms of the MRR@20. This confirms the importance of carefully tuning this central hyperparameter. As the results show, the differences between the best and worst MRR values can be more than 40% in some cases, simply by choosing a bad value for the embedding size. For some of the models, the difference between the minimum and maximum MRR values is less pronounced, but still in a range where researchers would report a substantial improvement over the state of the art when comparing models. We note that no clear pattern was found across the models that larger embedding sizes are consistently better than smaller ones. Furthermore, the effect size (η^2) is measured using the ANOVA test and found that its values for the DIGI and RSC15 datasets are 85.53% and 91.39% respectively, indicating the high variance between the mean MRR values of the GNN models.

Table 6. Results with varying embedding sizes, sorted based on difference

Models	MRR@20 mean ± std	Max MRR@20	Min MRR@20	Diff
RSC15 (1/64)				
GNRWW	0.237 ± 0.025	0.291	0.206	0.085
TAGAN	0.195 ± 0.021	0.226	0.143	0.083
FLCSP	0.160 ± 0.014	0.180	0.135	0.045
SR-GNN	0.238 ± 0.012	0.259	0.218	0.041
COTREC	0.277 ± 0.003	0.284	0.269	0.015
DIGI				
TAGAN	0.247 ± 0.033	0.294	0.164	0.130
COTREC	0.335 ± 0.010	0.343	0.309	0.034
FLCSP	0.206 ± 0.011	0.224	0.191	0.033
SR-GNN	0.313 ± 0.008	0.324	0.296	0.028
GNRWW	0.336 ± 0.005	0.346	0.327	0.019

[11] Note: the MRR@20 values of some GNN models are not the same as in Table 4 because we decreased the values of batch sizes to avoid memory issues, which in turn affects the accuracy of the models.

Random Seeds. The random seed used in an experiment is certainly not a hyperparameter to tune. Still, its choice can have notable impact on accuracy results [23,34]. We therefore conducted experiments to analyze to what extent this is true also for some of the examined GNN-based models. In these experiments, we selected random seed values between 100 to 10,000,000 by using a random function. In the shared repositories of the GNN models, random seed values between 2000 to 3000 can often been found, but there is no general pattern. In our experiments, we deliberately picked a larger range to examine if some more uncommon values (outliers) could have an unexpected effect. Again, we considered the RSC15 (1/64) and the DIGI dataset. We generated 100 random seed values for the RSC15 dataset, and 35 values for the larger DIGI dataset. The hyperparameters of the examined models were again left at their optimum values. The obtained results for MRR@20 when using different random seeds are shown in Table 7, again sorted by the difference between the minimum and the maximum observed value. Again, we can see that for some models and datasets, the obtained values using the same set of optimized hyperparameters vary strongly. For the RSC15 dataset and the FLCSP method, for example, the obtained accuracy results can be 30% lower than the best value simply because of a unlucky choice of the random seed value.[12] Clearly, the differences are not as extreme for all models and datasets and may be partly due to the large range of values that we explored. The distribution of the accuracy values Fig. 1 for the RSC15 (1/64) dataset. The detailed data of the entire experiment can be found in the online material.

Comparing the results of Table 7 and the results obtained when using the random seed values used in the original code shared by the authors (Table 4), we observe that often higher accuracy values can be achieved when exploring a large number of random seed values. For example, the optimal MRR@20 value for the FLCSP method on the RSC15 (1/64) dataset is 0.183. From the first row in Table 7 we can see that the worst MRR value was at 0.159 and the best one at 0.250, which can be seen as a substantial improvement, which is only obtained by changing the random seed. A similar observation can be made in other situations as well, e.g., considering the last row of Table 7, where the best MRR@20 value of GNRWW is at 0.352 for the DIGI dataset, whereas the best value in Table 4 was at 0.324. Note, however that in some cases the best MRR values obtained by exploring many (somewhat unusual) seed values are lower than when using more common choices as can be found in the shared code repositories, which means that these common ranges should be explored as well. Furthermore, η^2 is measured and found to be 97.79% and 89.56% for the DIGI and RSC15 datasets respectively, indicating the high variance between the mean MRR values of the GNN models.

[12] In this case, the random seed value leading to the worst MRR value was around 7,000,000. All detailed results can be found online.

Table 7. Experiments with random seeds, sorted based on difference

Models	MRR@20 mean ± std	Max MRR@20	Min MRR@20	Diff
RSC15 (1/64)				
FLCSP	0.192 ± 0.022	0.250	0.159	0.091
TAGNN	0.205 ± 0.011	0.232	0.175	0.057
GNRWW	0.273 ± 0.009	0.293	0.245	0.048
SR-GNN	0.249 ± 0.005	0.264	0.236	0.028
COTREC	0.283 ± 0.003	0.290	0.275	0.015
DIGI				
TAGNN	0.256 ± 0.015	0.281	0.228	0.052
FLCSP	0.203 ± 0.009	0.217	0.172	0.045
SR-GNN	0.322 ± 0.009	0.342	0.307	0.035
GNRWW	0.340 ± 0.006	0.352	0.327	0.025
COTREC	0.337 ± 0.003	0.343	0.329	0.014

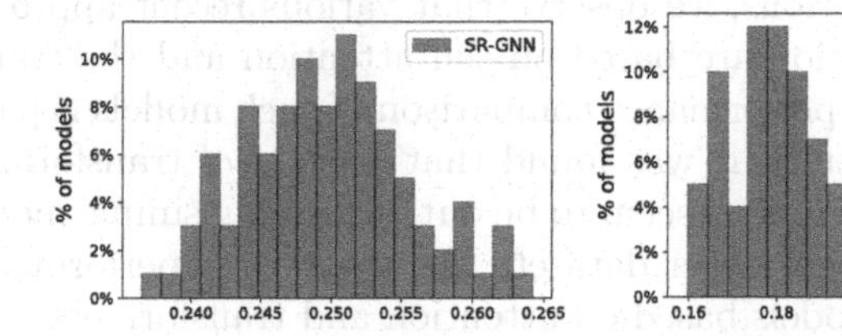

Fig. 1. Distribution of MRR@20 values for different random seeds (RSC15)

Tuning on Test Data. In a last experiment, we repeated the main experiment from Sect. 3, but this time tuned the hyperparameters of four training-efficient GNN models on the test set, which is unfortunately not an uncommon practice according to [29]. The results show that depending on the dataset an accuracy improvement by 1.5 to about 2.0% in terms of MRR@20 across models can be achieved. The strongest individual "improvement" was at 5.6%, which confirms the observation from [3] that such evaluation practice leads to overly optimistic assessments of the accuracy of a model. The detailed results of the experiment can be found in the online material.

4 Conclusion

Our systematic analysis of eight recent GNN-based models for session-based recommendation reveal that methodological problems that were identified several years ago still exist in this research area. Specifically, researchers tend to rely on experimental evaluations which mostly consider other recent neural models as baselines, but do not include simpler, yet often well-performing baselines in their comparisons. Considering again the 34 recent papers from top-tier outlets that we analyzed in our study, we find that none of them considered models like SFSKNN or VSTAN as baselines. Instead, trivial methods like popularity based ones or the sometimes poorly performing *item-KNN* baseline from [9] are considered. Only one paper [14] considered several non-neural baselines, but this was a particular study which compared the offline and online performance of various models based on the *session-rec* framework.

In sum, this leads to some worries regarding the true progress that is achieved in the field, in particular when we combine this finding with other potentially problematic factors that can influence the outcomes reported in research papers, as discussed in the additional analyses in this paper. On the more positive side, our analysis suggests that there is ample room for future improvements of session-based recommendation models. In particular, not too many works exist yet that consider the various types of side information, e.g., item category, price, that are available in some datasets. We believe that approaches based on Graph Neural Networks are indeed promising for such settings with rich data, even though the methods examined in our present study do not seem to exploit their full potential yet.

In terms of future directions, we observe that various recent approaches for session-based recommendation are based on self-attention and the transformer architecture, e.g., [4,11]. A performance comparison of such models is part of our future work. In a recent study, it was found that *sequential* transformer-based models like BERT4Rec [28] indeed seem to be outperforming simple models with a substantial gap, at least for larger datasets. An up-to-date performance comparison of *session-based* models based on attention and transformers is however still missing.

References

1. Anelli, V.W., Bellogín, A., Di Noia, T., Jannach, D., Pomo, C.: Top-N recommendation algorithms: a quest for the state-of-the-art. In: Proceedings of the 30th ACM Conference on User Modeling, Adaptation and Personalization, pp. 121–131 (2022)
2. Chamberlain, B., Clough, J., Deisenroth, M.: Neural embeddings of graphs in hyperbolic space. arxiv 2017. In: 13th International Workshop on Mining and Learning with Graphs, Held in Conjunction with KDD 2017 (2017)
3. Cremonesi, P., Jannach, D.: Progress in recommender systems research: Crisis? What crisis? AI Mag. **42**(3), 43–54 (2021)
4. Fan, X., Liu, Z., Lian, J., Zhao, W.X., Xie, X., Wen, J.R.: Lighter and better: low-rank decomposed self-attention networks for next-item recommendation. In: Proceedings of the 44th International ACM SIGIR Conference on Research and Development in Information Retrieval, pp. 1733–1737 (2021)
5. Ferrari Dacrema, M., Boglio, S., Cremonesi, P., Jannach, D.: A troubling analysis of reproducibility and progress in recommender systems research. ACM Trans. Inf. Syst. **39**(2), 1–49 (2021)
6. Ferrari Dacrema, M., Cremonesi, P., Jannach, D.: Are we really making much progress? A worrying analysis of recent neural recommendation approaches. In: Proceedings of the 2019 ACM Conference on Recommender Systems, Copenhagen (2019)

7. Garg, D., Gupta, P., Malhotra, P., Vig, L., Shroff, G.: Sequence and time aware neighborhood for session-based recommendations: STAN. In: Proceedings of the 42nd International ACM SIGIR Conference on Research and Development in Information Retrieval, pp. 1069–1072 (2019)
8. Hamilton, W.L., Ying, R., Leskovec, J.: Representation learning on graphs: methods and applications. IEEE Data Eng. Bull. **40**(3), 52–74 (2017)
9. Hidasi, B., Karatzoglou, A., Baltrunas, L., Tikk, D.: Session-based recommendations with recurrent neural networks. In: Proceedings of ICLR 2016 (2016)
10. Hidasi, B., Czapp, Á.T.: The effect of third party implementations on reproducibility. In: Proceedings of the 17th ACM Conference on Recommender Systems (2023)
11. Hou, Y., Hu, B., Zhang, Z., Zhao, W.X.: CORE: simple and effective session-based recommendation within consistent representation space. In: Proceedings of the 45th International ACM SIGIR Conference on Research and Development in Information Retrieval, pp. 1796–1801 (2022)
12. Jannach, D., Ludewig, M.: When recurrent neural networks meet the neighborhood for session-based recommendation. In: Proceedings of the 11th ACM Conference on Recommender Systems, Como, Italy (2017)
13. Jannach, D., Quadrana, M., Cremonesi, P.: Session-based recommendation. In: Ricci, F., Shapira, B., Rokach, L. (eds.) Recommender Systems Handbook (2021)
14. Kouki, P., Fountalis, I., Vasiloglou, N., Cui, X., Liberty, E., Al Jadda, K.: From the lab to production: a case study of session-based recommendations in the home-improvement domain. In: Proceedings of the 14th ACM Conference on Recommender Systems, pp. 140–149 (2020)
15. Lai, S., Meng, E., Zhang, F., Li, C., Wang, B., Sun, A.: An attribute-driven mirror graph network for session-based recommendation. In: Proceedings of the 45th International ACM SIGIR Conference on Research and Development in Information Retrieval, pp. 1674–1683 (2022)
16. Lin, J.: The neural hype and comparisons against weak baselines. In: ACM SIGIR Forum, vol. 52, pp. 40–51 (2019)
17. Liu, Q., Zeng, Y., Mokhosi, R., Zhang, H.: STAMP: short-term attention/memory priority model for session-based recommendation. In: Proceedings of the 24th ACM SIGKDD International Conference on Knowledge Discovery & Data Mining, pp. 1831–1839 (2018)
18. Ludewig, M., Jannach, D.: Evaluation of session-based recommendation algorithms. User Model. User-Adap. Inter. **28**, 331–390 (2018)
19. Ludewig, M., Mauro, N., Latifi, S., Jannach, D.: Performance comparison of neural and non-neural approaches to session-based recommendation. In: Proceedings of the 13th ACM Conference on Recommender Systems, pp. 462–466 (2019)
20. Ludewig, M., Mauro, N., Latifi, S., Jannach, D.: Empirical analysis of session-based recommendation algorithms: a comparison of neural and non-neural approaches. User Model. User-Adap. Inter. **31**, 149–181 (2021)
21. Makridakis, S., Spiliotis, E., Assimakopoulos, V.: Statistical and machine learning forecasting methods: concerns and ways forward. PLoS ONE **13**(3), e0194889 (2018)
22. Pang, Y., et al.: Heterogeneous global graph neural networks for personalized session-based recommendation. In: Proceedings of the Fifteenth ACM International Conference on Web Search and Data Mining, pp. 775–783 (2022)
23. Picard, D.: Torch. manual_seed (3407) is all you need: on the influence of random seeds in deep learning architectures for computer vision. arXiv preprint arXiv:2109.08203 (2021)

24. Rendle, S., Krichene, W., Zhang, L., Koren, Y.: Revisiting the performance of iALS on item recommendation benchmarks. In: Proceedings of the 16th ACM Conference on Recommender Systems, pp. 427–435 (2022)
25. Rendle, S., Zhang, L., Koren, Y.: On the difficulty of evaluating baselines: a study on recommender systems. arXiv preprint arXiv:1905.01395 (2019)
26. Shehzad, F., Jannach, D.: Everyone's a winner! On hyperparameter tuning of recommendation models. In: 17th ACM Conference on Recommender Systems (2023)
27. Song, W., Xiao, Z., Wang, Y., Charlin, L., Zhang, M., Tang, J.: Session-based social recommendation via dynamic graph attention networks. In: Proceedings of the Twelfth ACM International Conference on Web Search and Data Mining, pp. 555–563 (2019)
28. Sun, F., et al.: BERT4Rec: sequential recommendation with bidirectional encoder representations from transformer. In: Proceedings of the 28th ACM International Conference on Information and Knowledge Management, pp. 1441–1450 (2019)
29. Sun, Z., et al.: Are we evaluating rigorously? Benchmarking recommendation for reproducible evaluation and fair comparison. In: Proceedings of the 14th ACM Conference on Recommender Systems, pp. 23–32 (2020)
30. Valcarce, D., Bellogín, A., Parapar, J., Castells, P.: On the robustness and discriminative power of information retrieval metrics for top-N recommendation. In: Proceedings of the 12th ACM Conference on Recommender Systems, pp. 260–268 (2018)
31. Wang, H., Zeng, Y., Chen, J., Zhao, Z., Chen, H.: A spatiotemporal graph neural network for session-based recommendation. Expert Syst. Appl. **202**, 117114 (2022)
32. Wang, M., Ren, P., Mei, L., Chen, Z., Ma, J., De Rijke, M.: A collaborative session-based recommendation approach with parallel memory modules. In: Proceedings of the 42nd International ACM SIGIR Conference on Research and Development in Information Retrieval, pp. 345–354 (2019)
33. Wang, Z., Wei, W., Cong, G., Li, X.L., Mao, X.L., Qiu, M.: Global context enhanced graph neural networks for session-based recommendation. In: Proceedings of the 43rd International ACM SIGIR Conference on Research and Development in Information Retrieval, pp. 169–178 (2020)
34. Wegmeth, L., Vente, T., Purucker, L., Beel, J.: The effect of random seeds for data splitting on recommendation accuracy. In: Perspectives on the Evaluation of Recommender Systems Workshop (PERSPECTIVES 2023), Co-located with the 17th ACM Conference on Recommender Systems (2023)
35. Wu, S., Tang, Y., Zhu, Y., Wang, L., Xie, X., Tan, T.: Session-based recommendation with graph neural networks. In: Proceedings of the AAAI Conference on Artificial Intelligence, vol. 33, pp. 346–353 (2019)
36. Wu, Z., Pan, S., Chen, F., Long, G., Zhang, C., Philip, S.Y.: A comprehensive survey on graph neural networks. IEEE Trans. Neural Networks Learn. Syst. **32**(1), 4–24 (2020)
37. Xia, X., Yin, H., Yu, J., Shao, Y., Cui, L.: Self-supervised graph co-training for session-based recommendation. In: Proceedings of the 30th ACM International Conference on Information & Knowledge Management, pp. 2180–2190 (2021)
38. Xu, H., Yang, B., Liu, X., Fan, W., Li, Q.: Category-aware multi-relation heterogeneous graph neural networks for session-based recommendation. Knowl.-Based Syst. **251**, 109246 (2022)
39. Yu, F., Zhu, Y., Liu, Q., Wu, S., Wang, L., Tan, T.: TAGNN: target attentive graph neural networks for session-based recommendation. In: Proceedings of the 43rd International ACM SIGIR Conference on Research and Development in Information Retrieval, pp. 1921–1924 (2020)

40. Yuan, F., et al.: Future data helps training: modeling future contexts for session-based recommendation. In: Proceedings of The Web Conference 2020, pp. 303–313 (2020)
41. Zhang, Z., Wang, B.: Learning sequential and general interests via a joint neural model for session-based recommendation. Neurocomputing **415**, 165–173 (2020)
42. Zhang, Z., Wang, B.: Fusion of latent categorical prediction and sequential prediction for session-based recommendation. Inf. Sci. **569**, 125–137 (2021)
43. Zhang, Z., Wang, B.: Graph neighborhood routing and random walk for session-based recommendation. In: 2021 IEEE International Conference on Data Mining (ICDM), pp. 1517–1522 (2021)

A Reproducibility Study of Goldilocks: Just-Right Tuning of BERT for TAR

Xinyu Mao[1(✉)], Bevan Koopman[1,2], and Guido Zuccon[1]

[1] The University of Queensland, St. Lucia, Australia
{xinyu.mao,g.zuccon}@uq.edu.au
[2] CSIRO, Brisbane, Australia
bevan.koopman@csiro.au

Abstract. Screening documents is a tedious and time-consuming aspect of high-recall retrieval tasks, such as compiling a systematic literature review, where the goal is to identify all relevant documents for a topic. To help streamline this process, many Technology-Assisted Review (TAR) methods leverage active learning techniques to reduce the number of documents requiring review. BERT-based models have shown high effectiveness in text classification, leading to interest in their potential use in TAR workflows. In this paper, we investigate recent work that examined the impact of further pre-training epochs on the effectiveness and efficiency of a BERT-based active learning pipeline. We first report that we could replicate the original experiments on two specific TAR datasets, confirming some of the findings: importantly, that further pre-training is critical to high effectiveness, but requires attention in terms of selecting the correct training epoch. We then investigate the generalisability of the pipeline on a different TAR task, that of medical systematic reviews. In this context, we show that there is no need for further pre-training if a domain-specific BERT backbone is used within the active learning pipeline. This finding provides practical implications for using the studied active learning pipeline within domain-specific TAR tasks.

Keywords: Technology-Assisted Review (TAR) · Active Learning · Systematic Reviews

1 Introduction

Review tasks in professional domains often require screening of a large number of documents to ensure all evidence about a review topic is identified. This task is often associated with high recall retrieval (HRR); examples of such tasks include the compilation of systematic literature reviews, legal eDiscovery, and prior-art finding in patent applications [4,16,17]. The manual screening of a large number of candidate documents can be time-consuming and resource-intensive. To reduce the number of documents needing manual review, automated methods are used to identify the relevant documents in a given set, with the aim of achieving a targeted recall: a process known as technology-assisted review (TAR).

N. Goharian et al. (Eds.): ECIR 2024, LNCS 14611, pp. 132–146, 2024.
https://doi.org/10.1007/978-3-031-56066-8_13

Pre-trained language models such as BERT [3], T5 [20] and GPT [19] have exhibited state-of-the-art effectiveness in tasks such as general-domain search [14,24], question answering [11], or text summarization [15]. These language models follow the transformer architecture [25] and are able to model word semantics by performing a pre-training step, e.g., masked language modelling (MLM) [25].

In this work, we focus on reproducing the TAR pipeline of Yang et al. that exploits pre-trained language models [28]. Pre-trained language models [3,19,20] have exhibited state-of-the-art effectiveness across several tasks [11,14,15,24]. In the pipeline of Yang et al., the TAR task is modelled as a binary classification problem on a large dataset with categories and their pipeline uses active learning to continuously fine-tune BERT classifiers. A key aspect of the pipeline is the additional pre-training of the BERT backbone to the document collection of the TAR task, which is critical to obtain effectiveness improvements compared to a baseline logistic regression approach.

In this paper, we aim to reproduce and verify the findings of the original study and investigate whether the original conclusions generalise to other TAR tasks, and in particular that of a medical systematic review literature search [8–10]. The study of the pipeline's generalisability is not trivial because the TAR tasks considered in the original work crucially differ from those in medical systematic reviews for the following aspects, which are also visualised in Fig. 1:

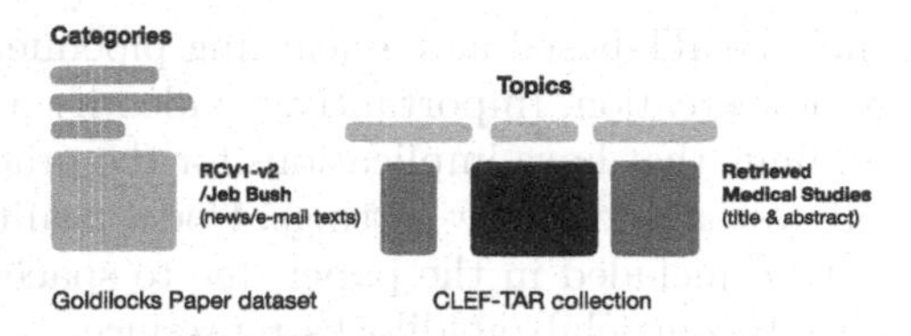

Fig. 1. Difference in task configuration between datasets.

- **Domain:** They contain biomedical literature and thus might require a more domain-specific approach.
- **Proposed task:** They are designed to evaluate ranking instead of classification, with the goal of ranking the set of documents associated with the topic in decreasing order of relevance.
- **Dataset composition:** They are composed of topics, each of which represents a systematic review, and a set of corresponding documents that are retrieved by a Boolean query and provided with only the title and abstract. TAR is performed separately and independently for each topic: this means that the TAR task is performed on different sources of documents as reference for each topic. This is unlike the datasets in the original work where the TAR task was performed once for each dataset. This is a crucial difference because (1) the size of the document set on which TAR is performed largely differs between the original and our setup, and (2) the nature of the classes assigned largely differ – very distinct classes in the original work vs. inclusion/exclusion classes related to topical relevance in our setup.

Our investigation provides insights into the reproducibility and generalisability of the original work by Yang et al., in particular in the context of applying

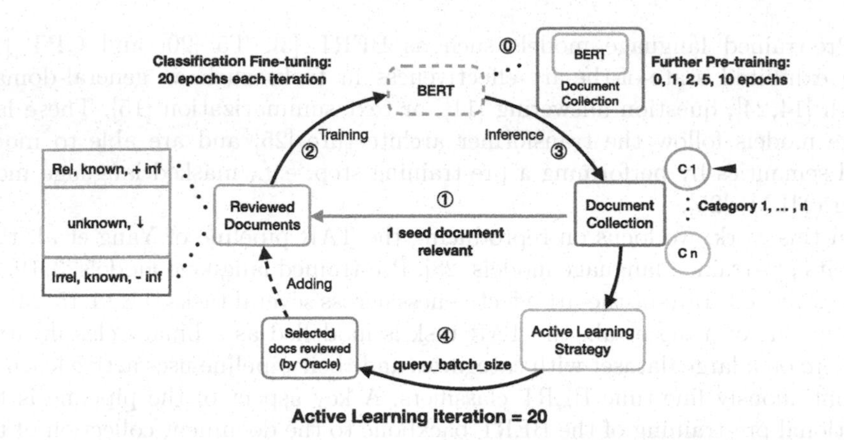

Fig. 2. The TAR workflow using active learning with BERT used by the Goldilocks paper.

their BERT-based active learning pipeline to the context of medical systematic review creation. Importantly, we identify a crucial difference in the experimental findings that have implications for the practical use of the pipeline within medical systematic review settings. Code, results and supporting material that could not be included in the paper due to space limitations can be found at https://github.com/ielab/goldilocks-reproduce.

2 Goldilocks: Just-Right Tuning of BERT

2.1 Overall Active Learning Pipeline

Yang et al. have proposed a pipeline that uses a BERT classifier to undertake a TAR task within an active learning workflow [28]. The pipeline is visualised in Fig. 2, and it consists of two key components: (1) Further Pre-training and; (2) Fine-tuning.

(1) Further Pre-training. Prior to the active learning process, a BERT pre-trained model undergoes additional pre-training using MLM on the entire target data. A key aspect of the original study was to investigate the effect that varying the number of epochs used in the further pre-training task has on the final model's effectiveness.

(2) Fine-tuning. The fine-tuning regime followed four steps:

i. *Initiating with seed document.* A seed document for a specific target category is specified to initiate the active learning process. A seed document is a document that is relevant to the category.

ii. *Classification fine-tuning.* BERT is fine-tuned with all the reviewed documents in the training set for a fixed number of epochs. The original paper demonstrated that fine-tuning using only newly labelled documents has worse performance. At the start, the reviewed documents only consist of the seed document. In subsequent iterations, more reviewed documents are added.
iii. *Scoring all remaining documents.* The fine-tuned model is used to score the remaining documents in the dataset. The scores are used by an active learning strategy (relevance feedback [12] or uncertainty sampling [22]) to select informative document samples to be reviewed/labelled by reviewers (known as the oracle). Already reviewed documents are excluded from selection.
iv. *Querying and labelling new documents and updating the training set.* The newly reviewed documents are added to the training set.

Steps ii, iii, and iv are repeated iteratively until all iterations are completed.

2.2 Key Findings from the Original Study

The key findings from the original study that we aim to reproduce are:

- Key to the effectiveness of the proposed pipeline is the existence of a Goldilocks (or 'just-right') epoch of further pre-training. This refers to the further pre-training step outlined in Sect. 2.1. The correct Goldilocks epoch depends on the dataset and task characteristics. An incorrect setting makes the BERT classifier underperform a baseline logistic regression, even on in-domain tasks.
- Further pre-training does not solve the issue of domain mismatch. Domain mismatch occurs when the data used for the pre-training does not match the data used at inference. Further pre-training on the target dataset is expected to address domain mismatch, but this is not the case when the target is out-of-domain, even when the correct Goldilocks epoch is identified.
- The computational costs associated with the proposed pipeline discourage its use because effectiveness improvements are only marginal compared to the logistic regression baseline.

3 Experiment Setup

We devise a set of experiments to address the following research questions (RQs):

RQ1. *To what extent the original results can be reproduced?*
RQ2. *How well do the original findings generalise to the task of screening in medical systematic review creation?*
RQ3. *Is there a domain-mismatch problem in the task of screening in medical systematic review creation, and can this be mitigated by adopting a BERT backbone specifically designed for the biomedical domain?*

3.1 Datasets

We use three groups datasets for our experiment: RCV1-v2 [13], Jeb Bush [5,21], both used in the original study [28], and the CLEF TAR collections [8–10]. The CLEF TAR consists of several datasets.

RCV1-v2 [13]: consists of newswire articles, and as such was regarded as in-domain for BERT by Yang et al.. The original study considered 658 categories, each containing no less than 25 documents, and selected 45 categories according to prevalence (rare, medium, common) and difficulty (hard, medium, easy) following previous works [26,27]. Then they downsampled to 160,833 documents (20%) for computational efficiency. We obtained a table of these 45 categories from Yang et al., and downsampled by ourselves with the same rate.

Jeb Bush [5,21]: consists of email texts related to specific local issues of a constituent, and because of this Yang et al. regarded this dataset as out-of-domain for BERT. The dataset contains 44 topics and 290,099 emails. We followed Yang et al. in de-duplicating these emails by comparing the md5 hash string of each email text, to obtain 274,124 unique emails. The emails are further randomly downsampled to 137,062 (50%) following the original study; this makes the dataset comparable in size to the downsampled RCV1-v2 dataset.

CLEF TAR Collections [8–10]: comprise six datasets – CLEF17 test (30 topics, total 117,557 documents), CLEF17 train (20 topics, total 149,404 documents), CLEF18 test (30 topics, total 218,484 documents), CLEF19 dta test (8 topics, total 30,521 documents), which include diagnostic test accuracy reviews, and CLEF19 intervention test (20 topics, total 41,996 documents), CLEF19 intervention train (20 topics, total 31,639 documents), which include intervention reviews. Each dataset contains a set of systematic review topics, along with documents (title and abstract of studies considered for inclusion in the systematic review) and inclusion/exclusion judgements. Although originally proposed to evaluate a ranking task (screening prioritisation), we adapt the datasets for classification (screening) similar to the RCV1-v2 and Jeb Bush datasets, so that we treat each topic as a separate classification task. We consider only the original splits of these datasets, which for each year include train and test splits, but we do not perform training on the portions labelled "train": instead we used them all for evaluation. This is because classifiers in these tasks are not trained across different reviews – they are instead trained "on-the-spot" on the specific systematic review under consideration, in an active learning setting.

We report further details on the datasets along with information on their pre-processing in the online repository.

3.2 Considered Methods: Baseline and BERT Model

We follow the original study in using a logistic regression (LR) classifier as the baseline model for comparison [28]; as in the original study, we also source the implementation from `scikit-learn`[1]. To replicate their BERT-based classifier, we acquire the common `bert-base-cased`[2] backbone from Hugging Face. The active learning implementation is based on the `libact`[3] library. We

[1] https://scikit-learn.org/stable/.
[2] https://huggingface.co/bert-base-cased.
[3] https://github.com/ntucllab/libact.

requested code from the original authors and obtained the BERT part, while we adapted related code for the baseline part, including tokenisation and training.

3.3 Evaluation Metrics

We report the same evaluation metrics as in the original study, including R-Precision, the uniform and expensive training cost [28].

Recall-Precision (R-Precision) calculates the proportion of relevant documents retrieved among the top R retrieved documents, where R is the total number of relevant documents (total recall) for a given category or topic: a recall target of 80% is common in eDiscovery; 95% in systematic reviews.

The review cost is characterised by the cost structure $(\alpha_p, \alpha_n, \beta_p, \beta_n)$, where: α_p and α_n are the unit cost of training a classifier on reviewed relevant and irrelevant documents respectively during the first phase; β_p and β_n are the unit cost of reviewing relevant and irrelevant unreviewed documents needed to attain the target recall post active learning training, i.e., in the second phase. The review cost is then computed as the cumulative product of the cost structure coefficients and the corresponding document numbers: $\alpha_p t_p + \alpha_n t_n + \beta_p m_p + \beta_n m_n$, where t_p and t_n are the number of relevant and irrelevant documents reviewed for training the classier at the first phase, while m_p and m_n are the number of remaining relevant and irrelevant documents to obtain the target recall in the second phase. Following the original study, we report the minimal total cost observed over the 20 active learning iterations and set the target recall to 80% [28]. We also consider two cost structures: (1,1,1,1) referred to as uniform, and (10,10,1,1) referred to as expensive (a cost of 10 is assigned to the training process, leaving the rest unchanged).

We report the statistical significance of the differences between methods using a paired t-test ($\alpha = 0.05$) with Bonferroni correction, as in the original study.

3.4 Parameters Setting and Experiment Environment

For executing the active learning process we use 3 HPC server nodes, each equipped with 3 NVIDIA A100 GPUs with 80 GB of memory per GPU for tokenization, further pre-training, fine-tuning, and inference. For the baselines using logistic regression, we run on a 48-core AMD CPU. The hyperparameters for BERT and logistic regression are set based on the original paper; further details are provided in the online documentation.

For the active learning pipeline, we use the same two sampling strategies as in the original paper, namely, relevance feedback and uncertainty sampling, with a query batch size of 200 for RCV1-v2 and Jeb Bush. However, for the CLEF collections, due to the large variation in the total number of documents for each topic, we set a moderate batch size value of 25; we did so informed by a previous work that executed activate learning for systematic reviews and considered a batch of 25 candidate documents as reasonable for the workflow [23].

The analysis was performed based on our understanding of the metrics, as there were no results files from the original authors to directly test and compare.

All of the results and analyses presented in this study are based on our own experiments using the materials described above.

4 Results

4.1 RQ1: Is the Goldilocks Finding Replicable?

Further Pre-training. Results related to RQ1 are reported in Table 1, where R-Precision values are obtained from the final active learning iteration of all categories in the two datasets. While the original study does not mention this being the case, it is reasonable to do so because of how they recorded the values: the R-Precision increases as the reviewed relevant documents are fixed and potentially more relevant documents are prioritised at the top position of the recording in next iterations.

Table 1. Results for RCV1-v2 and Jeb Bush dataset. In brackets for R-Precision: percentage difference between our results and those of the original study. Both uniform cost (Uni. Cost) and expensive training cost (Exp. Train.) values are presented as the relative cost difference between LR and BERT; the lower the value, the better. * indicates a statistically significant difference w.r.t. the baseline LR.

Dataset	Further Pre-training Epoch	R-Precision (↑)		Uni. Cost (↓)		Exp. Train. (↓)	
		Relevance	Uncertainty	Rel.	Unc.	Rel.	Unc.
In-domain RCV1-v2	LR	**0.754** (−4.32%)	**0.733** (−3.51%)	**1.000**	**1.000**	**1.000**	**1.000**
	0	0.688 (−8.51%)	0.740 (−2.08%)	*1.720	*2.059	*1.386	*1.516
	1	0.759 (+0.20%)	0.810 (+5.44%)	1.004	1.091	0.969	1.014
	2	0.771 (+1.52%)	0.822 (+7.32%)	0.935	1.053	0.935	0.987
	5	0.770 (+1.85%)	0.819 (+4.46%)	0.997	0.991	0.916	**0.946**
	10	**0.785** (+2.72%)	**0.838** (+9.52%)	**0.823**	**0.967**	**0.857**	0.952
Off-domain Jeb Bush	LR	**0.892** (−1.34%)	**0.866** (+1.04%)	**1.000**	**1.000**	**1.000**	**1.000**
	0	*0.528 (−27.07%)	*0.487 (−32.22%)	*6.471	*4.545	*3.609	*2.890
	1	*0.635 (−21.64%)	*0.569 (−30.21%)	*4.223	*3.239	*2.736	*2.317
	2	*0.678 (−16.54%)	*0.602 (−25.53%)	*3.439	2.691	*2.410	*1.976
	5	***0.706** (−12.90%)	***0.632** (−22.30%)	***2.748**	**2.509**	***2.079**	***1.858**
	10	*0.701 (−12.95%)	*0.600 (−26.35%)	*3.049	2.867	*2.199	*1.897

For RCV1-v2, the in-domain dataset, we achieved results that were largely on par or superior to those reported in the original study: the original values of R-Precision are shown in brackets in the table, along with the percentage change of our results w.r.t. the original. We find better results than the original when further pre-training occurs, regardless of the number of epochs. We instead found lower values than what was originally reported for the baseline and for the BERT classifier when no further pre-training is performed. With respect to the key findings, we identify that additional pre-training benefits the BERT model. In particular, the BERT model further pre-trained with 10 epochs showed the largest improvement in both R-Precision and uniform training cost. However, when uncertainty sampling is used under the expensive training cost, it showed a higher cost in terms of reviewing documents compared to the BERT model pre-trained for 5 epochs.

From our replication results for this dataset, we confirm the existence of a Goldilocks epoch, with which BERT can outperform the baseline model across all the metrics reported. In our case, this was found to be 10, contrasting with the original study (in which it was 5). Additionally, we noticed inconsistent performance of the active learning strategies on R-Precision and the two review costs. When comparing the top result under each strategy, uncertainty sampling outperformed relevance feedback on R-Precision but fell short on both uniform training cost and expensive training cost. While we were unable to replicate the original paper's precise values or their Goldilocks epoch as 5, which yielded optimal results in most settings according to their findings, our analysis offers complementary insights into the TAR workflow.

For Jeb Bush, the out-of-domain dataset, we find significant differences between the LR baseline and BERT. We also find that all our BERT results are sensibly lower than those in the original study, while for LR we only find a small difference (−1.34%) in R-Precision. Analysing our results, we observe that LR achieves high R-Precision values, and does not exhibit poor effectiveness in the out-of-domain dataset. Conversely, BERT shows large, significant losses compared to LR. For BERT, using 5 epochs for further pre-training provides the best results across all metrics – this is the Goldilocks epoch for this dataset. However, even at this Goldilocks epoch, BERT significantly trails behind LR in R-Precision, exhibiting an approximately 20% decrease across both active learning strategies.

In this case, our findings echo those of the original study: BERT, regardless of the extent of further pre-training, struggles to be an effective classifier on the Jeb Bush dataset. However, we reserve judgment on whether this is caused by the out-of-domain nature of the dataset later in our investigation.

The Goldilocks Epoch Varies Across Categories. The original study examined the impact of task characteristics on the Goldilocks epoch by examining the difficulty of categories in the RCV1-v2 dataset. We used the corresponding categories provided by the original authors, and we found the difficulty level is related to the number of relevant documents: 'Hard' means categories with less than 2000 relevant documents; 'Medium' refers to those with more than 2000 but fewer than 8000 documents, while 'Easy' denotes categories with more than 8000 relevant documents. Furthermore, each difficulty level is subdivided into three 'Prevalence' bins with further cut-off of relevant document numbers inside its range. As a result, each bin contains five distinct categories. Table 2 presents our reproduced results, showing the averaged relative cost differences compared to the baseline model for each category bin under the expensive training cost structure. Contrary to the original study, our results indicate higher values for some categories while having higher review costs compared to the baseline, marked with an upper arrow in the table. Since each bin only has five runs under each further pre-training epoch setting, we did not conduct statistical tests (neither did the original study).

Our results suggest that the BERT model is generally less effective in challenging categories for both relevance feedback and uncertainty sampling. Specifically, BERT is less effective in hard categories than in easy ones for relevance feedback, and it is not better than LR even with further pre-training in this case. In contrast, for uncertainty sampling, BERT provides an improvement on additional common prevalence under hard, on top of saving numbers of documents reviewed for training on all other easier categories. These findings suggest that a linear model may be more suitable for these scenarios, and we did not perform any special feature treatments to achieve these results. It is also important to note that we only randomly sampled the same number of documents as in the original study, so our results may differ due to the potentially different documents selected.

Table 2. Expensive Training (Exp. Train.) review costs for RCV1-v2 categories grouped in bins under the difficulty hierarchy. Values are the relative cost difference between the corresponding BERT models with different further pre-training epochs and the baseline LR model. Each bin under Prevalence contains 5 categories. The results are averaged over the five categories in each bin. ↑ shows results larger than both the original result and 1.

Difficulty	Prevalence	Relevance					Uncertainty				
		0	1	2	5	10	0	1	2	5	10
Hard	Rare	↑2.254	↑1.279	↑1.283	↑1.428	1.368	↑2.466	↑1.407	↑1.639	↑1.482	↑1.509
	Medium	↑2.229	↑1.982	↑2.206	↑2.326	1.592	↑1.893	↑1.289	↑1.204	↑1.422	↑1.294
	Common	↑1.949	↑1.462	↑1.519	↑1.292	1.381	↑1.050	0.734	0.729	0.655	0.883
Medium	Rare	↑1.339	0.814	0.799	0.706	0.728	↑1.557	0.937	0.879	0.866	0.814
	Medium	↑1.600	1.162	0.947	0.913	0.945	↑1.422	0.931	0.882	0.757	0.804
	Common	↑1.608	↑1.170	0.823	0.711	0.694	↑1.003	0.799	0.767	0.699	0.652
Easy	Rare	1.051	0.650	0.678	0.699	0.673	↑1.427	0.975	0.855	0.834	0.858
	Medium	1.348	0.881	0.894	0.822	0.862	↑1.513	↑1.038	↑1.031	0.965	0.976
	Common	0.756	0.694	0.610	0.543	0.505	↑1.078	0.898	0.886	0.837	0.833

Our findings also confirm that the trend of the Goldilocks epoch varies based on the difficulty and prevalence bins. However, our results do not reproduce the exact results of the original study. For example, we found that 2 epochs work best for the hard-medium bin instead of no further pre-training, and 1 epoch is the Goldilocks epoch for hard-rare, while the original study reported 5 epochs. These discrepancies suggest that the trend of Goldilocks is not dependent on the dataset and that it is difficult to determine a consistent Goldilocks epoch for further pre-training BERT.

Run-Time Analysis. In the original study, run-time was the only efficiency metric used. In our experiment, we observed an average run-time of 62 min for

RCV1-v2 and 56 min for Jeb Bush. The run-time includes BERT fine-tuning and inference on the dataset for one category, one further pre-training epoch setting and one active learning strategy. In contrast, LR took a mere 0.4 min per run. These times represent a substantial reduction from the original paper, which reported an average run-time of 18 h (or 1,080 min) per single active learning strategy run. After examining their code, we discovered that the original authors used fp32 precision computing throughout their entire experiment. Initially, we followed this method but found that it largely prolonged the run-time to about 3 to 4 h per run on our machine. Consequently, we switched to using fp16 precision, which considerably accelerated the process without noticeably affecting the results. Nonetheless, we agree with the original conclusion that when integrating BERT into the TAR workflow, its performance in terms of run-time cannot rival that of simpler baseline models like LR.

4.2 RQ2: Does the Goldilocks Finding Generalise?

Further Pre-training. Next, we consider our new results on the CLEF datasets, which are out-of-domain w.r.t. to the BERT backbone. Results are reported in Tables 3 and 4: the BERT-based model consistently outperforms the baseline in terms of R-Precision across almost all datasets, given the Goldilocks epoch and the corresponding active learning strategy. However, the specific pattern of the Goldilocks epoch becomes less distinct when focusing solely on R-Precision. Furthermore, it is noteworthy that the choice of active learning strategy influences the characteristics of the Goldilocks epoch. For example, CLEF 2017 train, CLEF 2018 test, and CLEF 2019 dta test exhibit identical epochs for both strategies, albeit with variations in the exact epoch number. Conversely, CLEF 2017 test and CLEF 2019 intervention test display divergent epochs under each strategy. In summary, we did not observe the 'domain mismatch' problem on CLEF TAR collections when adapting the BERT-base model to this task. We confirm that further pre-training works in TAR workflow as the performance can be improved when the number of further pre-training epochs increases among all the metrics.

Aside from investigating the Goldilocks epoch for the BERT-base model on the CLEF collections, we found that active learning strategies also affect final performance with the same epoch for further pre-training. Relevance feedback shows the tendency to identify more relevant documents during active learning and displays the least effort for screening the remaining documents after active learning has concluded. This is the same for review cost in the expensive training setting, as relevance feedback has the potential to rank unreviewed documents better for second-stage review with fewer reviewed documents to train. However, applying BERT requires reviewing more documents for training than the baseline LR model to obtain the same target recall, which is a significant drawback for practical applications.

Run-Time Analysis. For the CLEF collections, the run-time for each topic ranges from ≈2.75 min (topic CD010355 with 43 documents, a rarely small number across these collections) to 42 min (topic CD009263, with 79,782 documents) for each run (20 iterations, 2 active learning strategies). Inference time and training time depend on the number of documents associated with each topic. On topics with only a few documents, the cost is dictated by the training phase; in the last iteration (thus, the one that takes the longest) this is typically up to 20 s. On topics with a large number of documents, the cost is dictated by the inference phase; on the largest set of documents (CD009263), it takes ≈49 s for inference per iteration, while training time is up to 17 s.

4.3 RQ3: Can We Address the Domain-Mismatch Problem?

Since we have observed that the BERT-base model can work properly in TAR with further pre-training, we want to know what is the gap between such a paradigm and a domain-specific pre-trained model. To investigate this, we consider BioLinkBERT, the state-of-the-art BERT-like backbone across biomedical benchmarks such as BLURB and MedQA-USMLE [29]. Results when using the BioLinkBERT backbone with no further pre-training on the CLEF TAR datasets are reported in Tables 3 and 4. We find that this domain-specific backbone can largely outperform the LR baseline and the BERT model across almost all its tested further-pre-training settings: in the CLEF 2017–2018 collections, it can achieve ≈0.80 R-Precision, while it achieves up to 0.93 R-Precision in CLEF

Table 3. Results for CLEF 2017–2018 collections. * indicates statistical significance differences w.r.t. the baseline LR.

Collection	Further Pre-training Epoch	R-Precision (↑)		Uni. Cost (↓)		Exp. Train. (↓)	
		Relevance	Uncertainty	Rel.	Unc.	Rel.	Unc.
CLEF 2017 train	baseline	**0.734**	**0.603**	**1.000**	**1.000**	**1.000**	**1.000**
	0	*0.612	0.598	1.916	1.508	*2.233	*2.122
	1	*0.674	0.679	1.231	0.926	*1.665	*1.654
	2	0.697	0.669	1.147	0.935	*1.575	*1.605
	5	0.711	0.670	**0.974**	0.902	*1.500	*1.574
	10	**0.722**	**0.680**	0.989	**0.847**	***1.446**	*1.543
BioLinkBERT-ep0		***0.838**	***0.761**	**0.727**	**0.606**	**1.174**	***1.332**
CLEF 2017 test	baseline	**0.782**	**0.657**	**1.000**	**1.000**	**1.000**	**1.000**
	0	0.727	0.722	*1.879	*1.774	*2.120	*2.003
	1	0.756	0.738	*1.471	*1.448	*1.710	*1.820
	2	0.746	0.748	*1.588	*1.409	*1.687	*1.780
	5	**0.776**	0.728	*1.460	*1.561	***1.578**	*1.756
	10	0.776	**0.762**	**1.424**	**1.331**	*1.601	***1.750**
BioLinkBERT-ep0		**0.812**	***0.794**	**1.016**	**1.058**	***1.545**	***1.645**
CLEF 2018 test	baseline	**0.754**	**0.644**	**1.000**	**1.000**	**1.000**	**1.000**
	0	0.694	0.695	1.655	1.863	*2.174	*2.280
	1	0.725	0.725	1.485	1.617	*1.947	2.115
	2	0.720	0.722	1.246	1.776	*1.674	2.158
	5	0.729	0.729	1.148	1.454	*1.668	*1.934
	10	**0.747**	**0.750**	**1.067**	**1.021**	***1.596**	***1.585**
BioLinkBERT-ep0		**0.793**	***0.780**	**0.669**	**0.801**	**1.222**	***1.436**

2019. For review costs, we observe a concurrent improvement when compared to the common BERT backbone. In the uniform cost setting, the BioLinkBERT backbone, without additional pre-training, uses fewer or an equal number of reviewed documents compared to LR. While it shows significant improvement under the expensive training setting, it still falls short of LR. However, considering the time spent on further pre-training with the BERT-base model, and the effort in identifying the Goldilocks epoch (which is impractical to pinpoint in real-life scenarios), choosing an appropriate model emerges as a simple and adequate solution - if such a model exists. These findings suggest that using a tailored pre-trained model, like the BioLinkBERT backbone in the case of the CLEF collections considered here, can improve the effectiveness of the model in the target domain, and remove the effort required in finding the Goldilocks epoch. Moreover, our results highlight the importance of selecting an appropriate pre-trained backbone for domain-specific tasks.

5 Discussion

In our study, we experimented with both original and domain-specific BERT-based models (i.e. BERT-base and BioLinkBERT) as the backbone of the active learning pipeline for TAR. BioLinkBERT is a BERT model pre-trained on the PubMed abstract corpus with additional citation links. While it includes the standard MLM, BioLinkBERT replaces the next sentence prediction (NSP) task

Table 4. Results for CLEF 2019 collections. * indicates statistical significance differences w.r.t. the baseline LR.

Collection	Further Pre-training Epoch	R-Precision (↑)		Uni. Cost (↓)		Exp. Train. (↓)	
		Relevance	Uncertainty	Rel.	Unc.	Rel.	Unc.
CLEF 2019 dta test	baseline	**0.823**	**0.742**	**1.000**	**1.000**	**1.000**	**1.000**
	0	0.775	0.803	1.856	1.795	3.549	3.520
	1	0.802	0.793	1.787	1.736	3.349	3.573
	2	0.804	0.798	2.144	1.709	3.508	**3.279**
	5	**0.838**	**0.833**	1.788	1.820	**2.993**	3.352
	10	0.791	0.818	**1.523**	**1.542**	3.020	3.444
BioLinkBERT-ep0		**0.909**	**0.857**	**0.979**	**1.109**	***2.188**	***2.555**
CLEF 2019 intervention train	baseline	**0.913**	**0.843**	**1.000**	**1.000**	1.000	**1.000**
	0	0.898	0.891	1.206	1.254	*1.908	*1.834
	1	0.903	0.893	1.068	1.311	*1.997	*1.892
	2	0.912	0.919	**1.015**	0.976	*1.833	***1.834**
	5	**0.924**	0.920	1.029	0.950	***1.816**	*1.851
	10	0.913	**0.929**	1.079	**0.934**	*1.853	*1.869
BioLinkBERT-ep0		**0.939**	0.923	**0.902**	**0.620**	***1.596**	***1.617**
CLEF 2019 intervention test	baseline	**0.855**	**0.759**	**1.000**	**1.000**	1.000	**1.000**
	0	0.813	0.785	*1.678	*1.417	*2.486	*2.457
	1	0.851	0.813	*1.370	1.272	*2.185	*2.264
	2	0.859	**0.860**	*1.308	1.089	*2.131	*2.275
	5	**0.872**	0.852	**1.247**	**1.060**	*2.030	***2.224**
	10	0.820	0.799	1.342	1.285	***2.027**	*2.312
BioLinkBERT-ep0		**0.934**	***0.900**	**0.869**	**0.773**	***1.748**	***1.967**

with the document relation prediction (DPR) task. DRP is specifically designed to understand the relationships between documents, which could involve linking concepts. In our experiment with the domain-specific CLEF TAR collections, BioLinkBERT outperformed the BERT-base model without requiring any further pre-training. It also surpassed the baseline logistic regression (LR), which the BERT-base model may not outperform, even at its Goldilocks epoch.

Recently, models in the DeBERTa series, especially DeBERTa-v3 [6], have demonstrated superior performance compared to the BERT model in classification tasks. DeBERTa-v3 introduces a key modification by replacing MLM with replaced token detection (RTD), an ELECTRA-style pre-training task [1] aims to train a discriminator to distinguish whether a token in the text has been replaced by the generator. We experimented with both DeBERTa-base [7] and DeBERTa-v3-base models as the backbone for the CLEF TAR collections. However, these did not show more promising results compared to BERT-like backbones. The results will be updated in our GitHub repository.

6 Conclusions

In this paper, we investigated an active learning pipeline for TAR based on BERT models through the reproduction of previously published experiments, and extension to a new TAR context, that of screening of studies for the creation of medical systematic reviews.

Our reproduction of the original study was not fully successful: we were not able to obtain the same results, which we suggest are largely dataset-specific and pre-processing-specific. For instance, the optimal number of further pre-training epochs (i.e. the Goldilocks) in our experiments differs from that of the original paper. However, our reproduction confirms the general findings of the original paper: that while there exists a Goldilocks epoch that is best used for further pre-training BERT-based models for TAR in an active learning setting, this epoch is hard to determine a priori and it largely depends on the dataset.

Our study further sheds light on the 'domain-mismatch' problem highlighted by the original work. Despite the notable conceptual differences between the CLEF collections and the pre-training corpora used to obtain BERT, we show that the BERT base model can still outperform the baseline in these collections when the Goldilocks epoch is identified. Additionally, our experiments showcase the effectiveness of a domain-specific pre-trained model, BioLinkBERT, in the TAR workflow when medical systematic reviews are considered. Remarkably, this model surpasses both BERT set to the Goldilocks epoch and the logistic regression baseline, without the need for any further pre-training, thus saving a consistent amount of additional computation.

Our results also resonated with previous studies [2,12] on the influences of active learning strategies: relevance feedback can benefit logistic regression more than uncertainty sampling. However, when using BERT for TAR, the choice of active learning strategy is not consistent across metrics, particularly on the RCV1-v2 and Jeb Bush datasets. Interestingly, on the CLEF-TAR collections, instead, relevance feedback consistently yields better results on R-Precision and the expensive training cost.

Overall, our findings suggest that the search for the Goldilocks epoch is a laborious way of improving the effectiveness of BERT-based classifier models in TAR. Instead, we suggest that considering the task's characteristics and identifying an appropriate pre-trained BERT-like backbone may be a simpler and more effective way to achieve better effectiveness in TAR tasks. We also notice a recent study [18] has shown these pre-trained models can further improve the active learning process with zero-shot rankings, which indicates a promising trend in their application on professional retrieval tasks such as systematic review screening.

Acknowledgment. Xinyu Mao is supported by a UQ Earmarked PhD Scholarship and this research is funded by the Australian Research Council Discovery Projects programme ARC DP DP210104043.

References

1. Clark, K., Luong, M.T., Le, Q.V., Manning, C.D.: Electra: pre-training text encoders as discriminators rather than generators. arXiv preprint arXiv:2003.10555 (2020)
2. Cormack, G.V., Grossman, M.R.: Evaluation of machine-learning protocols for technology-assisted review in electronic discovery. In: Proceedings of the 37th International ACM SIGIR Conference on Research & Development in Information Retrieval, pp. 153–162 (2014)
3. Devlin, J., Chang, M.W., Lee, K., Toutanova, K.: BERT: pre-training of deep bidirectional transformers for language understanding. arXiv preprint arXiv:1810.04805 (2018)
4. Grossman, M.R., Cormack, G.V.: Technology-assisted review in e-discovery can be more effective and more efficient than exhaustive manual review. Richmond J. Law Technol. **17**(3), 11 (2011)
5. Grossman, M.R., Cormack, G.V., Roegiest, A.: TREC 2016 total recall track overview. In: TREC (2016)
6. He, P., Gao, J., Chen, W.: DeBERTaV 3: improving DeBERTa using electra-style pre-training with gradient-disentangled embedding sharing. arXiv preprint arXiv:2111.09543 (2021)
7. He, P., Liu, X., Gao, J., Chen, W.: DeBERTa: decoding-enhanced BERT with disentangled attention. arXiv preprint arXiv:2006.03654 (2020)
8. Kanoulas, E., Li, D., Azzopardi, L., Spijker, R.: CLEF 2017 technologically assisted reviews in empirical medicine overview. In: CLEF 2017 (2017)
9. Kanoulas, E., Li, D., Azzopardi, L., Spijker, R.: CLEF 2018 technologically assisted reviews in empirical medicine overview. In: CEUR Workshop Proceedings, vol. 2125 (2018)
10. Kanoulas, E., Li, D., Azzopardi, L., Spijker, R.: CLEF 2019 technology assisted reviews in empirical medicine overview. In: CEUR Workshop Proceedings, vol. 2380 (2019)
11. Karpukhin, V., et al.: Dense passage retrieval for open-domain question answering. arXiv preprint arXiv:2004.04906 (2020)

12. Lewis, D.D.: A sequential algorithm for training text classifiers: corrigendum and additional data. ACM SIGIR Forum **29**, 13–19 (1995). ACM New York, NY, USA
13. Lewis, D.D., Yang, Y., Russell-Rose, T., Li, F.: RCV1: a new benchmark collection for text categorization research. J. Mach. Learn. Res. **5**(Apr), 361–397 (2004)
14. Lin, J., Nogueira, R., Yates, A.: Pretrained transformers for text ranking: BERT and beyond. Synth. Lect. Hum. Lang. Technol. **14**(4), 1–325 (2021)
15. Liu, Y., Lapata, M.: Text summarization with pretrained encoders. arXiv preprint arXiv:1908.08345 (2019)
16. Lupu, M., Salampasis, M., Hanbury, A.: Domain specific search. In: Professional Search in the Modern World: COST Action IC1002 on Multilingual and Multifaceted Interactive Information Access, pp. 96–117 (2014)
17. Michelson, M., Reuter, K.: The significant cost of systematic reviews and meta-analyses: a call for greater involvement of machine learning to assess the promise of clinical trials. Contemp. Clin. Trials Commun. **16**, 100443 (2019). https://doi.org/10.1016/j.conctc.2019.100443, https://www.sciencedirect.com/science/article/pii/S2451865419302054
18. Molinari, A., Kanoulas, E.: Transferring knowledge between topics in systematic reviews. Intell. Syst. Appl. **16**, 200150 (2022)
19. Radford, A., Wu, J., Child, R., Luan, D., Amodei, D., Sutskever, I.: Language models are unsupervised multitask learners (2019)
20. Raffel, C., et al.: Exploring the limits of transfer learning with a unified text-to-text transformer. J. Mach. Learn. Res. **21**(140), 1–67 (2020)
21. Roegiest, A., Cormack, G.V., Clarke, C.L., Grossman, M.R.: TREC 2015 total recall track overview. In: TREC (2015)
22. Salton, G., Buckley, C.: Improving retrieval performance by relevance feedback. J. Am. Soc. Inf. Sci. **41**(4), 288–297 (1990)
23. Singh, G., Thomas, J., Shawe-Taylor, J.: Improving active learning in systematic reviews. arXiv preprint arXiv:1801.09496 (2018)
24. Tonellotto, N.: Lecture notes on neural information retrieval. arXiv preprint arXiv:2207.13443 (2022)
25. Vaswani, A., et al.: Attention is all you need. In: Advances in Neural Information Processing Systems, vol. 30 (2017)
26. Yang, E., Lewis, D.D., Frieder, O.: On minimizing cost in legal document review workflows. In: Proceedings of the 21st ACM Symposium on Document Engineering, pp. 1–10 (2021)
27. Yang, E., Lewis, D.D., Frieder, O., Grossman, D.A., Yurchak, R.: Retrieval and richness when querying by document. In: DESIRES, pp. 68–75 (2018)
28. Yang, E., MacAvaney, S., Lewis, D.D., Frieder, O.: Goldilocks: just-right tuning of BERT for technology-assisted review. In: Hagen, M., et al. (eds.) ECIR 2022. LNCS, vol. 13185, pp. 502–517. Springer, Cham (2022). https://doi.org/10.1007/978-3-030-99736-6_34
29. Yasunaga, M., Leskovec, J., Liang, P.: LinkBERT: pretraining language models with document links. arXiv preprint arXiv:2203.15827 (2022)

Optimizing BERTopic: Analysis and Reproducibility Study of Parameter Influences on Topic Modeling

Martin Borčin[1,2](✉) and Joemon M. Jose[1]

[1] University of Glasgow, Glasgow G12 8QQ, Scotland, UK
martin.borcin@gmail.com
[2] Gerulata Technologies, 841 04 Bratislava, Slovakia
https://www.gerulata.com

Abstract. This paper reproduces key experiments and results from the BERTopic neural topic modeling framework. We validate prior findings regarding the role of text preprocessing, embedding models and term weighting strategies in optimizing BERTopic's modular pipeline. Specifically, we show that advanced embedding models like MPNet benefit from raw input while simpler models like GloVe perform better with preprocessed text. We also demonstrate that excluding outlier documents from the topic model provides minimal gains. Additionally, we highlight that appropriate term weighting schemes, such as $\sqrt{TF}\text{-}BM25(IDF)$, are critical for topic quality. We manage to reproduce prior results and our rigorous reproductions affirm the effectiveness of BERTopic's flexible framework while providing novel insights into tuning its components for enhanced topic modeling performance. The findings offer guidance and provide insightful refinements and clarifications, serving as a valuable reference for both researchers and practitioners applying clustering-based neural topic modeling.

Keywords: BERTopic · Neural Topic modeling · Semantic Document Embeddings · Semantic Clustering · HDBCSAN · Term Weighting · *TF-IDF* · Topic Coherence · Topic Diversity

1 Introduction

1.1 Motivation for Reproducing BERTopic Experiments

Topic modeling is key in unsupervised text analysis, facilitating data exploration by uncovering latent topics. Topic modeling plays a pivotal role in information retrieval applications by automatically uncovering latent themes within vast text corpora, aiding in efficient document categorization and content recommendation. The recent advancements in the field of neural networks, especially the advent of advanced language models, have enhanced the extraction of nuanced textual semantics, which has driven significant improvements in topic modeling methods. BERTopic [4], a notable clustering-based framework introduced in

N. Goharian et al. (Eds.): ECIR 2024, LNCS 14611, pp. 147–160, 2024.
https://doi.org/10.1007/978-3-031-56066-8_14

2022, combines semantic embeddings and clustering for efficient topic identification. Its adaptability and effectiveness have driven widespread use[1].

However, the vast configuration space resulting from its flexibility poses a challenge. The initial BERTopic proposal offered limited experimental details and evaluations, leaving potential biases unaddressed and many configurations unexplored. A claim was made that BERTopic's performance would be augmented by enhancements in its sub-components like document-embedding models. Our study aims to both validate these claims and extend the original experiments. We explore a wider array of configurations to ascertain BERTopic's adaptability and robustness. This approach aims to strengthen BERTopic's position in neural topic modeling and offer empirical guidance for optimal configuration.

1.2 Model Overview

BERTopic [4], as a clustering-based neural topic modeling framework, consists of five main stages: (i) **Document Embedding**: BERTopic first generates semantic vector representations of input documents using pretrained language models like SBERT; (ii) **Dimensionality Reduction**: Techniques like UMAP reduce the dimensionality of the embeddings while preserving structure; (iii) **Clustering**: Algorithms like HDBSCAN group the embeddings into semantic clusters; (iv) **Vectorization**: Documents are tokenized and converted into sparse bag-of-words vectors; (v)**Topic Representation**: A term-weighting scheme (like *c-TF-IDF*) identifies salient words in each cluster to produce topic representations.

1.3 Objectives and Contributions

Specifically, this work aims to provide insight into and answer these research questions:

- RQ1: Reproduce the results of the original BERTopic evaluation [4].
- RQ2: Investigate impact of text preprocessing on BERTopic's topic quality. We compare topics from raw vs. preprocessed text embeddings to understand the role of input cleaning.
- RQ3: Assess how embedding model quality affects BERTopic's performance. In relation to RQ2, we hypothesize that more capable language models may be limited by heavy preprocessing in prior work.
- RQ4: Quantify the influence of ignoring outlier documents when building topic representations on topic coherence. We hypothesize outlier removal significantly improves coherence.
- RQ5: Evaluate how different term-weighting schemes impact quality of topic representations. We expect that choice of strategy substantially affects topic interpretability and hypothesize that a sub-linear transformation of term-frequency improves topic quality.

[1] To this day, the original proposal [4] has over 560 citations. See https://github.com/MaartenGr/BERTopic for its popular Python implementation (602 forks and 4.8k stars on GitHub).

2 Background and Related Work

Sia et al. [12] pioneered using pipelines of semantic embeddings, clustering, and term-weighting as an alternative to conventional topic models. They showed that pre-trained language model embeddings of vocabulary terms clustered using weighted K-means and frequency-based term re-ranking for cluster representation can provide faster, more interpretable topics than the standard methods. Following their findings, Thompson and Mimno [14] showed clustering token-level instead of vocabulary embeddings better captures contextual differences. Specifically, they found that contextual BERT embeddings performed best.

Angelov [1] developed Top2Vec, significantly improving on prior work by directly clustering document-level embeddings. Additional improvements resulted from the use of more modern and sophisticated components in the pipeline. Key advantages included improved dimensionality reduction due to UMAP's [7] superior structure preservation and improved clustering due to HDBSCAN's [6] automatic topic number estimation and noise exclusion. Top2Vec used cluster centroid proximity as the metric to select the most representative terms.

Grootendorst [4] criticised the term-selection strategies of the previous approaches for their assumptions about the (spherical) spatial structure of the clusters and instead proposed the BERTopic framework. It is similar to Top2Vec but with a novel term-weighting scheme avoiding any spatial assumptions, but relying on term frequencies within clusters instead. Another key benefit is BERTopic's flexibility and modularity - it supports virtually any embedding model, dimensionality reduction and clustering method, and term-selection strategy. Our work investigates BERTopic's customizability for optimal topic modeling.

Previously, Zhang et al. [15] conducted an empirical study through their CETopic framework to systematically evaluate clustering-based topic modeling variants. A key question was whether quality embeddings alone obviate sophisticated neural topic models. They concluded high-quality embeddings combined with effective term-weighting makes clustering pipelines viable alternatives, consistently outperforming neural topic models and traditional approaches.

Our work reproduces key aspects of these frameworks, providing guidance for optimal use in practice.

3 Experiments

3.1 General Experimental Setup

Datasets and Preprocessing. To perform our experiment, we utilize two common topic-modeling benchmarks. These datasets match the ones used in the original evaluation [4] as well as other works assessing clustering-based topic modeling techniques [15]:

- 20 Newsgroups[2] (20NG): 16863 newsgroup posts with mean length of 52.45 (st. dev. 141.50) words and 2124-term vocabulary, discussing diverse topics such as religion, politics, science, and technology. A well-established research benchmark.
- BBC News[3] [3]: 2,225 BBC articles from 2004–2005 with mean length 156.20 (st. dev. 87.94) words and 4614-term vocabulary. Discusses 5 broad topics: business, entertainment, politics, sport, and technology. Less varied than the 20 Newsgroups dataset and generally contains longer documents.

Prior studies used the OCTIS [13] library to evaluate their models. This library only provides datasets in their preprocessed form. However, in our experiments, we also need to work with their raw versions. Thus, instead of utilizing this library, we will preprocess the datasets independently. Our experiments differ with the original evaluation in this respect.

To generate clean documents following standard IR approaches, we preprocess using: (i) Tokenization with SpaCy[4]; (ii) Lemmatization; (iii) Stopword removal; (iv) Exclusion of punctuation, digits, and whitespace; (v) Removal of short (<3 chars.) and long (>14 chars.) tokens; (vi) Filtering of non-alphanumeric tokens; (vii) Removal of rare tokens (global freq. <0.5%); (viii) Discarding documents with <5 tokens.

Evaluation Measures. Evaluation of topic models at scale is a difficult task, as the interpretability of topics is a highly subjective measure. However, quantitative metrics that can serve as proxies for human judgement have been developed to enable automatic evaluation of topic models [5,8]. We assess the quality of topics generated by BERTopic as opposed to the model's inner document representations. This aligns with BERTopic's goal of producing interpretable semantic clusters and with the evaluation strategies employed by previous evaluation attempts. Specifically, we employ two complementary metrics:

- **Topic Coherence (C)** captures topical interpretability based on word co-occurrence statistics. We utilize two variants, C_V [11] and C_{NPMI} [5]. Both have shown high correlation with human judgement [11] and were used in evaluations of clustering-based topic models [4,15].[5] To ensure comparability with prior research, we compute term co-occurrence using the modeled dataset as the background dataset. This can bias the results towards higher coherence than would be expected if a large external dataset (e.g. Wikipedia) was used.

[2] http://qwone.com/~jason/20Newsgroups/.
[3] http://mlg.ucd.ie/datasets/bbc.html.
[4] https://spacy.io/usage/processing-pipelines#processing.
[5] We have used the Gensim library [16] to implement these measures, but added a slight technical modification. We avoid the re-computation of the term co-occurrence matrix between runs by pre-computing and reusing it. This allows for faster evaluation of outputs of multiple models on the same data.

- **Topic Diversity (TD)** measures lexical diversity across topics. It is computed as the percentage of unique terms across all topic representations [2]. Higher TD indicates more distinct topics with less word overlap.

We use C_{NPMI} and TD to exactly match the original evaluation setup [4]. In addition, we use C_V, which shown even higher correlation with human judgement than C_{NPMI} [11].

3.2 Effect of Embedding Models and Input Preprocessing

I this experiment, we attempt to reproduce the core experiment of the original evaluation [4]. We assess BERTopic's performance in relation to the quality of semantic document embeddings, influenced by both input preprocessing (RQ2) and the embedding model (RQ3). Examining the performance over both raw and preprocessed text, our study both reproduces and extends beyond prior, original evaluations that exclusively used preprocessed datasets, which contrasts the practical use-case of BERTopic. This approach aims to reveal insights into optimizing text inputs for varied embedding models, enhancing the understanding of their combined effects.

Table 1. Summary of model capabilities.

Model	Quality	Speed	Size
MPNet - *best*	69.57	2800	420
MiniLM - *default*	68.06	14200	80
GloVe - *baseline*	49.79	34000	420

Experimental Setup. The selected models are presented in Table 1. The Quality score is defined by the authors of the models as "Average performance on encoding sentences over 14 diverse tasks from different domains" and is in unknown units, but higher is better. Speed is in sent./sec. and size in MB.[6] We reproduce the original evaluation [4] by using the MPNet[7] and MiniLM[8] models. The latter serves as the current default embedding model in BERTopic. In addition, we also evaluate using the GloVe[9] model, which serves as a baseline due to its use in evaluating sentence encoders [10].

The original evaluation [4] used preprocessed versions of documents to generate embeddings using the MiniLM and MPNet models (and two other embedding models). We reproduce this setup and, in addition, we use another baseline model (GloVe), and also construct embeddings from the raw versions of documents, yielding a total of 6 sets of embeddings for each dataset (two types for each of the three models). We fit BERTopic to each embedding set and evaluate its performance across varying topic counts: 10, 20, 30, 40, 50, matching the

[6] For more information, see https://www.sbert.net/docs/pretrained_models.html.

[7] https://huggingface.co/sentence-transformers/all-mpnet-base-v2.

[8] https://huggingface.co/sentence-transformers/all-MiniLM-L6-v2.

[9] https://huggingface.co/sentence-transformers/average_word_embeddings_glove.6B.300d.

Table 2. The results for 'raw' document embeddings.

	20NG						BBC					
	C_V		C_{NPMI}		TD		C_V		C_{NPMI}		TD	
	mean	SD	mean	SD	mean	SD	mean	SD	mean	SD	mean	SD
MPNet	**0.747**	**0.023**	**0.184**	**0.011**	0.826	0.074	**0.770**	**0.012**	**0.189**	**0.014**	0.847	0.039
MiniLM	0.740	0.029	0.181	0.012	0.848	0.071	0.769	0.025	0.188	0.022	**0.854**	**0.023**
GloVe	0.746	0.023	0.182	0.012	**0.855**	**0.072**	0.731	0.018	0.163	0.015	0.832	0.032

Table 3. The results for 'preprocessed' document embeddings.

	20NG						BBC					
	C_V		C_{NPMI}		TD		C_V		C_{NPMI}		TD	
	mean	SD	mean	SD	mean	SD	mean	SD	mean	SD	mean	SD
MPNet	0.749	0.050	0.186	0.022	0.868	0.075	0.752	0.017	0.180	0.018	0.852	0.026
MiniLM	0.744	0.052	**0.188**	**0.023**	0.866	0.077	**0.766**	**0.014**	**0.190**	**0.014**	**0.874**	**0.032**
GloVe	**0.764**	**0.027**	**0.188**	**0.012**	**0.879**	**0.064**	0.746	0.018	0.175	0.018	0.825	0.035

original experimental setup. We re-run the pipeline 10 times for each topic count (the original experiment used 3 runs) and compare resulting topic quality using the coherence (C_V, C_{NPMI}) and diversity (TD) metrics.

To enable fair comparison to prior work, all other BERTopic components remain in default configuration. This includes using preprocessed versions of documents for the vectorization and topic representation stages, isolating the joint impact of the embedding model and text preprocessing. Finally, a two-way ANOVA determines the statistical significance of our findings.

Results. Tables 2 and 3 shows the results of the evaluation on embeddings of raw documents and preprocessed documents respectively. Input preprocessing generally improved most of the metrics. The MPNet model achieved the highest average coherence on raw documents, while GloVe and MiniLM performed best with preprocessed ones. ANOVA revealed the choice of embedding model significantly impacted all topic quality metrics ($p < 0.05$). Post-hoc analysis showed MiniLM and MPNet consistently positively influenced coherence compared to GloVe. There was a significant interaction effect between embedding model and document type on coherence. *This suggests the relationship between embedding model and topic quality depends on whether raw or preprocessed text is used as input.* While embedding model significantly affected diversity, post-hoc tests found no pairwise differences between models. Despite the statistical significance of these results, the pipeline achieved very similar performance across all examined conditions, with most differences between configurations falling within one standard deviation from the mean.

Discussion: *We managed to reproduce the results of the original evaluation (RQ1)* [4], although under our settings the model consistently achieved slightly higher performance. Moreover, the results align with prior work indicating pipeline's robustness to the choice of embedding model [4]. This enables flexibility for users prioritizing speed or scale. Contradicting the hypothesis that larger, more capable models were limited by input preprocessing in prior work, our investigation reveals that for the purposes of topic modeling, *all embedding models generally benefit from input preprocessing (RQ2)*.

The results confirm our hypothesis that *embedding model quality significantly impacts BERTopic's topic coherence and diversity (RQ3)*. However, the differences in metrics are relatively small (e.g. the largest difference in C_V is on the BBC dataset, between GloVe and MPNet, and is only 0.039). The advanced MPNet model achieved the highest raw text coherence indicating its sensitivity to the nuances of language. In contrast, the simpler, yet fast GloVe model performed well on preprocessed text, likely better capturing broad themes present in the datasets. Measurements also confirm that MiniLM's balance of quality regardless of input preprocessing and speed make it a sensible and robust default choice.

A potential limitation of this experiment is that BERTopic struggled to generate expected topic counts in some settings, which potentially skews the comparisons[10]. This is a known issue also reported by Zhang et al. [15].

3.3 Effect of Exclusion of Outlier Documents

During the clustering phase, by default, BERTopic leverages HDBSCAN and its ability to automatically exclude some datapoints from clusters as 'outliers'. The original proposal [4] expected that this feature would improve topic representations. In this section, we make a novel assessment of whether and how this capability affects topic quality (RQ4). We compare BERTopic's performance on portions of the dataset with varying amounts of 'outlier' documents included in the model. We also cross-check these results by fitting a baseline LDA model to the same datasets.

This 'inverse' experimental design reveals whether including more outliers degrades topic quality metrics. We expect a negative correlation between dataset proportion and coherence/diversity if eliminating outliers enhances performance. Comparing to LDA acts as a control to verify if potential benefits are specific to BERTopic's clustering approach and topic representation strategy or whether other topic modeling approaches could also benefit from the exclusion of outlier documents.

Experimental Setup. We first fit BERTopic on both evaluation datasets to identify 40 topics. We then iteratively decrease the proportion of documents labeled by HDBSCAN as noise using the default outlier reduction method built-in BERTopic. This assigns the 'outlier' documents into clusters based on the

[10] It was only able to generate at most 40 and 41 topics using the MiniLM and GloVe embeddings of the raw BBC dataset, and 41 using the GloVe embeddings of the preprocessed BBC dataset.

Table 4. The minimal proportion of dataset included in the topic model and corresponding measurements of topic quality averaged over 3 runs. Around 81% of the BBC dataset was already included in the model prior to the first reduction step, it had no effect.

	%	20NG			BBC		
		C_V	C_{NPMI}	TD	C_V	C_{NPMI}	TD
BERTopic	70	0.736	0.181	0.773	(0.796)	(0.206)	(0.818)
	80	**0.740**	**0.182**	0.779	**0.796**	**0.206**	**0.818**
	90	0.736	0.179	0.786	0.785	0.198	0.803
	99	0.731	0.175	**0.796**	0.779	0.196	0.802
LDA	70	0.596	0.084	0.869	0.463	−0.038	0.728
	80	**0.610**	**0.089**	**0.894**	**0.492**	**−0.016**	0.721
	90	0.601	0.086	0.875	0.478	−0.027	0.743
	99	0.597	0.087	0.893	0.470	−0.033	**0.755**

cosine distance of their *c-TF-IDF* representations. We dynamically adjust the 'threshold' parameter to achieve the desired proportions of outlier documents merged into clusters. In parallel, we use LDA[11] to model the same datasets while also excluding the same documents that were identified as outliers. We evaluate topic coherence (C_V, C_{NPMI}) and diversity (TD) of both LDA and BERTopic at outlier proportions of 30%, 20%, 10%, and 1%, averaging over 3 runs.

Results. BERTopic exhibited best coherence when 80% of dataset was retained (20% of data, marked as outlier, was removed). However, TD showed best performance for almost 100% of data. Consistently coherence measures (C) showed improvements on both datasets and on both models at 20% data reduction, however, TD performance showed variations. This suggests including outlier documents in clusters hurts BERTopic's performance.

Overall, not unexpectedly, BERTopic outperformed LDA, with wider score spreads on BBC. The only case where LDA scored higher than BERTopic was in TD on 20 Newsgroups. The two models disagree on the direction of influence on TD on the BBC dataset: BERTopic is negatively influenced, while LDA benefited from the inclusion of outlier documents.

For BERTopic on the BBC dataset, the differences in performance were found to be statistically significant, but not on 20 Newsgroups. For LDA, the differences were not significant, except for TD on the BBC dataset. Despite reaching optimal performance at 80%, Table 4 shows that changes in absolute scores were minimal (at most 0.03 points). Quality metrics remained stable across outlier proportions for both models.

[11] We use the default implementation provided by the Gensim library [16]: https://radimrehurek.com/gensim/models/ldamodel.html.

Discussion. Our results indicate that clustering-based topic modeling is robust to outlier inclusion, aligning with past work that used algorithms like K-Means that lack native outlier handling [12,14,15]. *We confirm assumptions that HDBSCAN's noise handling boosts the quality of topic representations* [4] *(RQ4)*. Both models peaked when around 80% of dataset was included, implying exclusion of outlier documents may help. However, its benefits seem dataset-dependent and do not dramatically improve absolute metrics. A limitation of our study is that around 81% of the BBC dataset were already included in clusters after the first run of BERTopic. Therefore, values reported for the 70% and 80% threshold for BBC are identical.

These findings are useful for practitioners constrained by run-time or those that already know the desired number of topics in advance.

3.4 Effect of Term-Weighting Schemes

The presence of stopwords in topic representations greatly degrades their interpretability. Stopwords do not contribute meaning and occupy space that could be filled with informative words. BERTopic offers four term-weighting strategies, initially aimed at reducing stopword concentration in topic representations from unpreprocessed documents. Our study evaluates their efficacy in this role.

In addition to that, prior work has shown that weighting strategies can significantly influence topic quality [15], but no prior quantitative analysis exists for the strategies employed by BERTopic. The original evaluation only mentions the default term-weighting strategy and does not discuss other strategies. We fill this gap, extending our investigation to their impact on topic quality when applied to preprocessed texts devoid of stopwords (RQ5).

Experimental Setup Let

- $tf_{t,c}$ be the frequency of term t in cluster c,
- $N_c = \sum_{t'} tf_{t',c}$ be the size of cluster c,
- $gf_t = \sum_{c'} tf_{t,c'}$ be the global frequency of term t,
- $\overline{N} = \frac{1}{C} \sum_{c'} N_{c'}$ be the average cluster size, where C is the total number of clusters.

Then the studied term-weighing strategies are

1. $c\text{-}TF\text{-}IDF : w_{t,c} = \frac{tf_{t,c}}{N_c} \cdot \log\left(1 + \frac{\overline{N}}{gf_t}\right)$
2. $c\text{-}\sqrt{TF}\text{-}IDF : w_{t,c} = \sqrt{\frac{tf_{t,c}}{N_c}} \cdot \log\left(1 + \frac{\overline{N}}{gf_t}\right)$
3. $c\text{-}\sqrt{TF}\text{-}(IDF) : w_{t,c} = \frac{tf_{t,c}}{N_c} \cdot \log\left(1 + \frac{\overline{N} - gf_t + 0.5}{gf_t + 0.5}\right)$
4. $c\text{-}\sqrt{TF}\text{-}BM35(IDF) : w_{t,c} = \sqrt{\frac{tf_{t,c}}{N_c}} \cdot \log\left(1 + \frac{\overline{N} - gf_t + 0.5}{gf_t + 0.5}\right)$

Table 5. Selection of topic representations generated using different term-weighting schemes. Using the default scheme often gives completely useless results.

Weighting strategy	% stopwords	Topic representation
$c\text{-}TF\text{-}IDF$	90	**the to** windows **it and is for with have that**
$c\text{-}TF\text{-}BM25(IDF)$	90	windows **it with have for is my on this and**
$c\text{-}\sqrt{TF}\text{-}IDF$	10	windows modem card monitor printer dos mouse problem driver **my**
$c\text{-}\sqrt{TF}\text{-}BM25(IDF)$	0	windows modem card monitor printer dos mouse problem driver fonts
$c\text{-}TF\text{-}IDF$	100	**is that of to the you not it and in**
$c\text{-}TF\text{-}BM25(IDF)$	90	**is that you not it of** god **are this be**
$c\text{-}\sqrt{TF}\text{-}IDF$	40	god atheists atheism belief **not** believe **that is you** moral
$c\text{-}\sqrt{TF}\text{-}BM25(IDF)$	10	god atheists atheism belief believe moral islam argument **not** atheist
$c\text{-}TF\text{-}IDF$	100	**the of and in to were they was that by**
$c\text{-}TF\text{-}BM25(IDF)$	70	**were they was in by** armenian **that their** israel armenians
$c\text{-}\sqrt{TF}\text{-}IDF$	30	armenian israel armenians turkish **were** jews israeli **they was** arab
$c\text{-}\sqrt{TF}\text{-}BM25(IDF)$	20	armenian israel armenians turkish jews **were** israeli arab jewish **they**
$c\text{-}TF\text{-}IDF$	90	**the of that** god **is to and in not he**
$c\text{-}TF\text{-}BM25(IDF)$	80	god **that he not** jesus **is as was his in**
$c\text{-}\sqrt{TF}\text{-}IDF$	10	god jesus sin jehovah christ lord father **he** elohim bible
$c\text{-}\sqrt{TF}\text{-}BM25(IDF)$	0	god jesus sin jehovah christ lord father elohim bible church

Initially, clusters are formed using the embeddings of raw documents. In the first part of the experiment, the topic representations are formed using all four term-weighting strategies over raw documents. Then the representations are evaluated by measuring the concentration of stopwords. In the second part, the focus shifts to assessing the impact of these strategies on topic quality, employing metrics C_V, C_{NPMI} and TD. In this part, representations are formed using preprocessed documents. For both parts of the experiment, topic representations are examined at thresholds of 10, 20, 30, 40, 50, 60, and 100 topics. To account for the variability introduced by stochastic elements in the pipeline, each experimental setting is repeated five times.

Table 6. The influence of term-weighting schemes on topic quality measures.

	20NG						BBC					
	C_V		C_{NPMI}		TD		C_V		C_{NPMI}		TD	
	mean	SD	mean	SD	mean	SD	mean	SD	mean	SD	mean	SD
$c\text{-}TF\text{-}IDF$	0.710	0.033	0.166	0.016	0.784	0.101	0.783	0.027	0.196	0.020	0.844	0.027
$c\text{-}TF\text{-}BM25(IDF)$	0.714	0.034	0.168	0.017	0.793	0.102	0.789	0.024	0.199	0.018	0.855	0.033
$c\text{-}\sqrt{TF}\text{-}IDF$	**0.729**	**0.051**	**0.173**	**0.025**	0.834	0.106	**0.807**	**0.014**	**0.201**	**0.012**	0.938	0.027
$c\text{-}\sqrt{TF}\text{-}BM25(IDF)$	0.728	0.054	0.172	0.027	**0.835**	**0.105**	0.806	0.017	**0.201**	**0.013**	**0.940**	**0.027**

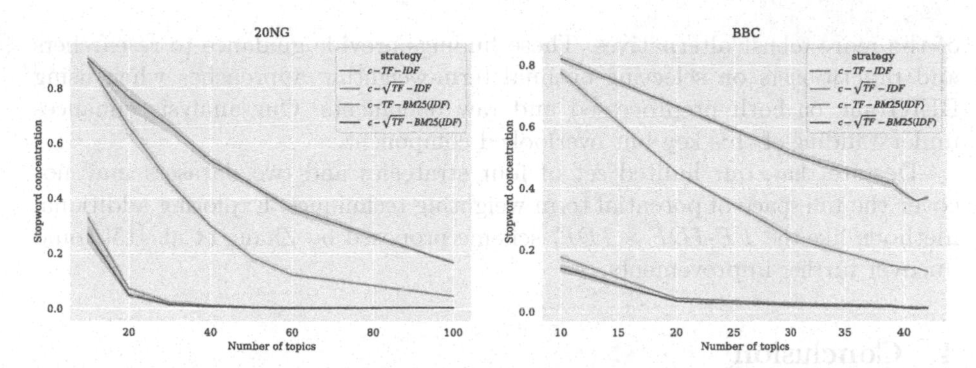

Fig. 1. Average percentage of topic representations that were stopwords on the 20NG (left) and BBC (right) datasets as a function of number of topics. BERTopic was not able to find more than 43 topics in the BBC dataset.

Results. Figure 1 shows the default *c-TF-IDF* scheme performed poorly, with very high stopword concentrations across the whole range of topic counts. Strategies that are using sub-linear frequency damping, namely $c\text{-}\sqrt{TF}\text{-}IDF$ and $c\text{-}\sqrt{TF}\text{-}BM25(IDF)$, substantially improved the scores, with some topics up to 90% points fewer stopwords compared to the default strategy (see Table 5).

Table 6 shows the results of the second half of the experiment. Very little difference between $c\text{-}\sqrt{TF}\text{-}IDF$ and $c\text{-}\sqrt{TF}\text{-}BM25(IDF)$ was observed. Both achieved higher coherence and diversity than the default strategy, with the former having slightly higher coherence and the latter slightly better diversity scores on both datasets. The $c\text{-}TF\text{-}BM25(IDF)$ scored second-lowest in a close tie with default $c\text{-}TF\text{-}IDF$, which achieved the worst performance across all datasets and evaluation metrics.

Discussion. The results of this study are consistent with previous literature, reinforcing the idea that term-weighting strategies play a critical role in the performance of clustering-based topic modeling approaches [9,12,15]. The sub-linear transformations of term-frequency (e.g. the square root) help reduce the impact of extremely frequent words, which improves topic quality. The schemes that employ such transformation showed exceptional resilience to stopwords while also improving coherence and diversity, outperforming their counterparts that lack this in both experiments. We also observe that this transformation is more effective in reducing stopword concentration than leveraging a BM25-inspired IDF term. This confirms our hypothesis that *term-weighting strategies have a significant effect on topic quality and damping term frequency enhances robustness and improves overall performance (RQ5).*

On the other hand, BERTopic's default strategy ($c\text{-}TF\text{-}IDF$), when used with raw input documents, suffered from extremely high stopword concentrations in topic representations, rendering practically useless. Additionally, it achieved the lowest topic quality metrics even if used with the preprocessed inputs. We would advise practitioners against its use and instead suggest they consider one of the more robust alternatives. These findings provide guidance to researchers and practitioners on selecting optimal term-weighting approaches when using BERTopic on both preprocessed and raw documents. Our analysis enhances understanding of this key but overlooked component.

Despite that, our limited set of four strategies and two datasets may not cover the full space of potential term weighting techniques. Exploring additional methods like the $TF\text{-}IDF \times IDF_i$ scheme proposed by Zhang et al. [15] could uncover further improvements.

4 Conclusion

This study systematically investigated the key components influencing BERTopic's topic generation performance. We studied the effect of embedding quality on topic quality and successfully reproduced the results of the original evaluation [4]. In addition, we studied the effect of input preprocessing, a factor overlooked in previous evaluations. Our insights reveal that even though preprocessing generally improves results, advanced models, notably MPNet, perform quite well with raw documents, whereas simpler embeddings like GloVe are advantageous with preprocessed data. The choice of embedding model and the presence of preprocessing do impact topic quality significantly.

Delving into noise handling, we identified that BERTopic's ability to automatically identify and exclude outlier documents is a notable feature and it was observed that the optimal performance was achieved when approximately 80% of the dataset is retained.

Notably, term-weighting strategies, especially the $\sqrt{TF}\text{-}BM25(IDF)$ approach, emerged as a critical determinant of topic interpretability, especially in cases of minimal input preprocessing. It was demonstrated that the current default term-weighting strategy lacks robustness to stopwords and that strategies that employ sub-linear damping of term-frequency produce higher-quality topic representations.

Collectively, this study reproduces and validates the results of core experiments form prior studies and, in addition, provides further insight into configuration of BERTopic, guiding both researchers and practitioners in harnessing its capabilities for optimal topic modeling results. Understanding BERTopic's configuration space strengthens the position of clustering-based topic models as a viable alternative to both standard and neural topic models. Their robustness, versatility and modularity makes them a valuable tool for diverse IR applications, and enables more accurate and useful results of unsupervised explorative analysis of large collections of documents.

References

1. Angelov, D.: Top2Vec: distributed representations of topics, August 2020. https://arxiv.org/abs/2008.09470, arXiv:2008.09470 [cs, stat]

2. Dieng, A.B., Ruiz, F.J.R., Blei, D.M.: Topic modeling in embedding spaces. Trans. Assoc. Comput. Linguist. **8**, 439–453 (2020). https://doi.org/10.1162/tacl_a_00325, https://aclanthology.org/2020.tacl-1.29
3. Greene, D., Cunningham, P.: Practical solutions to the problem of diagonal dominance in kernel document clustering. In: Proceedings of the 23rd International Conference on Machine Learning - ICML 2006, pp. 377–384. ACM Press, Pittsburgh, Pennsylvania (2006). https://doi.org/10.1145/1143844.1143892, https://portal.acm.org/citation.cfm?doid=1143844.1143892
4. Grootendorst, M.: BERTopic: neural topic modeling with a class-based TF-IDF procedure, March 2022. https://arxiv.org/abs/2203.05794, arXiv:2203.05794 [cs]
5. Lau, J.H., Newman, D., Baldwin, T.: Machine reading tea leaves: automatically evaluating topic coherence and topic model quality. In: Proceedings of the 14th Conference of the European Chapter of the Association for Computational Linguistics, pp. 530–539. Association for Computational Linguistics, Gothenburg, Sweden (2014). https://doi.org/10.3115/v1/E14-1056, https://aclweb.org/anthology/E14-1056
6. McInnes, L., Healy, J., Astels, S.: HDBSCAN: hierarchical density based clustering. J. Open Source Softw. **2**(11), 205 (2017). https://doi.org/10.21105/joss.00205, https://joss.theoj.org/papers/10.21105/joss.00205
7. McInnes, L., Healy, J., Melville, J.: UMAP: uniform manifold approximation and projection for dimension reduction, September 2020. https://arxiv.org/abs/1802.03426, arXiv:1802.03426 [cs, stat]
8. Newman, D., Lau, J.H., Grieser, K., Baldwin, T.: Automatic evaluation of topic coherence. In: Human Language Technologies: The 2010 Annual Conference of the North American Chapter of the Association for Computational Linguistics, pp. 100–108 (2010)
9. O'Callaghan, D., Greene, D., Carthy, J., Cunningham, P.: An analysis of the coherence of descriptors in topic modeling. Expert Syst. Appl. **42**(13), 5645–5657 (2015). https://doi.org/10.1016/j.eswa.2015.02.055, https://linkinghub.elsevier.com/retrieve/pii/S0957417415001633
10. Reimers, N., Gurevych, I.: Sentence-BERT: sentence embeddings using Siamese BERT-Networks, August 2019. https://arxiv.org/abs/1908.10084, arXiv:1908.10084 [cs]
11. Röder, M., Both, A., Hinneburg, A.: Exploring the space of topic coherence measures. In: Proceedings of the Eighth ACM International Conference on Web Search and Data Mining, pp. 399–408. ACM, Shanghai China, February 2015. https://doi.org/10.1145/2684822.2685324, https://dl.acm.org/doi/10.1145/2684822.2685324
12. Sia, S., Dalmia, A., Mielke, S.J.: Tired of topic models? Clusters of pretrained word embeddings make for fast and good topics too! October 2020. https://arxiv.org/abs/2004.14914, arXiv:2004.14914 [cs]
13. Terragni, S., Fersini, E., Galuzzi, B.G., Tropeano, P., Candelieri, A.: OCTIS: comparing and optimizing topic models is simple! In: Proceedings of the 16th Conference of the European Chapter of the Association for Computational Linguistics: System Demonstrations, pp. 263–270. Association for Computational Linguistics, Online (2021). https://doi.org/10.18653/v1/2021.eacl-demos.31, https://aclanthology.org/2021.eacl-demos.31
14. Thompson, L., Mimno, D.: Topic modeling with contextualized word representation clusters, October 2020. https://arxiv.org/abs/2010.12626, arXiv:2010.12626 [cs]

15. Zhang, Z., Fang, M., Chen, L., Namazi Rad, M.R.: Is neural topic modelling better than clustering? An empirical study on clustering with contextual embeddings for topics. In: Proceedings of the 2022 Conference of the North American Chapter of the Association for Computational Linguistics: Human Language Technologies, pp. 3886–3893. Association for Computational Linguistics, Seattle, United States (2022). https://doi.org/10.18653/v1/2022.naacl-main.285, https://aclanthology.org/2022.naacl-main.285
16. Řehůřek, R., Sojka, P.: Software framework for topic modelling with large corpora. In: Proceedings of the LREC 2010 Workshop on New Challenges for NLP Frameworks, pp. 45–50. ELRA, Valletta, Malta, May 2010

Does the Performance of Text-to-Image Retrieval Models Generalize Beyond Captions-as-a-Query?

Juan Manuel Rodriguez[1,2,5](✉), Nima Tavassoli[1], Eliezer Levy[3], Gil Lederman[3], Dima Sivov[3], Matteo Lissandrini[2,4], and Davide Mottin[1]

[1] Aarhus University, Aarhus, Denmark
davide@cs.au.dk
[2] Aalborg University, Aalborg, Denmark
jmro@cs.aau.dk
[3] Pnueli Lab, Tel Aviv Research Center Huawei Technologies, Tel Aviv, Israel
{eliezer.levy,gil.lederman,dima.sivov}@huawei.com
[4] University of Verona, Verona, Italy
matteo.lissandrini@univr.it
[5] UNCPBA-CONICET, Tandil, Argentina

Abstract. Text-image retrieval (T2I) refers to the task of recovering all images relevant to a keyword query. Popular datasets for text-image retrieval, such as Flickr30k, VG, or MS-COCO, utilize annotated image captions, e.g., "a man playing with a kid", as a surrogate for queries. With such surrogate queries, current multi-modal machine learning models, such as CLIP or BLIP, perform remarkably well. The main reason is the descriptive nature of captions, which detail the content of an image. Yet, T2I queries go beyond the mere descriptions in image-caption pairs. Thus, these datasets are ill-suited to test methods on more abstract or *conceptual queries*, e.g., "family vacations". In such queries, the image content is implied rather than explicitly described. In this paper, we replicate the T2I results on descriptive queries and generalize them to conceptual queries. To this end, we perform new experiments on a novel T2I benchmark for the task of conceptual query answering, called ConQA. ConQA comprises 30 descriptive and 50 conceptual queries on 43k images with more than 100 manually annotated images per query. Our results on established measures show that both large pretrained models (e.g., CLIP, BLIP, and BLIP2) and small models (e.g., SGRAF and NAAF), perform up to 4× better on descriptive rather than conceptual queries. We also find that the models perform better on queries with more than 6 keywords as in MS-COCO captions.

1 Introduction

Text-to-Image retrieval (T2I) aims to retrieve images that answer a user keyword query [3]. Recent methods based mainly on deep learning models such as CLIP [25], BLIP [19], and BLIP2 [18] achieve state-of-the-art performance on such a task by retrieving the relevant image 68% of the time. These models

N. Goharian et al. (Eds.): ECIR 2024, LNCS 14611, pp. 161–176, 2024.
https://doi.org/10.1007/978-3-031-56066-8_15

jointly learn vector embeddings from image-caption pairs, so the embedding of the image should be close to that of the caption describing that image.

Since the most adopted datasets are Flickr30k [33], VG [16], and MS-COCO [20], with only image-caption pairs, all evaluations implicitly treat captions as surrogate queries, overlooking that *a user query might not describe so explicitly the image content.* Moreover, these datasets contain at best N(captions)-to-1(image) T2I mapping but miss any 1-to-N mapping. This is again problematic for real applications in which a query usually aims to retrieve several images.

(a) *A pizza shown in an open box.*

(b) *A small propeller plane sitting on a field.*

Fig. 1. Examples of images, captions, and annotations in VG dataset.

For instance, existing T2I evaluations for Fig. 1b assume that image to be the only relevant image for the caption. Further, since they are trained only on mere descriptions of the images, they are unable to understand *conceptual queries* like "symbols of wars from the past" and potentially missing that image.

This paper focuses on **replicating** [1] **results in T2I** [10,18,19,25,35] under the lens of descriptive vs. conceptual queries. To this end, **we introduce a new dataset** for Conceptual Query Answering, ConQA, comprising both descriptive and conceptual queries with a length of three or four words. This length is similar to the average queries in Web search engines [14,22], and in the Google Trends for Google Images[1]. ConQA[2] comprises 80 queries on 43k images and over 100 annotated images per query using Amazon Mechanical Turk; 30 queries are descriptive, similar but shorter than those in previous datasets, and 50 queries are conceptual, an underrepresented type of queries in existing datasets.

Our extensive results highlight that state-of-the-art T2I models perform well when the text is a long image description, while the performance declines when the text is short or conceptual. These results confirm and extend those in a previous study [11] that highlighted deficiencies of T2I models, such as CLIP, in retrieving images containing a single object (e.g., a bird).

[1] Google Trends for Image search in 2023.
[2] The code and ConQA: https://github.com/AU-DIS/ConQA.

Contributions. In summary, we contribute with (i) A thorough replicability study of state-of-the-art T2I models validating previously reported results; (ii) A new dataset, ConQA, that extends MS-COCO and VG annotations with 1-to-N query-image pairs, providing a new benchmark for the T2I for both conceptual and descriptive queries; (iii) A reproducibility study on T2I models on ConQA short descriptive and conceptual queries. (iv) Important experimental findings showing the limitations of current T2I models those queries.

2 Related Work

Table 1. Main datasets for T2I and their characteristics. Annotation type: Caption (**C**), Object (**O**), Relationship (**R**), Segment (**S**), Attribute (**A**).

Dataset	Images	Relevant per Query	Conc. Queries	Annotation	Task
Flickr30k	31K	1	✘	C	Captioning
VG	108K	1	✘	O, A, R	Relationship Detection
MS-COCO	328K	1	✘	O, C, S	Obj. Det., Segmentation
GCC	3 300K	1	✘	C	Captioning, Retrieval
ConQA	43K	24.2	✔	O, C, A, R	Retrieval

Image retrieval aims to find images that match a query [3], whereby the query type reflects the task to solve. *Image-to-image* retrieval [32] returns images visually or semantically similar to a query image. *T2I* [13,31,34] returns images that match text queries or descriptions. *Hybrid* [36] retrieval finds images based on a combination of text, images, and additional annotations (e.g., semantic annotations). Here, we study the T2I task and describe current benchmarks' limitations to evaluate the T2I task.

Common Benchmarks in T2I. Two of the most popular datasets for evaluating image retrieval tasks are Flickr30k [33] and MS-COCO [20]. MS-COCO [20] consists of more than 328K images classified into 11 super-categories, e.g., "vehicle," divided into 91 categories e.g., "bicycle". MS-COCO is intended for object detection, segmentation, and captioning, as such images are paired with captions and annotated objects and segments. Similarly, Flickr30k [33] provides about 31K images crawled from Flickr; each image is associated with five captions. Flickr30k aims to analyze the semantic relationships and similarities between different captions. As such, MS-COCO and Flickr30k feature only one known correct image for each piece of text (caption or description). A more recent dataset for image captions is the Google Conceptual Caption (GCC) [29] that includes conceptual relations through synonyms and hypernyms. GCC consists of 3.3M images from the web, each with multiple captions, with 10.3 words on average. Yet, the image captions are quite detailed and descriptive. VG [16] is a dataset associating semantic annotations to objects in the image with a collection of 108*k* images from MS-COCO and YFCC100M [15] annotated with *scene* *graphs*. A *scene graph* is a graph with objects in the image as nodes and the relationships among them as edges, describing the image contents from the semantic point of view [21]. Images are manually annotated by more than 33 000 workers using crowdsourcing. VG images are still associated only with descriptive captions. Therefore, *none of the existing datasets are intended (nor appropriate) as a benchmark for the T2I task, since for a given textual clue we only have one image annotated as relevant (see Table* 1). Furthermore, these datasets implicitly encourage descriptive texts.

State-of-the-Art Models in T2I. The current best-performing methods for T2I are *vision-language models*, such as CLIP [25], BLIP [19], or BLIP2 [18]. Such models are large models with up to billions of parameters that require a large amount of training data and computational resources, e.g., some versions of CLIP require 500 V100 GPUs and 18 days of training. CLIP [25] employs a contrastive loss that promotes high similarity only for true image-caption pairs. BLIP [19] and BLIP-2 [18] are flexible models trained to perform several tasks. As CLIP, they are trained for the Image-Text Constastive (ITC) task, which allows fast image and text search by kNN search. Both models can also perform a fusion image-caption to estimate the probability that both are related, called Image-Text Matching (ITM). ITM is much slower than ITC, but it is more precise due to an attention mechanism that aligns the image and the text.

Besides large vision-language models, small *specialized models* [10,17,35] explicitly target T2I, often employing cross-attention mechanisms [7] to align words and image regions. SGRAF [10] leverages graph convolutional networks to model relationships between words and regions. NAAF [35] considers both matched and mismatched word-region pairs. These lightweight models train on consumer hardware. Yet, since such models are trained on MS-COCO or Flickr30k, it is unclear whether they generalize to more complex queries.

Diffusion Models. Diffusion models [27] are an orthogonal line of work aiming to solve a different task, namely generating images given textual descriptions. T2I retrieval can improve the results of such generative models [5] by including T2I models for similarity search as an initial filtering for image generation.

3 ConQA Creation

Despite recent advances in multi-modal representation learning, the replicability of the results and their generalizability when these methods are employed explicitly for the task of T2I have not been studied before. To provide a more robust test dataset for the T2I task, in this work, we design and collect new annotations for the popular VG [16] dataset to benchmark T2I methods. Then, we identify two query sets: descriptive and conceptual.

Image Selection. ConQA is a curated subset of VG 1.4. We select VG as it enriches MS-COCO with scene-graphs, which we use to select images with rich content. We filter images according to the following annotation quality criteria:

1. **Description**: The image should have one or more captions. Hence, we discarded the YFCC100M images with no caption, obtaining 91524 images from the MS-COCO subset of VG.
2. **Significance**: The image should be a non-trivial selection of scenes with at least two objects and one interaction. Notice that a picture of a single object, such as a car, can pass this filter if the graph annotates the parts of the objects and their relations, e.g., "license plates-on→bumper". We kept only images with *scene graphs* containing at least one edge, thus retaining 67 609 images.
3. **Coherence**: some VG edges contain noisy, erroneous, or meaningless annotations. In our case, we enforce that all relationships should be verbs and not contain nouns or pronouns. To detect this, we generated sentences for each edge as a concatenation of the words in the node labels and the relationship and applied Part of Speech tagging[3]. We remove all images with scene graphs that contain an edge not tagged as a verb or where the tag is not in an ad-hoc list of allowed non-verb keywords[4]. After filtering, we obtain 43 413 images.

Figure 1a depicts a discarded image as the *scene graph* has the edge "pizza has not been eaten yet", which contains the noun *pizza*. Figure 1b shows an image with an acceptable *scene graph*.

Query Generation. To ensure high-quality query-image relevance annotations, we devise a two-step approach.

The first step consists of generating queries for the dataset. Since there are no publicly available T2I query logs, the researchers created them. Six researchers, divided into three pairs, served as annotators to create both *conceptual* and *descriptive* queries. A descriptive query mentions some objects or actions in the image, such as "people chopping vegetables". While, a conceptual query does not mention specific objects or actions, but refers to a generic abstraction, e.g., "working class life". As a result, images with different objects can fit the query, e.g., a picture of an office or a factory ground can both be considered relevant. Notice that unlike GCC [29] that creates its "conceptual captions" by removing details of the original caption, e.g., removing proper names and object counts, or replacing a word by its hypernym, our conceptual queries are created independently of the images. As a result, while the GCC conceptual captions still describe the scene objects and relationships with fewer details than the original captions, *our conceptual queries do not prescribe a particular type of object or action in the scene.* The annotators also ensured that the queries are as realistic as possible. A post-hoc analysis confirms that the average number of words per query (3.4 words) is closer to the number of words reported for real query logs from Web search engines (2.56 words) [14] than the length of the average MS-COCO caption (11.26 words). The annotators generated 30 descriptive queries and 50 conceptual queries. After generating the queries, the annotators

[3] `en_core_web_sm` model in SpaCy [12].

[4] Allowed keywords: {top, front, end, side, edge, middle, rear, part, bottom, under, next, left, right, center, background, back, parallel}.

were tasked to search the dataset for images relevant to their query. To this end, the annotators used a prototype search engine using the "ViT-B/32" CLIP model [25]. The annotators could use ad-interim proxy or reformulated queries to find such relevant images. Anecdotally, we report that the annotators had to reformulate their queries several times before finding at least one relevant image. The annotators kept the queries for which they could define a non-empty initial relevant set.

Data Augmentation. We augment the initial set of images by adding the top-100 closest images in VG according to a pretrained ResNet152 model [9], meaning that the new images share visual features with the images in the original set. The purpose of this step is twofold, (1) evaluating a large amount of images, and (2) discriminating human- from machine-retrieved images.

Human Annotation. Finally, we set up a set of Human Intelligence Tasks (HITs) on Mechanical Turk (MTurk) [2], where each MTurk task consists of a query and 5 candidate images sampled from the augmented set of images per query. The workers are instructed to mark each image "Relevant" or "Irrelevant" for the given query[5]. They are also allowed to report "Unsure" when undecided. To reduce presentation bias, we randomize the order of images in each query. We also include validation tasks with control images to ensure a minimum quality in the annotation process, so workers failing 70% or more of validation tasks are excluded. Hence, we purposely present the worker images that the initial annotators manually tagged as completely irrelevant and only for descriptive queries. As a result, we employ 2 190 workers to tag 6.9*k* images with at least 3 workers annotating the same query-image pair. Each worker answered on average 2.5 HITs, 77% of workers performed at most 3 and only one performed 16 HITs.

Annotation Cost. We tag 100 images per query for a total of 80 queries. We show 5 images per HIT, and decided to obtain three judgments per query-image pair. Every time we show a descriptive query, one of the 5 images is replaced with a control image for that query. Therefore, to tag all the images for conceptual query we need 60 HITs and 75 for descriptive query. Since we pay USD 0.2 per HIT and Amazon charges us 10% of that, in total, the entire dataset tagging costs USD 1 155. Yet, since we run several preliminary versions of the HITs to validate the interface and the process before running the final task, the cost is about USD 1 500. Notice that this cost is only for Amazon Mechanical Turk, and it does not consider the time expended by the researchers defining the queries, analyzing the results of the tagging, and re-running HITs when needed.

4 ConQA Annotations and Content

After filtering the most inaccurate users, we obtain 6 941 images annotated as relevant/irrelevant for at least one query, and on average 100 images per query tagged as either relevant, non-relevant, or unsure by at least three workers. The queries have between 1 and 7 words with a median length of 3 words, and each

[5] An example is shown at https://benevolent-sawine-669286.netlify.app/.

has 100 annotated images as relevant/irrelevant by the workers. Finally, Fig. 2 shows that the images span several MS-COCO categories uniformly.

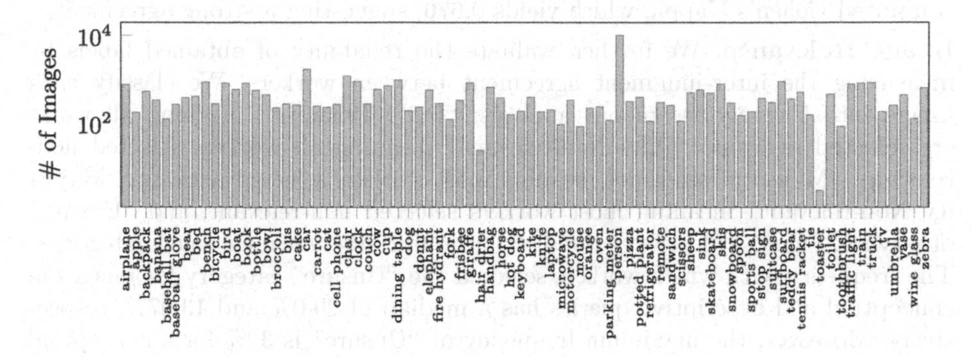

Fig. 2. Number of images per category in ConQA.

Table 2. Query depth in Wordnet hyponymy/hypernymy

Queries (#)	Depth		
	Min	Max	Average
ConQA Conceptual (50)	2.96 ± 1.94	6.73 ± 1.79	4.88 ± 1.57
ConQA Descriptive (30)	3.56 ± 2.30	8.03 ± 1.43	5.99 ± 1.35
MS-COCO captions (1.5M)	1.53 ± 0.88	9.72 ± 1.52	5.75 ± 0.93

Conceptual vs Descriptive Queries. We analyze the different abstraction levels of words in conceptual and descriptive queries. Hence, we measure the depth of the query words in the WordNet hyponym structure [23], which is a hierarchy where depth is related to concreteness as it goes from the most generic term [6] to the most concrete. We expect that the words in the conceptual queries have fewer nodes between them and a root concept, i.e. the most generic. Since queries have more than one word, Table 2 presents the average/minimum/maximum word depth averaged per query. The results show that conceptual query words are closer to the root node than descriptive query words. MS-COCO captions are similar to descriptive queries on average while the minimum and maximum values are, respectively, lower and higher than ConQA queries.

Validation Task. We study the annotation quality by analyzing the performance of the validation task. Figure 3 shows the correct/incorrect (unsure means the worker selected unsure as the answer) number of validation HITs with respect to the number of times a worker has been evaluated. Additionally, about one-third of the workers are not assigned any task with validation, but these are mostly workers who completed only 1 or 2 hits. Thus, it is reasonable to conclude that the vast majority of labels obtained are of high quality. To assess the inter-rater agreement between our evaluation and the workers' output, we computed Cohen's Kappa, which yields 0.676, suggesting a strong agreement.

Image Relevance. We further evaluate the reliability of obtained labels by measuring the inter-judgment agreement between workers. We classify each image into 5 categories in our analysis: "Fully Relevant" meaning all workers selected relevant, "Fully Non-relevant" meaning all workers selected non-relevant, "Majority Relevant" meaning most workers selected relevant, "Majority Non-relevant" meaning most workers selected non-relevant, and "Unsure" otherwise. Figure 4 illustrates the distribution of different mentioned categories. The frequency with which workers selected the "Unsure" category for images in conceptual and descriptive queries has a median of 20.0% and 13.97%, respectively. Moreover, the maximum frequency of "Unsure" is 32% for a conceptual query and 27.43% for a query in the descriptive category, showing that even humans tend to have difficulties when assigning relevance to some conceptual queries.

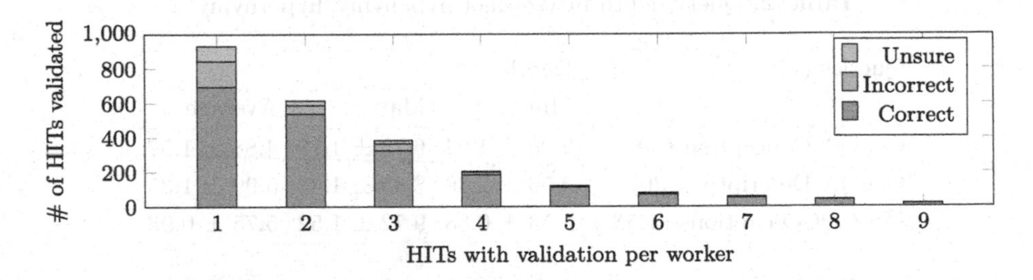

Fig. 3. Number of HITs with validation, only few control tasks failed.

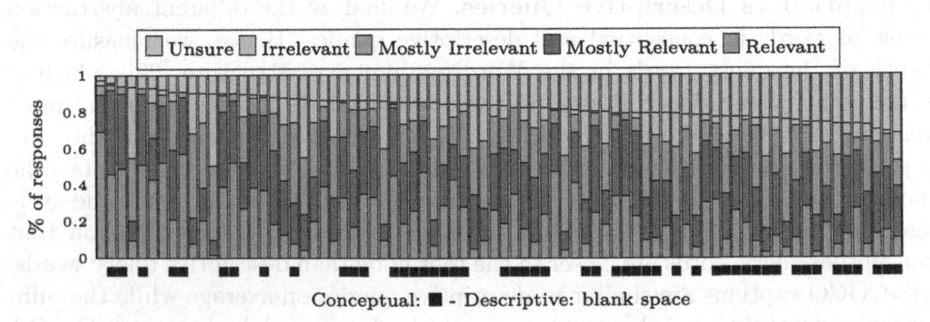

Fig. 4. Distribution of workers' responses per query.

5 Replicating and Generalizing T2I Experiments

We present the results of our experiments in two parts. First, we *reproduce* the T2I over MS-COCO-5K experiments as reported in the papers CLIP [25], BLIP [19], BLIP2 [18], SGRAF [10], and NAAF [35]. The second experiment aims at replicating the results in a zero-shot regime on the portion of the MS-COCO dataset for which we have annotations in ConQA. We additionally evaluate the models on ConQA conceptual and descriptive queries. We define reproducibility and replicability as they are defined by ACM in [1].

Models and Configurations. We evaluate three pretrained large vision-language models (LVMs) CLIP [25], BLIP [19], and BLIP2 [18], and two specialized T2I models, NAAF [35] and SGRAF [10]. In all the cases, we use the original implementations and pretrained weights. CLIP is trained on the WebImageText dataset (~400M images), while BLIP and BLIP2 are trained using a combination of datasets, including MS-COCO [20], VG [16], GCC [29], Conceptual [8], and SBU captions [24], resulting in more than 14M images. Further, LAION [28] is used as a source of noisy captions adding 100M images. For CLIP, we report the results for the ViT-L_14@336px model that attains the best results in the original paper [25]. For BLIP [19] and BLIP2 [18] we both use a version fine-tuned on MS-COCO for the reproducibility study (Sect. 5.1) and the pretrained version for the replicability experiment (Sect. 5.2). For BLIP and BLIP2, we use the ITC modality for obtaining an initial set of 128 images and then use ITM modality to re-rank them, as described in [18,19]. For SGRAF we use the pretrained model on MS-COCO for the reproducibility experiment (Sect. 5.1) and the model trained on Flickr30k for replicability (Sect. 5.2). For NAAF the only available pretrained model is on Flickr30k; for this reason we do not report results on reproducibility. The subscript FT next to each model's name indicates the version fine-tuned on MS-COCO.

Datasets. In the reproducibility analysis, we use MS-COCO-5K which comprises 5k images with 5 captions each used to test the model in the original papers [10,18,19,25,35]. For the replicability experiments, we report results on five datasets: CONC, DESC, MS-COCO-VG, GPT-CONC, and GPT-DESC. All datasets consist of the $6.3K$ images tagged by the MTurkers excluding the seed image annotated by the researchers. CONC and DESC are queries in ConQA in which an image is relevant for the query only if three MTurkers deem it relevant. MS-COCO-VG queries are the 5 MS-COCO captions for each image in the ConQA dataset. Note that while LVMs are trained with images from various sources, to the best of our knowledge they have not been exposed to the MS-COCO-VG queries tested in this work as we use the original hold-out set to test the models. The results in our reproducibility and replicability analyses seem to confirm this assumption. Finally, GPT-CONC and GPT-DESC are respectively the top-10 rephrasings of the CONC and DESC queries obtained using the GPT-J6B model [30] with prompt *"QUERY" can be rephrased as* to task the model to rephrase the query.

Measures. We report NDCG, R-precision, and Recall at 1, 5, and 10. NDCG evaluates the entire ranking by assigning a higher score to relevant images in the top positions. R-precision is the percentage of relevant images in the first R positions, where R is the number of relevant images for the query. For consistency with the T2I literature [10,17–19,25,35], here, we define Recall@K as the percentage of queries that return at least one relevant image within the top-K ranked images. We note that this definition assumes explicitly that each query (caption) can retrieve at most 1 relevant image. Our evaluation code adopts the ranx library [4], which follows TREC definitions, hence what we report as Recall@K here, it is called hit-rate@K in ranx.

5.1 Reproducing T2I Results

We first compare the experiments in the original papers and in the previous study [11]and highlight some differences among them. In the case of CLIP, the original paper presents a zero-shot experiment; while in the other cases the models (including BLIP and BLIP2) are fine-tuned on MS-COCO. Table 3 presents our reproduction of the T2I experiments as presented in the original works. The results on CLIP, BLIP_{FT} and BLIP2_{FT} are mostly consistent to those in the original papers [18,19,25]. Minor differences can be ascribed to technical differences in the platform as stated in the PyTorch documentation[6]. The most apparent difference is between our results and that of a previous evaluation [11] for the CLIP model that is depicted in the row "ECIR'23 Rep. CLIP" of Table 3. That work [11] reports on average 20% less Recall that originally reported for CLIP [25]. An inspection revealed that the ECIR'23 evaluation was conducted using the `ViT-L_14` in opposition to the `ViT-L_14@336px` used in the original CLIP paper and in our evaluation. Moreover, Sentence Transformer library [26] implementation was used in the ECIR'23 [11], while our experiments use the CLIP original implementation [25]. Therefore, we performed the experiment with `ViT-L_14` using both implementations. Our results show that the model and implementation differences do not account for the reported difference. Further analysis of the "ECIR'23 Rep. CLIP" code, shows that the authors used

Table 3. Results for T2I in MS-COCO-5K: our results vs. originally reported

	Reproduction			Originally reported		
	Recall@1	Recall@5	Recall@10	Recall@1	Recall@5	Recall@10
BLIP_{FT} [19]	65.1	86.3	91.8	65.1	86.3	91.8
BLIP2_{FT} [18]	67.9	87.5	92.4	68.3	87.7	92.6
CLIP [25]	37.1	61.6	71.5			
ECIR'23 Rep. CLIP [11]	21.9	40.2	49.8	37.8	62.4	72.2
CLIP ViT-L/14	36.5	61.0	71.1			
SGRAF_{FT} [10]	40.5	69.6	80.3	40.2	-	79.8

[6] https://pytorch.org/docs/stable/notes/randomness.html.

Table 4. Replicating results from ECIR'23 Rep. CLIP [11]

	Replication			ECIR'23 Rep.		
	Recall@1	Recall@5	Recall@10	Recall@1	Recall@5	Recall@10
CLIP ViT-L/14 [25]	21.9	40.7	49.9	21.9	40.2	49.8
ST-CLIP ViT-L/14 [26]	22.1	40.7	49.9			

an ad-hoc list of 20 252 images/captions from MS-COCO with only one caption per image instead of five captions per images, despite stating that they used the standard MS-COCO-5K [11]. Table 4 shows that we could replicate the results reported in [11] using the original CLIP implementation and the Sentence Transformer, called ST-CLIP in the table. Notice that, the MS-COCO based dataset

Table 5. Results for T2I Retrieval on Descriptive vs. Conceptual queries. The results with MS-COCO differ from those reported in previous work as we use 25% more testing examples and we do not fine-tune the models.

Method	Queries	NDCG	R-precision	Recall@1	Recall@5	Recall@10
BLIP2 [18]	MS-COCO-VG	**70.8**	**51.6**	**51.6**	**76.1**	**84.5**
	DESC	<u>48.0</u>	<u>15.3</u>	<u>20.7</u>	<u>51.7</u>	<u>62.1</u>
	GPT-DESC	40.2	9.5	12.8	32.4	43.4
	CONC	36.4	5.4	8.2	28.6	36.7
	GPT-CONC	32.1	3.4	3.9	16.1	25.3
BLIP [19]	MS-COCO-VG	**66.9**	**46.3**	**46.3**	**71.8**	**80.9**
	DESC	<u>46.2</u>	<u>15.3</u>	<u>20.7</u>	<u>58.3</u>	<u>62.1</u>
	GPT-DESC	39.3	10.4	13.4	32.8	48.6
	CONC	35.0	5.4	4.1	28.6	40.8
	GPT-CONC	31.6	3.8	3.1	14.5	22.0
CLIP [25]	MS-COCO-VG	**50.5**	**28.0**	**28.0**	<u>49.8</u>	<u>60.3</u>
	DESC	<u>47.8</u>	<u>16.5</u>	<u>20.7</u>	**58.6**	**65.5**
	GPT-DESC	40.2	10.1	12.1	34.5	45.2
	CONC	37.9	6.8	12.2	30.6	36.7
	GPT-CONC	32.4	3.7	5.3	15.5	24.3
NAAF [35]	MS-COCO-VG	<u>41.0</u>	**17.7**	**17.7**	**37.4**	**48.0**
	DESC	**41.5**	<u>10.6</u>	<u>13.8</u>	<u>34.5</u>	<u>44.8</u>
	GPT-DESC	36.7	7.3	9.0	26.6	33.8
	CONC	30.7	2.4	4.1	12.2	16.3
	GPT-CONC	29.2	1.9	3.7	8.2	12.2
SGRAF [10]	MS-COCO-VG	**39.6**	**16.5**	**16.5**	**35.6**	**46.1**
	DESC	<u>36.3</u>	<u>7.9</u>	<u>6.9</u>	<u>24.1</u>	<u>34.5</u>
	GPT-DESC	33.9	5.8	5.9	18.3	25.2
	CONC	28.4	1.3	0.0	8.2	10.2
	GPT-CONC	27.8	1.3	0.8	6.1	9.6

used in [11] (mistakenly labeled there as MS-COCO-5K), has 4x more images than MS-COCO-5K. Hence, the task is more challenging than in MS-COCO-5K as the task consist in ranking 20 252 images instead of only 5 000, which explains the difference in the results.

Table 6. Relative percentage improvement with p-value < 0.05 among pairs of queries and difference measures; values in bold: p-value < 0.01; X: not statistically significant; inf: the measure is 0 for the set of queries.

Method	Metrics	CONC DESC	CONC MS-COCO-VG	DESC MS-COCO-VG	GPT-CONC GPT-DESC	GPT-CONC MS-COCO-VG	GPT-DESC MS-COCO-VG
BLIP2 [18]	NDCG	**31.9**	**94.2**	**47.3**	**25.1**	**120.4**	**76.2**
	R-precision	**186.1**	**862.2**	236.3	**177.1**	**1412.1**	**445.7**
	Recall@1	X	**531.8**	**149.3**	**229.0**	**1230.1**	**304.2**
	Recall@5	81.0	**166.3**	**47.1**	**101.0**	**371.9**	**134.7**
	Recall@10	69	**130.0**	**36.1**	**71.7**	**233.9**	**94.5**
BLIP [19]	NDCG	**32.1**	91.1	**44.7**	**24.5**	**111.6**	**70.0**
	R-precision	181.5	**752.0**	X	**174.0**	**1117.7**	**344.3**
	Recall@1	406.9	**1034.3**	**123.8**	**339.3**	**1412.4**	**244.3**
	Recall@5	69	**151.2**	**48.7**	**126.1**	**395.4**	**119.1**
	Recall@10	52.1	98.3	**30.4**	**120.6**	**267.2**	**66.5**
CLIP [25]	NDCG	**26.2**	X	X	**23.9**	**55.9**	25.7
	R-precision	**142.9**	X	X	**173.5**	X	X
	Recall@1	X	**128.3**	X	**127.5**	**426.9**	**131.7**
	Recall@5	**91.5**	**62.5**	X	**122.3**	**220.8**	**44.3**
	Recall@10	**78.4**	**64.3**	X	**86.0**	**148.5**	**33.6**
NAAF [35]	NDCG	**35.3**	X	X	**25.6**	X	X
	R-precision	**333.4**	X	X	**295.1**	X	X
	Recall@1	X	**333.5**	X	**144.1**	**381.7**	**97.4**
	Recall@5	**181.6**	**205.1**	X	**225.3**	**357.6**	**40.7**
	Recall@10	**174.6**	**193.8**	X	**176.0**	**291.8**	**42.0**
SGRAFP [10]	NDCG	27.7	X	X	**22.1**	X	X
	R-precision	**504.7**	X	X	**335.8**	**1146.4**	X
	Recall@1	inf	inf	X	**618.1**	**1917.3**	**180.9**
	Recall@5	195.7	**335.7**	X	**198.5**	**480.9**	**94.6**
	Recall@10	**237.9**	**352.0**	X	**162.4**	**380.8**	**83.2**

Finally, we obtain results with SGRAF_{FT} slightly better than those reported in the original paper; our results refer to an improved model released by the authors in their code repository. As mentioned above, it is not possible to evaluate the NAAF model under the same conditions described in the paper [35].

5.2 Replicating and Generalizing T2I Results

We study the performance of T2I models under different query regimes. These results extend those of previous experimental evaluations [10,18,19,25], and pave the way to novel research in this field. Table 5 shows that the results vary significantly across query types.

Finding 1: The LVMs perform well in zero-shot MS-COCO-VG queries as opposed to specialized models. In zero-shot, the LVMs perform better than the small fine tuned NAAF and SGRAF. Yet, we see that the performance of the LVMs decreases in the ConQA queries compared to their original performance on MS-COCO-5K. This can be seen in BLIP, BLIP2 and CLIP Recall@1, 5, 10 in Table 3 with Table 5, e.g., BLIP2 Recall@1 on MS-COCO-5K is 67.9% and on MS-COCO-VG is 51.6%. Hence, while we conclude BLIP2 to be the best off-the-shelf method, and with superior zero-shot capabilities than small fine-tuned models, we still highlight difficulties in generalizing to other query sets.

Finding 2: The models are challenged by shorter queries. The models perform best on MS-COCO-VG queries that are similar to those employed in the model training. Yet, we observe a consistent performance drop on ConQA's queries, even those descriptive. Recall that ConQA has on average 24 relevant images per query and, despite such an easier case, most models do not identify relevant images in the top-5. This confirms our hypothesis that current models are less suited for shorter queries commonly found in web-search.
Finding 3: Current models struggle on conceptual queries. On conceptual queries that describe the content of the image in an abstract manner only, all models experience a significant setback up to 30% on all metrics compared to MS-COCO-VG queries. For example, Table 5 shows that BLIP2 Recall@1 on MS-COCO-VG is 51.6% and on conceptual queries is 8.2%.
Finding 4: Most of the results are statistically significant and exhibit a noticeable difference from MS-COCO-VG. We further report an experiment on the differences observed in Table 5. We perform a Mann-Whitney U test, a form of non parametric test to assess statistical significance of the alternative hypothesis that the model when retrieving queries in X got a lower score than those in Y. In our case, we compute differences for the following (X, Y) dataset pairs (Conc, MS-COCO-VG), (Desc, MS-COCO-VG), (GPT-Conc, GPT-Desc), (GPT-Conc, MS-COCO-VG), (GPT-Desc, MS-COCO-VG). We repeat the test for all measures and report the relative percentage improvement calculated as $(m_y - m_x)/m_x$ for a pair of measures (m_x, m_y). We report only the relative improvements for which the p-value < 0.05. The results in Table 5 confirm our findings. In particular, all models exhibit statistically significant differences in at least two metrics when comparing Conc with Desc and MS-COCO-VG queries. The Desc queries are also significantly different from those in MS-COCO-VG in the two largest models, BLIP and BLIP2. We evince that LVMs have a bias towards longer queries or image descriptions. Finally, we experience significant differences among GPT-paraphrased queries and MS-COCO-VG queries, showing that the difference is due to the content rather than the language.

6 Conclusions

We successfully reproduced and replicated results in the T2I task using LVMs, such as CLIP, BLIP and BLIP2, and specialized models trained on a specific dataset, such as SGRAF and NAAF. In particular, we obtained results within 1% of those reported for CLIP, BLIP, BLIP2, and SGRAF while, we could not reproduce NAAF due to the lack of pretrained model on MS-COCO. Moreover, we point out a limitation of a previous CLIP reproducibility study [11]. To perform a more systematic evaluation, we introduced ConQA, a dataset built on top of VG and MS-COCO datasets, that enriches the T2I task by adding conceptual queries that express the content of images in abstract terms. We found that small models, such as SGRAF and NAAF, are not able to generalize, even when training on Flickr30k and testing on MS-COCO. In contrast, BLIP and BLIP2 outperform CLIP even without fine-tuning. Furthermore, BLIP and

BLIP2 perform 1.5× better on MS-COCO-VG queries than ConQA descriptive queries. Hence, these models perform better on larger and more descriptive queries. Overall, we found out that all the models tend to perform better on descriptive queries rather than conceptual ones. Particularly, we found that some measures are 5× lower for conceptual than for ConQA's descriptive queries, and up to 10× than MS-COCO-VG queries. This shows a limitation of state-of-the-art models to generalize on conceptual queries. Our study provides evidence of unknown issues of current T2I approaches and paves the way to novel models that incorporate higher level abstractions to properly answer conceptual queries.

References

1. ACM (2020). Artifact review and badging - current. https://www.acm.org/publications/policies/artifact-review-and-badging-current
2. Aguinis, H., Villamor, I., Ramani, R.S.: MTurk research: review and recommendations. J. Manag. **47**(4), 823–837 (2021)
3. Alemu, Y., Koh, J.B., Ikram, M., Kim, D.K.: Image retrieval in multimedia databases: a survey. In: 2009 Fifth International Conference on Intelligent Information Hiding and Multimedia Signal Processing, pp. 681–689 (2009)
4. Bassani, E.: `ranx`: a blazing-fast Python library for ranking evaluation and comparison. In: Hagen, M., et al. (eds.) ECIR 2022. LNCS, vol. 13186, pp. 259–264. Springer, Cham (2022). https://doi.org/10.1007/978-3-030-99739-7_30
5. Blattmann, A., Rombach, R., Oktay, K., Müller, J., Ommer, B.: Retrieval-augmented diffusion models. In: Advances in Neural Information Processing Systems, vol. 35, pp. 15309–15324 (2022)
6. Bouras, C., Tsogkas, V.: W-kmeans: clustering news articles using WordNet. In: Setchi, R., Jordanov, I., Howlett, R.J., Jain, L.C. (eds.) KES 2010. LNCS (LNAI), vol. 6278, pp. 379–388. Springer, Heidelberg (2010). https://doi.org/10.1007/978-3-642-15393-8_43
7. Cao, M., Li, S., Li, J., Nie, L., Zhang, M.: Image-text retrieval: a survey on recent research and development. In: Proceedings of the Thirty-First International Joint Conference on Artificial Intelligence, IJCAI-2022, pp. 5410–5417 (7 2022). Survey Track
8. Changpinyo, S., Sharma, P., Ding, N., Soricut, R.: Conceptual 12M: pushing web-scale image-text pre-training to recognize long-tail visual concepts. In: Proceedings of the IEEE/CVF Conference on Computer Vision and Pattern Recognition (CVPR), pp. 3558–3568, June 2021
9. Chen, P., Liu, S., Jia, J.: Jigsaw clustering for unsupervised visual representation learning. In: Proceedings of the IEEE/CVF Conference on Computer Vision and Pattern Recognition (CVPR), pp. 11526–11535, June 2021
10. Diao, H., Zhang, Y., Ma, L., Lu, H.: Similarity reasoning and filtration for image-text matching. In: Proceedings of the AAAI Conference on Artificial Intelligence, vol. 35, pp. 1218–1226 (2021)
11. Hendriksen, M., Vakulenko, S., Kuiper, E., de Rijke, M.: Scene-centric vs. object-centric image-text cross-modal retrieval: a reproducibility study. In: Kamps, J., et al. (eds.) ECIR 2023. LNCS, vol. 13982, pp. 68–85. Springer, Cham (2023). https://doi.org/10.1007/978-3-031-28241-6_5
12. Honnibal, M., Montani, I., Van Landeghem, S., Boyd, A.: spaCy: Industrial-strength Natural Language Processing in Python, January 2022

13. Johnson, J., et al.: Image retrieval using scene graphs. In: 2015 IEEE Conference on Computer Vision and Pattern Recognition (CVPR), pp. 3668–3678 (2015)
14. Kacprzak, E., Koesten, L.M., Ibáñez, L.D., Simperl, E., Tennison, J.: A query log analysis of dataset search. In: Web Engineering, pp. 429–436 (2017)
15. Kalkowski, S., Schulze, C., Dengel, A., Borth, D.: Real-time analysis and visualization of the YFCC100m dataset. In: Workshop on Community-Organized Multimodal Mining: Opportunities for Novel Solutions, pp. 25–30, New York, NY, USA (2015)
16. Krishna, R., et al.: Visual genome: connecting language and vision using crowdsourced dense image annotations. Int. J. Comput. Vis. **123**, 32–73 (2017)
17. Lee, K.H., Chen, X., Hua, G., Hu, H., He, X.: Stacked cross attention for image-text matching. In: Ferrari, V., Hebert, M., Sminchisescu, C., Weiss, Y. (eds.) ECCV 2018. LNCS, vol. 11208, pp. 212–228. Springer, Cham (2018). https://doi.org/10.1007/978-3-030-01225-0_13
18. Li, J., Li, D., Savarese, S., Hoi, S.: BLIP-2: bootstrapping language-image pre-training with frozen image encoders and large language models (2023)
19. Li, J., Li, D., Xiong, C., Hoi, S.: BLIP: Bootstrapping language-image pre-training for unified vision-language understanding and generation. In: Proceedings of the 39th International Conference on Machine Learning. Proceedings of Machine Learning Research, vol. 162, pp. 12888–12900 (2022)
20. Lin, T.-Y., et al.: Microsoft COCO: common objects in context. In: Fleet, D., Pajdla, T., Schiele, B., Tuytelaars, T. (eds.) ECCV 2014. LNCS, vol. 8693, pp. 740–755. Springer, Cham (2014). https://doi.org/10.1007/978-3-319-10602-1_48
21. Lu, C., Krishna, R., Bernstein, M., Fei-Fei, L.: Visual relationship detection with language priors. In: Leibe, B., Matas, J., Sebe, N., Welling, M. (eds.) ECCV 2016. LNCS, vol. 9905, pp. 852–869. Springer, Cham (2016). https://doi.org/10.1007/978-3-319-46448-0_51
22. Luo, C., et al.: Query attribute recommendation at Amazon search. In: Proceedings of the 16th ACM Conference on Recommender Systems, RecSys 2022, pp. 506–508 (2022). https://doi.org/10.1145/3523227.3547395
23. Miller, G.A.: WordNet: a lexical database for English. Commun. ACM **38**(11), 39–41 (1995)
24. Ordonez, V., Kulkarni, G., Berg, T.: Im2Text: describing images using 1 million captioned photographs. In: Advances in Neural Information Processing Systems, vol. 24. Curran Associates, Inc. (2011)
25. Radford, A., et al.: Learning transferable visual models from natural language supervision. In: Proceedings of the 38th International Conference on Machine Learning. PMLR, vol. 139, pp. 8748–8763 (2021)
26. Reimers, N., Gurevych, I.: Sentence-BERT: sentence embeddings using Siamese BERT-networks. In: Proceedings of the 2019 Conference on Empirical Methods in Natural Language Processing, November 2019
27. Rombach, R., Blattmann, A., Lorenz, D., Esser, P., Ommer, B.: High-resolution image synthesis with latent diffusion models. In: Proceedings of the IEEE/CVF Conference on Computer Vision and Pattern Recognition (CVPR), pp. 10684–10695, June 2022
28. Schuhmann, C., et al.: LAION-400M: open dataset of clip-filtered 400 million image-text pairs (2021)
29. Sharma, P., Ding, N., Goodman, S., Soricut, R.: Conceptual captions: a cleaned, hypernymed, image alt-text dataset for automatic image captioning. In: Proceedings of the 56th Annual Meeting of the Association for Computational Linguistics, pp. 2556–2565 (2018)

30. Wang, B.: Mesh-Transformer-JAX: Model-Parallel Implementation of Transformer Language Model with JAX, May 2021. https://github.com/kingoflolz/mesh-transformer-jax
31. Wang, S., Wang, R., Yao, Z., Shan, S., Chen, X.: Cross-modal scene graph matching for relationship-aware image-text retrieval. In: Proceedings of the IEEE/CVF Winter Conference on Applications of Computer Vision (WACV) (2020)
32. Yoon, S., et al.: Image-to-image retrieval by learning similarity between scene graphs. AAAI **35**(12), 10718–10726 (2021)
33. Young, P., Lai, A., Hodosh, M., Hockenmaier, J.: From image descriptions to visual denotations: new similarity metrics for semantic inference over event descriptions. Trans. Assoc. Comput. Linguist. **2**, 67–78 (2014)
34. Yu, T., Fei, H., Li, P.: U-BERT for fast and scalable text-image retrieval. In: Proceedings of the 2022 ACM SIGIR International Conference on Theory of Information Retrieval, pp. 193–203 (2022)
35. Zhang, K., Mao, Z., Wang, Q., Zhang, Y.: Negative-aware attention framework for image-text matching. In: Proceedings of the IEEE/CVF Conference on Computer Vision and Pattern Recognition, pp. 15661–15670 (2022)
36. Zhao, Y., Song, Y., Jin, Q.: Progressive learning for image retrieval with hybrid-modality queries. In: Proceedings of the 45th International ACM SIGIR Conference on Research and Development in Information Retrieval, pp. 1012–1021 (2022)

Exploring the Nexus Between Retrievability and Query Generation Strategies

Aman Sinha, Priyanshu Raj Mall, and Dwaipayan Roy(✉)

Indian Institute of Science Education and Research Kolkata, Kalyani, India
{as18ms065,prm18ms118,dwaipayan.roy}@iiserkol.ac.in

Abstract. Quantifying bias in retrieval functions through document retrievability scores is vital for assessing recall-oriented retrieval systems. However, many studies investigating retrieval model bias lack validation of their query generation methods as accurate representations of retrievability for real users and their queries. This limitation results from the absence of established criteria for query generation in retrievability assessments. Typically, researchers resort to using frequent collocations from document corpora when no query log is available. In this study, we address the issue of reproducibility and seek to validate query generation methods by comparing retrievability scores generated from artificially generated queries to those derived from query logs. Our findings demonstrate a minimal or negligible correlation between retrievability scores from artificial queries and those from query logs. This suggests that artificially generated queries may not accurately reflect retrievability scores as derived from query logs. We further explore alternative query generation techniques, uncovering a variation that exhibits the highest correlation. This alternative approach holds promise for improving reproducibility when query logs are unavailable.

Keywords: Reproducibility · Retrievability · Empirical Study

1 Introduction

In the field of Information Retrieval (IR), the study of biases and fairness of retrieval models together with their effectiveness has been an active area of research [23,26]. Among the various metrics and methodologies developed, *retrievability* has emerged as a significant way of measuring how accessible a document is in a collection, given a retrieval system and a set of queries [8]. It reflects the degree to which the system favors certain documents while retrieval over others, regardless of their relevance. Retrievability can be used to evaluate the fairness, diversity, and coverage of retrieval systems, as well as to identify and analyze the factors that affect the accessibility of documents [7]. However,

A. Sinha and P.R. Mall—These authors contributed equally to this work.

N. Goharian et al. (Eds.): ECIR 2024, LNCS 14611, pp. 177–193, 2024.
https://doi.org/10.1007/978-3-031-56066-8_16

calculating the retrievability of a collection of documents is a complex task comprising several interconnected steps. Initially, a set of queries is selected based on which the retrievals are performed. Researchers have used query logs as well as artificially generated queries in this step. Subsequently, a retrieval model is configured with optimal parameters and the actual retrieval process follows, producing ranked document lists. Retrievability scores are then computed based on the rank of the documents in the retrieved lists. Finally, the computed scores are used to estimate "retrievability bias" by metrics like the Gini coefficient [27], Atkinson index [4], Theli index [30] or Palma index [39] borrowed from economics, to quantify bias and inequality (mainly in wealth) within a population.

Various query generation techniques employed at the initial stage can yield distinct query sets, potentially resulting in diverse retrievability scores for the same collection and retrieval system. Previous research has indicated that retrievability scores calculated using simulated queries differ from those obtained using actual queries extracted from query logs [45]. However, a comprehensive and systematic comparison of retrievability scores computed using various query simulation methods has yet to be undertaken. The discrepancy, if any, poses a challenge to the reproducibility of retrievability experiments as different researchers may employ different query generation methods, leading to diverse conclusions on the same collection. To investigate the potential impact on result reproducibility and experimental validation, this paper conducts a comparative analysis of computed retrievability scores using the most commonly adopted query generation techniques in the research community. The main contributions of this paper include:

- Providing an overview of query generation techniques commonly used in retrievability experiments documented in the literature.
- Conducting a comprehensive empirical assessment of the reproducibility of retrievability experiments employing different query generation techniques and test collections.

Additionally, we address the following research questions in this paper:

RQ1 - While keeping the other parameters constant, does varying the query simulation techniques lead to differences in retrievability scores, thus impacting the reproducibility of results?

RQ2 - Do these differences in retrievability scores exhibit consistency across various collections, thereby influencing the reproducibility of results?

The rest of the paper is organized as follows. In the next section, we provide essential background information and review previous research related to retrievability, with a particular emphasis on the prevalent query simulation techniques in this field. Section 3 delves into the specifics of our experimental setups before presenting the empirical result on benchmark test collections in Sect. 4. We conclude the paper in Sect. 5 highlighting the findings considering the issues with reproducing results, and suggesting some directions to pursue further study. The codes and the simulated queries are available from the following repository: https://mallpriyanshu.github.io/retrievability-survey/ecir2024.

2 Background and Motivation

2.1 Retrievability: A Measure of Accessibility

Retrievability is a measure that evaluates the ease with which a document can be retrieved in a specific configuration of an IR system. Azzopardi and Vinay [8] introduced the concept and formally defined the retrievability score for a document d in a collection D as $r(d)$ with respect to a given IR system. Typically, the calculation of retrievability involves the execution of the following sequence of steps [36]: *i*) query set generation, *ii*) system configuration and retrieval, *iii*) computation of retrievability scores, and *iv*) summarising the scores.

For an ideal calculation of retrievability scores, a comprehensive universe of all possible queries is needed, although obtaining this is practically unattainable. In practice, researchers commonly select from two alternatives: either using a sufficiently large query log containing diverse user queries [42,45] or simulating a set of queries from the collection that adequately represents its overall topic distribution [8,9,14,16]. Once an appropriate set of queries has been chosen, the second step is selecting a retrieval model with an optimal parameter setting. Prior research has extensively examined the impact of altering system configurations [8,15,50] where the authors conclude that changing the retrieval model can significantly impact the retrievability scores of a document collection. The retrievals are then performed with the set of queries, and the retrievability of a document d is computed in the third step as *the summation of the number of times d is retrieved within the top c position.* Mathematically, the retrievability $r(d)$ of a document d ($d \in D$) with respect to an IR system is computed using the formula in Eq. 1.

$$r(d) = \sum_{q \in Q} o_q \cdot f(k_{dq}, c) \quad (1)$$

Here, o_q indicates the likelihood of the query (usually considered uniform across query set) and k_{dq} is the rank of d for query q. Calculating $r(d)$ relies on the choice of a utility function $f(.)$, either cumulative or gravity-based. Previous studies have shown a high correlation between these two measures [47,48], and consequent research predominantly reports retrievability analysis using the cumulative-based function [13,21,45,49]. Defined as a binary function, the cumulative utility function $f(k_{dq}, c)$ is equal to 1 if document d is retrieved within the top c documents for query q; otherwise, it is considered 0. This utility function provides a straightforward interpretation of the retrievability score for each document. It simply corresponds to the number of times the document is retrieved within the top c ranks. Documents beyond the top c positions are disregarded, simulating a user who only examines the first c results. Therefore, the more queries that retrieve a document within rank c, the higher its retrievability score.

The final step involves summarizing the retrievability scores across the collection to assess existing bias. This quantification is commonly achieved using the

Gini coefficient, a measure frequently employed in economics and social sciences to assess inequality within a population [27]. Calculating the retrievability bias within a document collection, the Gini coefficient is computed as follows:

$$G = \frac{\sum_{i=1}^{N}(2i - N - 1) \cdot r(d_i)}{N \sum_{j=1}^{N} r(d_j)} \tag{2}$$

where N represents total number of documents in the collection ($|D|$). It provides a decimal value in the range $[0, 1]$. A Gini coefficient of zero denotes perfect equality, indicating that all documents in the collection have an equal retrievability score according to $r(d)$. Conversely, a Gini coefficient of one indicates total inequality, with only one document having $r(d) = |Q|$ while all other documents have $r(d) = 0$. Other inequality metrics, such as Atkinson index [4], Theli index [30], Palma index [39] etc., are also explored in [53] where the authors argue that various inequality metrics generally agree on which system and settings minimize the retrievability bias and the least inequality indicated by Gini coefficient is consistent with the other metrics discussed. Hence, the Gini coefficient G is commonly employed in subsequent works [8,16,18] to quantify the amount of bias in the distribution of the retrievability scores in a collection of documents.

To visually assess the retrievability distribution in a document collection, one can calculate retrievability scores for each document using Eq. 1 and plot the sorted values to create a *Lorenz curve*. This curve represents the cumulative score distribution of documents sorted by retrievability scores in ascending order. If the retrievability scores are evenly distributed, the Lorenz Curve will be linear; a skewed curve in contrast indicates a greater level of inequality or bias. The Lorenz curve is a concept commonly used in economics to depict wealth distribution, where a straight line signifies equal wealth distribution across a population, while a skewed curve represents inequality, as shown in Fig. 1.

Retrievability, and the underlying theory of retrievability, have found applications in various domains. For instance, it has been used in the development of inverted indexes to enhance the efficiency and performance of retrieval systems by capitalizing on terms that contribute to a document's retrievability [41]. Additionally, retrievability has been leveraged to investigate bias in search engines and retrieval systems on the web [6] and within patent collections [18], leading to improvements in system efficiency during pruning processes [54].

The initial and most critical step while calculating the retrievability of a collection of documents involves query generation, followed by retrieval using the generated queries, and finally, the computation of retrievability scores based on the appearance of documents within a rank cutoff. Prior research has extensively examined the impact of altering system configurations [8,9,15,45,50], modifying the nature of the function [48], and varying score summarization techniques [53]. However, there is limited empirical evidence concerning the influence of modifying query sets on the resultant retrievability scores. Filling the void, in this paper, we study the effect of various query-generation techniques on theretrievability

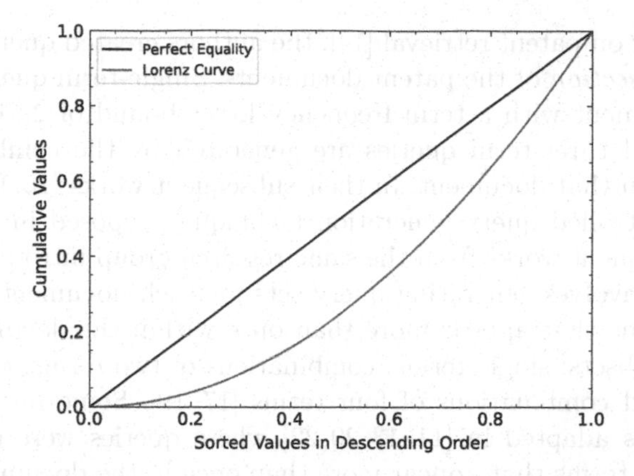

Fig. 1. A sample Lorenz Curve. The values belonging to the distribution in which the disparity is to be observed are arranged in a sorted manner along the X-axis.

scores of documents in a collection. We discuss some of the well-known techniques popular among researchers for query generation in the next section.

2.2 Query Generation

The initial step in a retrievability computation typically involves obtaining a set of all possible queries addressable by the collection (Q in Eq. 1). As per Eq. 1, the most ideal calculation of $r(d)$ would require an extensive set Q. In general, a query log would be the preferred option for Q; however, obtaining such a log can be difficult and often not feasible. Hence, researchers have inclined towards artificial simulation of the set Q from the collection. Following the inception of the concept of retrievability in [8], a popular choice among researchers in simulating Q based on term statistics in the collection itself [14–19,21,38]. The approach by Azzopardi and Vinay [8] in generating the query set was influenced by Callan and Connel [24]. The authors generated a set of queries, which included single-term queries, formed by selecting each term in the vocabulary that appeared at least 5 times, and bi-term queries, formed by selecting each bigram in the collection that appeared at least 20 times. The list of bigrams was truncated at 20 million, and each query was issued to the system to estimate its retrievability. Azzopardi and Bache later followed a similar methodology to generate query sets for the AP and WSJ collections by ranking the most frequent 100,000 bigrams in each collection [5]. The same approach was also adapted in [9] where the authors applied retrievability to quantify the accessibility of patent documents. Although the queries used in patent search are often longer than two words, the authors choose two-word queries because this means there will be "just one Boolean operator and it thus affords a comparison between the use of AND and OR within the various IR models". In the other studies on retrievability, the prevalent choice among researchers has been the adoption of a similar approach for query generation [47–53].

In their work on patent retrieval [14], the authors created queries using terms from the *claim section* of the patent documents. Single-term queries are formed from each document with a term frequency lower bound of 2. To make longer queries, two and three-term queries are generated by the combination of the frequent terms in that document. In their subsequent works [15,16], the authors employed a controlled query generation technique proposed in [31] for patent retrieval. Subsequent works from the same research group introduced a different approach that involves generating query sets for each document in the corpus. They select terms that appear more than once within the document, resulting in four query subsets: single terms, combinations of two terms, combinations of three terms, and combinations of four terms [17–19]. Some further truncation of query set was adapted in [11–13,20,22] where queries were generated with combinations of terms that appear more than once in the document. Stopwords and terms having document frequency $> 25\%$ of collection size were removed.

Assessing document retrievability in a collection necessitates a comprehensive set of queries that cover all aspects of the collection. Ideally, this would involve an extensive query log spanning a significant time period. However, acquiring such a log is often costly and unavailable, prompting researchers to resort to simulated queries. Despite the challenges, some research has been conducted using real user query logs acquired from certain closed domains. The authors in [44,45] use a set of real queries submitted to the National Library of the Netherlands, between March 2015 to July 2015. Additionally, to make the results comparable to [8], the authors create a set of simulated queries: they select the top 2 million unigrams and bigrams respectively as single term and two term queries. All the queries were finalized after stemming and removal of stopwords and operators. On comparing the characteristics of simulated queries to those formed by real user queries, the authors report significant differences regarding the composition, the number of unique terms used, and the presence of named entities: real user queries exhibited a considerably higher fraction of named entities compared to the simulated query set. Additionally, due to the considerably low number of real queries in contrast to the vast size of the collection (exceeding 102 million), the authors acknowledged the presence of a varying document-query ratio compared to those reported in [8]. Authors in [2] employ a transformer-based generative model docT5query [37] for generating queries and conclude that BERT-based rankers demonstrate less retrievability bias than user-given queries. Recently, Roy et al. [42] studied the retrievability distribution in an integrated retrieval system containing documents in multiple categories [42,43]. However, this work does not conduct a direct comparison of the retrievability scores (and their distributions) when computed using real and simulated query sets.

The contrasting nature of simulated queries, when compared to real queries, is briefly reported in [45]. The various techniques used for generating simulated queries illustrate the diversity in approaches taken at the initial stage of a retrievability analysis by different groups of researchers. This also highlights the absence of a well-defined structure for such an analysis paving a potential issue with reproducing results. It prompts the question: can we anticipate a similar

Table 1. Basic statistics of datasets used for empirical verification of the variation in the computed retrievability scores when different query sets are employed.

Dataset	# documents	Collection Type	# terms
TREC Robust	528,155	News articles	1,502,031
WT10g	1,692,096	Web pages	9,674,707
Wikipedia	6,584,626	Encyclopedia	18,797,260

distribution of retrievability scores when the query set is altered? Answering that question, in this paper, we present a systematic comparison of the retrievability scores computed based on several of these techniques together with real-user query logs. According to the observation in [45], the real queries contain a significantly higher fraction of named entities compared to the simulated query set. Inspired by this observation, we additionally introduce a rule-based query simulation technique for retrievability assessment, further contributing to the examination of this diversity in approaches. To maintain the primary narrative of the paper, we have detailed the method in the Appendix A.

3 Experimental Setup

As presented in Sect. 2, the set of queries employed in computing the retrievability plays an important role in the computation and our focus of this paper is to foresee whether alteration in this set can affect the final score, potentially leading to challenges in reproducing results for different collections and retrieval models. In this section, we present the datasets and the query sets that are used in this paper to verify this empirically with real data.

3.1 Datasets

We perform the experiments of the retrievability score distributions on two benchmark retrieval datasets as well as on a diverse encyclopedia of documents. Particularly, the TREC ad-hoc Robust collection [46] containing news articles, WT10g collection [29] with web pages, together with the English Wikipedia article dump of February, 2023 are employed in this study[1]. All the datasets are preprocessed with stopword removal following SMART stopword list[2] and stemmed using the Porter Stemmer before indexing them using Apache Lucene. Overall statistics of the datasets are presented in Table 1.

3.2 Simulated Queries

In order to foresee the variation, if any, in the retrievability distribution of documents in a collection, we use the following artificial query generation techniques:

[1] https://dumps.wikimedia.org/enwiki/.
[2] https://tinyurl.com/smart-stopword.

Table 2. Number of queries generated for each dataset using the individual query generation techniques. Note that, the simulated queries are generated from each dataset and applied in that particular dataset for the retrievability computation. AOL query set is filtered separately for each dataset as discussed in Sect. 3.3.

	Query Type	Number of queries		
		Robust04	WT10g	Wikipedia
SQ_1 [8]	Simulated	1061184	3902575	5320001
SQ_2 [13]	Simulated	6903196	18724509	84920098
SQ_3 [31]	Simulated	1644533	7346091	26272419
RSQ	Simulated	3900000	4500000	4550000
AOL [40]	Real Query	3885206	4302650	4550977

1. **Simulated Query set 1 (SQ_1):** Following the suggestion in [8], the first set of the queries with which the retrievals are performed is formed by a sampling method proposed in [24]. In this process, the terms undergo analysis and filtering, including stemming using the Porter stemmer and removing stopwords. Terms occurring more than 5 times in the collection are treated as single-term queries. Additionally, two-term queries are formed by pairing consecutively occurring terms that have a collection frequency of at least 20 times. These resulting bigrams are then ranked by occurrence, and the top two million are selected as the final set of two-term queries.
2. **Simulated Query set 2 (SQ_2):** The second technique is based on [13]. Specifically, queries are generated with the combinations of terms that appear more than once in the document. Stopwords and other terms with document frequency > 25% of collection size are removed. Queries up to 4 terms are formed using the boolean AND operator, with duplicate queries being removed.
3. **Simulated Query set 3 (SQ_3):** In this approach (based on [31]), a set of related documents is chosen as the source document set with clustering. Language models are defined for both the source documents and the corpus, with relative entropy used for comparison. Terms in the vocabulary are sorted based on their contribution to relative entropy, serving as a term discrimination scoring function. The most discriminating term initiates the query construction process, forming single-term and two-term queries for different query sets. The process continues by identifying next most informative term and generating subsequent queries until all terms have been used.
4. **Rule-based Simulated Query set (RSQ):** Building upon the insight in [45], that real queries contain a notably greater proportion of named entities compared to simulated queries, we introduce a rule-based query simulation technique for retrievability assessment. The approach is presented in Appendix A.

Table 3. Gini coefficient of Retrievability score distribution when computed using different query generation techniques. The highest and the lowest Gini values for each dataset are marked respectively with a superscript † and *.

Dataset	SQ_1	SQ_2	SQ_3	RSQ	AOL
Robust	0.3307	0.4556	0.4300	0.3052*	0.6032†
WT10g	0.5371	0.6391	0.6359	0.5009*	0.6541†
Wikipedia	0.5380	0.6210	0.6290	0.4820*	0.6798†

3.3 Real Queries

A majority of the works done in the domain of retrievability employ queries generated using simulation techniques. In this work, we use a general web search query log to make our study conclusive. Specifically, we utilize the AOL query log [40] that contains actual user query click data spanning from March 1st, 2006, to May 31st, 2006. It sparked significant controversy within the media, primarily due to privacy-related concerns [3,10,28]. Although the query log is more than a decade old, it is still widely used in recent literature [1,25,33–35,55].

To prepare the real query set for our experiments, the initial step involves extracting all unique queries from the AOL query log. Queries containing periods are eliminated to filter out website links. Given the distinct temporal scopes of our datasets and query sets, a conspicuous challenge arises in the form of topic and vocabulary mismatch. To mitigate this challenge as much as possible, we adopt an exclusionary approach: a query is selected if all of its terms are present in the corpus vocabulary. This strategic refinement serves to align the queries closely with the content of the target collection. Note that, the AOL queries are refined for each collection separately following the above exclusionary approach. Table 2 contains the individual count of queries finalized for each dataset.

3.4 Retrieval Model

In accordance with the insights and recommendations put forth in [8], we only use BM25 as the model for performing the retrieval. Specifically, we use the implementation available in Apache Lucene. To fine-tune the BM25 parameters, we utilize the TREC ad-hoc retrieval topic set [46], which is then applied on the Robust dataset for the retrievability study. Similarly, the TREC web topic sets [29] are employed for parameter tuning, and these adjusted parameters are subsequently applied to the WT10g and Wikipedia datasets.

4 Comparing Retrievability

4.1 Gini Coefficient and Lorenz Curve

We present the Gini coefficient for each dataset utilizing all five query sets in Table 3. The variation in Gini values is graphically presented in Fig. 2. In the

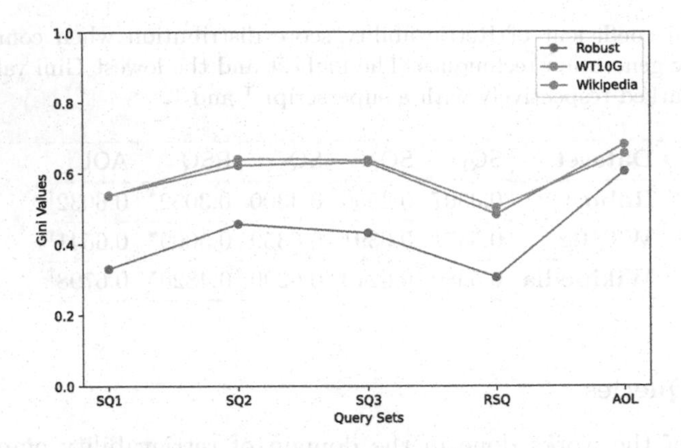

Fig. 2. Gini coefficient for each dataset when the retrievability scores are calculated using different query sets as discussed in Sect. 3.

table, the least disparity is observed across all the datasets when utilizing the rule-based simulation for query generation (denoted as RSQ in Table 3). The distributions of retrievability values are presented using Lorenz curve in Fig. 3. Furthermore, the original query simulation technique proposed by [8] (indicated by SQ_1 in Table 3) follows a similar trend. The other two simulation techniques (SQ_2 and SQ_3) exhibit higher disparity. In contrast, a significant disparity in retrievability scores across all the datasets is observed when the real-query log (AOL) is used. These results lead to the following conclusions:

- The web collection WT10g and the encyclopedia Wikipedia exhibit much more diverse retrievability scores as compared to the relatively smaller news collection TREC Robust across all the query sets.
- The approach in [8] selects terms from the collection based on their frequency, including non-informative queries (SQ_1). This method results in a more evenly distributed query generation covering a wide range of documents in the collection. As a result, it tends to produce lower Gini values compared to other query sets.
- Realistic queries generated using the rule-based approach (RSQ) result in even lower Gini coefficients across all datasets.
- AOL query logs mostly consist of known item searches rather than exploratory ones leading to a significant disparity in computed retrievability scores, with Gini coefficients consistently exceeding 0.6 across all datasets.

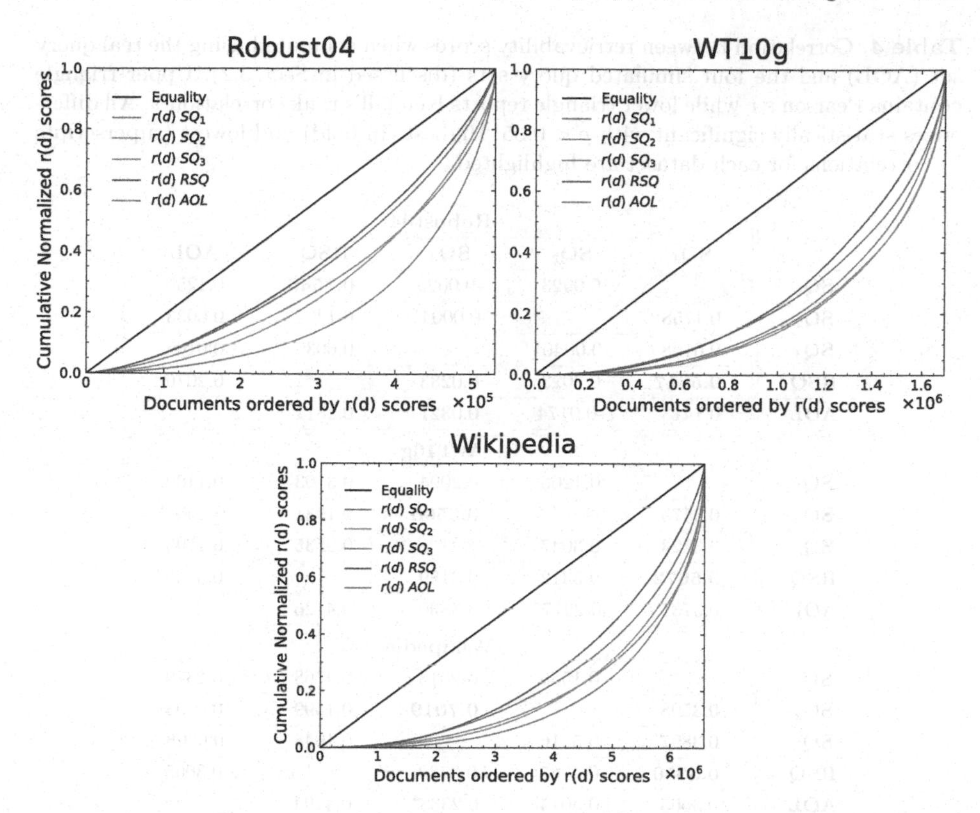

Fig. 3. Lorenz curve with retrievability computed with various query sets on different collections.

4.2 Comparing Correlation

Table 4 reports the degree of variation in retrievability scores when comparing the simulation-based query generation techniques (SQ_1, SQ_2, SQ_3, and RSQ) along with the AOL query log. We compute the Pearson's r between the distribution of each pair of retrievability scores computed using the different query sets and report in the upper half of Table 4. Kendall's τ is presented in the lower half of the table which is computed using the document rank arranged following their retrievability scores. The table shows negligible correlations between scores computed using almost any query set pairs. Interestingly, the highest correlations, in terms of both Pearson's r and Kendall's τ occur when using the rule-based technique RSQ (Appendix A). Further, the massive difference in the highest and lowest correlations validates that distinct query sets lead to divergent retrievability scores, affirming both **RQ1** and **RQ2**.

Table 4. Correlation between retrievability scores when computed using the real query set (AOL) and the four simulated query sets (discussed in Sect. 3.2). Upper triangle contains Pearson's r while lower triangle reports Kendall's rank correlation τ. All differences statistically significant with $p < 0.05$. Highest (in bold) and lowest (superscripts *) correlations for each dataset are highlighted.

	SQ_1	SQ_2	SQ_3	RSQ	AOL
			Robust04		
SQ_1		0.0923	-0.0025	**0.7549**	0.1250
SQ_2	0.1758		-0.0004*	0.1461	0.0033
SQ_3	0.0108	-0.0040*		0.0030	0.0165
RSQ	**0.5617**	0.3023	0.0233		0.2076
AOL	0.1413	-0.0174	0.0321	0.1811	
			WT10g		
SQ_1		0.1206	0.2094	0.3103	0.1100
SQ_2	0.3575		**0.5503**	0.1931	0.0900*
SQ_3	0.3223	0.3047		0.2535	0.1795
RSQ	**0.5672**	0.5312	0.3189		0.3135
AOL	0.3731	0.2977*	0.3830	0.4126	
			Wikipedia		
SQ_1		0.1586	0.2613	0.4768	0.2379
SQ_2	0.3508		**0.7019**	0.1599	0.0598
SQ_3	0.3897	0.5116		0.1048	0.0308*
RSQ	**0.5350**	0.4867	0.3739		0.3065
AOL	0.3603	0.3074	0.2338*	0.4194	

5 Conclusion

The examination of retrievability has predominantly relied on the use of simulated queries since its inception with few works using real query logs. However, a comprehensive comparative analysis of the computed values has been absent in the existing literature. Bridging this gap, in this paper, we examine the impact of five distinct and diverse query sets, ranging from simulated queries to actual user query logs, on the computed retrievability scores when the retrieval is performed using BM25.

Our findings on diverse datasets, including news, web, and Wikipedia collections expose substantial variations in computed retrievability scores across query sets, providing an affirmative response to **RQ1** and underscoring the reproducibility challenges associated with retrievability scores. Among the popular query simulation techniques for retrievability studies, the query set generated following the method in [8] (SQ_1) exhibits the least disparity in computed retrievability scores. In contrast, the AOL query log generates the highest disparity. Furthermore, the utilization of a rule-based simulation technique (RSQ) results in the least inequality in the retrievability scores among all. These observations reveal a high sensitivity of these scores with respect to the query generation methods employed. Our research demonstrates that the selection of a query set directly impacts computed retrievability scores, regardless of the nature of the collection. This provides a conclusive response to **RQ2**. These findings emphasize the need for standardization of query set construction in retrievability studies and contribute to a deeper understanding of the nuances of accessibility within the field of information retrieval. In previous research on retrievability, various studies utilized multiple retrieval models for a single query set per collection. As part of our future work, we will explore the impact of diverse retrieval models on the retrievability scores with different query sets.

A Appendix - A POS-Based Query Generation Technique

Identifying collocations in a text corpus typically involves counting co-occurring word pairs, revealing words that go beyond their individual meanings. Relying solely on the most frequent bigrams often yields uninteresting results, as many of them consist of function words, offering limited insights. To improve collocation quality, Justeson and Katz [32] introduced a simple yet effective heuristic. They apply a part-of-speech filter to candidate phrases, preserving patterns likely to represent genuine 'phrases' rather than random word combinations. This approach enhances the meaningfulness of the collocation identification process.

We apply this approach for query generation improving the relevance and effectiveness of the generated queries.

1. **Perform Part-of-Speech (POS) tagging**: Initially, we employ POS tagging on all the documents within the collection. This step assigns appropriate grammatical tags to each word, facilitating the subsequent identification of N-grams.
2. **Extract N-grams**: N-grams, where N represents the desired length of the word sequences, are extracted from the POS-tagged documents. In our case, we consider N to range from 1 to 4, enabling the identification of unigrams, bigrams, trigrams, and quadgrams.
3. **Select N-grams with 'query-like' POS tag patterns**: From the pool of extracted N-grams, we apply Justeson and Katz's [32] recommended POS tag patterns to filter and retain N-grams that exhibit patterns resembling queries. The specific POS tag patterns for each N-gram type are provided in Table 5. Tag patterns for Quadgrams are proposed by us heuristically from our observations.

Subsequently, the resulting list of N-grams is sorted in descending order based on their occurrence frequencies. To ensure a manageable and relevant set of queries, we truncate the list at specific thresholds. These thresholds are determined by drawing inspiration from the frequency distribution of queries found in the AOL query set [40]. We aim to maintain a proportional ratio between the selected N-grams and the query frequencies observed in the AOL real query set, preserving a close alignment with real-world query usage patterns.

Table 5. POS Tag rules for N-gram query generation

unigram	bigram	trigram	quadgram
noun	adj noun noun noun	adj adj noun adj noun noun noun adj noun noun noun noun noun adp noun	noun verb adp noun adj noun adj noun noun adp adj noun noun noun adp noun noun verb noun noun adv adj noun noun adj noun verb noun noun adj noun noun

References

1. Ahmad, W.U., Chang, K.W., Wang, H.: Context attentive document ranking and query suggestion. In: Proceedings of the 42nd International ACM SIGIR Conference on Research and Development in Information Retrieval, pp. 385–394. SIGIR'19, Association for Computing Machinery, New York, NY, USA (2019). https://doi.org/10.1145/3331184.3331246
2. Abolghasemi, A., Verberne, S., Askari, A., Azzopardi, L..: Retrievability bias estimation using synthetically generated queries. In: Proceedings of the First Workshop on Generative Information Retrieval - GenIR@SIGIR2023 held in conjunction with SIGIR 2023. GenIR@SIGIR2023 (2023). https://coda.io/@sigir/gen-ir/accepted-papers-17
3. Anderson, N.: The ethics of using aol search data. https://arstechnica.com/uncategorized/2006/08/7578/
4. Atkinson, A.B.: On the measurement of inequality. J. Econom. Theory **2**(3), 244–263 (1970). https://doi.org/10.1016/0022-0531(70)90039-6, https://www.sciencedirect.com/science/article/pii/0022053170900396
5. Azzopardi, L., Bache, R.: On the relationship between effectiveness and accessibility. In: Proceedings of the 33rd international ACM SIGIR conference on Research and development in information retrieval, pp. 889–890 (2010)
6. Azzopardi, L., Owens, C.: Search engine predilection towards news media providers. In: Proceedings of the 32nd international ACM SIGIR conference on Research and development in information retrieval, pp. 774–775 (2009)
7. Azzopardi, L., Vinay, V.: Accessibility in information retrieval. In: Macdonald, C., Ounis, I., Plachouras, V., Ruthven, I., White, R.W. (eds.) ECIR 2008. LNCS, vol. 4956, pp. 482–489. Springer, Heidelberg (2008). https://doi.org/10.1007/978-3-540-78646-7_46
8. Azzopardi, L., Vinay, V.: Retrievability: an evaluation measure for higher order information access tasks. In: Proceedings of the 17th ACM Conference on Information and Knowledge Management, pp. 561–570. CIKM '08, Association for Computing Machinery, New York, NY, USA (2008). https://doi.org/10.1145/1458082.1458157
9. Bache, R., Azzopardi, L.: Improving access to large patent corpora. Trans. Large Scale Data Knowl. Centered Syst. **2**, 103–121 (2010). https://doi.org/10.1007/978-3-642-16175-9_4
10. Barbaro, Michael; Zeller Jr, T.: A face is exposed for aol searcher no. 4417749. https://www.nytimes.com/2006/08/09/technology/09aol.html

11. Bashir, S.: Improving retrievability with improved cluster-based pseudo-relevance feedback selection. Expert Syst. Appl. **39**(8), 7495–7502 (2012). https://doi.org/10.1016/j.eswa.2012.01.041
12. Bashir, S.: Estimating retrievability ranks of documents using document features. Neurocomputing **123**, 216–232 (2014)
13. Bashir, S., Khattak, A.S.: Producing efficient retrievability ranks of documents using normalized retrievability scoring function. J. Intell. Inform. Syst. **42**, 457–484 (2014). https://doi.org/10.1007/s10844-013-0274-3
14. Bashir, S., Rauber, A.: Analyzing document retrievability in patent retrieval settings. In: Bhowmick, S.S., Küng, J., Wagner, R. (eds.) DEXA 2009. LNCS, vol. 5690, pp. 753–760. Springer, Heidelberg (2009). https://doi.org/10.1007/978-3-642-03573-9_63
15. Bashir, S., Rauber, A.: Identification of low/high retrievable patents using content-based features. In: Proceedings of the 2nd International Workshop on Patent Information Retrieval, pp. 9–16 (2009)
16. Bashir, S., Rauber, A.: Improving retrievability of patents with cluster-based pseudo-relevance feedback documents selection. In: Proceedings of the 18th ACM Conference on Information and Knowledge Management, pp. 1863–1866 (2009)
17. Bashir, S., Rauber, A.: Improving retrievability and recall by automatic corpus partitioning. In: Hameurlain, A., Küng, J., Wagner, R., Bach Pedersen, T., Tjoa, A.M. (eds.) Transactions on Large-Scale Data- and Knowledge-Centered Systems II. LNCS, vol. 6380, pp. 122–140. Springer, Heidelberg (2010). https://doi.org/10.1007/978-3-642-16175-9_5
18. Bashir, S., Rauber, A.: Improving retrievability of patents in prior-art search. In: Gurrin, C., He, Y., Kazai, G., Kruschwitz, U., Little, S., Roelleke, T., Rüger, S., van Rijsbergen, K. (eds.) ECIR 2010. LNCS, vol. 5993, pp. 457–470. Springer, Heidelberg (2010). https://doi.org/10.1007/978-3-642-12275-0_40
19. Bashir, S., Rauber, A.: On the relationship between query characteristics and ir functions retrieval bias. J. Am. Soc. Inform. Sci. Technol. **62**(8), 1515–1532 (2011)
20. Bashir, S., Rauber, A.: Automatic ranking of retrieval models using retrievability measure. Knowl. Inf. Syst. **41**, 189–221 (2014)
21. Bashir, S., Rauber, A.: Retrieval models versus retrievability. Current Challenges in Patent Information Retrieval, pp. 185–212 (2017)
22. Bashir, S., Rauber, A.: Retrieval models versus retrievability. In: Current Challenges in Patent Information Retrieval. TIRS, vol. 37, pp. 185–212. Springer, Heidelberg (2017). https://doi.org/10.1007/978-3-662-53817-3_7
23. Boratto, L., Faralli, S., Marras, M., Stilo, G. (eds.): Advances in Bias and Fairness in Information Retrieval. Springer Nature Switzerland (2023). https://doi.org/10.1007/978-3-031-37249-0
24. Callan, J., Connell, M.: Query-based sampling of text databases. ACM Trans. Inform. Syst. (TOIS) **19**(2), 97–130 (2001)
25. Dehghani, M., Zamani, H., Severyn, A., Kamps, J., Croft, W.B.: Neural ranking models with weak supervision. In: Proceedings of the 40th International ACM SIGIR Conference on Research and Development in Information Retrieval, pp. 65–74. SIGIR '17, Association for Computing Machinery, New York, NY, USA (2017). https://doi.org/10.1145/3077136.3080832
26. Ekstrand, M.D., Das, A., Burke, R., Diaz, F.: Fairness in information access systems. Foundations and Trends® in Information Retrieval **16**(1–2), 1–177 (2022). https://doi.org/10.1561/1500000079
27. Gini, C.: On the measure of concentration with special reference to income and statistics. Colorado College Publication, General Series **208**(1), 73–79 (1936)

28. Hafner, K.: Tempting data, privacy concerns; researchers yearn to use aol logs, but they hesitate. https://www.nytimes.com/2006/08/23/technology/23search.html
29. Hawking, D.: Overview of the TREC-9 web track. In: Voorhees, E.M., Harman, D.K. (eds.) Proceedings of The Ninth Text REtrieval Conference, TREC 2000, Gaithersburg, Maryland, USA, November 13–16, 2000. NIST Special Publication, vol. 500–249. National Institute of Standards and Technology (NIST) (2000). http://trec.nist.gov/pubs/trec9/papers/web9.pdf
30. Johnston, J.: H. Theil. economics and information theory. Econom. J. **79**(315), 601–602 (09 1969). https://doi.org/10.2307/2230396
31. Jordan, C., Watters, C., Gao, Q.: Using controlled query generation to evaluate blind relevance feedback algorithms. In: Proceedings of the 6th ACM/IEEE-CS Joint Conference on Digital Libraries, pp. 286–295 (2006)
32. Justeson, J.S., Katz, S.M.: Co-occurrences of antonymous adjectives and their contexts. Computational Linguistics **17**(1), 1–20 (1991). https://aclanthology.org/J91-1001
33. Kang, Y.M., Liu, W., Zhou, Y.: Queryblazer: efficient query autocompletion framework. In: Proceedings of the 14th ACM International Conference on Web Search and Data Mining, pp. 1020–1028. WSDM '21, Association for Computing Machinery (2021). https://doi.org/10.1145/3437963.3441725
34. Ma, Z., Dou, Z., Bian, G., Wen, J.R.: Pstie: time information enhanced personalized search. In: Proceedings of the 29th ACM International Conference on Information & Knowledge Management, pp. 1075–1084. CIKM '20, Association for Computing Machinery (2020). https://doi.org/10.1145/3340531.3411877
35. MacAvaney, S., Macdonald, C., Ounis, I.: Reproducing personalised session search over the aol query log. In: Hagen, M., Verberne, S., Macdonald, C., Seifert, C., Balog, K., Nørvåg, K., Setty, V. (eds.) ECIR 2022. LNCS, vol. 13185, pp. 627–640. Springer, Cham (2022). https://doi.org/10.1007/978-3-030-99736-6_42
36. McLellan, C.: The relationship between retrievability bias and retrieval performance. Ph.D. thesis, University of Glasgow, UK (2019). https://ethos.bl.uk/OrderDetails.do?uin=uk.bl.ethos.775857
37. Nogueira, R., Lin, J.: From doc2query to doctttttquery. In: Online preprint 6 (2019). https://github.com/castorini/docTTTTTquery
38. Noor, S., Bashir, S.: Evaluating bias in retrieval systems for recall oriented documents retrieval. Int. Arab J. Inform. Technol. (IAJIT) **12**(1) (2015)
39. Palma, J.G.: Homogeneous middles vs. heterogeneous tails, and the end of the 'inverted-u': the share of the rich is what it's all about. Cambridge working papers in economics, Faculty of Economics, University of Cambridge (2011). https://EconPapers.repec.org/RePEc:cam:camdae:1111
40. Pass, G., Chowdhury, A., Torgeson, C.: A picture of search. In: Proceedings of the 1st International Conference on Scalable Information Systems, pp. 1-es. InfoScale '06, Association for Computing Machinery (2006). https://doi.org/10.1145/1146847.1146848
41. Pickens, J., Cooper, M., Golovchinsky, G.: Reverted indexing for feedback and expansion. In: Proceedings of the 19th ACM International Conference on Information and Knowledge Management, pp. 1049–1058 (2010)
42. Roy, D., Carevic, Z., Mayr, P.: Studying retrievability of publications and datasets in an integrated retrieval system. In: Proceedings of the 22nd ACM/IEEE Joint Conference on Digital Libraries. JCDL '22, Association for Computing Machinery (2022). https://doi.org/10.1145/3529372.3530931

43. Roy, D., Carevic, Z., Mayr, P.: Retrievability in an integrated retrieval system: an extended study. Int. J. Digital Libr. (Apr 2023). https://doi.org/10.1007/s00799-023-00363-4
44. Traub, M.C., Samar, T., van Ossenbruggen, J., Hardman, L.: Impact of crowd-sourcing ocr improvements on retrievability bias. In: Proceedings of the 18th ACM/IEEE on Joint Conference on Digital Libraries, pp. 29–36. JCDL '18, Association for Computing Machinery (2018). https://doi.org/10.1145/3197026.3197046
45. Traub, M.C., Samar, T., Van Ossenbruggen, J., He, J., de Vries, A., Hardman, L.: Querylog-based assessment of retrievability bias in a large newspaper corpus. In: 2016 IEEE/ACM Joint Conference on Digital Libraries (JCDL), pp. 7–16. IEEE (2016)
46. Voorhees, E.M.: Overview of the TREC 2004 robust track. In: Voorhees, E.M., Buckland, L.P. (eds.) Proceedings of the Thirteenth Text REtrieval Conference, TREC 2004, Gaithersburg, Maryland, USA, November 16–19, 2004. NIST Special Publication, vol. 500–261. National Institute of Standards and Technology (NIST) (2004). http://trec.nist.gov/pubs/trec13/papers/ROBUST.OVERVIEW.pdf
47. Wilkie, C., Azzopardi, L.: An initial investigation on the relationship between usage and findability. In: Serdyukov, P., et al. (eds.) ECIR 2013. LNCS, vol. 7814, pp. 808–811. Springer, Heidelberg (2013). https://doi.org/10.1007/978-3-642-36973-5_90
48. Wilkie, C., Azzopardi, L.: Relating retrievability, performance and length. In: Proceedings of the 36th International ACM SIGIR Conference on Research and Development in Information Retrieval, pp. 937–940 (2013)
49. Wilkie, C., Azzopardi, L.: Best and fairest: an empirical analysis of retrieval system bias. In: de Rijke, M., et al. (eds.) Advances in Information Retrieval, pp. 13–25. Springer International Publishing, Cham (2014)
50. Wilkie, C., Azzopardi, L.: Efficiently estimating retrievability bias. In: de Rijke, M., et al. (eds.) Advances in Information Retrieval, pp. 720–726. Springer International Publishing, Cham (2014)
51. Wilkie, C., Azzopardi, L.: A retrievability analysis: Exploring the relationship between retrieval bias and retrieval performance. In: Proceedings of the 23rd ACM International Conference on Conference on Information and Knowledge Management, pp. 81–90 (2014)
52. Wilkie, C., Azzopardi, L.: Query length, retrievability bias and performance. In: Proceedings of the 24th ACM International on Conference on Information and Knowledge Management, pp. 1787–1790. CIKM '15, Association for Computing Machinery, New York, NY, USA (2015). https://doi.org/10.1145/2806416.2806604
53. Wilkie, C., Azzopardi, L.: Retrievability and retrieval bias: a comparison of inequality measures. In: Hanbury, A., Kazai, G., Rauber, A., Fuhr, N. (eds.) ECIR 2015. LNCS, vol. 9022, pp. 209–214. Springer, Cham (2015). https://doi.org/10.1007/978-3-319-16354-3_22
54. Zheng, L., Cox, I.J.: Document-oriented pruning of the inverted index in information retrieval systems. In: 2009 International Conference on Advanced Information Networking and Applications Workshops, pp. 697–702. IEEE (2009)
55. Zhu, Y., et al.: Contrastive learning of user behavior sequence for context-aware document ranking. In: Proceedings of the 30th ACM International Conference on Information & Knowledge Management, pp. 2780–2791. CIKM '21, Association for Computing Machinery (2021). https://doi.org/10.1145/3459637.3482243

Reproducibility Analysis and Enhancements for Multi-aspect Dense Retriever with Aspect Learning

Keping Bi[1,2](✉), Xiaojie Sun[1,2], Jiafeng Guo[1,2](✉), and Xueqi Cheng[1,2]

[1] CAS Key Lab of Network Data Science and Technology, ICT, CAS, Beijing, China
{bikeping,sunxiaojie21s,guojiafeng,cxq}@ict.ac.cn

[2] University of Chinese Academy of Sciences, Beijing, China

Abstract. Multi-aspect dense retrieval aims to incorporate aspect information (e.g., brand and category) into dual encoders to facilitate relevance matching. As an early and representative multi-aspect dense retriever, MADRAL learns several extra aspect embeddings and fuses the explicit aspects with an implicit aspect "OTHER" for final representation. MADRAL was evaluated on proprietary data and its code was not released, making it challenging to validate its effectiveness on other datasets. We failed to reproduce its effectiveness on the public MA-Amazon data, motivating us to probe the reasons and re-examine its components. We propose several component alternatives for comparisons, including replacing "OTHER" with "CLS" and representing aspects with the first several content tokens. Through extensive experiments, we confirm that learning "OTHER" from scratch in aspect fusion is harmful. In contrast, our proposed variants can greatly enhance the retrieval performance. Our research not only sheds light on the limitations of MADRAL but also provides valuable insights for future studies on more powerful multi-aspect dense retrieval models. Code will be released at: https://github.com/sunxiaojie99/Reproducibility-for-MADRAL.

Keywords: Multi-aspect Retrieval · Dense Retrieval · Aspect Learning

1 Introduction

Standing on the shoulders of pre-trained language models (PLMs) [5,22], dense retrieval models have exhibited impressive performance in the first stage of information retrieval [6,7,11,14]. Most dense retrieval models concentrate on unstructured textual data, while much less attention has been paid to structured item retrieval such as product search and people search. These scenarios have a wide population of users and the aspect information like brand (e.g., "Apple") and affiliation (e.g., "Microsoft") can be pivotal to enhance relevance matching. Nonetheless, it remains largely unexplored how to effectively integrate these aspects within dense retrieval models.

N. Goharian et al. (Eds.): ECIR 2024, LNCS 14611, pp. 194–209, 2024.
https://doi.org/10.1007/978-3-031-56066-8_17

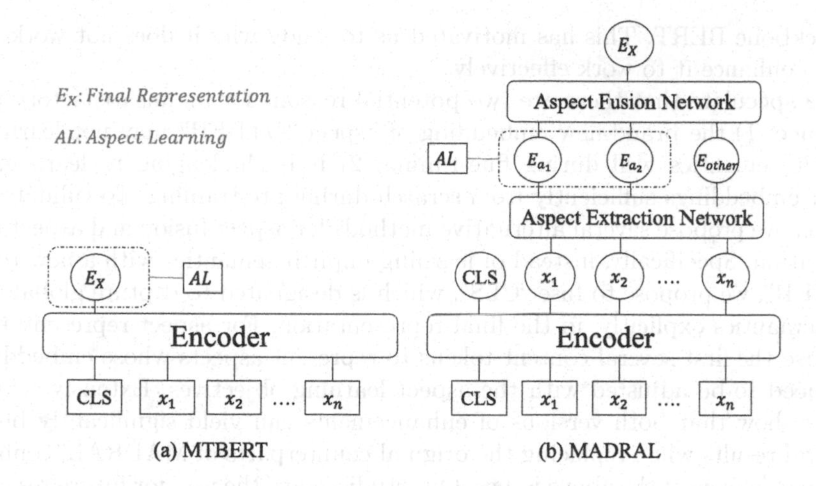

Fig. 1. Two multi-aspect dense retrieval models proposed by Kong et al. [10].

Recently, Kong et al. [10] initiated such a study by proposing a Multi-Aspect Dense Retriever with Aspect Learning, named MADRAL, and a simpler yet competitive baseline MTBERT. As illustrated in Fig. 1, MADRAL has three major components, i.e., aspect extraction, aspect learning, and aspect fusion, to produce the final representation E_X. Specifically, this model employs an aspect extraction network to extract extra aspect embeddings alongside the initial BERT parameters and conducts aspect learning by predicting the value IDs of an aspect (e.g., the ID of "Beauty" in the vocabulary of the product category). Notably, a special aspect "OTHER" is included to capture the implicit semantics that the explicit aspects cannot cover. Then for relevance matching, these aspect embeddings are integrated using an aspect fusion network to produce the final query/item representation. In contrast, MTBERT only conducts aspect learning on the CLS token, which is also used for relevance matching. Both models significantly outperform the original BERT and MADRAL can achieve much more compelling performance. The framework of MADRAL is insightful for the research on multi-aspect dense retrieval.

Although claimed to be effective, MADRAL has been experimented on proprietary data (i.e., Google shopping) that is not accessible to the public. The code of MADRAL has not been released either, which makes it even harder to reproduce the experimental results in [10]. Since Google shopping data has aspect information of both queries and items, it is also unknown whether MADRAL will be effective on other datasets of different properties. We have tried to reproduce its performance on the public MA-Amazon data, which has large-scale real-world queries and multiple aspect information associated with the items, but surprisingly find that MADRAL[1] has significantly worse performance than

[1] The authors have not provided their code upon our request but verified our implementation of MADRAL.

its backbone BERT. This has motivated us to study why it does not work and how to enhance it to work effectively.

We speculate that there are two potential reasons for its unsatisfactory performance: 1) the brand-new embedding of aspect "OTHER" may not learn the implicit semantics well during fine-tuning; 2) it is challenging to learn extra aspect embeddings sufficiently from scratch during pre-training. To validate the reasons, we propose several alternative methods for aspect fusion and aspect representation. Specifically, instead of learning implicit semantics with a new token "OTHER", we propose to fuse "CLS", which is designated to capture global content semantics explicitly, in the final representation. For aspect representation, we reuse the first several content tokens to represent aspects whose embeddings only need to be adjusted with the aspect learning objectives. Extensive experiments show that both versions of enhancements can yield significantly better retrieval results when replacing the original counterparts of MADRAL, confirming the existence of the above issues. Our studies pave the way for future research on this topic that uses MADRAL as a benchmark and also provide valuable insights into the development of more powerful multi-aspect dense retrievers.

2 Related Work

Dense Retrieval. Dense retrieval models typically adopt a bi-encoder architecture, which encodes a query and a document into two vectors and uses a similarity function like a dot product to measure their relevance. Karpukhin et al. [8] have explored pre-trained language models (PLMs) for information retrieval by using the BERT as the encoder and training it with in-batch negatives. This achieves superior performance compared to the models before the PLM era. Subsequently, researchers delved into various fine-tuning techniques to improve dense retrieval such as mining hard negatives [16,26], distilling the knowledge from cross encoders [23], and representing documents with multi-vector representations [9,13,28].

Most research efforts on dense retrieval have been spent on unstructured text until recently Kong et al. [10] proposed an effective method MADRAL that incorporates the structured aspect information of queries or items into the dense retrievers. This work leverages a typical way of injecting the aspect information [1,2] to the item representation, i.e., predicting the values associated with the aspects as an auxiliary training objective. Following this paradigm, Sun et al. [21] studied how to capture fine-grained semantic relations between different aspect values. As MADRAL [10] is the first multi-aspect dense retriever and adopts a typical manner of modeling the aspects, our reproducibility study on it will pave the way for future research that uses MADRAL as a benchmark in this direction.

Multi-field Retrieval. It has been a longstanding research topic on how to effectively utilize multi-field information such as titles, keywords, and descriptions in documents. The earliest attempt can date back to BM25F [18]. More recently, Liu et al. [12] explored the incorporation of multi-field information into the relevance models. Prior to the advent of PLMs, researchers have investigated leveraging the document fields in neural ranking models [3,4,27]. For example, Zamani et al. [27] proposed to aggregate field-level representations using a matching network and trained the model with field-level dropout. There has been ongoing research on the utilization of multi-field information [19,24] after PLMs have become the dominant retriever backbone. For instance, Shan et al. [19] proposed to leverage field-level cross interactions between queries and items as an auxiliary fine-tuning objective to improve retrieval performance. Sun et al. [20] treated item aspects as text strings and proposed a pre-training method to enhance the retriever.

Although the aspects can be simply treated as fields, multi-field retrievers only focus on the document side and cannot handle the case that query aspects are also available. Moreover, aspects and fields have some essential differences: fields are comprised of unstructured text that has infinite semantic space, whereas an aspect is defined by a finite set of values, serving as the aspect annotations. Consequently, multi-aspect and multi-field retrieval face distinct challenges. MADRAL [10] has been experimented on the data having aspects for both queries and items, and it was not compared to any baselines that treat aspects as fields. Since we use the public MA-Amazon dataset that only has item aspects, we also include a straightforward baseline that uses aspects as fields and concatenates the aspect texts, i.e., BIBERT-CONCAT in Sect. 7.

3 Preliminaries of Multi-aspect Dense Retrievers

Task Definition. In multi-aspect dense retrieval, queries and candidate items can have multiple aspects such as brand, color, and category. Given a query q or item i, each of its associated aspects a has a finite vocabulary of value set, denoted as V_a, and an embedding lookup table $T_a \in \mathbb{R}^{|V_a| \times H}$, where each value ID maps to an H-dimensional vector. Suppose that the aspect set is A containing k aspects, i.e., $A = a_1, a_2, \cdots, a_k$, their corresponding annotated value sets are $\mathcal{A}_{a_1}, \mathcal{A}_{a_2}, \cdots, \mathcal{A}_{a_k}$. The content tokens of q or i (that can include titles and descriptions) are denoted as $X = x_1, x_2, \cdots, x_n$. A multi-aspect dense retrieval model aims to learn effective representations of q and i by incorporating the aspect information and capturing the content semantics so that their similarities can reflect their relevance.

Multi-aspect Dense Retrievers with Aspect Learning. As shown in Fig. 1 and 2, typical multi-aspect dense retrievers [10] usually have three major components: 1) *Aspect Representation*, that either declares extra aspect embeddings (e.g., in MADRAL) or reuses the "CLS" token (e.g., in MTBERT) to capture the aspect information; 2) *Aspect Learning*, that injects the aspect-value information into the aspect representation by predicting its associated value IDs during pre-training (may also be beneficial in fine-tuning); 3) *Aspect Fusion* that merges the learned aspect representations into the final query/item representation for

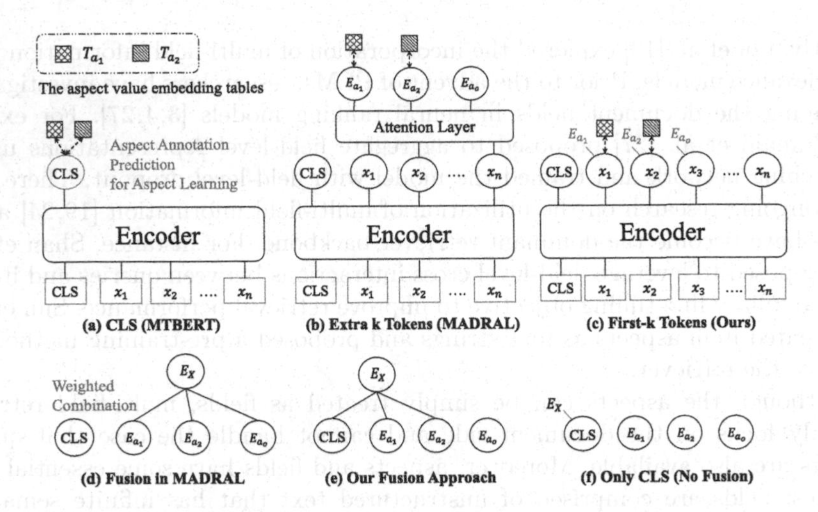

Fig. 2. The upper figures illustrate the aspect representation and learning of MTBERT, MADRAL, and our variant. The lower figures show aspect fusion methods to yield the final representation E_X. The weighted combination can be CLS gating or presence weighting. a_o denotes the special aspect "OTHER".

relevance matching during fine-tuning. In MTBERT, all the aspect learning is conducted on the "CLS" token, so no additional fusion is needed.

In the next two sections, we elaborate on the component variants of aspect representation and aspect fusion we study. Since queries and items have the same learning process, we only use items in the illustration for brevity.

4 Component Variants: Aspect Representation

To effectively incorporate the aspect-value information into an item, reasonable aspect representation is pivotal. Figure 2(a), 2(b), and 2(c) show the three variants of aspect representations we compare.

Reusing CLS Token (MTBERT). MTBERT [10] reuses the "CLS" token to conduct aspect learning (see Fig. 2(a)). It naturally injects aspect annotations of an item into the "CLS" token that also captures content semantics. The aspect information can be learned on top of "CLS", which can be a decent starting point. However, it compresses the information from multiple aspects and the content into a single token without weighting mechanisms. So, it does not differentiate the importance of each information source to relevance ranking, which could yield suboptimal retrieval results.

Declaring k Extra Embeddings (MADRAL). MADRAL [10] represents aspects with extra embeddings (Fig. 2(b)), that are computed based on the attention over the encoded content tokens $Encoder(X) = Encoder(x_1, x_2, \cdots, x_n)$, where *Encoder* can be any transformer-based encoders like BERT. The aspect embeddings E_A, stacked from $E_{a_1}, E_{a_2}, \cdots, E_{a_k}$, is computed as follows:

$$\begin{aligned} E_A &= Attention(QW^Q, Encoder(X)W^K, Encoder(X)W^V), \\ Attention(Q, K, V) &= softmax(\frac{QK^T}{\sqrt{H}})V. \end{aligned} \tag{1}$$

Attention is the multi-head attention function involving Q, K, V in the standard transformer [25]. Q is the set of aspect embeddings in this case. In this way, each aspect has its own representation and can be dedicated to its own learning. The influence of each aspect on the final representation can be automatically learned. However, in MADRAL, these new parameters are only learned from the aspect learning objectives introduced in Sect. 6, which could be challenging to learn them well from scratch, especially when there are not many aspect annotations.

Reusing First-k Content Tokens. Instead of training extra k aspect embeddings from scratch, we propose an alternative approach that reuses the encoder output of the first k content tokens, i.e., $x_1, x_2, \cdots, x_k$, to represent the k-associated aspects (shown in Fig. 2(c)). In MADRAL, the embeddings of content tokens are loaded from a pre-trained BERT and also updated by the masked language model (MLM) loss when ingesting the local corpus. Hence, they are learned more sufficiently and can serve as better starting points than brand-new extra tokens. The tokens at the beginning of the content are usually important to represent the content semantics. Guiding these tokens with the aspect learning loss is a way that not only binds the aspect information to content tokens like MTBERT does but also differentiates the influence of each aspect on the final representation like MADRAL does. So, it can leverage the advantages from both perspectives.

5 Component Variants: Aspect Fusion

During relevance matching, the aspect embeddings are fused into a single item embedding E_X, i.e.,

$$E_X = \sum_{a \in A} w_a E_a. \tag{2}$$

MTBERT does not have the phase of aspect fusion and uses the CLS token directly for relevance matching. MADRAL [10] has three fusion networks: weighted sum, CLS-gating, and presence weighting. Since the first one does not perform well [10], we only study the latter two. We will elaborate from two perspectives: the weighting mechanism and the objects to fuse.

5.1 Weighting Mechanism

CLS Gating. The encoded embedding of "CLS" is projected to $|A|$ (i.e., the number of aspects) logits, with a linear layer: $Linear(E_{CLS}) \in \mathbb{R}^{|A|}$, and the softmax weight computed for each logit is used as the final weight. In other words, taking one logit $\gamma_a \in Linear(E_{CLS})$ for example, $w_a = \frac{e^{\gamma_a}}{\sum_{a' \in A} e^{\gamma_{a'}}}$.

Presence Weighting. This mechanism computes the weight of each aspect according to its presence probability in the item, i.e., $w_a = P(I_a)\gamma_a$, where I_a indicates that item I has annotated values for a, $P(I_a) = Sigmoid(E_a)$, and γ_a is a learned parameter. $P(I_a)$ is learned with a cross-entropy loss based on whether an aspect has associated values in an item, which will be described in Sect. 6.

5.2 Objects to Fuse

The objects to fuse into the final item representation have a huge impact on retrieval performance, which we will show in the experiments. For both CLS-Gating and presence weighting, besides the original objects MADRAL fuses, we propose an alternative approach for fusion that slightly revises the fusion objects and can greatly enhance the performance. We introduce both ways as follows:

Aspect and Implicit Token ("OTHER"). In MADRAL, for both CLS-Gating and Presence Weighting, besides the standard aspects like brand and color, it adds a special aspect "OTHER" to capture the important information that may not be included in the explicit aspects. No aspect learning is conducted on this special aspect and it is supposed to learn implicit semantics automatically. If we denote the special aspect "OTHER" as a_o, the k elements in the set A in Equation (2) becomes:

$$A = \{a_1, a_2, \cdots, a_{k-1}, a_o\}. \tag{3}$$

The fusion weights and the embedding of a_o need to be learned during fine-tuning. When there is not sufficient training data with relevance labels to fine-tune the retriever, it could be difficult for the model to learn them well, especially a_o, which inevitably harms the retrieval performance.

Aspect and Explicit Token ("CLS"). We believe that an effective item representation should capture both content semantics and the aspect information of the item. Rather than fusing with the embedding of "OTHER" that learns implicit information, we use the "CLS" token that captures the global item semantics explicitly as a pseudo aspect in the fusion. In other words, the set A in Equation (2) becomes:

$$A = \{a_1, a_2, \cdots, a_{k-1}, CLS\}. \tag{4}$$

Then, only fusion weights need to be learned during fine-tuning and the objective is clear: balancing the effects of content and aspects on the final representation for relevance matching.

Only CLS (No Aspect Fusion). Though we have pre-trained the aspect embeddings, during relevance matching (or fine-tuning), we do not fuse these aspect embeddings but instead use the embedding of "CLS" as the item embedding. In this way, the aspect learning process conducted on the extra k embeddings can be considered as purely multi-task learning that could guide the underlying parameters in the encoder to a better optimum. Then, the content tokens also carry some aspect information and the final "CLS" embedding could be a better representation for relevance matching.

6 Aspect Learning and Overall Training Objectives

Aspect Prediction (AP). The typical way of learning aspect embeddings is to predict the annotated value IDs of the aspects [1,2]. MADRAL [10] also adopts this method. Take an arbitrary aspect a for instance, given its ground-truth annotation set $\mathcal{A}_a$ and its global value set V_a, the loss function is:

$$\mathcal{L}_{AP}^{a} = -\sum_{v^+ \in \mathcal{A}_a} log \frac{exp(E_a \cdot E_{v^+})}{\sum_{v \in V_a} exp(E_a \cdot E_v)}, \quad (5)$$

where $E_v / E_{v^+} \in \mathbb{R}^H$ is the aspect value embedding from a's embedding lookup table T_a. This is the major loss function to learn aspect embeddings.

Aspect Presence Prediction (APP). MADRAL [10] also proposes the loss of predicting whether an item has a valid value for a certain aspect as a part of presence weighting (introduced in Sect. 5.1). The APP loss is essentially a binary classification loss:

$$\mathcal{L}_{APP}^{a} = -y_a \log(P(I_a)) - (1 - y_a) \log(1 - P(I_a)), \quad (6)$$

where $y_a = 0$ when $\mathcal{A}_a = \varnothing$ and $y_a = 1$ otherwise. Since this objective can also guide aspect embedding training even if not used in the weighting function, we include it as an aspect learning objective, supplementary to the AP loss.

Pre-training Objective. The backbone model is pre-trained on the local corpus to adapt the model parameters and learn the aspect embeddings. The overall pre-training objectives consist of the masked language model (MLM) loss and the aspect learning objectives that can be scaled with λ_p, i.e.,

$$\mathcal{L}_{pretrain} = \mathcal{L}_{MLM} + \lambda_p \sum_{a \in A \setminus \{\text{OTHER}\}} (\mathcal{L}_{AP}^{a} + \mathcal{L}_{APP}^{a}). \quad (7)$$

Note that $\mathcal{L}_{APP}^{a}$ is optional and only takes effect when needed.

Fine-tuning Objective. Similarly, the loss for fine-tuning has relevance loss, and the aspect learning loss (with optional $\mathcal{L}_{APP}^{a}$) controlled by λ_f, i.e.,

$$\mathcal{L}_{finetune} = \mathcal{L}_{REL} + \lambda_f \sum_{a \in A \setminus \{\text{OTHER}\}} (\mathcal{L}_{AP}^{a} + \mathcal{L}_{APP}^{a}), \quad (8)$$

where $\mathcal{L}_{REL}$ is a standard softmax cross-entropy loss that uses the relevant items and in-batch negative samples for training as in [11].

7 Experimental Settings

Multi-aspect Amazon ESCI Dataset (MA-Amazon). The MA-Amazon dataset [20] has 482k products with the aspects of "brand", "color", and "category" besides their titles and descriptions. Only items have aspect information. The coverage of brand, color, and category of levels 1-2-3-4 on the items are 94%, 67%, and 87%-87%-85%-71%, respectively. The relevance dataset for fine-tuning has 17k, 3.5k, and 8.9k real-world queries for training, validation, and testing, respectively. For each query, the relevance dataset provides an average of 20.1 items, each accompanied by relevance judgments - "Exact", "Substitute", "Complement", and "Irrelevant". As in [17,20], we treat *Exact* as positive instances and the other judgments as negative during training and for recall calculation. Although MA-Amazon does not have query aspects as the private Google shopping data does, it is public and has large-scale real-world queries with relevance judgments. We are unaware of other such public datasets, so we only conduct experiments on MA-Amazon.

Methods for Comparison. We compare the MADRAL variants with various dense retrieval baselines, some incorporating aspect information and others not. Besides the baselines the original MADRAL compares, we also include a baseline that uses the aspects as text strings rather than conducting aspect classification for its learning. The baselines are: *BIBERT*: A typical bi-encoder retriever and the backbone of MADRAL, using BERT's `CLS` token for query and item encoding. It is pre-trained with $\mathcal{L}_{MLM}$ in Eq. (7) and fine-tuned with $\mathcal{L}_{REL}$ in Eq. (8); *Condenser*: An advanced pre-trained model for textual dense retrieval. It enhances the `CLS` embedding during pre-training by connecting middle-layer tokens to the top layers; *BIBERT-CONCAT*: A straightforward method that concatenates the text strings of aspect annotations with item content, adds an indicator token before each aspect, and also uses $\mathcal{L}_{MLM}$ for pre-training and $\mathcal{L}_{REL}$ for fine-tuning; *MTBERT*: A multi-task model based on BIBERT proposed in [10], reusing CLS for aspect prediction alongside MLM during pre-training. The MADRAL variants are: *MADRAL-ori*: The best original MADRAL model [10] introduced in Sect. 4 & 5.1; *MADRAL-en-v1*: Our first enhanced version of MADRAL-ori that only refines it with the best aspect fusion method in Sect. 5. They only differ during fine-tuning; *MADRAL-en-v2*: The second enhanced version of MADRAL-ori that incorporates the change in version 1 and the best aspect representation in Sect. 4.

Evaluation Metrics. We use recall (R) and normalized discounted cumulative gain (NDCG) as evaluation metrics. Specifically, we report R@100, R@500, NDCG@10, and NDCG@50. As in [17,20], the gains of E, S, C, and I judgments are set to 1.0, 0.1, 0.01, and 0.0, respectively. We conducted two-tailed paired t-tests ($p < 0.05$) to check statistically significant differences.

Implementation Details. Since MADRAL does not release code, we have implemented all the MADRAL variants and the baseline methods on our own to ensure consistent experimentation details and fair comparisons. **Pre-training.** For all methods, we share the encoder for both queries and items to promote knowledge sharing. In particular, we pre-train the models on the products in MA-Amazon to acquire the shared encoder for subsequent fine-tuning. In line with prior research [14,15], we initialize all BERT components using Google's public checkpoint and employ the Adam optimizer with the linear warm-up technique. The learning rate and pre-training epoch are set to 1e-4 and 20 respectively. We accommodate a maximum token length of 156 and employ MLM mask ratios of 0.15. For the scaling coefficient of AP and APP objectives, i.e., λ_p in Eq. 7, we slightly tune it from 0 to 0.5. Based on the validation results, it is set to 0.1. **Fine-tuning.** For both datasets, we fine-tune all the models for 20 epochs. Following the previous work [8], we include a hard negative sample for each query besides in-batch negatives. We use a learning rate of 5e-6 and a batch size of 64. The maximum token lengths are set to 32 for queries and 156 for items. λ_f in Eq. 8 is scanned from {0,0.05,0.1,0.2,0.3}, and the best value is 0 according to evaluation results. We conduct fine-tuning on the pre-trained model checkpoints every two epochs and select the best-performing one on the validation set.

8 Experimental Results

8.1 Comparisons Between MADRAL Variants and Baselines

The results of the baselines, the original MADRAL, and our two variants of enhanced MADRAL are shown in Table 1. Among the baselines, we find that better pre-trained models (i.e., Condenser) have better performance, and incorporating the aspect information (MTBERT and BIBERT-CONCAT) can boost the retrieval performance. Note that the method that considers aspects as text strings, i.e., BIBERT-CONCAT, achieves competitive performance compared to methods that use aspects as auxiliary training objectives. For instance, the performance of MTBERT is better than BIBERT-CONCAT at lower positions (R@500) but worse at higher positions (NDCG@10,50). This indicates that treating aspects as text strings is also an effective approach to leverage these aspects, as observed in [19,20]. Yet, careful learning strategies are required when concatenating the aspect strings with the original content, especially on the query side [20]. It is also promising to study how to combine the two ways of using aspects (i.e., as text strings and by conducting associated value ID prediction) [21]. Further discussions on leveraging aspects as strings are beyond the scope of this paper and interested users can refer to [19–21] for more information.

When we compare the performance of the MADRAL variants, it is surprising that the best results the original MADRAL can achieve (i.e., with presence weighting) are still worse than the baselines by a large margin. We attribute this to the insufficient learning of the special aspect - "OTHER" during fine-tuning. In contrast, the best variants we propose to enhance MADRAL in terms

Table 1. Comparisons between various retrievers. †, ‡, and ∗ indicate significant improvements over BIBERT, BIBERT-CONCAT, and MTBERT, respectively. The best overall and baseline results are underlined and bold.

Method	R@100	R@500	NDCG@10	NDCG@50
BIBERT	0.6075	0.7795	0.3148	0.3929
Condenser	0.6091	0.7801	0.3191	0.3960
BIBERT-CONCAT	0.6137	0.7814	<u>0.3223</u>	<u>0.4005</u>
MTBERT	<u>0.6139</u>†	<u>0.7849</u>†‡	0.3183†	0.3969†
MADRAL-ori	0.5016	0.7121	0.2086	0.2823
MADRAL-en-v1	0.6159†	0.7892†‡∗	0.3220†∗	0.4003†∗
MADRAL-en-v2	**0.6219**†‡∗	**0.7922**†‡∗	**0.3291**†‡∗	**0.4076**†‡∗

of the objects to fuse and aspect representation (denoted as "-en-v1" and "-en-v2" respectively) can significantly boost the retrieval performance. MADRAL-en-v1 uses "CLS" instead of "OTHER" and performs significantly better than MTBERT regarding almost all the metrics. It is also better than BIBERT-CONCAT at lower positions and similar at higher positions. Besides replacing "OTHER" with "CLS", MADRAL-en-v2 uses the first k content tokens to represent the aspects instead of declaring extra k aspect tokens. These two changes together lead to significantly better performance than all the baselines, showing that the proposed variants are effective ways of enhancements.

8.2 Comparisons of Fusion Methods

If we only modify the fusion methods of MADRAL, only fine-tuning will be affected and the changes are relatively small. We introduce our studies on this part before aspect representation variants that incur more changes. Table 2 shows how the fusion objects affect the retrieval performance when equipped with different aspect learning objectives (AP only or AP plus APP) during pre-training

Table 2. Comparisons of various aspect fusion methods. The best results are in bold. All the methods that use either explicit token "CLS" in fusion (see Sect. 5.2) or do not conduct fusion (see Sect. 5.1) are significantly better than fusion with implicit token "OTHER". † indicates significant improvements over the BIBERT.

Method	R@100	R@500	NDCG@10	R@100	R@500	NDCG@10
Aspect Fusion	Explicit Token("CLS")			Implicit Token("OTHER")		
CLS-Gating	0.6090	0.7832†	0.3158	0.4743	0.7002	0.1809
APP+CLS-Gating	0.6118†	0.7836†	**0.3194**†	0.4717	0.6973	0.1744
PresenceWeighting	**0.6148**†	**0.7878**†	0.3184†	0.5016	0.7121	0.2086
NoAspectFusion	0.6120†	0.7835†	0.3172†	–	–	–
APP+NoAspectFusion	0.6117†	0.7832†	0.3188†	–	–	–

and multiple weighting mechanisms during fine-tuning. We can see that for both CLS-gating and presence weighting, using the token "CLS" to explicitly capture content semantics can boost the performance over learning implicit semantics with the token "OTHER". This confirms our speculation that learning the embedding of "OTHER" sufficiently from scratch during fine-tuning can be challenging, which may be a less severe issue on the Google Shopping data in [10] than on the MA-Amazon data as it has more training data for fine-tuning.

When we do not conduct aspect fusion and only use CLS as the final representation during fine-turning, the performance is also significantly better than BIBERT and competitive with the aspect fusion methods with "CLS" in it. It indicates that the aspect learning objectives (AP and APP) during pre-training are beneficial for the backbone model's parameters. It again confirms that introducing the special token "OTHER" during fine-tuning and training it from scratch is the reason that harms model performance. A side observation is that using the aspect presence prediction objective as an auxiliary pre-training task can improve some metrics a little, e.g., NDCG@10, showing that it does not have to be paired with the presence weighting mechanism when training MADRAL.

8.3 Comparisons of Aspect Representation

We compare different ways of aspect representation in the model architecture in Table 3. The reported numbers are based on their best fusion methods, i.e., presence weighting for extra k and CLS gating for the rest. Similar to Table 2, we show the performance of both using "OTHER" and "CLS" in aspect fusion. It is obvious that reusing the first k content tokens as aspect representations outperforms declaring extra k embeddings in the original MADRAL. Since reusing existing tokens will not face the issue of insufficient learning of brand-new parameters, it is not surprising that "first k" can be a better aspect representation option. Another interesting finding is that the gap between using "CLS" and "OTHER" for reusing content tokens is much smaller than "Extra k". This means that based on the pre-trained model that attaches the aspect learning with existing content tokens, it reduces the difficulty for the model to act effectively by learning an implicit embedding "OTHER" during fine-tuning.

Table 3. Study of aspect representation approach. "br", "co", and "ca" are short for "brand", "color", and "category". The best results are in bold. † and ‡ indicates significant improvements over BIBERT and the "Extra k" method, respectively.

Method	"CLS" in Fusion			"OTHER" in Fusion		
	R@100	R@500	NDCG@10	R@100	R@500	NDCG@10
Extra k	0.6148†	0.7878†	0.3184†	0.5016	0.7121	0.2086
First k (br,co,ca)	0.6216†‡	0.7919†‡	0.3285†‡	0.6087‡	0.7839†‡	0.3157‡
First k (ca,co,br)	**0.6219**†‡	**0.7922**†‡	**0.3291**†‡	0.6132†‡	0.7859†‡	0.3179†‡
Random k	0.6188†‡	0.7874†	0.3230†‡	0.6081‡	0.7836†‡	0.3113‡

We also study whether the positions of the content tokens that the aspects are mapped to would affect model performance. By using the first 3 tokens to represent different aspects, i.e., (brand, color, category) and (category, color, brand), we do not see significant differences when CLS is in the fusion while the latter is better when "OTHER" is in the fusion. Since "category" is the most important aspect [10,20], it seems that mapping it to a higher position can help the model be stable when learning a new representation during fine-tuning. We also mapped the aspects to random k positions and conducted aspect learning, denoted as "Random k". As smaller positions are often more important in representing an item, we limited the random selection to within the first twenty positions. The performance of "Random k" (i.e., positions 3, 9, 15, and 7 as "brand", "color", "category", and "OTHER" when needed) is lower than "First k" but still beats "Extra k". It implies that reusing positions at higher positions for aspect representation would be better, which is not surprising since the beginning tokens are usually more important than the others.

Table 4. Accuracy@3 of the predicted values using the aspect embeddings learned after pre-training and fine-tuning.

Method	Category		Brand		Color	
	pre-train	fine-tune	pre-train	fine-tune	pre-train	fine-tune
MTBERT(al = 0)	0.9728	0.1118	0.9727	0.0004	0.7843	0.0060
MTBERT(al = 0.05)	0.9728	0.4758	0.9727	0.0066	0.7843	0.2367
MADRAL-en-v1(al = 0)	0.9712	0.8786	0.9811	0.8251	0.7725	0.2008
MADRAL-en-v1(al = 0.05)	0.9712	0.9664	0.9811	0.9649	0.7725	0.7333

8.4 Effect of AL Coefficient λ_f in Fine-Tuning

In the original paper [10], the aspect learning objectives during fine-tuning are helpful for the retrieval performance but we find that they would harm both MTBERT and MADRAL. Figure 3 and Table 4 show the retrieval performance and the accuracy of aspect prediction when using different coefficients of aspect learning during fine-turning. From Table 4, we observe that a small amount of aspect prediction (AP) loss during fine-tuning will boost the AP accuracy. However, the one with higher AP accuracy has worse retrieval performance, as shown in Fig. 3. The more AP is used, the more retrieval performance will drop. This implies that the learning objectives between relevance matching and aspect prediction guide the model in different directions.

8.5 Effect of Aspect Annotation Amount

We divide the aspect annotations of the item into three partitions and gradually include more aspect annotations for aspect learning during pre-training.

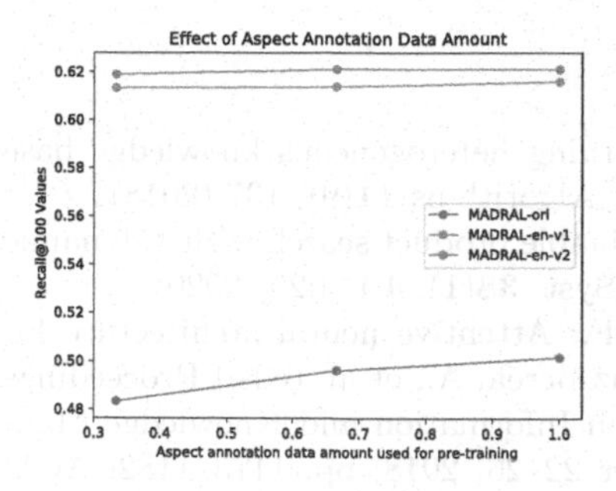

Fig. 3. Effect of AL Coefficient λ_f

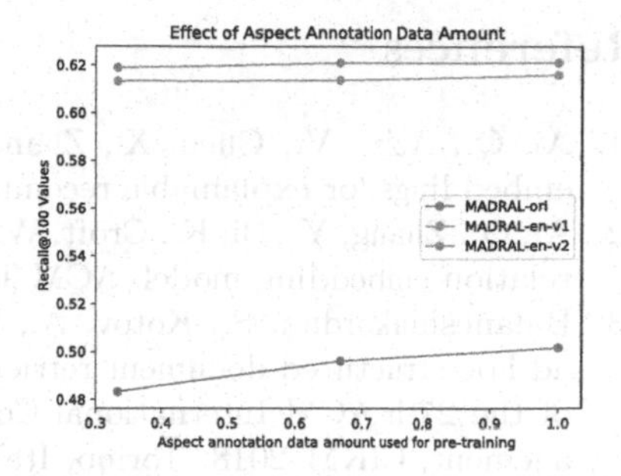

Fig. 4. Effect of Annotation Amount

The MLM loss on the entire corpus is always used for pre-training. Figure 4 illustrates the effect of aspect annotation amount on the original MADRAL and our two versions of enhancements. For MADRAL-ori, more annotations help the model perform better (i.e., from 0.4833 to 0.5016), which is consistent with our speculation that the extra k aspect embeddings require more aspect annotations for sufficient learning. When "CLS" replaces "OTHER" during fusion (in MADRAL-en-v1), more annotations bring fewer benefits (i.e., from 0.6133 to 0.6159), indicating that this fusion manner requires fewer aspect annotations to act effectively. When the first k content tokens are adjusted by aspect learning (in MADRAL-en-v2), the recall at 100 saturates with 2/3 aspect annotations. It shows that when aspect learning is used for refining existing important content tokens (first k), even fewer aspect annotations are needed to act effectively.

9 Conclusion

In conclusion, this paper presents a critical examination of the first multi-aspect dense retrieval model, MADRAL. Observing its failure on the public MA-Amazon data, we conduct a thorough investigation into MADRAL's components of aspect representation and fusion. We propose several alternative approaches for each component and compare them with their original counterparts. We find that it has a detrimental effect on retrieval performance to learn implicit semantics with the special aspect "OTHER". In contrast, the proposed variants, including replacing "OTHER" with "CLS" (that represents the overall content semantics explicitly) and representing aspects with the first few content tokens, have demonstrated significant improvements in retrieval performance.

Acknowledgments. This work was funded by the National Natural Science Foundation of China (NSFC) under Grants No. 62302486, the Innovation Project of ICT CAS under Grants No. E361140, the CAS Special Research Assistant Funding Project, the Lenovo-CAS Joint Lab Youth Scientist Project, and the project under Grants No. JCKY2022130C039.

References

1. Ai, Q., Azizi, V., Chen, X., Zhang, Y.: Learning heterogeneous knowledge base embeddings for explainable recommendation. Algorithms **11**(9), 137 (2018)
2. Ai, Q., Zhang, Y., Bi, K., Croft, W.B.: Explainable product search with a dynamic relation embedding model. ACM Trans. Inf. Syst. **38**(1), 4:1–4:29 (2020)
3. Balaneshinkordan, S., Kotov, A., Nikolaev, F.: Attentive neural architecture for ad-hoc structured document retrieval. In: Cuzzocrea, A., et al. (eds.) Proceedings of the 27th ACM International Conference on Information and Knowledge Management, CIKM 2018, Torino, Italy, October 22–26, 2018, pp. 1173–1182. ACM (2018)
4. Choi, J.I., Kallumadi, S., Mitra, B., Agichtein, E., Javed, F.: Semantic product search for matching structured product catalogs in e-commerce. CoRR abs/2008.08180 (2020)
5. Devlin, J., Chang, M., Lee, K., Toutanova, K.: BERT: pre-training of deep bidirectional transformers for language understanding. CoRR abs/1810.04805 (2018)
6. Gao, L., Callan, J.: Condenser: a pre-training architecture for dense retrieval. In: Moens, M., Huang, X., Specia, L., Yih, S.W. (eds.) Proceedings of the 2021 Conference on Empirical Methods in Natural Language Processing, EMNLP 2021, Virtual Event / Punta Cana, Dominican Republic, 7–11 November, 2021, pp. 981–993. Association for Computational Linguistics (2021)
7. Gao, L., Callan, J.: Unsupervised corpus aware language model pre-training for dense passage retrieval. In: Muresan, S., Nakov, P., Villavicencio, A. (eds.) Proceedings of the 60th Annual Meeting of the Association for Computational Linguistics (Volume 1: Long Papers), ACL 2022, Dublin, Ireland, May 22–27, 2022, pp. 2843–2853. Association for Computational Linguistics (2022)
8. Karpukhin, V., Oguz, B., Min, S., Wu, L., Edunov, S., Chen, D., Yih, W.: Dense passage retrieval for open-domain question answering. CoRR abs/2004.04906 (2020)
9. Khattab, O., Zaharia, M.: Colbert: efficient and effective passage search via contextualized late interaction over BERT. In: Huang, J.X., et al., (eds.) Proceedings of the 43rd International ACM SIGIR conference on research and development in Information Retrieval, SIGIR 2020, Virtual Event, China, July 25–30, 2020, pp. 39–48. ACM (2020)
10. Kong, W., et al.: Multi-aspect dense retrieval. In: Zhang, A., Rangwala, H. (eds.) KDD '22: The 28th ACM SIGKDD Conference on Knowledge Discovery and Data Mining, Washington, DC, USA, August 14–18, 2022,m pp. 3178–3186. ACM (2022)
11. Lin, J., Nogueira, R., Yates, A.: Pretrained Transformers for Text Ranking: BERT and Beyond. Morgan & Claypool Publishers, Synthesis Lectures on Human Language Technologies (2021)
12. Liu, B., Lu, X., Kurland, O., Culpepper, J.S.: Improving search effectiveness with field-based relevance modeling. In: Proceedings of the 23rd Australasian Document Computing Symposium, ADCS 2018, Dunedin, New Zealand, December 11–12, 2018, pp. 11:1–11:4. ACM (2018)
13. Luan, Y., Eisenstein, J., Toutanova, K., Collins, M.: Sparse, dense, and attentional representations for text retrieval. Trans. Assoc. Comput. Linguistics **9**, 329–345 (2021)
14. Ma, X., Guo, J., Zhang, R., Fan, Y., Cheng, X.: Pre-train a discriminative text encoder for dense retrieval via contrastive span prediction. In: Amigó, E., Castells, P., Gonzalo, J., Carterette, B., Culpepper, J.S., Kazai, G. (eds.) SIGIR '22: The

45th International ACM SIGIR Conference on Research and Development in Information Retrieval, Madrid, Spain, July 11–15, 2022, pp. 848–858. ACM (2022)
15. Ma, X., Guo, J., Zhang, R., Fan, Y., Ji, X., Cheng, X.: Prop: Pre-training with representative words prediction for ad-hoc retrieval. Proceedings of the 14th ACM International Conference on Web Search and Data Mining (2021)
16. Qu, Y., et al.: Rocketqa: An optimized training approach to dense passage retrieval for open-domain question answering. In: Toutanova, K., et al., (eds.) Proceedings of the 2021 Conference of the North American Chapter of the Association for Computational Linguistics: Human Language Technologies, NAACL-HLT 2021, Online, June 6–11, 2021, pp. 5835–5847. Association for Computational Linguistics (2021)
17. Reddy, C.K., et al.: Shopping queries dataset: a large-scale ESCI benchmark for improving product search. CoRR abs/2206.06588 (2022)
18. Robertson, S., Zaragoza, H., Taylor, M.: Simple bm25 extension to multiple weighted fields. In: Proceedings of the thirteenth ACM international conference on Information and knowledge management, pp. 42–49 (2004)
19. Shan, H., Zhang, Q., Liu, Z., Zhang, G., Li, C.: Beyond two-tower: attribute guided representation learning for candidate retrieval. In: Proceedings of the ACM Web Conference 2023, pp. 3173–3181 (2023)
20. Sun, X., et al.: Pre-training with aspect-content text mutual prediction for multi-aspect dense retrieval. arXiv preprint arXiv:2308.11474 (2023)
21. Sun, X., et al.: A multi-granularity-aware aspect learning model for multi-aspect dense retrieval (2023)
22. Sun, Y., et al.: ERNIE: enhanced representation through knowledge integration. CoRR abs/1904.09223 (2019). http://arxiv.org/abs/1904.09223
23. Tahami, A.V., Ghajar, K., Shakery, A.: Distilling knowledge for fast retrieval-based chat-bots. In: Huang, J.X., Chang, Y., Cheng, X., Kamps, J., Murdock, V., Wen, J., Liu, Y. (eds.) Proceedings of the 43rd International ACM SIGIR conference on research and development in Information Retrieval, SIGIR 2020, Virtual Event, China, July 25–30, 2020, pp. 2081–2084. ACM (2020)
24. Ueda, A., Santos, R.L.T., Macdonald, C., Ounis, I.: Structured fine-tuning of contextual embeddings for effective biomedical retrieval. In: Diaz, F., Shah, C., Suel, T., Castells, P., Jones, R., Sakai, T. (eds.) SIGIR '21: The 44th International ACM SIGIR Conference on Research and Development in Information Retrieval, Virtual Event, Canada, July 11–15, 2021, pp. 2031–2035. ACM (2021)
25. Vaswani, A., et al.: Attention is all you need. In: Advances in Neural Information Processing Systems 30 (2017)
26. Xiong, L., et al.: Approximate nearest neighbor negative contrastive learning for dense text retrieval. In: 9th International Conference on Learning Representations, ICLR 2021, Virtual Event, Austria, May 3–7, 2021. OpenReview.net (2021)
27. Zamani, H., Mitra, B., Song, X., Craswell, N., Tiwary, S.: Neural ranking models with multiple document fields. In: Chang, Y., Zhai, C., Liu, Y., Maarek, Y. (eds.) Proceedings of the Eleventh ACM International Conference on Web Search and Data Mining, WSDM 2018, Marina Del Rey, CA, USA, February 5–9, 2018, pp. 700–708. ACM (2018)
28. Zhang, S., Liang, Y., Gong, M., Jiang, D., Duan, N.: Multi-view document representation learning for open-domain dense retrieval. In: Muresan, S., Nakov, P., Villavicencio, A. (eds.) Proceedings of the 60th Annual Meeting of the Association for Computational Linguistics (Volume 1: Long Papers), ACL 2022, Dublin, Ireland, May 22–27, 2022, pp. 5990–6000. Association for Computational Linguistics (2022)

Measuring Item Fairness in Next Basket Recommendation: A Reproducibility Study

Yuanna Liu[1](✉), Ming Li[2], Mozhdeh Ariannezhad[3], Masoud Mansoury[1,4], Mohammad Aliannejadi[1], and Maarten de Rijke[1]

[1] University of Amsterdam, Amsterdam, The Netherlands
{y.liu8,m.mansoury,m.aliannejadi,m.derijke}@uva.nl
[2] AIRLab, Amsterdam, The Netherlands
m.li@uva.nl
[3] Booking.com, Amsterdam, The Netherlands
mozhdeh.ariannezhad@booking.com
[4] Discovery Lab, Elsevier, Amsterdam, The Netherlands

Abstract. Item fairness of recommender systems aims to evaluate whether items receive a fair share of exposure according to different definitions of fairness. Raj and Ekstrand [26] study multiple fairness metrics under a common evaluation framework and test their sensitivity with respect to various configurations. They find that fairness metrics show varying degrees of sensitivity towards position weighting models and parameter settings under different information access systems. Although their study considers various domains and datasets, their findings do not necessarily generalize to next basket recommendation (NBR) where users exhibit a more repeat-oriented behavior compared to other recommendation domains. This paper investigates fairness metrics in the NBR domain under a unified experimental setup. Specifically, we directly evaluate the item fairness of various NBR methods. These fairness metrics rank NBR methods in different orders, while most of the metrics agree that repeat-biased methods are fairer than explore-biased ones. Furthermore, we study the effect of unique characteristics of the NBR task on the sensitivity of the metrics, including the basket size, position weighting models, and user repeat behavior. Unlike the findings in [26], Inequity of Amortized Attention (IAA) is the most sensitive metric, as observed in multiple experiments. Our experiments lead to novel findings in the field of NBR and fairness. We find that Expected Exposure Loss (EEL) and Expected Exposure Disparity (EED) are the most robust and adaptable fairness metrics to be used in the NBR domain.

1 Introduction

Fairness of information access systems is increasingly drawing attention [8]. Such systems are not only required to have high accuracy, but they should also be fair to both users and providers. Usually, items are exposed to users as ranked lists based on a *relevance* or *utility* score. The *exposure* of items influence users' browsing, clicking, and

M. Ariannezhad—Work done when the author was a member of AIRLab at the University of Amsterdam.

N. Goharian et al. (Eds.): ECIR 2024, LNCS 14611, pp. 210–225, 2024.
https://doi.org/10.1007/978-3-031-56066-8_18

purchase behavior [26]. However, items may not receive fair exposure due to algorithmic or data biases [22]. To understand different notions and aspects of fairness, many fairness definitions and metrics have been proposed. Raj and Ekstrand [26] unify several fair ranking metrics under a common evaluation framework and compare them empirically using different information access tasks, viz. book recommendation and scholarly article retrieval and re-ranking. The authors focus on provider-side group fairness of ranked lists and design a sensitivity analysis to evaluate the robustness of fairness metrics w.r.t. position weighting models and parameter settings. Importantly, they find that *fairness metrics show different patterns of sensitivity for different search and recommendation tasks* — this means that the lessons learned in one search and recommendation scenario usually cannot be directly and completely transferred to another domain.

Item Fairness and Next Basket Recommendation. Recently, there has been increased interest in the task of next basket recommendation (NBR) [3]. NBR models users' preferences based on a sequence of historical sets of items (baskets, playlists, reading lists, ...) and then predicts, for each user, a set of items they are likely to purchase, listen to, or read next. The NBR task is important because it is relevant across a broad range of domains, from e-commerce and travel to education and entertainment. We are interested in item fairness in the context of NBR: is the exposure assigned by an NBR system fairly distributed and do the main findings from [26] generalize to the NBR task? In particular, we focus on the following findings from [26] and examine their validity in the context of NBR:

- Metrics usually disagree on the orderings of methods; we expect to find similar patterns for NBR baselines.
- For the FairTREC full retrieval task, changing the size of the ranked list has no impact on the performance of fairness measurements. But the fairness metrics Attention-Weighted Rank Fairness (AWRF) [27], IAA, EEL, EED, and Expected Exposure Relevance (EER) are stable on the GoodReads recommendation task, while Demographic Parity (logDP), Exposed Utility Ratio (logEUR), and FAIR [34] change with different-sized ranked lists and even alter the relative order of recommendation algorithms. We change the basket size and focus on the magnitude of the change of metric values and whether the fairness order of NBR methods will be affected.
- When used with different position weighting models, fairness metrics show different degrees of robustness. Position weighting models influence metric values and algorithms' ordering, especially for EEL and Realized Utility Ratio (logRUR). For the NBR scenario, we experiment with the document-based click-through rate model (DCTR) [5], which means each position has the same exposure in a list. We expect to identify the most unstable metrics.

Item Fairness and Repeat Behavior. Going beyond the above generalizability questions, there is a specific characteristic of NBR that may affect item fairness in unknown ways: in many NBR scenarios, users display a significant amount of repeat behavior, whereby they purchase or consume the same item multiple times. Some users mostly purchase repeat items (i.e., items that they consumed before) as part of their next basket, while others tend to explore items (items that they never bought before). Ekstrand et al. [7] suggest we need to look at metric values across different user groups rather than at the population as a whole. Following this advice, we design experiments to assess item fairness across different user groups with varying degrees of repeat purchase behavior.

To the best of our knowledge, we are the first to study fairness in next basket recommendation. We focus on four research questions:

(RQ1) What is the fairness ranking of NBR methods? Are the lessons observed in [26] valid for the NBR task?
(RQ2) Which fairness metrics are more robust to basket size?
(RQ3) Which fairness metrics are more robust to position weighting models for NBR?
(RQ4) How does repeat purchase behavior affect item fairness?

2 Related Work

Next Basket Recommendation. NBR systems make personalized product recommendations to users based on their historical baskets. Repeat purchase behavior is a prominent pattern of NBR, which makes it different from typical recommendation domains, such as movie recommendation [11]. Since the models designed for recommending repeat items and explore items vary significantly, recent work proposes more targeted task settings and algorithms [14, 17, 18] to address the unique challenges in this domain by modeling the repeat behavior of the users while trying to optimize the explore behavior of the users.

A significant amount of research concentrates on employing neural networks to learn representations for sequences of baskets. The effectiveness of recurrent neural networks (RNNs) in sequential modeling has led to their application in NBR. Yu et al. [32] propose a dynamic recurrent basket model, feeding a series of basket representations to the recurrent architecture to obtain the dynamic representation of a user. Le et al. [16] model correlation information to augment the representation of basket sequence. Yu et al. [33] propose a model based on graph neural networks (GNNs) for temporal sets prediction, where sets are constructed as weighted graphs and a graph convolutional network is applied to capture relationships among elements in each set. Ariannezhad et al. [2] analyze users' repeat consumption behavior and propose a repeat consumption-aware neural network for NBR.

However, recent nearest neighbor-based methods show more effective performance and efficiency than neural network-based baselines on NBR. Faggioli et al. [9] propose recency-aware, user-wise popularity and incorporate it into both user- and item-based collaborative approaches. Hu et al. [13] integrate temporal dynamics into personalized item frequency and then use a user-KNN method to make predictions. Naumov et al. [24] improve on [13] by considering time intervals between interactions.

Li et al. [19] reproduce NBR methods in a unified experimental setting and propose a new angle to evaluate the performance obtained from repeat items and explore items, respectively. According to whether a method tends to recommend repeat items or explore items, it is called repeat-biased or explore-biased. Our work reproduces several representative NBR methods in a unified experimental setting and focuses on measuring item fairness and exploring the impact of repeat behavior on fairness.

Item Fairness in Recommender Systems. Fairness research raises challenges for information access systems characterized by (i) a multi-stakeholder nature, (ii) a rank-based problem setting, (iii) the requirement of personalization in many cases, and (iv) the role of user feedback [8]. Wang et al. [30] collect fairness definitions in the recommendation literature and provide views of classifying fairness issues. Consumer fairness cares about whether users receive comparable recommendation quality from the system [21]. In contrast, provider fairness focuses on how to assign reasonable exposure to each document, content provider, or group [26]. The allocation criteria can be with reference to a distribution [27,31,34] or proportional to merit [6].

Some metrics have primarily undergone testing using small and/or synthetic datasets, and they encounter challenges in handling complex real-world information access applications where incomplete data and extreme cases happen [26]. Raj and Ekstrand [26] implement experiments on book recommendation and scholarly article retrieval tasks and test the sensitivity of fairness metrics towards parameter setting. Kowald et al. [15] reproduce the analyses of [1] to investigate how popularity bias causes unfairness for both long-tail items and low-mainstream users in the context of music recommendation. We can observe that the conclusions about the fairness evaluation in one domain cannot be completely transferred to another domain.

NBR has distinguishing characteristics, i.e., repeat items figure prominently amongst the recommended results and make a considerable contribution to accuracy [19]. There is no prior work studying fairness in the context of NBR. Li et al. [20] argues that the frequency bias harms the fairness of NBR system. To fill this gap, we reproduce the fairness evaluation and sensitivity experiments of [26] to (i) see if the patterns they found can be generalized to NBR domain, and (ii) select robust metrics suitable for NBR.

3 Reproducibility Setup

3.1 Problem Formulation

Our experiments concern two main parts: NBR and fairness evaluation. Firstly, in next basket recommendation, we denote D as the item set and Q as the user set. A basket is a set of items $B = [d_1, d_2, \ldots, d_m]$, where $d_i \in D$ denotes an item from the basket B. Items have no temporal order and hold equal significance in a basket. For each user $q \in Q$, there is a sequence of historical purchase baskets $B^q = [B^q_1, B^q_2, \cdots, B^q_n]$, where B^q_i indicates the i-th basket purchased by the user. The goal of NBR is to predict

Table 1. Notation used in the paper; adapted from [26].

$d \in D$	item	$q \in Q$	user
L	ranked list of N items (predicted basket)	$L^{-1}(i)$	the item in position i of basket L
$L(d)$	rank of item d in L	$y(d\|q)$	relevance of d to q
$\hat{y}(d\|q)$	predicted relevance of d to q	$G(L)$	group alignment matrix for items in L
G^+	popular group	G^-	unpopular group
$\mathbf{a}_L$	exposure vector for items in L	ϵ_L	the exposure of groups in L $(G(L)^T\mathbf{a}_L)$

the next basket L for each user. Then, we evaluate the item fairness of predicted baskets among all users using the fair ranking metrics in Sect. 3.4. Specifically, the predicted basket L can be seen as a ranked list where the ranking is based on the user-specific relevance score $\hat{y}(d \mid q)$ predicted by each NBR method. The goal of item fairness is to measure whether the exposure is fairly distributed among the groups according to a specific principle. The notation used in this paper is summarized in Table 1.

3.2 Datasets

Following [2,12,25], we use three publicly available datasets for our experiments: (i) Instacart,[1] which contains a sample of over three million grocery orders from users. Items belonging to the same order form a basket. (ii) Dunnhumby,[2] which includes household-level transactions over two years from 2,500 households. (iii) TaFeng,[3] which contains transaction data from a Chinese grocery store from November 2000 to February 2001. We treat all transactions of a user within a day as a basket.

For each dataset, we remove users with fewer than three baskets and items bought fewer than five times [2]. Due to the large number of baskets in Instacart, the calculation of some methods exceeds the memory, therefore we randomly sampled 20,000 users from Instacart before filtering [24]. Table 2 shows the statistics of three datasets after preprocessing. Avg. repeat ratio refers to the average proportion of repeat items in the ground truth baskets [19]. We split each dataset following [2,9,24]. The training baskets consist of all baskets of users except the last one. For users who have more than 50 baskets in the training data, we only consider their last 50 baskets in the training set [19]. The last baskets of all the users are split into 50% validation set and 50% test set.

3.3 NBR methods

We select the following representative NBR methods with open-source code, including frequency-based, nearest neighbor-based, and deep learning-based methods. According to the classification of NBR methods in [19], G-TopFreq and Dream are explore-biased methods (the recommended baskets are skewed towards explore items averaged over all users), which are also popularity-based methods. Other methods are repeated-biased

Table 2. Statistics of datasets after preprocessing.

Dataset	#Users	#Items	#Baskets	Avg. #baskets/user	Avg. #items/basket	Avg. repeat ratio
Instacart	19,210	29,399	305,582	15.91	10.06	0.60
Dunnhumby	2,482	37,162	107,152	43.17	10.07	0.43
TaFeng	10,182	15,024	82,387	8.09	6.14	0.21

[1] https://www.kaggle.com/c/instacart-market-basket-analysis/data.

[2] https://www.dunnhumby. com/source-files/.

[3] https://www.kaggle.com/datasets/chiranjivdas09/ta-feng-grocery-dataset.

Table 3. Summary of fairness metrics; adapted from [26].

Category	Metrics	Goal	Binomial	More Fair
Equal opportunity	logEUR	Exposure proportional to relevance	✓	○
	logRUR	Click-through rate proportional to relevance	✓	○
	IAA	Exposure proportional to predicted relevance	×	↓
	EEL,EER	Exposure matches ideal (from relevance)	×	EEL ↓, EER ↑
Statistical parity	EED	Exposure well-distributed	×	↓
	logDP	Exposure equal across groups	✓	○

methods (the recommended baskets are skewed towards repeat items). We cover various types of NBR method and study their fairness performance.

Frequency-Based Methods. (i) G-TopFreq recommends the most popular k items across all historical purchases to all users. (ii) P-TopFreq counts the k products with the highest frequency in each user's historical purchase records. (iii) GP-TopFreq firstly recommends personal history most popular items and then fills the empty slot with global most popular items.

Nearest Neighbor-Based Methods. (i) TIFUKNN [13] integrates temporal dynamics modeled by time-decayed weights to generate user representations. Then, the target user representation and its nearest neighbors are combined to make the prediction. (ii) UP-CF@r [9] incorporates recency-aware user-wise popularity in a collaborative filtering framework.

Deep Learning-Based Methods. (i) Dream [32] is an RNN-based model that learns a dynamic representation of a user and captures global sequential features among baskets. (ii) DNNTSP [33] constructs weighted graphs to learn basket-level element relationships. The attention mechanism is used to learn the temporal dependencies of sets and elements. Static and dynamic representations are fused by a gated updating mechanism. (iii) ReCANet [2] focuses on repeat consumption, combines user-item representations with historical consumption patterns, and models temporal signals by LSTM layers.

3.4 Fair Ranking Metrics

Raj and Ekstrand [26] unify several fair ranking metrics in a common framework and notation. They clarify the limitation of these metrics when applying to practical information access systems, and test the robustness of these metrics towards design and parameter choices. We follow the notation and fairness implementation of this paper. The fairness metrics shown in Table 3 are selected for the following reasons: (i) The predicted basket can be formulated as a ranking L. These metrics are well-known fairness metrics for rankings. (ii) These metrics cover two categories of fairness definitions: *statistical parity* (aimed at ensuring comparable exposure among groups) and *equal opportunity* (aimed at promoting equal treatment based on merit or utility, regardless of group membership) [26]. (iii) Since fair exposure is unlikely to be satisfied in any single ranking [4] and it is more practical to pursue fair exposure of items to overall users, we only consider the fairness of multiple rankings for NBR setting.

Assume $\pi(L \mid q)$ is a user-dependent distribution and $\rho(q)$ is a distribution over users, overall rankings among all the users follow the distribution of $\rho(q)\pi(L \mid q)$. $\epsilon_L = G(L)^T \mathbf{a}_L$ is the group exposure within a single ranking. Its expected value $\epsilon_\pi = E_{\pi\rho}[\epsilon_L]$ is the group exposure among all the rankings.

Equal Opportunity. Singh and Joachims [28] propose two ratio-based metrics. Exposed Utility Ratio (EUR) quantifies the deviation from the objective that the exposure of each group is proportional to its utility $Y(G)$:

$$\mathrm{EUR} = \frac{\epsilon_\pi (G^+) / Y(G^+)}{\epsilon_\pi (G^-) / Y(G^-)}. \tag{1}$$

Realized Utility Ratio (RUR) [28] models actual user engagement, the click-through rates for the groups $\Gamma(G)$ are proportional to their utility:

$$\mathrm{RUR} = \frac{\Gamma(G^+) / Y(G^+)}{\Gamma(G^-) / Y(G^-)}. \tag{2}$$

Biega et al. [4] propose Inequality of Amortized Attention (IAA), which takes the L_1 norm of the difference between cumulative exposure and cumulative system-predicted relevance $\hat{y}(d|q)$ for each group. For consistency, we normalize predicted relevance scores to be in the same range as exposure values:

$$\mathrm{IAA} = \|\epsilon_\pi - \hat{Y}\|_1. \tag{3}$$

Diaz et al. [6] define the target exposure ϵ^* as the expected exposure under the ideal policy. Expected Exposure Loss (EEL) is the distance between expected exposure and target exposure:

$$\mathrm{EEL} = \|\epsilon_\pi - \epsilon^*\|_2^2 = \|\epsilon_\pi\|_2^2 - 2{\epsilon_\pi}^T \epsilon^* + \|\epsilon^*\|_2^2. \tag{4}$$

EEL can be decomposed into EER $= 2{\epsilon_\pi}^T \epsilon^*$ (measuring the alignment of exposure and relevance) and Expected Exposure Disparity (EED).

Statistical Parity. EED [6] measures the inequality in exposure distribution across groups:

$$\mathrm{EED} = \|\epsilon_\pi\|_2^2. \tag{5}$$

Demographic Parity (DP) [28] measures the ratio of average exposure given to the two groups:

$$\mathrm{DP} = \epsilon_\pi (G^+) / \epsilon_\pi (G^-). \tag{6}$$

Following [26], DP is reformulated as $\mathrm{logDP} = \log(\epsilon_\pi(G^+) + 10^{-6}) - \log(\epsilon_\pi(G^-) + 10^{-6})$ to address the empty-group problem and enhance interpretability. logEUR and logRUR are defined in the same way.

Table 4. Position weighting models for computing $\mathbf{a}_L$; adapted from [26].

Metric	Model	Formula	Parameters
IAA	Geometric	$\gamma(1-\gamma)^{L(d)-1}$	patience γ
logDP, logEUR, logRUR	Logarithmic	$1/\log_2 \max\{L(d), 2\}$	–
EER, EED,EEL	RBP [23]	$\gamma^{L(d)}$	patience γ
EER, EED,EEL	Cascade	$\gamma^{L(d)-1} \prod_{j\in[0,L(d))}[1-\phi(y(L^{-1}(j) \mid y))]$	patience γ stopping probability function ϕ
–	DCTR	$1/\lvert L(d)\rvert$	–

3.5 Position Weighting Models

Since users are likely to pay decreasing attention to lower-ranked items (position bias) [4], position weighting models are required when computing exposure [6]. These metrics explicitly represent the position weighting model as a position weight vector $\mathbf{a}_L$, as shown in Table 4 [26]. In NBR, many works [14,29,32] treat recommended results as ranked lists and evaluate NBR methods based on ranking metrics, e.g. normalized discounted cumulative gain (NDCG). Therefore, it is required to take the position of the recommended items into account when applying fair ranking metrics in NBR.

3.6 Implementation Details

In order to evaluate group fairness using the above metrics, following previous work [10], the division of items is based on item popularity, i.e., the number of purchases in the historical baskets of all users. We define the top 20% items with the highest purchase as popular group G^+ and the remaining 80% of the items as unpopular group G^-. The default basket size is set to 10. For all NBR baselines, we perform a grid search based on the hyperparameter ranges given in the original papers to find the optimal hyperparameters using the validation set. For TIFUKNN, the number of nearest neighbors k is tuned on $\{100, 300, 500, 900, 1100, 1300\}$, the number of groups m is chosen from $\{3, 7, 11, 15, 19, 23\}$, the within-basket time-decayed ratio r_b and the group time-decayed ratio r_g are selected from $\{0.1, 0.2, 0.3, 0.4, 0.5, 0.6, 0.7, 0.8, 0.9, 1\}$, and the fusion weight α is tuned on $\{0, 0.1, 0.2, 0.3, 0.4, 0.5, 0.6, 0.7, 0.8, 0.9, 1\}$. For UP-CF@r, recency window r is tuned on $\{1, 5, 10, 25, 100, \infty\}$, locality q is tuned on $[1, 5, 10, 50, 100, 1000]$, and asymmetry α is tuned on $\{0, 0.25, 0.5, 0.75, 1\}$. For Dream and DNNTSP, the item embedding size is tuned on $\{16, 32, 64, 128\}$. For ReCANet, user embedding size and item embedding size are tuned on $\{16, 32, 64, 128\}$. We run each method 5 times with 5 fixed random seeds to eliminate the random initialization effect and report the average results. Following [13,19,33], we use three accuracy metrics. Recall assesses the capacity to retrieve all relevant items. NDCG measures ranking quality, which considers sequence order by giving the lower-ranked items a discount. Personalized Hit Ratio (PHR) computes the proportion of predicted baskets that include at least one item from the ground truth basket. We release our code and hyperparameters at https://github.com/lynEcho/NBR-fairness.

4 Experiments and Results

Fairness Valuation. To answer **RQ1**, we measure item fairness of the recommendation results obtained by each NBR method on Instacart, Dunnhumby, and TaFeng. We

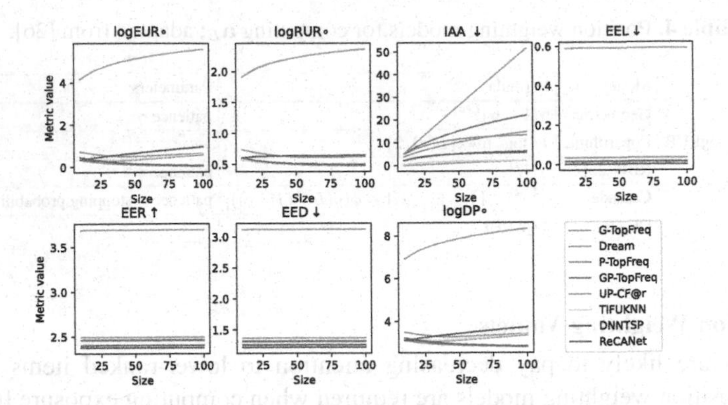

Fig. 1. Performance of different NBR methods in terms of item fairness with varying basket size. Each plot represents one item fairness metric.

implement the fairness metrics following the configurations from their original papers. Table 5 compares the NBR methods in terms of accuracy and fairness metrics. We make the following observation: (i) In most cases, repeat-biased methods exhibit more effectiveness in Recall, NDCG, and PHR, compared to explore-biased methods. (ii) Consistent with observations in [26], our experimental results lead to different orderings of the NBR methods when ranked using the fairness metrics. For instance, on Instacart, EEL indicates DNNTSP and P-TopFreq as the top two methods, whereas logRUR ranks TopFreq and GP-TopFreq as the top two. (iii) Metrics that consider equal opportunity, namely, logEUR, logRUR, and EEL deem repeat-biased methods fairer than the explore-biased ones (G-TopFreq and Dream). Surprisingly, according to EER, G-TopFreq and Dream are fairer than the repeat-biased methods. Both G-TopFreq and Dream only recommend popular items, which is unfair. (iv) Statistical-parity metrics, namely, EED and logDP, agree that repeat-biased methods are fairer than explore-biased methods. Because statistical parity aims at ensuring equal exposures across groups, however G-TopFreq and Dream only recommend popular items.

Basket Size. To answer **RQ2**, we change the basket size from 10 to 100 for each dataset. Figure 1 shows the variation of fairness values as basket size changes on Instacart.[4] It can be observed that: (i) IAA shows the highest sensitivity to basket size, as we see the highest increase in its value as the basket size grows. We relate it to the fact that as more items are considered in the list, it leads to more accumulation of gaps between position weights and predicted relevance. And the fairness order of P-TopFreq and G-TopFreq even changes. This differs from [26], where IAA exhibits stable behavior in the GoodReads recommendation task. (ii) For logDP, logEUR, and logRUR, we observe a mixed behavior for different basket sizes. As basket size increases, DNNTSP, GTop-Freq, and Dream become less fair in terms of all three metrics; PTop-Freq and ReCANet become fairer. GP-TopFreq, UF-CF@r, and TIFUKNN, however, exhibit

[4] We observe a similar trend on the Dunnhumby and TaFeng datasets. Because of space limitations, we only report the results on the Instacart dataset.

Table 5. Overall performance comparison of NBR methods.

Data-set	Method	Accuracy			Equal opportunity					Statistical parity	
		Recall↑	NDCG↑	PHR↑	logEUR∘	logRUR∘	IAA↓	EEL↓	EER↑	EED↓	logDP∘
Instacart	G-TopFreq	0.0704	0.0817	0.4600	4.1803	1.9140	3.8532	0.5899	3.6884	3.1169	6.9148
	Dream	0.0704	0.0817	0.4600	4.1803	1.9143 (0.0001)	0.1466 (0.0053)	0.5895	3.6879	3.1160 (0.0001)	6.9148
	P-TopFreq	0.3143	0.3339	0.8447	0.3419	0.5756	4.9492	0.0138	2.4561	1.3086	3.1112
	GP-TopFreq	0.3150	0.3343	0.8460	0.3616	0.5764	5.1071	0.0144	2.4633	1.3164	3.1313
	UP-CF@r	0.3377	0.3582	0.8586	0.7217	0.7383	5.1405	0.0351	2.4951	1.3688	3.4917
	TIFUKNN	0.3456	0.3657	0.8639	0.4275	0.5800	0.4227	0.0154	2.4095	1.2635	3.1972
	DNNTSP	0.3295 (0.0002)	0.3434 (0.0005)	0.8581 (0.0001)	0.3458 (0.0171)	0.6050 (0.0104)	2.1511 (0.0462)	0.0020 (0.0004)	2.3760 (0.0070)	1.2166 (0.0075)	3.1154 (0.0171)
	ReCANet	0.3490 (0.0001)	0.3699 (0.0001)	0.8668 (0.0003)	0.4588 (0.0184)	0.6069 (0.0056)	2.1972 (0.0691)	0.0173 (0.0004)	2.4000 (0.0015)	1.2559 (0.0019)	3.2283 (0.0184)
Dunnhumby	G-TopFreq	0.0897	0.0798	0.3795	6.2326	3.5166	2.8829	1.0020	3.1503	3.2310	6.6211
	Dream	0.0896 (0.0003)	0.0759 (0.0002)	0.3873 (0.0013)	6.2326	3.5511 (0.0020)	0.3248 (0.0035)	0.9940	3.1427	3.2154	6.6211
	P-TopFreq	0.1628	0.1562	0.5399	0.9611	1.6879	4.6402	0.4039	2.7369	2.2195	3.2579
	GP-TopFreq	0.1628	0.1562	0.5399	0.9620	1.6879	4.6480	0.4039	2.7369	2.2196	3.2588
	UP-CF@r	0.1699	0.1639	0.5536	1.2771	1.6287	4.4460	0.4279	2.7297	2.2364	3.5630
	TIFUKNN	0.1763	0.1683	0.5729	1.6677	1.8252	0.9255	0.6236	2.8502	2.5525	3.9346
	DNNTSP	0.0871 (0.0019)	0.0792 (0.0016)	0.4303 (0.0048)	0.5513 (0.0268)	1.7653 (0.0506)	0.1037 (0.0363)	0.2545 (0.0126)	2.7432 (0.0130)	2.0765 (0.0255)	2.8581 (0.0263)
	ReCANet	0.1730 (0.0011)	0.1625 (0.0010)	0.5655 (0.0018)	1.0334 (0.0204)	1.6296 (0.0334)	0.4417 (0.0127)	0.4775 (0.0075)	2.7926 (0.0066)	2.3489 (0.0139)	3.3279 (0.0197)
TaFeng	G-TopFreq	0.0812	0.0893	0.2565	5.4916	2.4948	2.9267	1.0932	3.1283	3.3155	7.6939
	Dream	0.0888 (0.0032)	0.0924 (0.0011)	0.2798 (0.0059)	5.4916	2.5259 (0.0072)	0.6672 (0.0055)	1.0929 (0.0001)	3.1280 (0.0001)	3.3150 (0.0002)	7.6939
	P-TopFreq	0.1087	0.0983	0.3608	0.4059	0.8820	5.5137	0.2735	2.7269	2.0944	2.4192
	GP-TopFreq	0.1183	0.1018	0.3740	0.4357	0.8986	5.6991	0.2775	2.7335	2.1050	2.4492
	UP-CF@r	0.1406	0.1222	0.4325	1.6427	1.4053	4.7278	0.6896	2.9401	2.7237	3.6597
	TIFUKNN	0.1618	0.1419	0.4697	1.1420	1.3862	0.8630	0.6299	2.8835	2.6074	3.1573
	DNNTSP	0.1483 (0.0004)	0.1239 (0.0004)	0.4482 (0.0011)	1.4265 (0.0645)	1.5422 (0.0444)	0.3508 (0.0332)	0.4712 (0.0246)	2.8531 (0.0121)	2.4183 (0.0367)	3.4427 (0.0647)
	ReCANet	0.1287 (0.0002)	0.1188 (0.0002)	0.4195 (0.0005)	0.9450 (0.0045)	1.2147 (0.0080)	0.1644 (0.0098)	0.4343 (0.0072)	2.8093 (0.0042)	2.3376 (0.0114)	2.9597 (0.0045)

↑ indicates that larger value means better fairness; ↓ indicates that smaller value means better fairness; ∘ means that the closer the value is to 0, the better the fairness. The number in each cell is mean, and the number in parentheses is standard deviation. The missing standard deviation represents 0.0000.

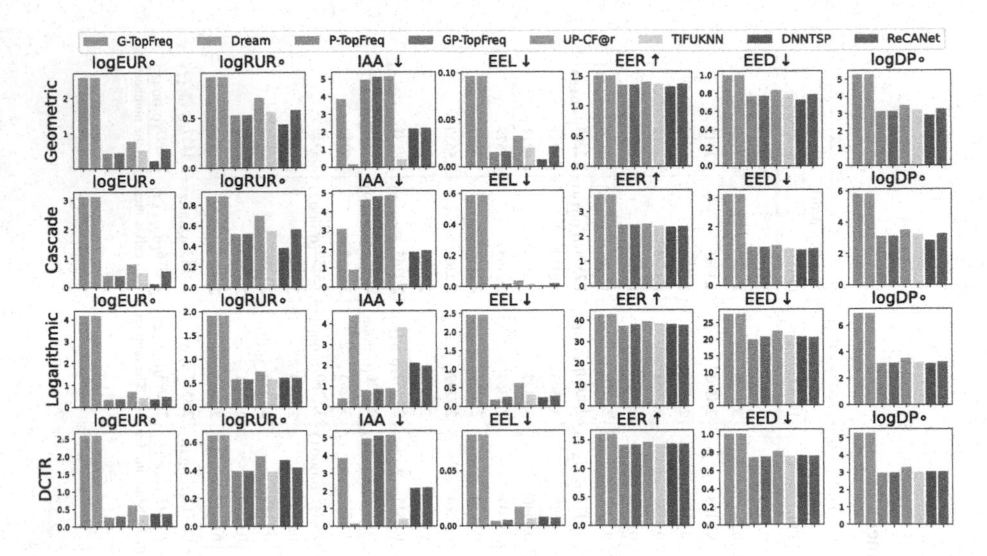

Fig. 2. Item fairness with varying position weighting models. Each column represents an item fairness metric. Each row represents a position weighting model.

non-monotonic changes; and these metrics reorder the NBR methods. (iii) In line with [26], we see that EEL, EED, and EER are stable with varying basket sizes.

Position Weighting Models. To answer **RQ3**, we perform experiments with different position weighting models (in Table 4) for each fairness metric and report the results of Instacart[5] in Fig. 2. The parameters are assigned to the default values: patience parameter $\gamma = 0.5$, and a stopping probability of 0.5.[6] We summarize our observations below: (i) Different from [26], we observe high sensitivity of IAA on the three datasets for different position weighting models. (ii) Except for IAA, other metrics maintain the order of NBR methods across different position weighting models. Also, some slight ordering adjustments happen among methods with similar fairness values. For example, for logEUR in Fig. 2, the relative fairness order of TIFUKNN and DNNTSP is changed when the position weight model is changed from Logarithmic to DCTR. (iii) We find that the exposure values for each position weighting model is quite different. Since these metrics capture the gap between exposure and relevance, this could explain why their values change significantly in Fig. 2 when using different position weighting models.

Repeat Purchase Behavior. To answer **RQ4**, we group users into five subgroups based on their repeat consumption behavior. Here, *repeat ratio* is defined as the proportion of repeat items in the ground-truth basket for each user [19]. Hence, we create five

[5] We observe a similar trend on the Dunnhumby and TaFeng datasets. Because of space limitations, we only report the results on the Instacart dataset.

[6] The Geometric and Rank-biased precision (RBP) share the same formula under this parameter setting. Therefore, we only report the results obtained by the Geometric weighting model for fairness metrics.

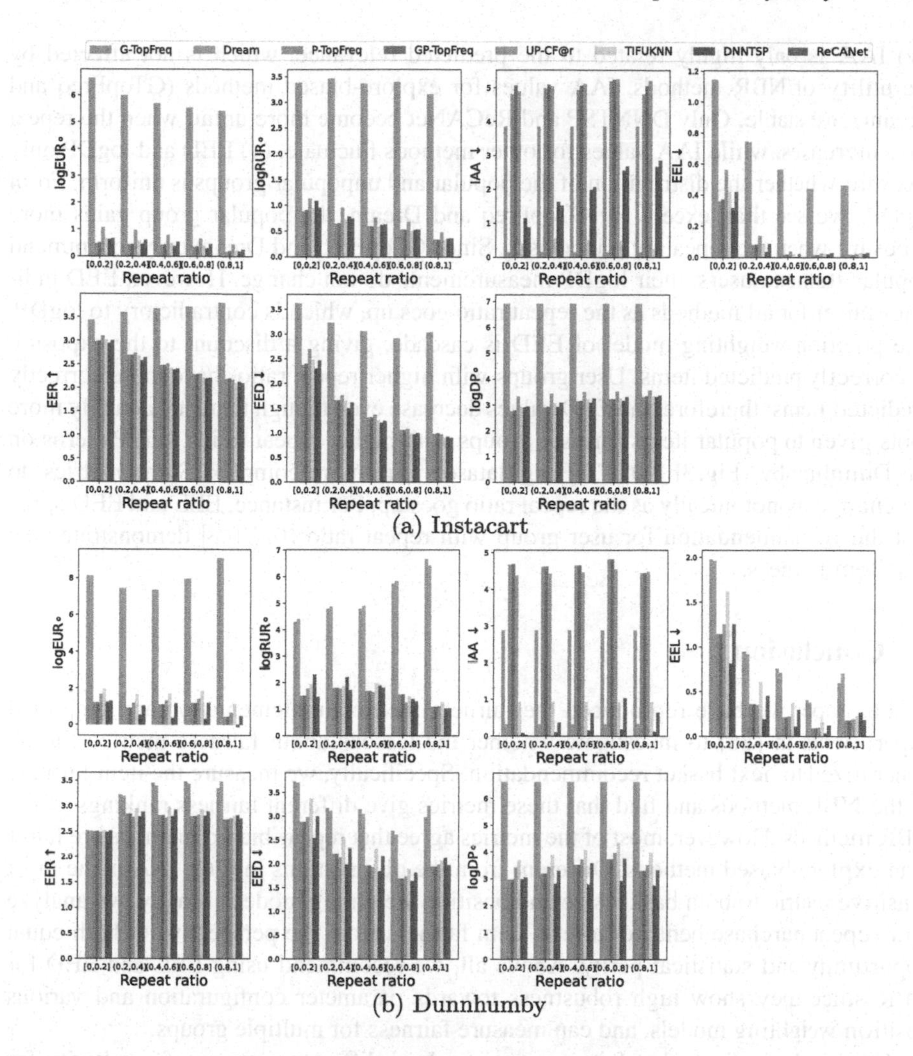

Fig. 3. Item fairness under different repeat ratios.

subgroups of users with a repeat ratio of $[0, 0.2]$, $(0.2, 0.4]$, $(0.4, 0.6]$, $(0.6, 0.8]$, and $(0.8, 1.0]$, respectively. We calculate fairness metrics for each subgroup to test the sensitivity of fairness metrics to different degrees of repeat purchase behavior and report the results in Fig. 3. We can observe that: (i) The fairness metrics show clear differences for different user groups, indicating that repeat purchase behavior does affect item fairness. (ii) The pattern on the Instacart dataset is relatively clear (Fig. 3a). logEUR, logRUR, and EEL consider user groups with higher repeat ratios fairer because they depend highly on the utility of the NBR methods. As NBR utility increases for higher repeat ratios [19], fairness for higher repeat ratios increases too. (iii) EER also belongs to the notion of equal opportunity, but these repeat-biased methods become more unfair as the repeat ratio increases, which is inconsistent with logEUR, logRUR, and EEL.

(iv) IAA is only highly related to the predicted relevance, which is not affected by the utility of NBR methods. IAA values for explore-biased methods (GTopFreq and Dream) are stable. Only DNNTSP and ReCANet become more unfair when the repeat ratio increases, while IAA values for other methods fluctuate. (v) EED and logDP only measure whether the distribution of the popular and unpopular groups is uniform. From logDP, we see that, except for GTopFreq and Dream, the popular group gains more exposure when the repeat ratio increases. Since GTopFreq and Dream only recommend popular items to users, their logDP measurements do not change. However, EED indicates fairer for all methods as the repeat ratio goes up, which is contradictory to logDP. The position weighting model of EED is cascade, giving a discount to the exposure of correctly predicted items. User groups with higher repeat ratios have more correctly predicted items; therefore, the EED values decrease even though there are actually more spots given to popular items for user groups with higher repeat ratios. (vi) Patterns on the Dunnhumby (Fig. 3b) and TaFeng datasets[7] are more complex. Some metrics do not change monotonically as the repeat ratio goes up. For instance, EEL and EED agree that the recommendation for user group with repeat ratio $(0.6, 0.8]$ demonstrates the best item fairness.

5 Conclusion

In this paper we have reproduced the fairness metrics implementation and empirical experiments in [26] to investigate whether the lessons about fairness metrics can be generalized to next basket recommendation. Specifically, we measure the item fairness of the NBR methods and find that these metrics give different fairness rankings of the NBR methods. However, most of the metrics agree that repeat-biased methods are fairer than explore-biased methods. Different from the observations in [26], IAA is the most sensitive metric to both basket size and position weighting models. Finally, we analyze how repeat purchase behavior affects item fairness from the perspective of both equal opportunity and statistical parity. Above all, we recommend using EEL and EED for NBR since they show high robustness towards parameter configuration and various position weighting models, and can measure fairness for multiple groups.

Our work confirms that fairness metrics show different patterns of sensitivity for different information access systems due to the characteristics of the scenarios and subtle differences in metrics implementation. For the sake of rigor, we suggest testing the sensitivity of fairness metrics for every specific scenario, following the evaluation framework used by us and by Raj and Ekstrand [26] to ensure the employed metrics are reliable.

This paper is the first attempt to study item exposure fairness in NBR domain, which provides a reference for the optimization direction of subsequent NBR methods. We have mainly considered group fairness grouped by popularity, but it is worth examining other ways of grouping, such as by brand or category. From a practical perspective, equal opportunity is a safe and reasonable optimization goal, however, we cannot conclude whether statistical parity is applicable to NBR since it is not reasonable to assign

[7] The pattern on the TaFeng dataset is similar to that on the Dunnhumby dataset. Because of space limitations, we report the results on the TaFeng dataset in the repository.

equal exposure to popular and unpopular groups in practical grocery shopping scenarios. Nevertheless, future work should investigate the ideal exposure distribution of the two groups.

Acknowledgments. We thank our reviewers for their valuable feedback.

This research was supported by the China Scholarship Council under grant nrs. 202206290080 and 20190607154, Ahold Delhaize, the Hybrid Intelligence Center, a 10-year program funded by the Dutch Ministry of Education, Culture, and Science through the Netherlands Organisation for Scientific Research, https://hybrid-intelligence-centre.nl, project LESSEN with project number NWA.1389.20.183 of the research program NWA ORC 2020/21, which is (partly) financed by the Dutch Research Council (NWO), and the FINDHR (Fairness and Intersectional Non-Discrimination in Human Recommendation) project that received funding from the European Union's Horizon Europe research and innovation program under grant agreement No 101070212.

All content represents the opinion of the authors, which is not necessarily shared or endorsed by their respective employers and/or sponsors.

References

1. Abdollahpouri, H., Mansoury, M., Burke, R., Mobasher, B.: The unfairness of popularity bias in recommendation. In: 13th ACM Conference on Recommender Systems, RecSys 2019 (2019)
2. Ariannezhad, M., Jullien, S., Li, M., Fang, M., Schelter, S., de Rijke, M.: ReCANet: a repeat consumption-aware neural network for next basket recommendation in grocery shopping. In: Proceedings of the 45th International ACM SIGIR Conference on Research and Development in Information Retrieval, pp. 1240–1250 (2022)
3. Ariannezhad, M., Li, M., Jullien, S., de Rijke, M.: Complex item set recommendation. In: SIGIR 2023: 46th international ACM SIGIR Conference on Research and Development in Information Retrieval, pp. 3444–3447, ACM (July 2023)
4. Biega, A.J., Gummadi, K.P., Weikum, G.: Equity of attention: amortizing individual fairness in rankings. In: The 41st International ACM SIGIR Conference on Research and Development in Information Retrieval, pp. 405–414 (2018)
5. Craswell, N., Zoeter, O., Taylor, M., Ramsey, B.: An experimental comparison of click position-bias models. In: Proceedings of the 2008 international conference on web search and data mining, pp. 87–94 (2008)
6. Diaz, F., Mitra, B., Ekstrand, M.D., Biega, A.J., Carterette, B.: Evaluating stochastic rankings with expected exposure. In: Proceedings of the 29th ACM International Conference on Information & Knowledge Management, pp. 275–284 (2020)
7. Ekstrand, M.D., Carterette, B., Diaz, F.: Distributionally-informed recommender system evaluation. ACM Transactions on Recommender Systems (2023)
8. Ekstrand, M.D., Das, A., Burke, R., Diaz, F.: Fairness in information access systems. Found. Trends Inf. Retr. **16**(1–2), 1–177 (2022)
9. Faggioli, G., Polato, M., Aiolli, F.: Recency aware collaborative filtering for next basket recommendation. In: Proceedings of the 28th ACM Conference on User Modeling, Adaptation and Personalization, pp. 80–87 (2020)
10. Ge, Y., et al.: Towards long-term fairness in recommendation. In: Proceedings of the 14th ACM international conference on web search and data mining, pp. 445–453 (2021)
11. Goyani, M., Chaurasiya, N.: A review of movie recommendation system: limitations, survey and challenges. ELCVIA: Electron. Lett. Comput. Vision Image Anal. **19**(3), 0018–37 (2020)

12. Hu, H., He, X.: Sets2sets: learning from sequential sets with neural networks. In: Proceedings of the 25th ACM SIGKDD International Conference on Knowledge Discovery & Data Mining, pp. 1491–1499 (2019)
13. Hu, H., He, X., Gao, J., Zhang, Z.L.: Modeling personalized item frequency information for next-basket recommendation. In: Proceedings of the 43rd International ACM SIGIR Conference on Research and Development in Information Retrieval, pp. 1071–1080 (2020)
14. Katz, O., Barkan, O., Koenigstein, N., Zabari, N.: Learning to ride a buy-cycle: a hyper-convolutional model for next basket repurchase recommendation. In: Proceedings of the 16th ACM Conference on Recommender Systems, pp. 316–326 (2022)
15. Kowald, D., Schedl, M., Lex, E.: The unfairness of popularity bias in music recommendation: a reproducibility study. In: Jose, J.M., et al. (eds.) Advances in Information Retrieval: 42nd European Conference on IR Research, ECIR 2020, Lisbon, Portugal, April 14–17, 2020, Proceedings, Part II, pp. 35–42. Springer International Publishing, Cham (2020). https://doi.org/10.1007/978-3-030-45442-5_5
16. Le, D.T., Lauw, H.W., Fang, Y.: Correlation-sensitive next-basket recommendation. In: the Twenty-Eighth International Joint Conference on Artificial Intelligence, pp. 10–18 (2019)
17. Li, M., Ariannezhad, M., Yates, A., de Rijke, M.: Masked and swapped sequence modeling for next novel basket recommendation in grocery shopping. In: Proceedings of the 17th ACM Conference on Recommender Systems, pp. 35–46 (2023)
18. Li, M., Ariannezhad, M., Yates, A., de Rijke, M.: Who will purchase this item next? Reverse next period recommendation in grocery shopping. ACM Trans. Recomm. Syst.**1**(2), Article 10 (June 2023)
19. Li, M., Jullien, S., Ariannezhad, M., de Rijke, M.: A next basket recommendation reality check. ACM Trans. Inform. Syst. **41**(4), 1–29 (2023)
20. Li, X., et al.: Mitigating frequency bias in next-basket recommendation via deconfounders. In: 2022 IEEE International Conference on Big Data (Big Data), pp. 616–625, IEEE (2022)
21. Li, Y., Chen, H., Fu, Z., Ge, Y., Zhang, Y.: User-oriented fairness in recommendation. In: Proceedings of the Web Conference 2021, pp. 624–632 (2021)
22. Li, Y., et al.: Fairness in recommendation: a survey. ACM Transactions on Intelligent Systems and Technology (2022)
23. Moffat, A., Zobel, J.: Rank-biased precision for measurement of retrieval effectiveness. ACM Trans. Inform. Syst. (TOIS) **27**(1), 1–27 (2008)
24. Naumov, S., Ananyeva, M., Lashinin, O., Kolesnikov, S., Ignatov, D.I.: Time-dependent next-basket recommendations. In: European Conference on Information Retrieval, pp. 502–511, Springer (2023)
25. Qin, Y., Wang, P., Li, C.: The world is binary: contrastive learning for denoising next basket recommendation. In: Proceedings of the 44th International ACM Sigir Conference on Research and Development in Information Retrieval, pp. 859–868 (2021)
26. Raj, A., Ekstrand, M.D.: Measuring fairness in ranked results: an analytical and empirical comparison. In: Proceedings of the 45th International ACM SIGIR Conference on Research and Development in Information Retrieval, pp. 726–736 (2022)
27. Sapiezynski, P., Zeng, W., E Robertson, R., Mislove, A., Wilson, C.: Quantifying the impact of user attention fair group representation in ranked lists. In: Companion proceedings of the 2019 World Wide Web Conference, pp. 553–562 (2019)
28. Singh, A., Joachims, T.: Fairness of exposure in rankings. In: Proceedings of the 24th ACM SIGKDD International Conference on Knowledge Discovery and Data Mining, pp. 2219–2228 (2018)
29. Sun, W., Xie, R., Zhang, J., Zhao, W.X., Lin, L., Wen, J.R.: Generative next-basket recommendation. In: Proceedings of the 17th ACM Conference on Recommender Systems, pp. 737–743 (2023)

30. Wang, Y., Ma, W., Zhang, M., Liu, Y., Ma, S.: A survey on the fairness of recommender systems. ACM Trans. Inform. Syst. **41**(3), 1–43 (2023)
31. Yang, K., Stoyanovich, J.: Measuring fairness in ranked outputs. In: Proceedings of the 29th International Conference on Scientific and Statistical Database Management, pp. 1–6 (2017)
32. Yu, F., Liu, Q., Wu, S., Wang, L., Tan, T.: A dynamic recurrent model for next basket recommendation. In: Proceedings of the 39th International ACM SIGIR conference on Research and Development in Information Retrieval, pp. 729–732 (2016)
33. Yu, L., Sun, L., Du, B., Liu, C., Xiong, H., Lv, W.: Predicting temporal sets with deep neural networks. In: Proceedings of the 26th ACM SIGKDD International Conference on Knowledge Discovery and Data Mining, pp. 1083–1091 (2020)
34. Zehlike, M., Bonchi, F., Castillo, C., Hajian, S., Megahed, M., Baeza-Yates, R.: FA*IR: a fair top-k ranking algorithm. In: Proceedings of the 2017 ACM on Conference on Information and Knowledge Management, pp. 1569–1578 (2017)

Query Generation Using Large Language Models

A Reproducibility Study of Unsupervised Passage Reranking

David Rau(✉) and Jaap Kamps

University of Amsterdam, Amsterdam, The Netherlands
{d.m.rau,kamps}@uva.nl

Abstract. Existing passage retrieval techniques predominantly emphasize classification or dense matching strategies. This is in contrast with classic language modeling approaches focusing on query or question generation. Recently, Sachan et al. introduced an Unsupervised Passage Retrieval (UPR) approach that resembles this by exploiting the inherent generative capabilities of large language models. In this replicability study, we revisit the concept of zero-shot question generation for re-ranking and focus our investigation on the ranking experiments, validating the UPR findings, particularly on the widely recognized BEIR benchmark. Furthermore, we extend the original work by evaluating the proposed method additionally on the TREC Deep Learning track benchmarks of 2019 and 2020. To enhance our understanding of the technique's performance, we introduce novel experiments exploring the influence of different prompts on retrieval outcomes. Our comprehensive analysis provides valuable insights into the robustness and applicability of zero-shot question generation as a re-ranking strategy in passage retrieval.

Keywords: LLMs based Query Generation · Neural Ranking Models · Language Modeling Framework for IR

1 Introduction

Text ranking stands at the beginning of several downstream NLP tasks, such as open-domain question answering (Open-QA). In this context, the objective is to retrieve relevant passages that may contain the answer. Text ranking is a rapidly evolving research area that has recently undergone drastic performance improvements by employing neural ranking models based on large-language models (LLMs). Typically, these models require extensive fine-tuning specifically tailored to the ranking task, frequently employing large training sets [2,3] on MS-Marco [1].

A useful way to classify ranking approaches is based on whether these are dense or sparse, and whether these are supervised or unsupervised [7]. While neural models dominate the supervised approaches, up to now unsupervised approaches tend to be traditional lexical systems, either traditional IR models

N. Goharian et al. (Eds.): ECIR 2024, LNCS 14611, pp. 226–239, 2024.
https://doi.org/10.1007/978-3-031-56066-8_19

based on text statistics, or classic dense IR models such as LSI. Given the earlier success of the statistical modeling framework in information retrieval [9], our main motivation in this paper is to explore how effective closely related unsupervised approaches based on LLMs are.

With the recent availability of various LLMs with billions of parameters like T0 [13], BLOOM [14], OPT [18], Llama [17], there lies an opportunity for further performance enhancement by utilizing these even bigger models for text retrieval. Nevertheless, fine-tuning becomes progressively expensive with increasing model size, prompting a surge of interest in applying these models in an unsupervised way. Sachan et al. [12] were the first to utilize the generative power of these LLMs proposing an unsupervised query generation method for re-ranking, coined as Unsupervised Passage Re-ranker (UPR). UPR utilizes the log-likelihood of the query conditioned on a passage for re-ranking without requiring specific training. This approach inspired follow-up research, exploring various techniques such as pairwise and listwise approaches [4,10,11,19].

In this study aimed at replicating Sachan et al. [12] work, we reexamine the UPR concept, with a specific focus on ranking experiments on the popular BeIR benchmark [16], and validate findings presented at EMNLP 2022. Our goal is to shed light on the practicality and reproducibility of this approach, further advancing our understanding of zero-shot question generation for re-ranking using LLMs, particularly through instruction prompts.

In this work, we formulate and address the following research questions:

- **RQ1**: *Are the results of UPR on BeIR reproducible?*

We partially reproduce (Different team, different experimental setup) the core re-ranking experiments of the original work. We find the approach to be computationally very expensive limiting our reproducibility results to only a subset of the datasets. Next, we reduce the number of passages to be re-ranked and repeat the first experiment.

- **RQ2**: *Does shallower re-ranking impact performance negatively?*

While investigated in the original paper on a single dataset we extend their experiment to the entire BeIR and come, in contrast to the original work, to the conclusion that re-ranking a smaller set can improve performance.

After, we investigate whether unifying both retrieval stages using a query likelihood model (QL) instead of BM25 as an initial retriever leads to a positive interaction between the two QL models.

- **RQ3**: *Does initial retrieval using QL improve re-ranking performance?*

Our results suggest the opposite, interactions between BM25 and UPR seem to be more beneficial for performance.

We then extend to work of Sachan et al. [12] by evaluating UPR two new test sets of TREC DL containing more fine-grained NIST judgments.

- **RQ4**: *Does UPR achieve similar performance gains over BM25 on TREC DL?*

Our results answer this positively. Although the original paper is focusing on question answering tasks, experiments on TREC DL confirm the viability of the model for ranking.

Finally, we investigate the robustness of the performance by experimenting with different instruction prompts for the LLM.

- **RQ5**: *What is the impact of the prompt on the re-ranking performance?*

We find the model overall to be robust in many cases to the prompts, however, also find that small changes can have a negative impact on performance.

Main Contributions. Our main contributions are: (i) We conduct a replicability study of zero-shot question generation (UPR) focusing on re-ranking. (ii) We extend an ablation study of the impact of the passage candidate size to the entire BeIR (iii) We complement the original work by evaluating UPR on two new testsets, TREC Deep Learning 2019 and 2020, which present the golden standard to evaluate re-ranking with fine-grained relevance judgments. (iii) We investigate the impact of different prompts on the retrieval outcomes. And finally, (iv) we provide our codebase to facilitate the reproducibility of all results presented.

The code to reproduce the results of this paper can be found under: https://github.com/davidmrau/upr_reproducibility_ecir24.

This paper is structured as follows: First, in §2 we give an overview of the reproduced method UPR. §3 we detail our experimental setup. Next, in §4 we present our experiments and results. Finally, in §5 we conclude this reproducibility study.

2 Method

In this section, we detail the approach proposed by Sachan et al. [12]. The approach follows the classical re-ranking pipeline. First, a retriever retrieves a set of passages, which is then re-ranked by a more complex model. We will detail the pipeline in the following subsections, following the original notation.

2.1 Retriever

The retriever provides an initial set of candidate passages that can later be re-ranked. For this, the retriever ranks all passages $\mathcal{D} = \{d_1, \cdots, d_M\}$ according to the given question q. This yields a sub-set of relevant passages $\mathcal{Z} \subset \mathcal{D}$ containing possible answers to the question q. The retriever provides the most K relevant passages $\mathcal{Z} = \{z_1, \cdots, z_K\}$.

2.2 Unsupervised Passage Re-ranking (UPR)

It is the task of the re-ranker to re-score the set $\mathcal{Z}$ candidate passages ranking the most relevant passages at the top. While retrievers are required to be efficient in potentially reranking millions of passages the focus for re-rankers is to provide more fine-grained ranking scores. For each passage $z_i \in \mathcal{Z}$ a relevance score is computed as $p(z_i|q)$. The approach proposed by [12] introduced a straightforward yet insightful ranking approach that leverages the inherent generative capabilities of large language models. They utilize a pre-trained large language model (LLM) to estimate the conditional probability of generating the query q given the text passage z. In this specific case, the LLM is only trained on the next token prediction task learning to model language. The model is therefore not trained for the specific task of query generation and is therefore *unsupervised.* In more detail, applying the Bayes' rule to the conditional probability $p(z_i|q)$ yields:

$$\log p(z_i|q) = \log p\left(q \mid z_i\right) + \log p\left(z_i\right) + c, \tag{1}$$

where $p(z_i)$ denotes the prior of the passage and a common constant for all z_i. The authors make the simplifying assumption of a uniform assumption for all passages z_i. Therefore we obtain:

$$\log p(z_i|q) \propto \log p\left(q \mid z_i\right), \forall z_i \in \mathcal{Z} \tag{2}$$

The LLM parameterized by Θ is then used to calculate the average log-likelihood over query terms. as follows:

$$\log p(q|z_i) = \frac{1}{|q|} \sum_t \log p\left(q_t \mid q_{<t}, z_i; \Theta\right) \tag{3}$$

To make sure the generated query stays close to the actual query teacher forcing is applied through the entire input. For re-ranking the negative log-likelihood is then used as a relevance score to rank passages. An example of the model input is given in the following:

> "*Passage:* {}. *Please write a question based on this passage. Question:* {}", where {} is replaced with the content of the passage/query respectively.

3 Experimental Setup

In this section, we detail our experimental setup consisting of datasets, retrievers, and the pre-trained LLM.

3.1 Datasets

In this section we detail the used datasets. The original work evaluates UPR on the BeIR Benchmark. We extend the evaluation to two TREC DL testsets.

BeIR Benchmark. BeIR Benchmark is a popular test suite for zero-shot evaluation it consists of various datasets of which only a subset is publicly available. Each dataset contains relevance judgments, queries, and evidence passages. The datasets span different retrieval tasks such as question answering, fact-checking, etc., and stem from different domains. We rely on the publicly available datasets provided on the HuggingFace Hub. For two datasets we find different numbers of queries than reported in the original. For CQA-Dupstack 12,569 instead 13,145 and for Nfcorpus 308 instead 323.

TREC Deep Learning Testsets. The TREC Deep Learning Track testsets are the golden standard for evaluating re-ranking models. In contrast to the publicly available BeIR datasets TREC DL testsets contain fine-grained relevance judgements provided by NIST accessors and are of high quality. We use the passage retrieval task testsets of TREC DL 2019/2020 [2,3]. The queries are based on real user search queries from the Bing search engine.

3.2 Retriever

BM25. BM25 is a lexical-based retriever that is based on TF-IDF. We follow the original work and use the publicly available Pyserini Python toolkit [8] for initial retrieval with BM25 using the default parameters.

Query Likelihood (QL). The query likelihood retrieval model is a probabilistic retriever that assesses the relevance of passage to a user's query based on the likelihood that the passages would generate the query. It calculates the probability of observing the query terms in a passage and ranks passages by their estimated likelihood of being relevant to the query. Again, we use the publicly available Pyserini Python toolkit [8] for initial retrieval with QL with the default parameters.

3.3 Pre-trained Large-Language Model

T0_3B. Following the original work, we employ a T5-based model with 3 billion parameters. T5 employs an encoder-decoder architecture, where the encoder processes input text and the decoder generates output text, allowing it to excel in tasks such as text generation, translation, and summarization. We utilize the publicly available model available on HuggingFace hub[1] under "bigscience/T0_3B". To accelerate inference we use the model in half-precision.

4 Experiments and Results

In this section, we answer the research questions posed in Sect. 1. We organized the experiments in subsections of which each will answer one research question.

[1] https://huggingface.co/.

First, in §4.1 we focus on the reproducibility of the re-ranking results [12] using UPR on the BeIR Benchmark. Second, in §4.2, we repeat an ablation study changing the experimental setup by (i) UPR re-ranking a smaller number of passages, and (ii) evaluating the ablation instead of on only one dataset in the original paper on the entire BeIR Benchmark. Then, in §4.3 we investigate the impact of the performance when using a query likelihood model for the initial retrieval as well. In §4.4 we test UPR on two new testsets, namely TREC DL 2019 and 2020. Finally, in §4.5 we test the robustness of the model to the instruction prompt.

4.1 Reproducibility: Unsupervised Passage Re-ranking

We are interested in whether the main re-ranking results on BeIR in the original work can be reproduced. In this section, we aim to reproduce the passage re-ranking results on the BeIR Benchmark. **RQ1**: *Are the results of UPR on BeIR reproducible?*

To address RQ1, we reproduce the best-performing method presented in Table 6 of the original paper [12], namely re-ranking the top-1,000 passages retrieved by BM25 with *UPR* leveraging the T0 3B Language model. A more detailed report on the BeIR Benchmark can be found in Appendix A.4 in the original paper, which will serve as a reference for this reproducibility study. Following the original paper we report NDCG@10 and Recall@100 averaged over queries.

While the authors shared their code[2], running the BeIR benchmark is not supported. Previous reproducibility studies have [5,6,15] shown how different pre-processing can have grave impacts on the reported performance, therefore we believe in the importance of this reproducibility study to confirm the results independently. To this end, we independently re-implement UPR from scratch[3], including the support of running the BeIR Benchmark out-of-the-box. We further add the support of CQADupStack to the HuggingFace hub, which is currently not available. While the retriever scores are replicated (different team, same experimental setup) the UPR scores are reproduced (different team, different experimental setup).

Contrary to the original paper we report conistently non-capped recall scores across datasets (the original paper reports capped recall scores for trec-covid). The original work does not specify a max. input length for UPR or whether special tokens should be added. We choose a max. length of 512 and add special tokens.

Results. Due to the computational complexity of re-ranking the top-1,000 results with such a large LLM, we could only reproduce results for a limited subset of the datasets in BeIR. To put this into context, just for evaluating MS-Marco re-ranking ≈ 7M passages (6,980 queries x 1,000 passages) with max.

[2] https://github.com/DevSinghSachan/unsupervised-passage-reranking.
[3] https://github.com/davidmrau/upr_reproducibility_ecir24.

Table 1. Reproduced Unsupervised passage re-ranking results on the BEIR benchmark. #Q and # denote the number of queries and evidence passages, respectively. We show results for BM25 top 1,000 and re-ranking the same with the T0-3B language model using UPR. Δ indicates the performance difference relative to the original results [12]. Results – could not be reproduced due to the computational complexity. The average is only over the reported datasets, excluding the non-capped recall values.

Dataset	#Q	#E	NDCG@10		Recall@100	
			BM25 (Δ)	re-ranked (Δ)	BM25 (Δ)	re-ranked (Δ)
Scifact	300	5K	66.5	70.6 (+0.3)	90.8	94.8 (+0.6)
Scidocs	1000	25K	15.8	-	35.6	-
Nfcorpus	308	3.5K	34.1 (+1.6)	36.3 (+1.5)	26.2 (+1.2)	27.4 (−0.6)
FIQA-2018	648	57K	23.6	-	53.9	-
Trec-covid	50	0.2M	65.6 (+0.1)	68.0 (−0.8)	11.4	12.7 (*)
Touche-2020	49	0.4M	36.7 (−0.1)	22.3 (+1.7)	53.8	46.6 (+0.9)
NQ	3452	2.7M	32.9	-	76.0	-
MS-MARCO	6980	8.8M	22.8	-	65.8	-
HotpotQA	7405	5.2M	60.3	-	74.0	-
ArguAna	1406	8.7K	41.4 (+9.9)	-	94.3	-
CQADupStack	12569	0.5M	29.9	-	60.6	-
Quora	10000	0.5M	78.9	-	97.3	-
DBPedia	400	4.6M	31.3	36.3 (+0.9)	39.8	48.4 (−4.9)
Fever	6666	5.4M	75.3	-	93.0	-
Climate-Fever	1535	5.4M	21.3	-	43.6	-
Average			42.4 (+0.8)	46.7 (+0.7)	64.6 (+0.1)	54.3 (−1.0)

* non-capped

batch size 196 on a single A100 40 GB would take 130 h. Since the number of queries is the defining parameter for this experiment we choose all datasets with < 500 queries-namely Scifact, NFcorpus, Trec-covid, touche-2020, and DBPedia. In the next section, we will provide results for all datasets in BeIR using the top-100 passages instead.

Our results reproducing the results of the original work can be found in Table 1. The table shows NDCG@10 and Recall@100 for the initial retrieval of top-1,000 passages using BM25. Δ indicates the difference from the originally reported results.

We first investigate the performance of our BM25 retriever. Our replicated scores match the scores for most datasets reported in the original paper macro average NDCG@10 42.4 vs 43.2 and Recall@100 64.6 vs 64.7. Note that the macro average of the re-ranking presented in Table 1 are not directly comparable with the original work, as we average only over the scores present in the table (excluding non-capped recall). Only for datasets Nfcorpus, Trec-covid, Touche-2020, and ArguAna, we observe minimal score variations except for

ArguAna which deviates strongly (+20% NDCG@10). The original work reports NDCG@10 31.5 while we observe a score of 41.4. Since the experimental setup is the same it is not clear where the score differences originate (our scores match the official reported scores using Pyserini[4]).

Considering UPR, the main interest in this reproducibility experiment, we observe small score differences. The original paper reports a macro average for NDCG@10 46.7 vs 47.4 and Recall@100 54.3 vs 53.4. These performance differences are likely rooted in the score deviations for BM25.

Answer to RQ1. Our first research question was: *Are the results of UPR on BeIR reproducible?* Due to the computational complexity of the approach, only a subset of the results could be reproduced. While observing small score differences for BM25 and UPR, for the experiments that we could reproduce, we were overall able to validate the claims presented in the original work regarding the passage re-ranking of UPR on the BeIR benchmark.

4.2 Shallow Re-ranking with UPR

The computational complexity of re-ranking in the previous experiment allowed us to only reproduce results for a subset of BeIR. In this section, we seek to answer **RQ2**: *Does shallower re-ranking impact performance negatively?* We repeat the experiment of the previous section but reduce the number of passages to be re-ranked from the top-1000 to top-100. This not only enables us to assess UPR across all the originally reported datasets using our limited computational resources but also facilitates an analysis of the influence of re-ranking a smaller subset of the initially retrieved passages with UPR, significantly reducing GPU runtime. The original paper explores a similar experiment in Sect. 4.2.3, only on the NQ dataset. We complement the original paper by validating their result on the entire BeIR Benchmark using NDCG@10 as a metric, as the initial results are of most interest. We omit reporting Recall@100 as it remains constant re-ranking the top-100.

Results. We present our results in Table 2. The UPR NDCG@10 scores for re-ranking are directly comparable to Table 1 and Table 10 in the original work. We observe that re-ranking only the top-100 passages yields a performance of NDCG@10 46.7 on average over all datasets compared to 44.0 in the original work, being 2,7 points higher. These results suggest that re-ranking a smaller number of passages using UPR can improve performance, indicating that UPR is overestimating the relevance of some passages that are ranked lower than top-100.

Answer to RQ2. Our second research question was: *Does shallower re-ranking impact performance negatively?* We find that re-ranking the top-100 passages

[4] https://castorini.github.io/pyserini/2cr/beir.html.

Table 2. Reproduced Unsupervised passage re-ranking results on the BEIR benchmark. #Q and # denote the number of queries and evidence passages, respectively. We show results for BM25 top 100 and re-ranking the same with the T0-3B language model using UPR. Δ indicates the performance difference relative to the original results [12].

Dataset	#Q	#E	NDCG@10	
			BM25 (Δ)	UPR (re-ranked) (Δ)
Scifact	300	5K	66.5	70.5 (+ 0.2)
Scidocs	1000	25K	15.8	17.6 (+ 0.6)
Nfcorpus	308	3.5K	34.1 (+1.6)	36.4 (+ 1.6)
FIQA-2018	648	57K	23.6	41.5 (− 2.9)
Trec-covid	50	0.2M	65.6 (+0.1)	76.0 (+ 7.2)
Touche-2020	49	0.4M	36.7 (−0.1)	22.4 (+ 1.8)
NQ	3452	2.7M	32.9	45.6 (+ 0, 2)
MS-MARCO	6980	8.8M	22.8	29.4 (− 0.8)
HotpotQA	7405	5.2M	60.3	70.7 (− 2.6)
ArguAna	1406	8.7K	41.4 (+9.9)	50.0 (+12.8)
CQADupStack	12569	0.5M	29.9	37.8 (− 3.8)
Quora	10000	0.5M	78.9	83.5 (+ 0.4)
DBPedia	400	4.6M	31.3	35.5 (− 0.1)
Fever	6666	5.4M	75.3	67.1 (+ 0.8)
Climate-Fever	1535	5.4M	21.3	16.0 (+ 4.3)
Average			42.4 (+0.8)	46.7 (+ 1.8)

using UPR does not have a negative impact on performance. In contrast, we observe the performance to increase for Trec-covid over re-ranking a very deep pool (top-1000) of candidate passages. Reproducing the ablation study of the original work on the entire BeIR (instead of only NQ) suggests re-ranking a shallower pool of candidate passages to be sufficient if not beneficial (for getting the top results right), which is the opposite of what the original work suggests.

4.3 UPR with Query Likelihood Retriever

The original paper utilizes both sparse retrievers (BM25) and dense retrievers (Contriever) for the initial retrieval process. In both cases, this creates a discrepancy between the first-stage and the second-stage rankers. This leaves unifying first- and second-stage retrieval by using query likelihood models for both stages unexplored. We fill this gap by repeating the experiment but using a Query Likelihood model as a retriever.

In this section, we seek to answer **RQ3**: *Does initial retrieval using QL improve re-ranking performance?* Employing query likelihood-based rankers for both stages might lead to positive interaction effects between them, as similar

Table 3. Unsupervised passage re-ranking results on the BEIR benchmark for the Query Likelihood retriever of top 100 and re-ranking the same with the T0-3B language model using UPR. #Q and #E denote the number of queries and evidence passages, respectively.

Dataset	#Q	#E	NDCG@10	
			QLD	UPR (re-ranked)
Scifact	300	5K	66.5	70.9
Scidocs	1000	25K	14.9	17.1
Nfcorpus	323	3.5K	33.5	36.6
FIQA-2018	648	57K	20.5	39.7
Trec-covid	50	0.2M	54.0	75.3
Touche-2020	49	0.4M	49.8	25.2
NQ	3452	2.7M	29.3	44.7
MS-MARCO	6980	8.8M	20.8	29.4
HotpotQA	7405	5.2M	58.4	67.0
ArguAna	1406	8.7K	36.1	49.1
CQADupStack	13145	0.5M	24.0	36.1
Quora	10000	0.5M	64.8	81.7
DBPedia	400	4.6M	27.6	34.3
Fever	6666	5.4M	71.3	66.8
Climate-Fever	1535	5.4M	20.7	16.2
Average			40.1	44.7

Table 4. Unsupervised passage re-ranking results for TREC DL 2019 and 2020 for BM25 and re-ranked with the T0-3B language model using UPR.

Dataset	#Q	#E	NDCG@10		MRR@10	
			BM25	UPR (re-ranked)	BM25	UPR (re-ranked)
TREC DL 2019	43	8.8M	50.6	61.1	70.2	70.9
TREC DL 2020	54	8.8M	48.0	61.7	65.3	74.4

passages might be promoted by both models. Again, limited by the computational complexity re-rank only the top-100 passages.

Results. We report the performance in Table 3 in NDCG@10. We find both the initial ranking as well as the re-ranking using a query likelihood-based model to perform worse than when using BM25 as an initial retriever (comparing results with Table 2).

Answer to RQ3. Our third research question was: *Does initial retrieval using QL improve re-ranking performance?* Our results suggest that unifying first- and second-stage retrieval using a query likelihood-based model does not lead to an improved re-ranking performance.

4.4 UPR on TREC DL

In this section we investigate **RQ4**: *Does UPR achieve similar performance gains over BM25 on TREC DL?* While the original paper explores various different datasets mostly focussing on passage retrieval for QA, the main testsets for passage re-ranking TREC DL are not reported, and is unclear whether the performance gains translate to the fine-grained high-quality relevance labels of the NIST accessors. We first retrieve the top-100 passages using BM25 and then re-rank them using UPR. We show the most popular metrics for TREC DL NDCG@10 and MRR@10.

Results. The results for TREC DL 2019/2020 can be found in Table 4. We observe UPR comfortably outperforming BM25 for both measures, except for TREC DL 2019 MRR@10 is only marginally higher for the re-ranked passages.

Answer to RQ4. Our fourth research question was: *Does UPR achieve similar performance gains over BM25 on TREC DL?* Re-ranking using UPR is effective and able to outperform BM25 by a large margin, similar to the claims of the original work. Our results on two new datasets validate the claims of the original work that re-ranking using UPR can improve over traditional unsupervised models (over 22% on TREC DL 2019, and 18% on 2020), even though with a smaller margin 10% on BeIR.

4.5 Robustness of UPR to Instruction Prompts

While the original work briefly mentions instruction prompt selection in A.2. they do not provide ablations to different prompts or indicate the robustness of the performance to different prompts. We investigate this with **RQ5**: *What is the impact of the prompt on the re-ranking performance?*

For this experiment, we systematically evaluate the robustness UPR using a variety of different prompts changing the instruction component's order, removing instruction terms, lowercasing, and paraphrasing. Similar to the previous experiment we retrieve top-100 candidates using BM25 and re-rank with UPR. We measure the performance on the TREC DL 2020 testset in NDCG@10.

Results. The prompts alongside the performance can be found in Table 5. We observe that overall the model is robust against changes to the instruction prompts. However, we observe unexpected performance drops for exchanging the term "question" for "query" (see Table 5 *paraphrasing* 7.) as well as removing the instructions entirely (see *removing instruction terms* 3.). Shortening the instructions to a minimum (*removing instruction terms* 5.) seems to be an effective way of prompting for UPR.

Table 5. Robustness of UPR to changes in the instruction prompts reported on TREC DL 2020.

Prompt	NDCG@10
default:	
1. *Passage:* {}. *Please write a question based on this passage. Question:* {}	61.74
changing instruction component order:	
1. *Please write a question based on the passage. Passage:* {}. *Question:* {}	61.10
paraphrasing:	
1 *Passage:* {}. *Please write a question based on the previous input. Question:* {}	62.03
2 *Text:* {}. *Please write a question based on this text. Question:* {}	61.78
3 *Passage:* {}. *Please write a question that this passage could answer. Question:* {}	60.27
4 *We are using zero-shot question generation to re-rank passages of the generated question. For this based on the likelihood please write a question based on this passage. Passage:* {}. *Question:* {}	60.98
5 *Passage:* {}. *Please generate a question based on this passage. 4. Question:* {}	61.69
6 *Passage:* {}. *Please output a question based on this passage. Question:* {}	60.59
7. *Passage:* {}. *Please write a query based on this passage. Query:* {}	54.57
removing instruction terms:	
1 *Passage:* {}. ~~*Please*~~ *Write a question based on this passage. Question:* {}	60.55
2 ~~*Passage:*~~ {}. ~~*Please write a question based on this passage.*~~ *Question:* {}	53.02
3 *Passage:* {}. ~~*Please write a question based on the passage.*~~ *Question:* {}	56.97
4 ~~*Passage:*~~ {}. *Please write a question based on this passage. Question:* {}	61.49
5. *Passage:* {}. ~~*Please*~~*Write* ~~*a*~~ *question* ~~*based on this passage.*~~ *Question:* {}	61.74
lowercasing:	
1 *passage:* {}. *please write a question based on this passage. question:* {}	60.88

Answer to RQ5. Our fifth research question was: *What is the impact of the prompt on the re-ranking performance?* While the model overall seems to be fairly robust against the proposed instruction prompt manipulations we find very subtle changes in the prompt such as exchanging single instruction terms with synonyms can lead to a large performance drop.

5 Discussion and Conclusions

In this reproducibility study, we have examined Unsupervised Passage Reranking using an LLM. We investigated the *reproducibility* of the zero-shot question generation results reported on the BeIR. To this end, we re-implemented UPR, and this way allow others to reproduce and experiment with UPR on BeIR, as well as TREC DL 2019/2020. For the datasets of BeIR that could be we evaluated, we found the original results to be reproducible, even though with small score differences, likely caused by different retrieval scores by BM25. We complemented the original work with several ablation experiments. First, we examined the impact of the size of the passage candidate set, and came, in contrast to the original work, to the conclusion, that re-ranking a smaller set of initially retrieved passages can improve performance using UPR while saving

GPU inference time. Second, we investigated the impact of unifying first- and second-stage retrieval using a query likelihood-based model for both and found it to compromise performance. Third, we test UPR also on two IR datasets (TREC DL 2019/2020) validating the effectiveness claims made by the original authors. Finally, we systematically test UPR for robustness to changes in the instruction prompt. We observe the performance of UPR to be very robust to most changes in the instruction prompt, however, also discovered how minimal changes can lead to a large drop in performance. Further, we found a minimal instruction of "Write question" to be sufficient.

Overall, our findings demonstrate the viability completely unsupervised neural information retrieval models. Any modern LLM can be used simply as a "language model" and estimate query likelihood in similar ways to the classic language modeling framework [9]. This opens up a new line of research in unsupervised neural IR models that complements the dominant focus on supervised neural IR models [7]. There are several benefits of pursuing such models. First, these models are applicable to very large language models where training is prohibitively expensive or impossible. Second, as unsupervised neural models clearly outperform traditional lexical approaches, they also present more realistic baselines for supervised ranking models. Third, there is conceptual and theoretical interest in unifying the statistical language modeling framework with current LLMs. Fourth, the controllable way of the unsupervised query generation model is by design avoiding hallucination, one of the main open problems of using LLMs for IR.

Acknowledgments. This research is funded in part by the Netherlands Organization for Scientific Research (NWO CI # CISC.CC.016).Views expressed in this paper are not necessarily shared or endorsed by those funding the research.

References

1. Bajaj, P., et al.: MS MARCO: a human generated machine reading comprehension dataset. CoRR (2016)
2. Craswell, N., Mitra, B., Yilmaz, E., Campos, D.: Overview of the TREC 2020 deep learning track. CoRR abs/2102.07662 (2021). https://arxiv.org/abs/2102.07662
3. Craswell, N., Mitra, B., Yilmaz, E., Campos, D., Voorhees, E.M.: Overview of the TREC 2019 deep learning track. CoRR abs/2003.07820 (2020). https://arxiv.org/abs/2003.07820
4. Dai, Z., et al.: Promptagator: few-shot dense retrieval from 8 examples. arXiv preprint arXiv:2209.11755 (2022)
5. Hendriksen, M., Vakulenko, S., Kuiper, E., de Rijke, M.: Scene-centric vs. object-centric image-text cross-modal retrieval: A reproducibility study. In: Advances in Information Retrieval: 45th European Conference on Information Retrieval, ECIR 2023, Dublin, Ireland, April 2–6, 2023, Proceedings, Part III, p. 68–85, Springer-Verlag, Berlin, Heidelberg (2023), ISBN 978-3-031-28240-9, https://doi.org/10.1007/978-3-031-28241-6_5
6. Lajewska, W., Balog, K.: From baseline to top performer: A reproducibility study of approaches at the trec 2021 conversational assistance track. In: Advances in Information Retrieval: 45th European Conference on Information Retrieval, ECIR 2023,

Dublin, Ireland, April 2–6, 2023, Proceedings, Part III, p. 177–191, Springer-Verlag, Berlin, Heidelberg (2023), ISBN 978-3-031-28240-9, https://doi.org/10.1007/978-3-031-28241-6_12
7. Lin, J., Ma, X.: A few brief notes on deepimpact, coil, and a conceptual framework for information retrieval techniques. CoRR abs/2106.14807 (2021). https://arxiv.org/abs/2106.14807
8. Lin, J., Ma, X., Lin, S.C., Yang, J.H., Pradeep, R., Nogueira, R.: Pyserini: a Python toolkit for reproducible information retrieval research with sparse and dense representations. In: Proceedings of the 44th Annual International ACM SIGIR Conference on Research and Development in Information Retrieval (SIGIR 2021), pp. 2356–2362 (2021)
9. Ponte, J.M., Croft, W.B.: A language modeling approach to information retrieval. In: Croft, W.B., Moffat, A., van Rijsbergen, C.J., Wilkinson, R., Zobel, J. (eds.) SIGIR '98: Proceedings of the 21st Annual International ACM SIGIR Conference on Research and Development in Information Retrieval, August 24–28 1998, Melbourne, Australia, pp. 275–281, ACM (1998), https://doi.org/10.1145/290941.291008
10. Pradeep, R., Sharifymoghaddam, S., Lin, J.: Rankvicuna: Zero-shot listwise document reranking with open-source large language models. arXiv preprint arXiv:2309.15088 (2023)
11. Qin, Z., et al.: Large language models are effective text rankers with pairwise ranking prompting. arXiv preprint arXiv:2306.17563 (2023)
12. Sachan, D., et al..: Improving passage retrieval with zero-shot question generation. In: Proceedings of the 2022 Conference on Empirical Methods in Natural Language Processing, pp. 3781–3797, Association for Computational Linguistics, Abu Dhabi, United Arab Emirates (Dec 2022), https://doi.org/10.18653/v1/2022.emnlp-main.249, https://aclanthology.org/2022.emnlp-main.249
13. Sanh, V., et al.: Multitask prompted training enables zero-shot task generalization. arXiv preprint arXiv:2110.08207 (2021)
14. Scao, T.L., et al.: Bloom: a 176b-parameter open-access multilingual language model (2023)
15. Schütz, M.: Disinformation detection: Knowledge infusion with transfer learning and visualizations. In: Advances in Information Retrieval: 45th European Conference on Information Retrieval, ECIR 2023, Dublin, Ireland, April 2–6, 2023, Proceedings, Part III, p. 468–475, Springer-Verlag, Berlin, Heidelberg (2023), ISBN 978-3-031-28240-9, https://doi.org/10.1007/978-3-031-28241-6_54
16. Thakur, N., Reimers, N., Rücklé, A., Srivastava, A., Gurevych, I.: BEIR: a heterogeneous benchmark for zero-shot evaluation of information retrieval models. In: Thirty-fifth Conference on Neural Information Processing Systems Datasets and Benchmarks Track (Round 2) (2021). https://openreview.net/forum?id=wCu6T5xFjeJ
17. Touvron, H., et al.: Llama: Open and efficient foundation language models (2023)
18. Zhang, S., et al.: Opt: open pre-trained transformer language models (2022)
19. Zhuang, H., et al.: Rankt5: fine-tuning t5 for text ranking with ranking losses. In: Proceedings of the 46th International ACM SIGIR Conference on Research and Development in Information Retrieval, pp. 2308–2313 (2023)

IR for Good Papers

Absolute Variation Distance: An Inversion Attack Evaluation Metric for Federated Learning

Georgios Papadopoulos(✉), Yash Satsangi, Shaltiel Eloul, and Marco Pistoia

Global Technology Applied Research, JPMorgan Chase, New York, USA
georgios.papadopoulos@jpmorgan.com

Abstract. Federated Learning (FL) has emerged as a pivotal approach for training models on decentralized data sources by sharing only model gradients. However, the shared gradients in FL are susceptible to inversion attacks which can expose sensitive information. While several defense and attack strategies have been proposed, their effectiveness is often evaluated using metrics that may not necessarily reflect the success rate of an attack or information retrieval, especially in the context of multidimensional data such as images. Traditional metrics like the Structural Similarity Index (SSIM), Peak Signal-to-Noise Ratio (PSNR), and Mean Squared Error (MSE) are typically used as lightweight metrics, assume only pixel-wise comparison, but fail to consider the semantic context of the recovered data. This paper introduces the Absolute Variation Distance (AVD), a lightweight metric derived from total variation, to assess data recovery and information leakage in FL. Unlike traditional metrics, AVD offers a continuous measure for extracting information in noisy images and aligns closely with human perception. Our results combined with a user experience survey demonstrate that AVD provides a more accurate and consistent measure of data recovery. It also matches the accuracy of the more costly and complex Neural Network based metric, the Learned Perceptual Image Patch Similarity (LPIPS). Hence it offers an effective tool for automatic evaluation of data security in FL and a reliable way of studying defence and inversion attacks strategies in FL.

Keywords: Federated learning · Differential privacy · Evaluation metrics

1 Introduction

In the age of Large Models, with size of billions of parameters, data play a crucial role for their continuous development. Therefore, the availability of large amounts of data for their training and fine-tuning is critical. Traditionally, data was concentrated in centralized repositories, but the increased awareness of privacy and the decentralized nature of information generation from devices and multiple regional data centres has necessitated a more nuanced approach. In this regard, Federated Learning (FL) enables models to learn from a multitude of decentralized edge

N. Goharian et al. (Eds.): ECIR 2024, LNCS 14611, pp. 243–256, 2024.
https://doi.org/10.1007/978-3-031-56066-8_20

devices or servers holding local data samples, obviating the need to exchange raw data. The standard FL configuration is achieved with a central aggregator node that exchanges gradients to train a centralised model [12]. Particularly, at each training step t, a client node receives neural network model weights, W_t, from an aggregator server and calculates loss l with local data (x_t, y_t) for a batch, B, which generates gradients with respect to the model weights:

$$\Delta W_t = -\frac{\gamma}{B} \sum_{b<B} \frac{\partial l(F_{W_t}(x_{t,b}, y_{t,b}))}{\partial W_t}, \tag{1}$$

where F_{W_t} is a neural network parameterized by W_t. The gradients are typically averaged in the server with a rate, γ. Because of their flexibility and the client anonymity that they offer, FL models have been deployed in a variety of real-world applications [14,17,23].

However, the gradients, ΔW_t, shared by the client are vulnerable to inversion attacks instigated by a malicious eavesdropper that can expose the original sensitive data. Existing literature on inversion attacks [1,6,24,26,28] have shown that these inversion attacks can be highly successful, with potentially recovering large batch of data after several rounds of learning, and at a pixel resolution [6]. In general, these attacks more or less follow the same paradigm, generate a dummy dataset (usually images) and then use a loss function with priors distributions [1] as regulators or with additional generative models to minimise the loss between the FL model and the dummy gradients. The success of the inversion attacks prevents FL from becoming a fully trustful framework for distributed training.

To mitigate inversion attacks in FL, several defence strategies were proposed to reduce the leakage of information [3,8,20,21]. These include, data transformation from the client side [9], homomorphic encryption techniques [15], data sanitation methods [28], and defense strategies originating from Differential Privacy approaches (DP) [4]. In FL, defence techniques can be achieved predominantly by adding noise, but also by truncating the data, mixing the gradients and batch of gradients shared [5,27,28]. All techniques inherent, inevitably, a potential loss in model training performance.

One important step towards the development of robust defence strategies and the prevention and understanding of such attacks is by assessing their success. Therefore, an inversion attack is studied experimentally by comparing the information revealed from the recovered input data to that of the original dataset [8]. There exist a few metrics that are commonly used to measure the reconstruction quality, with the most popular being the structural similarity index (SSIM) [22], the peak signal-to-noise ratio (PSNR) [2], the learned perceptual image patch similarity (LPIPS) [25], and the mean squared error (MSE).

In this paper, we show how the aforementioned metrics are insufficient to properly assess the success of an inversion attack on the gradients of an FL model. Especially, for cases that the FL model generates multidimensional outputs that contain contextual information like in images. The reason is that these metrics assume spatial independence when comparing the recovered image to the original one. Therefore, in many cases they do not capture the semantic context, and fail

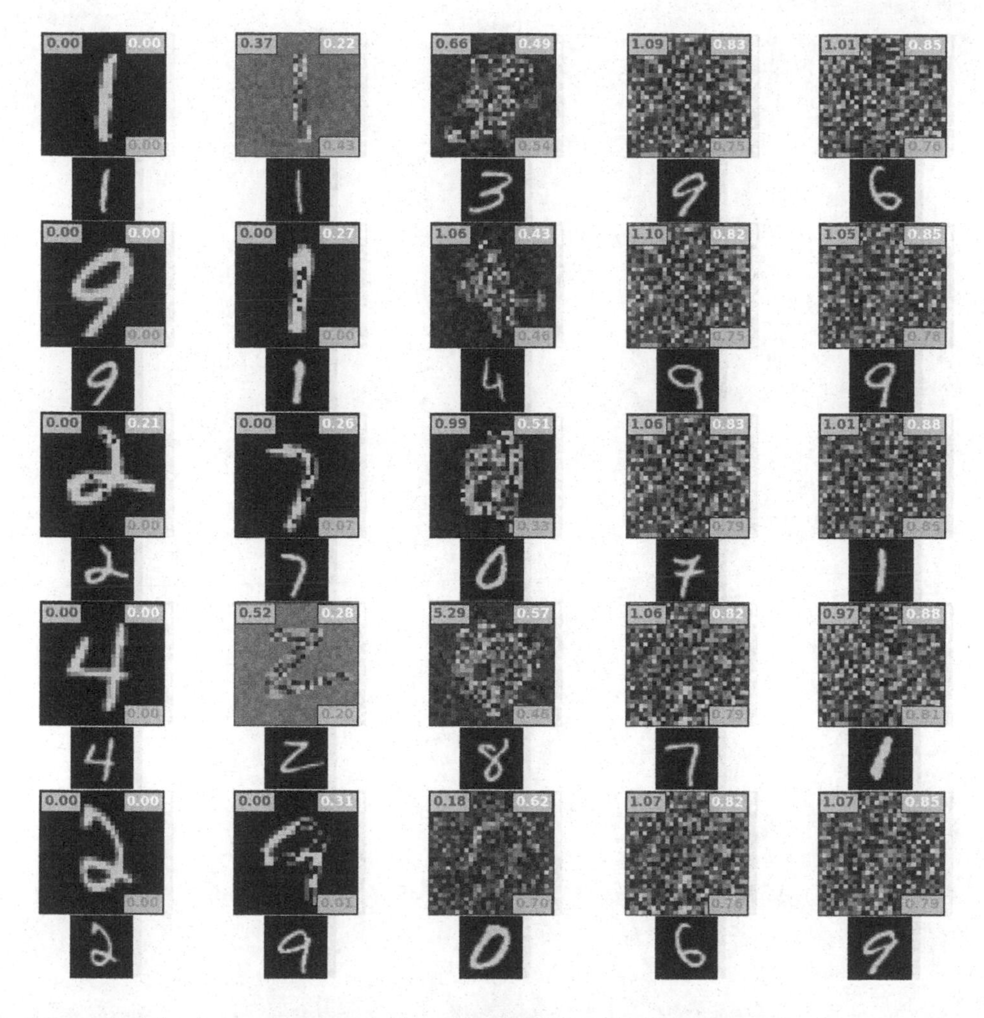

Fig. 1. Random recovered vectors from the MNIST dataset. It shows the images sorted by the AVD metric. The smaller figures represent the original images associated with image above. The figures include complex images, such as row 1 & column 3, that is a combination of digits 4 and 3.

to measure accurately the minimal information revealed out of a noisy image. For example, a common defense mechanism for an FL model is to add noise to an image that depicts a number from the MNIST dataset. An attack on the gradients of this model may recover an approximate image with the same number but considerably different background colour (see Fig. 1, row 4 & column 2). Metrics, such as SSIM, PSNR, and MSE fail to provide a consistent and accurate result and indicate this attack as unsuccessful (in the image example $MSE = 0.52$, which is considered high). That is, because of the use of Euclidean distance-per-pixel measure, they miss the fact that the attacker has recovered the most important element, the actual number; regardless of how noisy or changed

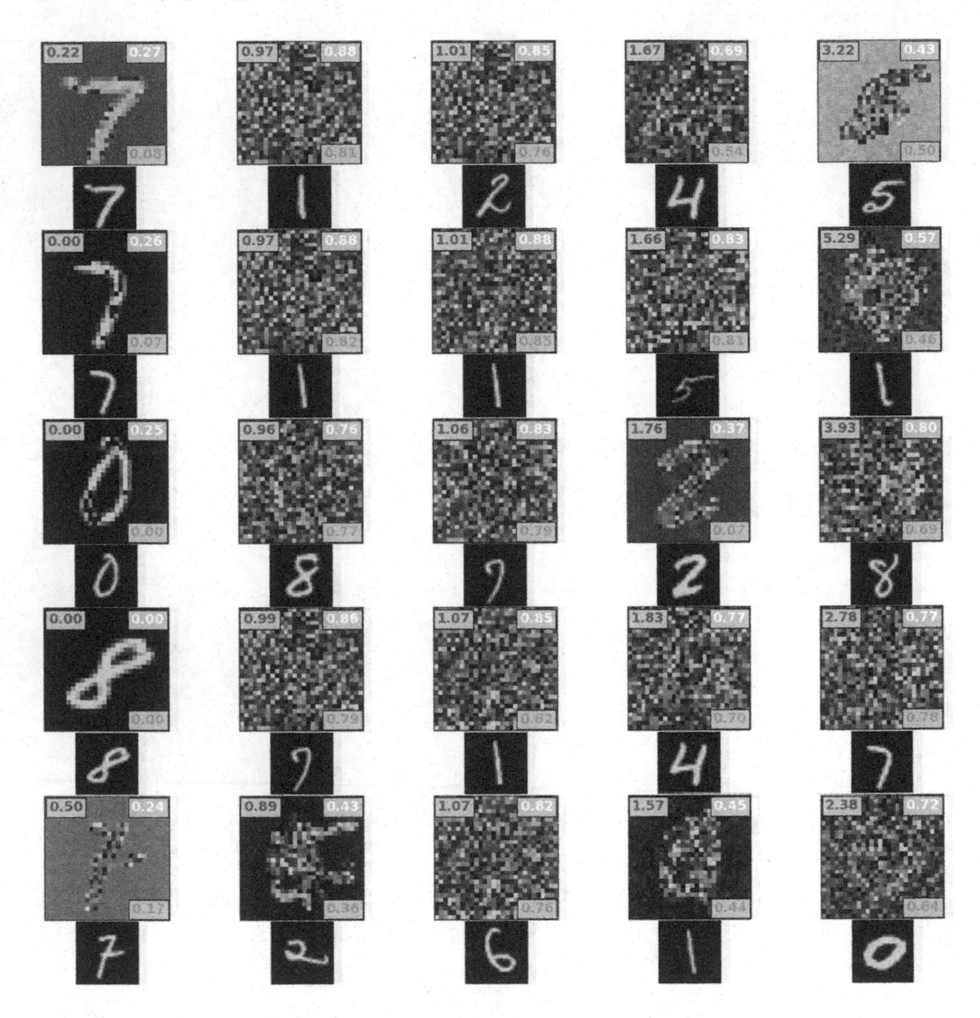

Fig. 2. Random recovered vectors from the MNIST dataset. It shows the images sorted by the MSE metric. The smaller figures represent the original images associated with image above. The figures include complex images, such as row 3 & column 4, that is a combination of digits 2 and 8.

the background of the image is. Therefore, these metrics discard the contextual information in the image that a human's vision would have otherwise recognized, e.g. edges and points of interests (another example is 1, row 2 & column 3, where $MSE = 1.06$ but a human would have read the number). It becomes even a larger challenge to use these metrics as a mechanism to approve data for FL in real-time. Since the development of attacks models are mostly empirical and data dependant, it is plausible to have an automatic verification (e.g. as a smart contract/client service) to assess the security of data by applying brute-force attacks before submission of gradients. For that purpose, a reliable and lightweight metric is needed.

This problem has not gone unnoticed and efforts to address these challenges have resulted in proposal of specialized metrics that may be computationally expensive, for example, Learned Perceptual Image Patch Similarity (LPIPS) from [25]. The authors use the power of deep neural networks (DNN) to create LPIPS that is aligned with human perception metric. A major downside is that LPIPS is a complex and a computationally costly metric that is difficult to interpret due to the underlying DNN themselves that may require training for new data.

This paper introduces a new distance metric to assess data recovery, the *Absolute Variation Distance* (AVD). It is derived from total variation and in contrast to standard methods (MSE, SSIM), it offers a continuous metric for measuring information recovered in noisy images. Furthermore, we show via a user study that AVD is highly correlated with human perception, but at the same time it is computationally more efficient and interpretable compared to LPIPS. Our results show that recovery of data is more visible as AVD decreases in a continuous manner. In contrast the MSE metric for MNIST fluctuates drastically when the image is not completely clear or a blend, and can obtain various values similar or higher than the MSE for the pure noise input.

Table 1. Types of gradient inversion attacks employed in our study.

Attack Name	Main Objective Function	Description
2-norm	g^{l2} (3)	Euclidean distance and initial label determination
Angle & var	g^{ang} + TV (4)	[6] proposed to leverage cosine similarity, total variation (TV) and initial label determination
Angle & var & Orth_regulators	g^{ang} + TV + Orth	Cosine distance with orthogonal regulator for the input + initial label determination. [16]

2 Absolute Variation Distance

In this paper we developed AVD, a variant of total variation metric [18], which is a more suitable indicator to compare the spatial gradient of the recovered image and source image. Given two images v^{source} and v^{target}, we define AVD between them as following:

$$\begin{aligned} \text{AVD}(v^{source}, v^{target}) = \\ \|(|\nabla v^{source}| - |\nabla v^{target}|)\| + \\ \|(|\nabla^2 v^{source}| - |\nabla^2 v^{target}|)\| \end{aligned} \tag{2}$$

where $\nabla v = \frac{dv}{di} + \frac{dv}{dj}$ is the spatial gradient and $\nabla^2 v = \frac{d^2 v}{di^2} + \frac{d^2 v}{dj^2}$ is the second order gradient. Here we treat the image as a 2-D array with values $v(i, j)$. Therefore, because AVD measures distance in gradient space it allows to consider boundaries and edges in images which are a common discriminator in visual recognition, whilst the gradient of noise remains as noise. Note that this work

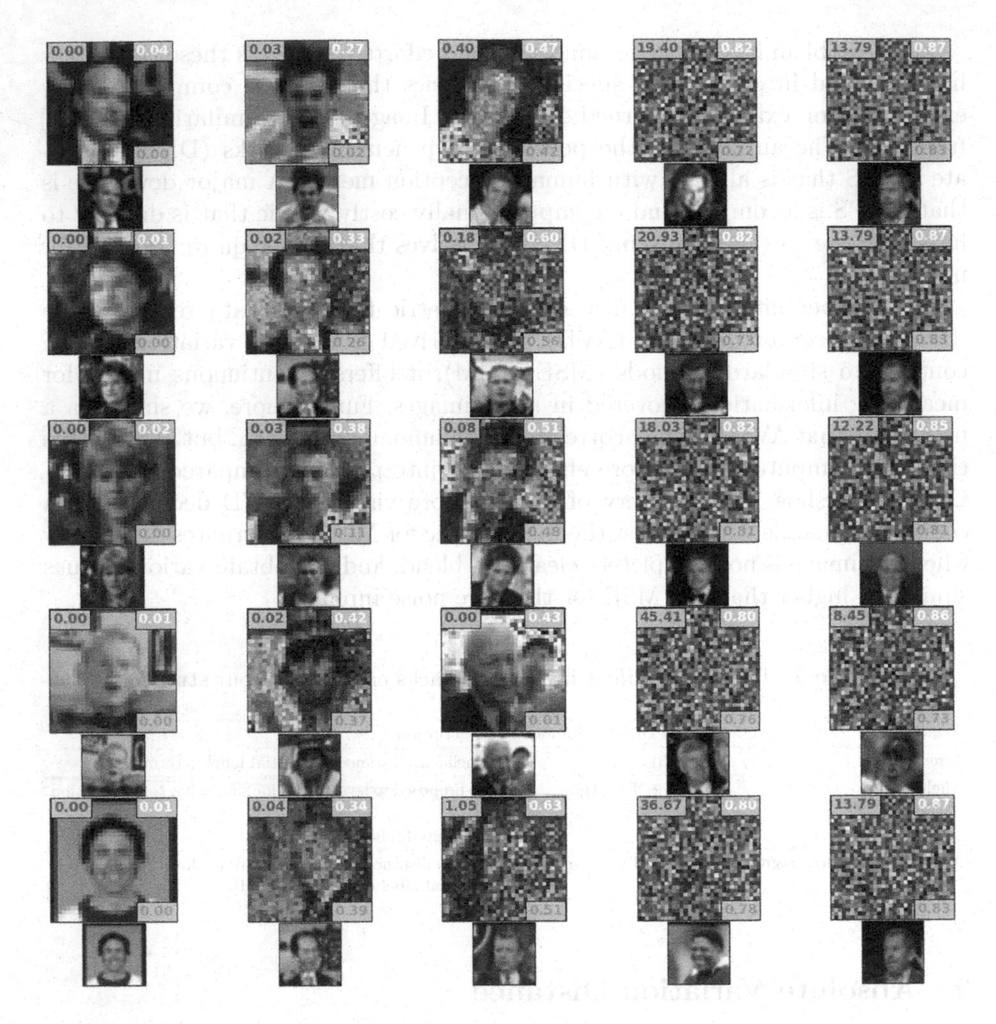

Fig. 3. Random recovered vectors from the LFW face dataset. It shows the images sorted by the AVD metric. The smaller figures represent the original images associated with image above.

which includes the contribution of the second order spatial derivative improves the previous observation on Absolute Variation Distance [5], and verified here with a user study.

2.1 Inversion Attack Algorithms

In the context we are examining (which is often the case), the gradient inversion attack is executed by selecting proxy model variables x'_t, y'_t based on the model $F'(x'_t, y'_t)$ as mentioned in [5]. The attack involves reducing the difference between the gradients calculated by the proxy model $\Delta W'_t$ and the original

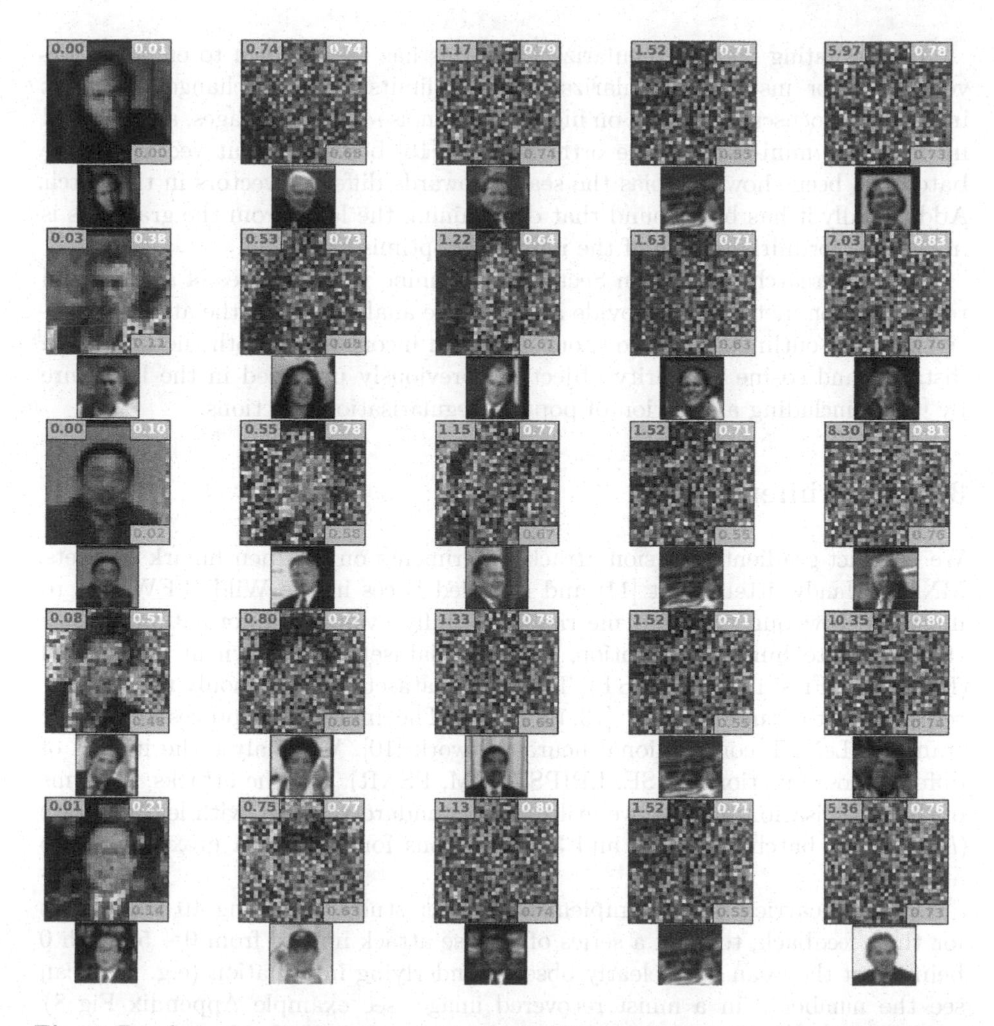

Fig. 4. Random recovered vectors from the LFW face dataset. It shows the images sorted by the MSE metric. The smaller figures represent the original images associated with image above.

model's gradients. A common approach for this is to minimize the norm of the difference between these gradients:

$$g^{l2}(x'_t, y'_t) = \min||\Delta W'_t - \Delta W_t|| \tag{3}$$

This method aims to find a model $F'(x'_t, y'_t)$ that approximates the gradient vector size seen by the client. However, further empirical research suggests that using cosine distance leads to better convergence, as indicated by [6]:

$$g^{ang}(x'_t, y'_t) = \min 1 - \frac{\langle \Delta W'_t, \Delta W_t \rangle}{||\Delta W'_t|| \cdot ||\Delta W_t||} \tag{4}$$

Incorporating various regularization terms has been shown to enhance convergence. For instance, regularization that limits extreme changes in input images and focuses the search on high-quality, noise-reduced images, as discussed in [6,24]. In mini-batches the orthogonality [16] between input vectors in the batch has been shown to bias the search towards different vectors in the batch. Additionally it has been found that determining the label from the gradients is important for initialisation of the numerical optimisation [24].

In our research detailed in Sect. 4, we examine various types of attacks and regularization methods to provide an extensive analysis about the attack's effectiveness. As outlined in Table 1, our approach incorporates both the Euclidean distance and cosine similarity objectives previously proposed in the literature by [6,28] including a selection of popular regularisation functions.

3 Experiments

We conduct gradient inversion attack experiments on two benchmark datasets, MNIST Handwritten Digit [11] and Labelled Faces in the Wild (LFW) [7], to illustrate how our proposed metric successfully evaluates information leakage that aligns to human perception. The two datasets are shown in Figs. 3 & 4 (LFW) and Figs. 1 & 2 (MNIST). These two datasets are commonly used among researchers to study attacks [13,19,26,28]. The images are processed by the standard LeNET convolutional neural network [10]. We analyse the impact of different loss functions (MSE, LPIPS, SSIM, PSNR). For the attacks, in terms of the optimisation scheme, we utilized the standard LFBGS, with learning rate (lr) of 0.05, batch size of 4, and 300 iterations for running a proxy model to attack.

We also carried out a complementary user study by asking 10 individuals for their feedback, to rank a series of inverse attack images from $0 - 5$. With 0 being that they can very clearly observe underlying information (e.g. they can see the number 9 in a mnist recovered image, see example Appendix Fig. 8), to 5 that they cannot extract any useful information (e.g. the image is pure noise). We randomly generated 6 groups, with 100 images each (total 600 images); specifically, three 10×10 frames of LFW images and three 10×10 frames of MNIST images (see a similar version to the 5×5 Appendix, Figs. 7 and 8 for an example of LFW and MNIST). The images were not ranked by noise level/clarity, but they were randomly allocated into the 10×10 frame. After the individuals ranked them then we averaged the scores they gave for each of the 100 images and we compared it to the score each image achieved from MSE, AVD, and LPIPS metrics. We used Pearson correlation (ρ) as a measure of similarity. The results can be studied in the heatmap in Fig. 6.

	Noisy Reference	Noisy Reference	Noisy Reference	Noisy Reference	Reference	Noisy Reference
AVD	0.00	0.62	0.50	0.48	0.40	0.80
MSE	0.00	0.51	1.32	1.05	0.92	0.60
SSIM	0.99	0.09	0.13	0.12	0.18	0.12
PSNR	50.77	15.45	14.88	13.63	15.81	17.76
LPIPS	0.00	0.26	0.49	0.26	0.34	0.82

Fig. 5. Comparison table of the most widely used metrics in FL for evaluation of inversion attacks (MSE, SSIM, PSNR, LPIPS), plus our own novel metric, the AVD. Each column compares the metrics between a noisy (attack generated) image and its reference (original) image. We included a wide array of examples, from no noise (column 1), to a mixture with two references (column 4), and complete noise (column 5). The SSIM and PSNR metrics have an opposite direction from the rest. When a generated noisy image has no noise, column 1, and matches perfectly the reference then SSIM gets a max value of 1 and PSNR gets a high value -no bounds-. Whereas, AVD, MSE, and LPIPS get a min value of 0, with higher values indicating miss match between reference and noisy image. What we observe from this table is the consistency that the AVD and LPIPS measure the noisy images with regards to the original reference. With 0 being for images that have no difference from the reference and the highest value (around .80) for completely noisy images.

4 Results

For Fig. 1 and Fig. 2 we ran the experiment for the MNIST dataset. The Fig. 1 sorts the images by AVD value. For comparison, the Fig. 2 sorts the images by the MSE. The smaller images represent the original MNIST data. Low score ranking represents that the attack was successful and the image matches the original. High score ranking shows that the attack was unsuccessful and the generated attack images are very noisy with no observable pattern. From Fig. 2, row one and columns four and five, the $MSE_{1,4} = 1.67$ and $MSE_{1,5} = 3.22$. These images clearly have a pattern, a user might be able to infer that the number of the last image relates to the number 5. So the attacker can extract private information from an FL model. But, according to the MSE these images are more private (noisy) when compared against images two and three from the same row ($MSE_{1,2} = 0.97$ and $MSE_{1,2} = 1.01$). On the other hand, our metric AVD captures these irregularities, with $AVD_{1,4} = 0.69$ and $AVD_{1,5} = 0.43$ being lower than $AVD_{1,2} = 0.88$ and $AVD_{1,3} = 0.85$. The AVD scores also agree with the LPIPS benchmark in these images, which indicates the AVD follows the human perception to evaluate the success of an inversion attack. In the LFW dataset, Figs. 3 and Fig. 4, we can observe the same phenomenon when using the MSE as a score to evaluate FL inversion attacks. In Fig. 5 we compare different inversion attack metrics, including PSNR, SSIM, MSE, LPIPS, and AVD. The LPIPS and AVD results are consistent and agree very well with human consensus; they attribute the lowest value ($LPIPS = 0$ & $AVD = 0$) to

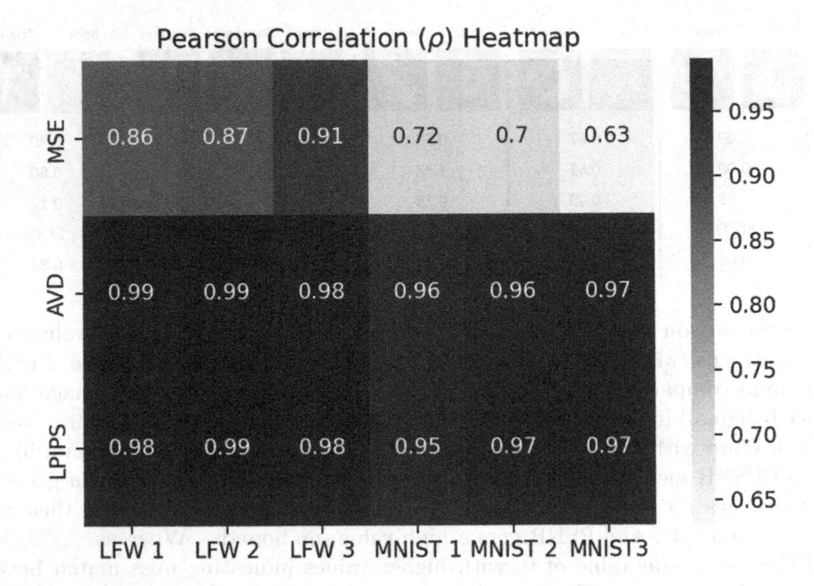

Fig. 6. Pearson correlation ρ heatmap between the average ranking score (0–5) of 10 people for each image, and the MSE, LPIPS, AVD scores of these images. A ranking score of 0 means that the human can perceive very clear information in the image and a rank of 5 means the image is pure noise. The x-axis shows 6 groups (LFW1, LFW2, LFW3, MNIST1, MNIST2, MNIST3) of images. Each group has a 10×10 frame, 100 images in each frame. The y-axis shows the the three inversion attack metrics. The correlation is between the metrics (y-axis) and the average ranking score by the users. For more details refer to Sect. 3. In this matrix we only report the MSE, AVD, and LPIPS. The reason being that SSIM and PSNR do not defer much from the MSE results in terms of interpretability.

column one that the two images are identical, and the largest value at column five ($LPIPS = 0.82$ & $AVD = 0.80$), where the generated image is just noise.

In our final experiment, we contacted a qualitative survey amongst 10 people, Fig. 6. When we evaluate inversion attacks in multidimensional data that exhibit strong intercorrelation amongst the data-points, such as images, then a contextual interpretation of the image is imperative for an accurate evaluation. Therefore, the similarity metric should be able to showcase human-like perception. Our survey results further support the quantitative analysis that we conducted in Fig. 5 and show that the AVD is highly correlated ($\rho \geq 0.96$) with how a human would have recognised information from an attack generated image. For the LFW group of images, the MSE had a correlation between $0.86 \leq \rho \leq 0.91$ with the average human score. On the other hand, for the same images, the LPIPS and the AVD achieved very high levels of correlation, $0.98 \leq \rho \leq 0.99$. For the MNIST group, the MSE showed correlation between $0.63 \leq \rho \leq 0.72$. The AVD though retained consistently high levels of correlation with the human score, $0.96 \leq \rho \leq 0.97$. It seems that the mixing of numbers and the change of the background that we observed in the MNIST examples (Figs. 1 and 5) "confuse" the MSE score and drive the results further away from human perception, reducing its accuracy.

5 Conclusion

In this paper, we have addressed a significant challenge in the field of FL - the evaluation of the success of inversion attacks and the effectiveness of defense strategies. Traditional metrics such as the SSIM, PSNR, and MSE have been shown to be insufficient for accurately assessing the success of these attacks, particularly in the context of multidimensional outputs like images. These metrics, which assume spatial independence, fail to consider the semantic context of the recovered data, leading to potentially misleading evaluations.

To overcome these limitations, we introduced the AVD, a metric for assessing data recovery and information leakage in FL. Derived from total variation, AVD offers a continuous measure for extracting information in noisy images, aligning closely with human perception. It is computationally more efficient and mathematically more interpretable than the LPIPS, a deep learning-based metric. The quantitative experiments demonstrated that AVD provides a more accurate and consistent measure of data recovery, thereby offering a more reliable tool for evaluating defense strategies against inversion attacks in FL. Also, the survey that we contacted amongst 10 people, asking them to rank random generated images by scoring the success of the recovered image, showed that human perception had high correlation with the AVD scores

By providing a more accurate measure of data recovery, AVD allows researchers to better understand the effectiveness of their defense strategies and to develop robust FL evaluation of data security. We hope that our work will inspire further advancements in the field of FL and contribute to the development of more secure and reliable distributed learning systems.

Disclaimer. This paper was prepared for informational purposes by the Global Technology Applied Research center of JPMorgan Chase & Co. This paper is not a product of the Research Department of JPMorgan Chase & Co. or its affiliates. Neither JPMorgan Chase & Co. nor any of its affiliates makes any explicit or implied representation or warranty and none of them accept any liability in connection with this paper, including, without limitation, with respect to the completeness, accuracy, or reliability of the information contained herein and the potential legal, compliance, tax, or accounting effects thereof. This document is not intended as investment research or investment advice, or as a recommendation, offer, or solicitation for the purchase or sale of any security, financial instrument, financial product or service, or to be used in any way for evaluating the merits of participating in any transaction.

Appendix

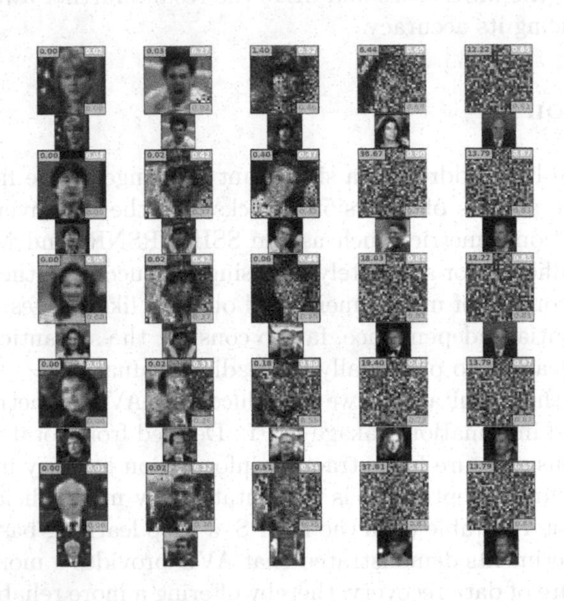

Fig. 7. Random recovered vectors from LFW datasets, column-wise sorted via the AVD.

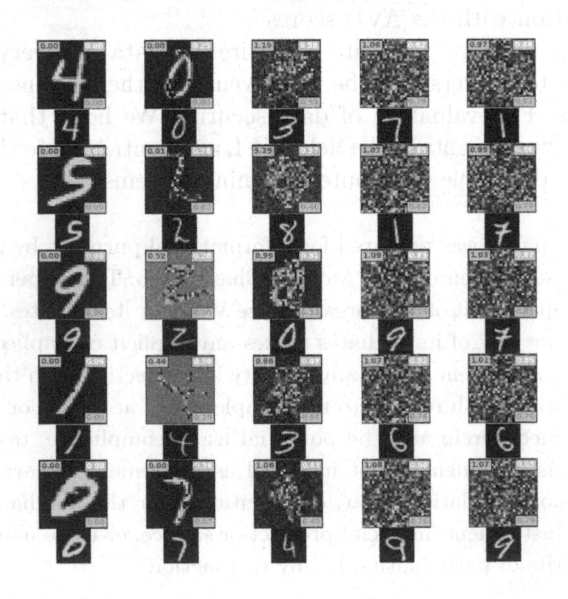

Fig. 8. Random recovered vectors from MNIST datasets, column-wise sorted via the AVD.

References

1. Balunović, M., Dimitrov, D.I., Staab, R., Vechev, M.: Bayesian framework for gradient leakage (2022)
2. Cahn, C.: A note on signal-to-noise ratio in band-pass limiters. IRE Trans. Inform. Theor. **7**(1), 39–43 (1961). https://doi.org/10.1109/TIT.1961.1057616
3. Chen, Y., Gui, Y., Lin, H., Gan, W., Wu, Y.: Federated learning attacks and defenses: a survey (2022)
4. Dwork, C.: Differential privacy. In: Bugliesi, M., Preneel, B., Sassone, V., Wegener, I. (eds.) Automata, Languages and Programming, pp. 1–12. Springer Berlin Heidelberg, Berlin, Heidelberg (2006). https://doi.org/10.1007/11787006_1
5. Eloul, S., Silavong, F., Kamthe, S., Georgiadis, A., Moran, S.J.: Mixing gradients in neural networks as a strategy to enhance privacy in federated learning. In: Proceedings of the IEEE/CVF Winter Conference on Applications of Computer Vision (WACV), pp. 3956–3965 (January 2024)
6. Geiping, J., Bauermeister, H., Dröge, H., Moeller, M.: Inverting Gradients - How easy is it to break privacy in federated learning? In: Advances in Neural Information Processing Systems 33: Annual Conference on Neural Information Processing Systems 2020, NeurIPS 2020(December), pp. 6–12, 2020. virtual (2020). https://proceedings.neurips.cc/paper/2020/hash/c4ede56bbd98819ae6112b20ac6bf145-Abstract.html
7. Huang, G.B., Ramesh, M., Berg, T., Learned-Miller, E.: Labeled faces in the wild: a database for studying face recognition in unconstrained environments. Tech. Rep. 07–49, University of Massachusetts, Amherst (October 2007)
8. Huang, Y., Gupta, S., Song, Z., Li, K., Arora, S.: Evaluating gradient inversion attacks and defenses in federated learning. In: Beygelzimer, A., Dauphin, Y., Liang, P., Vaughan, J.W. (eds.) Advances in Neural Information Processing Systems (2021). https://openreview.net/forum?id=0CDKgyYaxC8
9. Huang, Y., Song, Z., Li, K., Arora, S.: Instahide: Instance-hiding schemes for private distributed learning (2021)
10. LeCun, Y., et al.: Handwritten Digit Recognition with a Back-Propagation Network. In: Advances in Neural Information Processing Systems. vol. 2. Morgan-Kaufmann (1990), https://proceedings.neurips.cc/paper/1989/file/53c3bce66e43be4f209556518c2fcb54-Paper.pdf
11. LeCun, Y., Cortes, C., Burges, C.: Mnist handwritten digit database. ATT Labs. http://yann.lecun.com/exdb/mnist 2 (2010)
12. McMahan, B., Moore, E., Ramage, D., Hampson, S., Arcas, B.A.y.: Communication-efficient learning of deep networks from decentralized data. In: Proceedings of the 20th International Conference on Artificial Intelligence and Statistics. Proceedings of Machine Learning Research, vol. 54, pp. 1273–1282, PMLR (20–22 Apr 2017), https://proceedings.mlr.press/v54/mcmahan17a.html
13. Melis, L., Song, C., Cristofaro, E.D., Shmatikov, V.: Exploiting unintended feature leakage in collaborative learning. In: 2019 IEEE Symposium on Security and Privacy, SP 2019, San Francisco, CA, USA, May 19–23, 2019, pp. 691–706. IEEE (2019). https://doi.org/10.1109/SP.2019.00029
14. Nguyen, D.C., et al.: Federated learning for smart healthcare: a survey. ACM Comput. Surv. (CSUR) **55**(3), 1–37 (2022)
15. Phong, L.T., Aono, Y., Hayashi, T., Wang, L., Moriai, S.: Privacy-preserving deep learning via additively homomorphic encryption. IEEE Trans. Inf. Forensics Secur. **13**(5), 1333–1345 (2018). https://doi.org/10.1109/TIFS.2017.2787987

16. Qian, J., Nassar, H., Hansen, L.K.: Minimal model structure analysis for input reconstruction in federated learning (2021)
17. Rieke, N., et al.: The future of digital health with federated learning. npj Digital Med. **3**(1) (Sep 2020). https://doi.org/10.1038/s41746-020-00323-1
18. Rudin, L.I., Osher, S., Fatemi, E.: Nonlinear total variation based noise removal algorithms. In: Proceedings of the Eleventh Annual International Conference of the Center for Nonlinear Studies on Experimental Mathematics: Computational Issues in Nonlinear Science: Computational Issues in Nonlinear Science, pp. 259–268. Elsevier North-Holland Inc, USA (1992)
19. Shokri, R., Stronati, M., Song, C., Shmatikov, V.: Membership inference attacks against machine learning models. In: 2017 IEEE Symposium on Security and Privacy, SP 2017, San Jose, CA, USA, May 22–26, 2017, pp. 3–18. IEEE Computer Society (2017). https://doi.org/10.1109/SP.2017.41
20. Sikandar, H.S., Waheed, H., Tahir, S., Malik, S.U.R., Rafique, W.: a detailed survey on federated learning attacks and defenses. Electronics 12(2) (2023). https://doi.org/10.3390/electronics12020260, https://www.mdpi.com/2079-9292/12/2/260
21. Wainakh, A., et al.: Federated learning attacks revisited: a critical discussion of gaps, assumptions, and evaluation setups (2022)
22. Wang, Z., Bovik, A., Sheikh, H., Simoncelli, E.: Image quality assessment: from error visibility to structural similarity. IEEE Trans. Image Process. **13**(4), 600–612 (2004). https://doi.org/10.1109/TIP.2003.819861
23. Yang, Q., Liu, Y., Chen, T., Tong, Y.: Federated machine learning: concept and applications. ACM Trans. Intell. Syst. Technol. **10**(2) (jan 2019). https://doi.org/10.1145/3298981, https://doi.org/10.1145/3298981
24. Yin, H., Mallya, A., Vahdat, A., Alvarez, J., Kautz, J., Molchanov, P.: See through gradients: image batch recovery via GradInversion. In: 2021 IEEE/CVF Conference on Computer Vision and Pattern Recognition (CVPR), pp. 16332–16341 (2021). https://doi.org/10.1109/CVPR46437.2021.01607
25. Zhang, R., Isola, P., Efros, A.A., Shechtman, E., Wang, O.: The unreasonable effectiveness of deep features as a perceptual metric (2018)
26. Zhao, B., Mopuri, K.R., Bilen, H.: idlg: Improved deep leakage from gradients (2020)
27. Zhao, Y., et al.: Local differential privacy based federated learning for internet of things (2020)
28. Zhu, L., Liu, Z., Han, S.: Deep leakage from gradients. In: Advances in Neural Information Processing Systems. vol. 32. Curran Associates, Inc. (2019). https://proceedings.neurips.cc/paper/2019/file/60a6c4002cc7b29142def8871531281a-Paper.pdf

Ranking Distance Metric for Privacy Budget in Distributed Learning of Finite Embedding Data

Georgios Papadopoulos(✉), Yash Satsangi, Shaltiel Eloul, and Marco Pistoia

Global Technology Applied Research, JPMorgan Chase,New York, USA
georgios.papadopoulos@jpmorgan.com

Abstract. Federated Learning (FL) is a collective of distributed learning paradigm that aims to preserve privacy in data. Recent studies have shown FL models to be vulnerable to reconstruction attacks that compromise data privacy by inverting gradients computed on confidential data. To address the challenge of defending against these attacks, it is common to employ methods that guarantee data confidentiality using the principles of Differential Privacy (DP). However, in many cases, especially for machine learning models trained on unstructured data such as text, evaluating privacy requires to consider also the finite space of embedding for client's private data. In this study, we show how privacy in a distributed FL setup is sensitive to the underlying finite embeddings of the confidential data. We show that privacy can be quantified for a client batch that uses either noise, or a mixture of finite embeddings, by introducing a normalised rank distance (d_{rank}). This measure has the advantage of taking into account the size of a finite vocabulary embedding, and align the privacy budget to a partitioned space. We further explore the impact of noise and client batch size on the privacy budget and compare it to the standard ϵ derived from Local-DP.

Keywords: Federated Learning · Differential Privacy · Privacy metric

1 Introduction

Federated learning (FL) [5] emerged as a collective of paradigms for distributed learning models without sharing data from local clients. In its common setup, a central ML model is trained by aggregating parameter updates (gradients) computed on local databases or edge-devices; thereby, avoiding transfer of private data to a central server. However, FL is still prone to reconstruction attacks, which can recover large training samples from the ML models by inverting gradient updates [24,25]. To mitigate this privacy challenge in distributed learning on private data, defense strategies originating mainly from Differential Privacy (DP) [10] have been proposed to reduce the leakage of information from shared gradients [13,24,25].

DP technique relies on the information recovery principal that information can be protected by considering a noise mechanism operate on the data [10].

N. Goharian et al. (Eds.): ECIR 2024, LNCS 14611, pp. 257–269, 2024.
https://doi.org/10.1007/978-3-031-56066-8_21

In FL, this can be achieved by adding noise to the shared gradients or input data, which inevitably leads to a potential loss in model performance [10]. DP gives a formal definition of an ϵ-differential private mechanism that provide a theoretical boundary for trade-off between privacy and performance of the model. Furthermore, this definition allows practitioners to evaluate the overall privacy of data against inversion attacks [23]. The privacy of a mechanism can be measured using Local Differential Privacy (LDP), as commonly employed [8,24] in FL. According to LDP, a mechanism M with domain $\mathcal{X}$ and range $\mathcal{Y}$ is ϵ- locally differential private if for all $\mathcal{S} \subseteq Y$

$$\Pr[M(x) \in \mathcal{S}] \leq e^{\epsilon} \Pr[M(x') \in \mathcal{S}], \tag{1}$$

where $x, x' \in \mathcal{X}$.

Equation 1 can be used to compute ϵ given a mechanism, which can operate on randomised gradients or dataset, to obtain a measure of privacy in typical machine learning task, such as in computer vision or signal processing classifications [12,16].

For FL models trained on text embeddings, graphs, tabular data, or categorical unstructured data, the embeddings of raw data are typically finite (for example, a small vocabulary). Finite vocabulary embeddings are particularly important in industries to obtain efficient training, for example medical documents or financial and legal documents are sometimes limited in scope compared to general text such as a book or a newspaper. Note, that these type of documents are usually the ones that contain highly confidential information that need to be protected (patient medical data or banking client data). For context, the English dictionary contains approximately 170,000 in-use words [4], whereas the Oxford Concise Medical Dictionary contains 12,000 words [3], the Merriam-Webster Legal Dictionary contains 10,000 words [1], and the Investopedia Financial Terms Dictionary contains 13,000 words [2]. Moreover, for many cases of machine learning model, reducing the vocabulary is a common practice to obtain effective training fit with a manageable training dataset and model size.

In such models, the inversion attack is potentially done on embedding vectors [15,19], where the task of the attacker is to invert to an embedding within the vocabulary. For a small vocabulary that simply means that a gradient inversion attack can be successful in recovering the exact data even without reconstructing accurately the input embedding, but be sufficiently 'near' the embedding [14]. Specifically, depending on the attacker's budget, they can infer the n-th nearest embedding to the recovered vector embedding to recover valuable information. This is especially true for text or natural language data, since it is sufficient to recover a small fraction of sporadic words to extract valuable information. This poses a unique challenge in quantifying the privacy in FL with the typical LDP setup using Eq. 1. LDP in its standard application may bound the probability of reconstructing the original vector embeddings, but it does not necessarily reason about recovery of information when the inversion attack reconstructs embedding that are 'sufficiently' near the original embedding, especially when the original vocabulary is finite and small.

In this paper we propose a distance metric for measuring the privacy budget in FL settings, named d_{rank}. The measure relates the privacy budget to the probability of an attacker recovering the original data from an inversion attack in FL setups of textual data. We show that by using d_{rank} measure, we can control the privacy of an FL system by changing the batch size and the added noise, as commonly done in FL. We compare d_{rank} to ϵ-LDP. Our analysis shows that in many cases ϵ fails to accurately represent the level of privacy and overestimates it against inversion attacks, especially as the vocabulary size is decreased. On the other hand, d_{rank} shows a better representation of the privacy budget, and can be used in FL to assess tolerance against inversion attacks or to optimise the batch and noise size.

2 Related Work

Previous papers on inversion attacks show that it is possible to recover private textual data of clients in a FL setup [19,24], directly as opposed to inference or attribution attacks. Typically, in FL setup, DP is assessed for image benchmarks [6] as a defense against such inversion attacks, but not for words embeddings from vocabulary as discussed here. Recently, [7,15,22] examined sentence privacy and developed the Truncated Exponential Mechanism (TEM) that considers the sensitivity of the embedding space and takes into account the density of the space around the input sentences/words (token neighbours). [22] expands the work into sentence level privacy (ϵ-SentDP), and [18] focuses on using linear programming to construct a mechanism that achieves the optimal utility. These works however do not propose a metric that can assess privacy in FL setup but instead they propose novel mechanisms that might be ϵ-differential private.

Following that, [17] discussed how to estimate privacy in trained embedding of text by considering the context of text on trained embedding. Their study is originated from d-χ-privacy technique [9], allowing distance distribution function to also be considered in Eq. 1, where ϵ is multiplied by the distance function $d(x, x')$, which describes the distribution density of the database, sometimes is also referred as Metric DP. [15] applies this approach to the domain of words embedding to capture the contextual similarity between words in trained embeddings. They were able to relate lower values of ϵ, by adding noise to the embedding which results in less-alike words. Hence to estimate privacy against inference attack from contextual meaning of a recovered text in noisy embedding. Here as opposed of inference attack, we consider the common recovery attack in FL, where an attacker can guess the right embedding just by searching nearby embeddings, regardless the context of the word, either by assuming untrained or trained embeddings, and any categorised data.

Overall, previous works of [7,15,22] focus on trained embeddings, so to define the distance function in neighbour Euclidean space and how the embeddings cluster in that space. Their interest is not necessarily in FL applications and mainly to avert inference attacker by exploiting semantic affinity in embeddings.

Our research, on the other hand, focuses on evaluating the privacy of embeddings against inversion attack in FL system. Specifically, we measure rank distance between untrained embeddings and their space size (vocabulary), as an FL attacker does not necessarily knows or requires the text context. We demonstrate that the d_{rank} measure allows us to manage privacy by adjusting the combination of batch size and the amount of noise introduced, which are a common practice in FL.

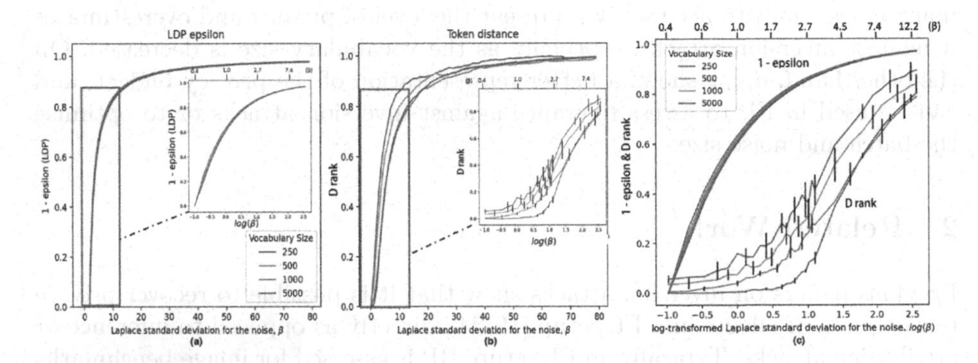

Fig. 1. Laplace noise only experiment. Larger values (close to 1) show that the data are more private. The (a) figure shows the $1 - \epsilon$ parameter as a function of standard deviation of Laplace noise, b. The $1 - \epsilon$ formula is derived in the Appendix. This example, explores the behaviour of ϵ, for different vocabulary sizes (250, 500, 1000, 5000), when the batch size is fixed at 1 and b varies (the batch size $|B|$ from Eq. 2). The figure (b) shows the token rankings ratio, (d_{rank}, Eq. 5), as a function of standard deviation of Laplace noise, b. Both plots show the results for up to $b = 80$ or $b = 13$ for the zoomed-in versions of the plots. The $x-$axis of the zoomed-in plots shows the $log(b)$ at the bottom and the b at the top. The results are the average values from a 5-times run experiment. The vertical lines in the middle zoomed-in version represent the 1-standard deviation error from the experiments we ran. The figure (c) is a combination of the two zoomed-in plots. The top curved lines are the 1 - ϵ and the bottom lines are the token rankings ratio. Ratio in the y-axis refers to the normalised $1 - \epsilon$ and d_{rank} $\in [0, 1]$.

3 Local Differential Privacy for Inversion Attacks on Text Data

In this section we describe the problem setting and mechanism we are interested in and then compute the ϵ for the mechanism to tackle inversion attacks for a client's batch of data that can be mapped to a finite embedding set in a typical FL setup. Then we define d_{rank} as a metric to quantify privacy in this setup and we subsequently provide an algorithm for computing d_{rank}.

Let $I = \{0, 1, \ldots, N-1\}$, we start with a vocabulary $V = \{T_t\}$ for $t \in I$ of N tokens T_t. We define a document C with a finite number of sentences (S_l), where $l \in \{0, \ldots, L-1\}$ is the index of the sentence and a sentence is a collection of tokens.

We assume an embedding function ϕ that takes as input a token and creates a random embedding vector $\mathbf{x}_t = \phi(T_t)$, where $\mathbf{x}$ is real-valued m-dimensional input vector. These embedding vectors (or a collection of them) can be used as input to train a machine learning model locally and share updates, $f : \mathcal{X} \rightarrow \mathbb{R}^r$, where r is the output dimension size. Generally, these updates can be computed over batches of data or tokens $B \subseteq I$. Given f, we can define a Laplace mechanism $\mathbf{M}$ on a batch where we add Laplace distribution $Lap(0, b)$ to the output of f average over the batch. The Laplace is a widely used distribution in the DP literature, the benefits over Gaussian or others have been extensively studied [11,20].

$$\mathbf{M}(B, f, b) = (\sum_{t \in B} \frac{f(\mathbf{x}_t)}{|B|}) + Lap(0, b) \tag{2}$$

We can then show (proof in Appendix) that for two different batches $B, B' \subseteq I$ the probability that the above mechanism yields same output y is less than $\exp(\frac{|\Delta f|}{br|B|})$, where

$$\Delta f = \left|\left|\Big(\sum_{t \in B} f(\mathbf{x}_t)\Big) - \Big(\sum_{t' \in B} f(\mathbf{x}_{t'})\Big)\right|\right|$$

and

$$\frac{\Pr(\mathbf{M}(B, f, b)) = y)}{\Pr(\mathbf{M}(B', f, b)) = y)} \leq \exp(\frac{\Delta f}{br|B|}) \tag{3}$$

While this provides a way to bound the probability of reconstruction of original $\mathbf{x}$ or B, here, the ϵ does not consider the embedding partition. Meaning that even when adding small noise, $b + \delta$, an attacker that inverses $\mathbf{M}$, may still obtain the correct embedding by searching the nearest embedding in the set of all possible embeddings, whilst ϵ will be varied continuously due to the added noise. This partition of the embedding space becomes significant as the vocabulary get smaller as we show in the results.

To tackle for this challenge, we suggest a distance function that accounts for the closest neighbours of embeddings (or tokens). Given $\mathbf{x}$ we rank the all possible $\mathbf{x}'$ according to the $L2$ or Euclidean norm, that is, we create an ordered list $R = \{t_0, t_1, \ldots t_{N-1}\}$ such that

$$\forall i < j \in I, \| f(\mathbf{x}) - f(\mathbf{x}_{t_i}) \| < \| f(\mathbf{x}) - f(\mathbf{x}_{t_j}) \|$$

Given the mechanism $\mathbf{M}$ defined before, we can compute the rank of output of $\mathbf{M}$ (denote by $\mathbf{M}$) in the list R, that is,

$$D_{rank}(\mathbf{M}, \mathbf{x}) = \arg\min_{i \in I} : ||\mathbf{M} - f(\mathbf{x})|| \geq ||f(\mathbf{x}) - f(\mathbf{x}_{t_i})||, \tag{4}$$

where t_i are the tokens in ordered list R. Given a sentence S of Z tokens and corresponding embeddings $(\mathbf{x}_0, \mathbf{x}_1 \ldots \mathbf{x}_{Z-1})$, we take the average to compute

$$\langle D_{rank} \rangle_S = \frac{\sum_{z=0}^{Z-1} D_{rank}(\mathbf{M}, \mathbf{x}_z)}{Z},$$

Finally, we suggest that the privacy budget can be estimated for embedded dataset and finite vocabulary as:

$$d_{rank} = 1 - \frac{2\langle D_{rank} \rangle}{N} \tag{5}$$

This is the expected d_{rank} for a mechanism $\mathbf{M}$ normalised to the expected rank of an average random guess of the embedding. d_{rank} is bounded to 1, when the outcome of $\mathbf{M}$ is fully preserving privacy, as the expected rank is equivalent to the expected value of the random guess, and bounds to 0, when the output of the mechanism is similar to the output of the input vector. The d_{rank} hence provides how successful an attack can be on embeddings, as the rank partitioned the space uniformly on the embeddings but discards non-uniform distribution of the embeddings, for example in case of trained embedding. This measure is in particular relevant to inverse attacks in FL models where the data are inversion is due to locality of an embedding and without inference i.e. finding similarity between words or sentences. This measure does not formulate the relation between the noise, batch, and its privacy budget as shown for DP, but we compute this relations as provided in Algorithm 1, using sampling of the data to get the expected values and for each data batch sharing of the client.

4 Experiments

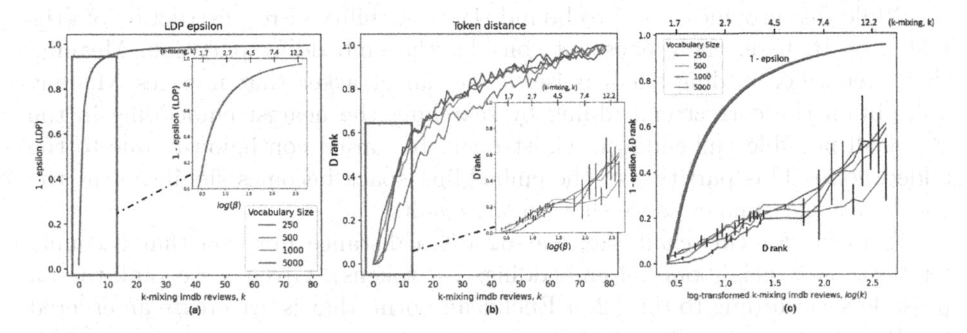

Fig. 2. Mixing-k + Laplace noise ($b = 1$ constant). Larger values (close to 1) show that the data are more private. The figure (a) shows the $1 - \epsilon$ parameter as a function of k-mixing tokens (k here refers to the batch size $|B|$ from Eq. 2). The $1 - \epsilon$ formula is derived in the Appendix. This example, explores the behaviour of ϵ, for different vocabulary sizes (250, 500, 1000, 5000), when k is changing and the Laplace noise $b = 1$ remains constant. The figure (b) shows the token rankings ratio, (d_{rank}, Eq. 5) normalised to $[0, 1]$, as a function of k-mixing tokens. Both plots show the results for up to $k = 80$ or $k = 13$ for the zoomed-in versions of the plots. The $x-$axis of the zoomed-in plots shows the $log(k)$ at the bottom and the k at the top. The results are the average values from a 5-times run experiment. The vertical lines in the middle zoomed-in version represent the 1-standard deviation error from the experiments we ran. The figure (c) is a combination of the two zoomed-in plots. The top curved lines are the 1 - ϵ and the bottom lines are the token rankings ratio. Ratio in the y-axis refers to the normalised $1 - \epsilon$ and $d_{rank} \in [0, 1]$.

We perform an experiment on a typical sentences Database [21]. The experiment is split into two groups. One group, contains noise generated from a Laplace distribution, with location $\mu = 0$ and a variable scale, $b \in (0, 80]$. Specifically, we ran the experiment 5 times, each time we chose a random sentence (we name it sentence, S_l in Sect. 3 and in Algorithm 1) and we transform it into collection of embeddings. The embeddings come from a random non-contextual (not trained) model from PyTorch (default *nn.Embedding* model). For each run, we incrementally increased the scale of the Laplace noise that we add to the token, from $b = 0.0001$ to $b = 80$ (Eq. 2). Then, we averaged the outputs ($1 - \epsilon$ and d_{rank}) of the 5 experiments for each noise level.

Similarly, for a batch B of size k ($k = |B|$), we expand the experiments by adding consecutive k sentences on the initial token, plus a fixed standard Laplace ($\mu = 0, b = 1$) noise. A detailed description of the process can be found in Algorithm 1. We follow the same process for different vocabulary sizes (250, 500, 1,000, and 5,000 words).

5 Results

In Fig. 1a, we plot the values of the normalised 1 - ϵ $[0, 1]$ as a function of the different levels of the Laplace standard deviation (b) while keeping $k = 1$ fixed. Larger values (close to 1) show that the data are more private. We observe how the 1 - ϵ values experience a steep increase even from small values of b and then slows down and eventually plateaus around $b = 20$. On the other hand, Fig. 1b shows a more accurate measure of privacy by using our ranking distance d_{rank} from Eq. 5. The reason for this discrepancy between the different metrics of how they interpret the privacy in embeddings is what we described theoretically in Sect. 3. The ϵ-parameter does not take into account adjacent embeddings.

The zoomed-in version (Fig. 1c) offers more context on this plot (x-axis logged transformed). Using the LDP ϵ as a measure of privacy can be potentially misleading in a finite embedding dataset. We observe that for a scale of $b = 0.6$ then the $1 - \epsilon = 0.2$ (a lower $1 - \epsilon$ epsilon offers weaker privacy) and for a value of $b = 1$ then $1 - \epsilon = 0.6$. But when we compare the $1 - \epsilon$ results against the normalised distance (d_{rank}), it is obvious that the ϵ-parameter does not capture the true privacy of the data. At $b = 0.6$ the normalised distance is $d_{rank} = 0.13$ and for a value of $b = 1$ the normalised distance is $d_{rank} = 0.18$. Furthermore, the ϵ cannot capture the differences amongst the various vocabulary sizes. The normalised distance between tokens shows a very clear difference in the privacy efficiency when we take into account the size of the vocabulary. In fact, the data reach the maximum private capacity only after we increase the scale of the Laplacian noise to values of $b = 7$ and above compared to $b = 2$ that the ϵ suggests.

For the next experiment, we create a batch mechanism as described in Sect. 4, Algorithm 1 and Eq. 2, by mixing together k embeddings and adding Laplace noise. As before, Fig. 2, validates the theoretical assumptions from Sect. 3. In Fig. 2c, we observe that the discrepancy in the quality of privacy between the

Algorithm 1. Privacy Mechanism

input review $\{S_l\}, l = \{0, \ldots, L-1\}$; batch size k; vocabulary V
set S_0 as the sentence we want to protect
  Let Z be the number of tokens in S_0 and $S_0 = \{t_1, t_2, \ldots, t_Z\}$ where t_z are the indexes of tokens in V.
  Create weights: $\{w_{num}, w_{den}\} : k = \frac{w_{num}}{w_{den}}$
  for $z \in \{0, \ldots, Z-1\}$ **do**
    Compute embedding $\mathbf{x}_{t_z} = \phi(T_{t_z})$.
    Perturb embedding to obtain $\hat{\mathbf{x}}_{t_z} = \mathbf{x}_{t_z} + Lap(0, 1)$
    $\mathbf{x}_{batch} := 0$
    for $l \in \{0, \ldots, w_{num} - 1\}$ **do**
      Sample embedding $\mathbf{x}_{S_l, t_z}$
      $\mathbf{x}_{batch} += \mathbf{x}_{S_l, t_z}$
    end for
    $\hat{\mathbf{x}} := w_{den}\hat{\mathbf{x}}_{t_z} + \mathbf{x}_{batch}$
    Obtain perturbed $D_{rank,z}$ from vocabulary V. Eq. 4
  end for
  Get the average D_{rank} of the sentence S_0

$$\langle D_{rank} \rangle = \frac{\sum_{z=0}^{Z-1} D_{rank,z}}{Z} \tag{6}$$

two normalised values ($1 - \epsilon$ and $d_{rank} \in [0, 1]$) is even greater that what it was in Fig. 1. Again, for a mixing value of $k = 4.5$ the $1 - \epsilon = 0.8$, whereas in reality, the $d_{rank} = 0.2$. Therefore, a higher k enables greater depth, and thus a higher level of privacy. Similarly to Fig. 1, the difference amongst the various vocabulary sizes is more clear with the d_{rank} metric.

Overall, from Appendix, Fig. 3 shows why d-rank is an important metric for differential privacy levels and captures privacy in embeddings better than epsilon. At Laplace noise level $\beta = 1$ the epsilon gives us a ratio of 0.6 out of maximum value of 1. With 0.6 the user might consider the embeddings to be "secured" in a sense that the attack party cannot infer the original sentence. But from the Fig. 3 example we observe that the majority of the tokens in the sentence have been unchanged, noise of that level does not really protect the embeddings, especially apparent for larger vocabularies (5, 000+ tokens). The d-rank draws a more realistic picture with level around 0.1. We need to reach an epsilon level of 0.9 (or d-rank around 0.4) for the embedding to be masked by noise. Similarly, for Fig. 4, for a mixing approach, the epsilon of mixing one extra sentence has a ratio of 0.5, whereas d-rank is 0.1; the sentence has retained most of the original tokens. Yet, is important to point out that the mixing approach offers, even for lower levels, a more robust privacy compared to just using noise. Comparing the sentences after adding noise/mixing between (appendix) Fig. 3 and Fig. 3 we observe that mixing makes more challenging for an attacker to infer the original sentence.

6 Conclusion

This work addresses the evaluation of privacy in the context of finite embedding spaces, such as those encountered in text, tabular data, and small vocabulary datasets. Traditional LDP measures can overestimate privacy for finite size word embeddings, particularly in the case of inversion attacks in Federated-Learning. This is primarily due to the fact that LDP does not consider how privacy changes with the size of the vocabulary or with a finite space vocabulary.

To address this limitation, we propose a ranking distance for the privacy budget, named d_{rank}. This measure takes into consideration a finite embedding space and relates the privacy budget to the probability of an attacker guessing the correct embedding from an inversion attack in FL setups of textual data. Our analysis demonstrates that d_{rank} offers a more accurate representation of the privacy budget in finite vocabulary spaces, and can be used in FL to assess tolerance against inversion attacks as function of added noise and batch size.

The introduction of d_{rank} aims to provide a more accurate and reliable tool for evaluating privacy in the context of finite embedding spaces. By allowing practitioners to better understand and manage their privacy budgets, d_{rank} can contribute to the development of more secure and reliable FL systems.

A Appendix

Starting with LHS of 3 and substituting probability density function of Laplace distribution we get

$$\begin{aligned}
\frac{\Pr(\mathbf{M}(B,f,b))=y)}{\Pr(\mathbf{M}(B',f,b))=y)} &= \Pi_i \frac{\exp\big(-\frac{|y_i - \frac{\sum_t f(\mathbf{x}_t)_i}{|B|}|}{br}\big)}{\exp\big(-\frac{|y_i - \frac{\sum_{t'} f(\mathbf{x}_{t'})_i}{|B|}|}{br}\big)} \\
&= \Pi_i \frac{\exp\big(-\frac{|\frac{|B|y_i - \sum_t f(\mathbf{x}_t)_i}{|B|}|}{br}\big)}{\exp\big(-\frac{|\frac{|B|y_i - \sum_{t'} f(\mathbf{x}_{t'})_i}{|B|}|}{br}\big)} \\
&= \Pi_i \frac{\exp\big(-\frac{||B|y_i - \sum_t f(\mathbf{x}_t)_i|}{|B|br}\big)}{\exp\big(-\frac{||B|y_i - \sum_{t'} f(\mathbf{x}_{t'})_i|}{|B|br}\big)} \\
= \Pi_i \exp\big(&\frac{||B|y_i - \sum_{t'} f(\mathbf{x}_{t'})_i| - ||B|y_i - \sum_t f(\mathbf{x}_t)_i|}{|B|br}\big) \\
&\leq \Pi_i \exp\big(\frac{|\sum_{t'} f(\mathbf{x}_{t'})_i - \sum_t f(\mathbf{x}_t)_i|}{br|B|}\big) \\
&= \exp(\sum_i \big(\frac{|\sum_{t'} f(\mathbf{x}_{t'})_i - \sum_t f(\mathbf{x}_t)_i|}{br|B|}\big) \\
&= \exp(\epsilon)
\end{aligned}$$

The first inequality follows from triangle inequality and in the last step ϵ is substituted for its definition.

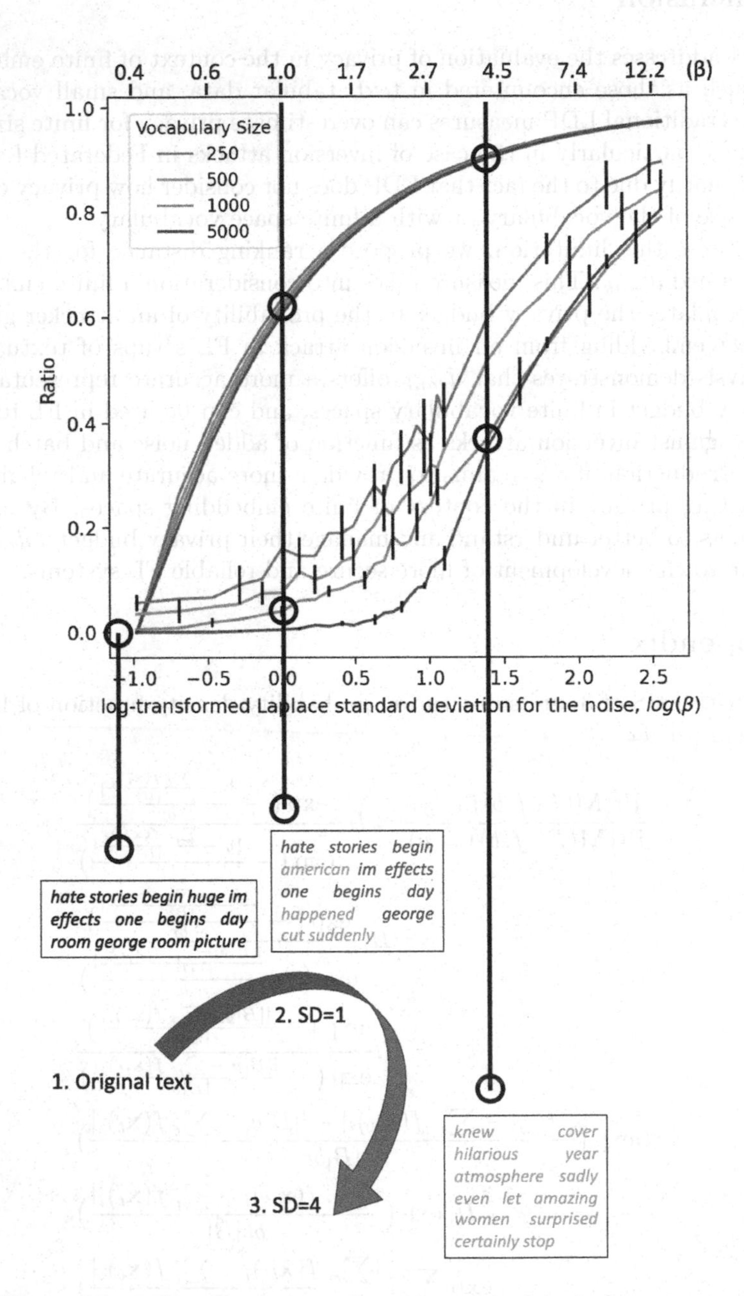

Fig. 3. From Fig. 1 The figure shows how the original text changes as we increase the Laplace noise we add to the embeddings and it maps the texts to the epsilon and d-rank. For the example we use a vocabulary of 1,000 tokens.

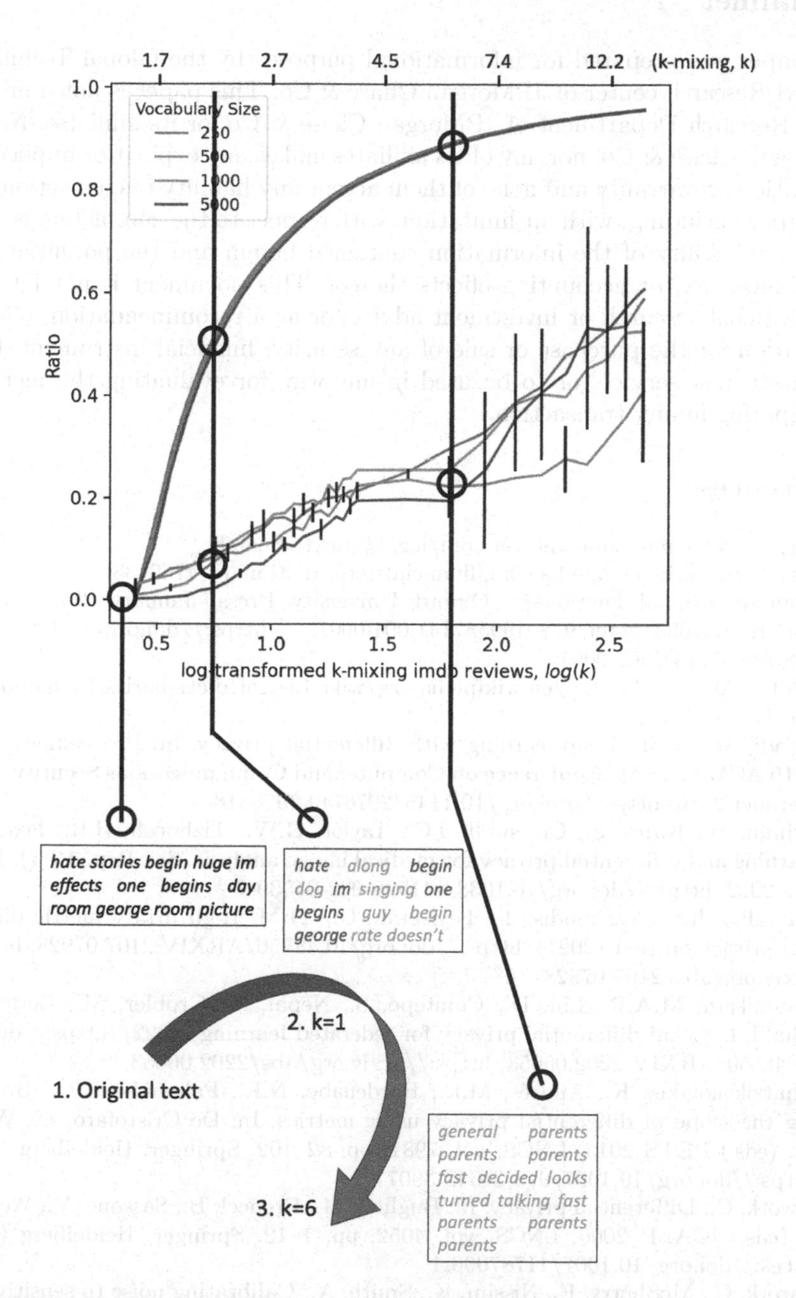

Fig. 4. From Fig. 2 The figure shows how the original text changes as we increase the size of the batches (k) we add to the embeddings and it maps the texts to the epsilon and d-rank. For the example we use a vocabulary of 1,000 tokens.

Disclaimer

This paper was prepared for informational purposes by the Global Technology Applied Research center of JPMorgan Chase & Co. This paper is not a product of the Research Department of JPMorgan Chase & Co. or its affiliates. Neither JPMorgan Chase & Co. nor any of its affiliates makes any explicit or implied representation or warranty and none of them accept any liability in connection with this paper, including, without limitation, with respect to the completeness, accuracy, or reliability of the information contained herein and the potential legal, compliance, tax, or accounting effects thereof. This document is not intended as investment research or investment advice, or as a recommendation, offer, or solicitation for the purchase or sale of any security, financial instrument, financial product or service, or to be used in any way for evaluating the merits of participating in any transaction.

References

1. https://www.merriam-webster.com/legal#::text=Search
2. https://www.investopedia.com/financial-term-dictionary-4769738
3. Concise Medical Dictionary. Oxford University Press, January 2010. https://doi.org/10.1093/acref/9780199557141.001.0001, https://doi.org/10.1093/acref/9780199557141.001.0001
4. (Oct 2023), https://en.wikipedia.org/wiki/List_of_dictionaries_by_number_of_words
5. Abadi, M., et al.: Deep learning with differential privacy. In: Proceedings of the 2016 ACM SIGSAC Conference on Computer and Communications Security. ACM, October 2016. https://doi.org/10.1145/2976749.2978318
6. Adnan, M., Kalra, S., Cresswell, J.C., Taylor, G.W., Tizhoosh, H.R.: Federated learning and differential privacy for medical image analysis. Sci. Rep. **12**(1), February 2022. https://doi.org/10.1038/s41598-022-05539-7
7. Carvalho, R.S., Vasiloudis, T., Feyisetan, O.: TEM: High utility metric differential privacy on text (2021). https://doi.org/10.48550/ARXIV.2107.07928, https://arxiv.org/abs/2107.07928
8. Chamikara, M.A.P., Liu, D., Camtepe, S., Nepal, S., Grobler, M., Bertok, P., Khalil, I.: Local differential privacy for federated learning (2022). https://doi.org/10.48550/ARXIV.2202.06053, https://arxiv.org/abs/2202.06053
9. Chatzikokolakis, K., Andrés, M.E., Bordenabe, N.E., Palamidessi, C.: Broadening the scope of differential privacy using metrics. In: De Cristofaro, E., Wright, M. (eds.) PETS 2013. LNCS, vol. 7981, pp. 82–102. Springer, Heidelberg (2013). https://doi.org/10.1007/978-3-642-39077-7_5
10. Dwork, C.: Differential privacy. In: Bugliesi, M., Preneel, B., Sassone, V., Wegener, I. (eds.) ICALP 2006. LNCS, vol. 4052, pp. 1–12. Springer, Heidelberg (2006). https://doi.org/10.1007/11787006_1
11. Dwork, C., McSherry, F., Nissim, K., Smith, A.: Calibrating noise to sensitivity in private data analysis. In: Halevi, S., Rabin, T. (eds.) TCC 2006. LNCS, vol. 3876, pp. 265–284. Springer, Heidelberg (2006). https://doi.org/10.1007/11681878_14
12. Elkordy, A.R., Zhang, J., Ezzeldin, Y.H., Psounis, K., Avestimehr, S.: How much privacy does federated learning with secure aggregation guarantee? (2022). https://doi.org/10.48550/ARXIV.2208.02304, https://arxiv.org/abs/2208.02304

13. Eloul, S., Silavong, F., Kamthe, S., Georgiadis, A., Moran, S.J.: Enhancing privacy against inversion attacks in federated learning by using mixing gradients strategies (2022)
14. Feyisetan, O., Aggarwal, A., Xu, Z., Teissier, N.: Research challenges in designing differentially private text generation mechanisms (2020). https://doi.org/10.48550/ARXIV.2012.05403, https://arxiv.org/abs/2012.05403
15. Feyisetan, O., Balle, B., Drake, T., Diethe, T.: Privacy- and utility-preserving textual analysis via calibrated multivariate perturbations (2019). https://doi.org/10.48550/ARXIV.1910.08902, https://arxiv.org/abs/1910.08902
16. Gafni, T., Shlezinger, N., Cohen, K., Eldar, Y.C., Poor, H.V.: Federated learning: a signal processing perspective. IEEE Sign. Process. Mag. **39**(3), 14–41 (2022). https://doi.org/10.1109/msp.2021.3125282
17. Hu, L., Habernal, I., Shen, L., Wang, D.: Differentially private natural language models: Recent advances and future directions (2023)
18. Imola, J., Kasiviswanathan, S., White, S., Aggarwal, A., Teissier, N.: Balancing utility and scalability in metric differential privacy. In: UAI 2022 (2022). https://www.amazon.science/publications/balancing-utility-and-scalability-in-metric-differential-privacy
19. Klymenko, O., Meisenbacher, S., Matthes, F.: Differential privacy in natural language processing the story so far. In: Proceedings of the Fourth Workshop on Privacy in Natural Language Processing. Association for Computational Linguistics (2022). https://doi.org/10.18653/v1/2022.privatenlp-1.1, https://doi.org/10.18653
20. Koufogiannis, F., Han, S., Pappas, G.J.: Optimality of the Laplace mechanism in differential privacy (2015). https://doi.org/10.48550/ARXIV.1504.00065, https://arxiv.org/abs/1504.00065
21. Maas, A.L., Daly, R.E., Pham, P.T., Huang, D., Ng, A.Y., Potts, C.: Learning word vectors for sentiment analysis. In: Proceedings of the 49th Annual Meeting of the Association for Computational Linguistics: Human Language Technologies. pp. 142–150. Association for Computational Linguistics, Portland, Oregon, USA (Jun 2011), https://aclanthology.org/P11-1015
22. Meehan, C., Mrini, K., Chaudhuri, K.: Sentence-level privacy for document embeddings (2022). https://doi.org/10.48550/ARXIV.2205.04605, https://arxiv.org/abs/2205.04605
23. Wei, K., et al.: Federated learning with differential privacy: algorithms and performance analysis. IEEE Trans. Inf. Forensics Secur. **15**, 3454–3469 (2020). https://doi.org/10.1109/TIFS.2020.2988575
24. Zhao, Y., et al.: Local differential privacy based federated learning for Internet of Things (2020)
25. Zhu, L., Liu, Z., Han, S.: Deep leakage from gradients. In: Wallach, H., Larochelle, H., Beygelzimer, A., d'Alché-Buc, F., Fox, E., Garnett, R. (eds.) Advances in Neural Information Processing Systems. vol. 32. Curran Associates, Inc. (2019). https://proceedings.neurips.cc/paper_files/paper/2019/file/60a6c4002cc7b29142def8871531281a-Paper.pdf

Experiments in News Bias Detection with Pre-trained Neural Transformers

Tim Menzner[1] and Jochen L. Leidner[1,2(✉)]

[1] Information Access Research Group, Coburg University of Applied Sciences, Friedrich-Streib-Straße 2, 96459 Coburg, Germany

[2] Department of Computer Science, University of Sheffield, Regents Court, 211 Portobello, Sheffield S1 4DP, UK

leidner@acm.org

Abstract. The World Wide Web provides unrivalled access to information globally, including factual news reporting and commentary. However, state actors and commercial players increasingly spread biased (distorted) or fake (non-factual) information to promote their agendas.

We compare several large, pre-trained language models on the task of sentence-level news bias detection and sub-type classification, providing quantitative and qualitative results.

Our findings are to be seen as part of a wider effort towards realizing the conceptual vision, articulated by Fuhr et al. [10], of a "nutrition label" for online content for the social good.

Keywords: media bias · propaganda detection · content quality · content nutrition label · text analysis · metadata enrichment · pre-trained language models · natural language processing · information retrieval

1 Introduction

The triad of media bias, propaganda and fake news is threatening democracy: the media, which have been called the "fourth estate" of a democratic system, are needed to inform citizens of the events of the world, of the response of politicians that they elected, and of potential scandals to keep the political system honest. Democracy may even be seen as facing an existential threat in scenarios where most citizens do not consume news about daily events from reliable sources, but from online commercial Websites or social media platforms whose business model are more aligned with propagating controversies than balanced news, because such a behavior promotes attention, and in an attention society,

The authors gratefully acknowledge the funding provided by the Free State of Bavaria under its "Hitech Agenda Bavaria". We would also like to thank Michael Reiche for help with annotation, the MBICS team for sharing their dataset as well as three anonymous reviewers for feedback which improved our paper. All views are the authors' and do not necessarily reflect the views of any funders or affiliated institutions.

N. Goharian et al. (Eds.): ECIR 2024, LNCS 14611, pp. 270–284, 2024.
https://doi.org/10.1007/978-3-031-56066-8_22

increased attention and stay time translates to higher advertising revenues. Consequently, in response substantial research has been conducted to – manually or automatically – determine the bias of news publishers, publications or individual news articles, story verification and fake news detection, and spotting signs of propaganda. Our own project biasscanner.org aims to add to these efforts.

Contribution. In line with the social importance of this agenda, we present the first experiments to evaluate and compare the efficacy of a range of the latest generation of pre-trained language models, namely OpenAI GPT-3, GPT-4 and Meta Llama2, on the task of sentence-level news bias detection and sub-type classification, in both zero-shot and some fine-tuned settings.

Any evidence obtained by these models can then be used by search engines as part of their ranking model, or post-hoc by users to filter out highly biased material upon user request.[1]

2 Related Work

2.1 News Bias: The Phenomenon

Media bias [12,13,19,24] has a long tradition of being investigated, and often, readers are aware of the political leanings of a newspaper, online or in paper form. However, the increasing use of the Internet as a target for information warfare and propaganda has led to an increase of various types of bias, fake news and other undesirable phenomena. Evidence also suggests [29] that fake news spreads faster than balanced news. DellaVigna et al. [9] focused on the effect of news bias and voting patterns, suggesting a significant increase in republican votes in towns where Fox News entered the cable market in 2000. Groeling [12] presents a survey of the literature covering partisan bias. However, studying political bias in U.S. news reporting, Budak et al. [5] found that "news outlets are considerably more similar than generally believed. Specifically, with the exception of political scandals, major news organizations present topics in a largely nonpartisan manner, casting neither Democrats nor Republicans in a particularly favorable or unfavorable light."

2.2 Automatic Detection of News Bias

After early pioneering work on bias from economics [13], Arapakis et al. [1] labeled 561 articles along 14 quality dimensions including subjectivity. Yano, Resnik and Smith [31] manually annotated sentence-level partisanship bias. A system to detect sentiment in news articles was developed by Zhang et al. [32] while Baumer et al. [3] focused on detecting framing language. Bhuiyan et al. [4] compare crowdsourced and expert assessment criteria for credibility on statements about climate change. Chen et al. [6] demonstrated that incorporating

[1] We do not believe in automatic censorship. Citizens should be able to see bias material on request.

second-order information, such as the probability distributions of the frequency, positions, and sequential order of sentence-level bias, can enhance the effectiveness of article-level bias detection, especially in cases where relying solely on individual words or sentences is insufficient. Recasens et al. [22] exploit the Wikipedia edit history based on "violations of neutral point of view" to obtain training data, and theypresent a logistic regression model based on a lexicon and various linguistic features. Lim, Jatowk and Yoshikawa [20] use a method based on 16 simple features to identify bias-carrying terms: they first cluster news articles by their words at the document level, identifying similar event stories as belonging to the same cluster, and then find contrasting paragraph/sentence pairs, one from insider and one from outside a given cluster in the hope of obtaining representants of two quite different perspectives. Lim and colleagues are also to be credited for first exploiting the value of rare (= high-IDF [28]) words as bias-bearing.

Media bias datasets with different focus where released by various groups [1,16,25,26]. MBIC, a media bias identification corpus, which we also use here, was introduced by Spinde et al. [27]. Horne et al. [16] released a larger dataset annotated for political partisanship bias, but without grouping articles by event, which makes apples-to-apples comparison harder; Chen et al. [7] addressed this issue by resorting to another corpus sampled from the website https://allsides.com, which includes human labels by U.S. political orientation (on the ordinal scale $\{LL, L, C, R, RR\}$); they also present an ML model to flip the orientation to the opposite one. Spinde et al. published a dataset containing biased sentences and evaluated detection techniques on it [25,26]. Hube and Fetahu [17] were the first to extract biased words from Wikipedia. They used crowdsourcing to have humans annotate bias instances with a low reported Fleiss $\kappa = 0.35$ intercoder agreement, suggesting the high subjectivity of the task as defined. Conrad et al. [8] focused on content mining to measure credibility of authors on the Web; credibility is orthogonal but related to bias. Ghanem et al. [11] analyze an interesting way to distinguish between real/credible news and fake news by looking at the distribution of affective words within the document. Spinde et al. [26] tried to extract a lexicon of bias workds from ordinary news. The Ph.D. thesis of Hamborg [14] provides a frame-oriented bias analysis technique. Hamborg et al. [15] provided a recent and interdisciplinary literature review to suggest methods how bias could be bias detection could be automated. See [2] for a good survey of the related area of automated and explainable fact checking from an NLP perspective.

Our work is perhaps closest in spirit to that of Wessel et al. [30], who evaluated transformer techniques on detecting nine different types of bias across 22 selected datasets, which they obtained and merged (= their MBIB dataset). In contrast to our work, that paper used a different dataset (MBIB≠MBIC) and they did not compare to the any of the transformer types we used, so their findings are not directly comparable. To the best of our knowledge, no previous work has compared OpenAI GPT-3.5, GPT-4 and Meta Llama2 with each other or a fine-tuned version of one of them, for the news media bias detection task.

3 Data

We evaluated using the MBIC dataset [27], which consists of 1,700 statements that contain various occurrences of media bias, annotated by 10 different judges. The statements were taken from eight different US-American news outlets: The HuffPost, MSNBC, and AlterNet as examples of left-oriented media; The Federalist, Fox News, and Breitbart to represent right-wing media outlets; and two sources from the center: USA Today and the Reuters news agency.

After removing those statements for which annotators could not arrive at a final decision, 1,551 statements (1,018 biased, 533 non-biased) were left for us to use.

4 Methods

We conducted three lines of experiments to assess quantitatively and qualitatively the ability of three of the latest-generation large pre-trained languge models, GPT-3.5, GPT-4 and Llama2 (Table 1), when employed to classify sentences from English news with respect to whether or not they may express any type of bias, and if so, which type.

Table 1. Methods Used in Our Experiments

Method 1	Open AI, GPT-3.5	one/zero-shot + prompt engineering, called via API
Method 2.	Open AI, GPT-3.5	prompt engineering, fine-tuned by us, via API
Method 3.	Open AI, GPT-4	one-shot, via API
Method 4.	Meta Llama2	one-shot, on premise

All of these were evaluated using MBIC. When iterating through the dataset for evaluation, two different modes were used. In the first mode, sentences were not evaluated individually, but in batches joined together in groups of ten sentences, separated by a line break. This was done for the sole benefit of minimizing the number of API calls, as the model had not to be prompted for each single sentence. Also, this might be considered a more natural case for bias detection, as it models the process of bias identification in longer texts. As the cleaned dataset contained 1,551 sentences, which can not be divided by 10 without a remainder, the last sentence was removed from the dataset in this mode. The results for this evaluation can be found in Table 2 below (Sect. 5).

In the second mode (results shown in Table 3 below), sentences were evaluated *individually*, independent of preceding or following sentences, to see whether batching unrelated sentences in the first line of experiments may have harmed performance of the models. The language model's instruction prompt underwent iterative development[2] to guarantee the production of consistent and high-

[2] The text of the various prompts used here is too long to include it in this paper; we will share them in the GitHub repository at (Anonymized) upon acceptance of this paper.

quality results. It incorporates a precise delineation of each sought-after bias category (linguistic bias, text-level context bias, reporting-level context bias, cognitive bias, hate speech, fake news, racial bias, gender bias, and political bias), with the definitions given by [30], along with an illustrative example demonstrating the desired JSON output structure. Subsequently, the model's output underwent a post-processing and filtering phase to mitigate potential errors before flagging biased sentences within the content.

For each bias-flagged sentence, the model was also asked to offer insights into the bias type, a detailed explanation, and a quantified bias strength score. The development of the prompt followed evaluated best practice techniques, like asking the model to assume an expert role, breaking the task into smaller sub tasks to complete step by step or providing an example output for one shot prompting [21]. While the evaluations presented in Table 2 and Table 3. were all conducted with the same prompt, several additional variations were tested for the experiments presented in Table 4, to gain insights in the effect of different prompt engineering techniques and parameter changes on the quality of results. All evaluations unless stated otherwise were conducted with a temperature of 0.0 to keep the model's outputs more consistent. Evaluations in Table 2 and Table 3 were conducted using GPT-3.5-turbo-16k, two different fine-tuned GPT-3.5-turbo models (called Variant A and B, respectively in the results tables below), GPT-4 and Llama2-70b-chat-awq. The first fine-tuned model, for the experiment presented in Table 2, was fine-tuned with 50 examples, each consisting of the system prompt, the article built from 10 sentences and a desired model output for this case (FT Variant A). While the sentences considered biased could be taken directly from the dataset, the information about bias type and bias score (which was not relevant to the result of the evaluation, but was needed to stay in format), was filled in using GPT-4. The 500 sentences (50 batches a 10 sentences) used for fine-tuning were removed from the dataset before the evaluation of the model. For the experiments presented in Table 3, fine-tuning was conducted in a similar manner, but using 50 held-out examples comprising *individual* sentences (FT Variant B). GPT-3.5-turbo-16k served as the foundational model for all the tests in Table 4. This decision allowed us to conveniently evaluate the effects of different prompt and parameter modifications. It is important to note that different models may produce varying responses to these changes.

In the third line of experiments, next to the described prompt, different variations were evaluated. One version where a (fictional) example given in the original prompt was removed, to see how significant the influence of such an example can be. In a different approach, we excluded the definitions of text-level context bias, reporting-level context bias, and cognitive bias from the prompt. These concepts heavily rely on context, which the dataset we are using cannot provide, as explained in Sect. 3. In another scenario, we completely omitted the explanation of different bias types. This was done to assess the model's performance when relying solely on its internal knowledge of bias. For yet another different experiment, we implemented a second round of prompt engineering techniques. This involved making subtle modifications to enhance clarity in instructing the model on the specific steps to follow, along with replacing certain words with more precise or commonly used expressions. In a separate run, we leveraged the model's bias score to filter out sentences in which only a mild bias was detected. Specifically, we removed all sentences with a bias score less than 0.6 from the results. For another experiment, we set no custom system prompt but instead appended the text that would have gone there at the start of the user message. While using the system message for prompting the basic behaviour of the model is considered best practice, there had initially been discussion about it not having the same weight as the user message [23]. Finally, in a final experiment, we set the model temperature from 0.0 to GPT's default of 0.7, to test a variant that uses more influence of randomness.

5 Evaluation

In this section, we report the results of our experimental findings.

5.1 Quantitative Evaluation

Table 2 shows the results for the comparison of GPT-3.5, both *as is*, as well as versus our fine-tuned variant of it, with GPT-4 and Llama2. In this first set of experiments, we posed the news text to be classified to the models in groups of ten sentences at a time without overlap. Our fine-tuned version of GPT-3.5 outperformed all other models in terms of F1 through a substantial increase of Precision by 14%, traded for a drop in Recall by 16%. Overall, GPT-4 has the highest precision (84%) in absolute terms, but its small lead of 2% over the fine-tuned GPT-3.5 is not worth the added energy, cost and memory for almost all applications, in particular given that its Precision is also lower.

Table 2. Evaluation Results for GPT-3.5-turbo-16k, GPT-3.5-turbo fine-tuned, GPT-4 and Llama2-70b-chat-awq, on 10 sentence blocks. Best results are highlighted in bold.

Model	TP	FP	FN	TN	F1-Score	Recall	Precision
GPT-3.5	965	460	53	72	0.790	**0.948**	0.677
GPT-3.5 (FT Variant A)	442	100	119	189	**0.802**	0.788	0.816
GPT-4	739	138	279	394	0.780	0.726	**0.843**
Llama2	579	241	439	291	0.630	0.569	0.706

Table 3 shows the results for a different set of experiments, in which sentences were posed individually to the models, leaving out the formation of 10-sentence blocks, in order to assess the degree of influence the mixture of different biased and unbiased sentences in a longer text has on the transformer models' ability to make the right decisions. The model's decision about the bias level of a sentence might not be absolute but relative to the bias exhibited by other sentences included in the same prompt. To our surprise, GPT-3.5 did quite well on individual sentences, slightly better than on groups of sentences even, whereas fine-tuning appears to benefit from longer texts, since the performance of the model trained on single sentences, dropped substantially (-14%). GPT-4 improved in the absence of other sentences, from 78% to 82% in terms of F1.

Table 3. Evaluation Results for GPT-3.5-turbo-16k, GPT-3.5-turbo fine-tuned, GPT-4 and Llama2-70b-chat-awq, on individual sentences. Best results are highlighted in bold.

Model	TP	FP	FN	TN	F1-Score	Recall	Precision
GPT-3.5	943	370	75	163	0.809	0.926	0.718
GPT-3.5 (FT Variant B)	527	82	455	437	0.662	0.537	0.865
GPT-4	1003	415	15	118	0.823	0.985	0.707
Llama2	1014	525	2	10	0.793	**0.998**	0.659

In addition, we conducted an extra set of experiments with GPT-3.5, in which we varied prompts and parameter settings. The results are shown in Table 4.

As these results show, providing an example within the prompt only had a minor effect on the outcome, whereas removing the definitions for context dependent bias criteria led to a small shift towards fewer total positives and slightly more negatives, while keeping the F1-Score consistent. Restructuring the prompt to substitute unclear words and to work out the division into individual work steps more clearly, led to an increased F1-score and small gains in precision, while the recall performance dropped at the same time. As expected, filtering out results where the bias score was below 0.6, led to a notable increase in precision and a notable decrease in recall. Interestingly, giving the model no definition of bias and different bias types at all, resulted in the best F1-score, with a higher precision and a lower recall. In this setting, the model had to come up with its own definition of bias types. Most of the times, it selected a generic categorisation like "negative bias" (21%), "positive bias" (4%) or just "bias" (15%). However, it also identified more specific bias categories like "political bias" (15%), "loaded language" (12%), "spin" (7%), "emotional bias" (2%), "gender bias" (2%) or "bias by omission" (1%). Some of these are identical or resemble the categories used in our prompt. In many cases, the model further appears to use different spellings or different wording potentially describing the same thing like "omission" (1%) additionally to the already mentioned "bias by omission" or "sexism" (0.4%) next to the discussed "gender bias". The model also came up with rather obscure and specific bias categories like "bias against conspiracy theories" or "bias against Bernie Sanders", respectively exactly one time.

In comparison, our fine tuned GPT 3.5 model, which was the best performing model in the 10 batch mode, managed to stick to the defined bias categories nearly every time, spread as following: Political bias (57%), linguistic bias (13%), reporting-level context bias (8%), text-level context bias (8%), gender bias (5%), racial bias (4%), fake news (0.2%). The remaining percentages were split on cases where the model named two different biases at the same time (3%) or introduced own categories like "economic bias" or "generational bias", which could also all be found in the results of the run without definitions in the prompt (1%).

Furthermore, using a larger user message instead of submitting a non-changing basic instruction as a system message, which is considered best practice, might increase recall and decrease precision. Ultimately, setting the temperature of the model back to the GPT default of 0.7 up from the otherwise used 0.0, decreased the performance on all evaluated metrics. While these tests indicate that certain modifications to prompt and technique can have an influence on the performance, it is important to bear in mind that the non deterministic nature of large language models might influence the results in one or the other direction.

Table 4. Evaluation results for GPT-3.5-turbo-16k with different prompt variants and parameter settings, on 10 sentence blocks. Best results are highlighted in bold.

Model	TP	FP	FN	TN	F1-Score	Recall	Precision
GPT-3.5 Base	965	460	53	72	0.790	0.948	0.677
No Example in Prompt	966	468	52	64	0.788 ▼	**0.949** ▲	0.674 ▼
No Contextual/Cognitive	938	426	80	106	0.788 ▼	0.921 ▼	0.688 ▲
Restructured Prompt	956	439	62	93	0.792 ▲	0.939 ▼	0.685 ▲
Bias Score Threshold	632	241	273	110	0.711 ▼	0.698 ▼	**0.724** ▲
No Bias Definition	894	344	124	188	**0.793** ▲	0.878 ▼	0.722 ▲
No System Prompt	986	486	32	46	0.792 ▲	0.969 ▲	0.670 ▼
High Temperature	931	446	87	86	0.778 ▼	0.915 ▼	0.676 ▼

All of these numbers above are related to the binary +/-BIAS sentence-level classification, for which MBIC contained the binary gold labels. Our modes also output sub-categories, for which no gold label was available, so we drew a random sample of $N = 100$ sentences. Three human annotators ($K = 3$) independently judged the model output either as "right" or "wrong". The result, per category and overall, is shown in Table 5. Using automatic creation of a silver dataset via majority voting between the three human judge's decision, the overall accuracy of the model was $A = 87\%$.

5.2 Qualitative Evaluation

Overall, we are satisfied with the quality pre-trained transformers can detect instances of news bias on *sentence granularity* and classify them into sub-categories. Some examples of what could be considered a high quality output, are

Table 5. Evaluation Results for the Bias Subcategory Evaluation GPT 3.5 (FT Variant A) model (random sample $N = 100, K = 3$, silver label through majority voting)

Bias Sub-Type	Correct	Incorrect	Total	Accuracy
Political bias	56	7	63	88.89%
Racial bias	5	1	6	83.33%
Gender bias	3	1	4	75.00%
Text-level context bias	5	1	6	83.33%
Linguistic bias	11	1	12	91.67%
Reporting-level context bias	7	0	7	100.00%
Sub-type violation (hallucination)	n/a	2	2	n/a
Total (all classes)	87	13	100	87.00%

presented in Example 1. We are less satisfied with the level of our understanding of some sub-classes of news bias and the degree of subjectivity that is still visible from the various gold data sets available today. Examples for questionable model assessments can be found in Example 2.

Example 1. Examples for high quality model decisions

sentence text	In October, Breitbart London reported that French police are allegedly actively assisting migrants to avoid detection by UK Border Force, telling them when and when not to attempt to cross the channel, according to a people smuggler
gold label	Biased
assessment by model	GPT-3.5 FT
bias_type	Reporting-level context bias
bias_score	0.5
bias_description	The sentence cites Breitbart London as the source of the information, which is known for its conservative bias. This could be seen as a bias in the choice of source, potentially influencing the readerś perception of the reported event.
sentence text	Football supports Trump in its promotion of racial division, the crushing of dissent, and the spread of misinformation, inequality, and brutality
dataset label	Biased
assessment by model	Llama2
bias_type	Political bias
bias_score	0.9
bias_description	This sentence directly implies that football, by extension the NFL, supports Trump and his policies, which is a highly controversial and debated topic. The use of the phrase'promotion of racial division' and'spread of misinformation' implies a strong negative bias against Trump and the NFL, and could be seen as taking a political stance.

sentence text	On one hand, naive teenagers who signed off their financial futures to leftist, anti-American institutions would catch a break, The college cartel screwed you; now here's a government waiver to make you forever grateful to the Democrats
gold label	Biased
assessment by model	GPT-4
bias_type	Linguistic bias
bias_score	0.8
bias_description	The use of the term'naive teenagers' and'leftist, anti-American institutions' introduces a negative judgement and bias against those who support college debt forgiveness and the institutions that provide higher education. The sentence also implies that the Democrats' motive for the policy is to make people'forever grateful' to them, suggesting manipulation.
sentence text	Male and female are the two sexes - but transgender advocates now claim more than 50 "genders" - including agender, bigender, gender fluid, gender non-conforming, genderqueer, intersex, two-spirit, etc., plus "cis-gender" for the 99+ percent of people who are normal, non-transgender people
gold label	Biased
assessment by model	GPT-4
bias_type	Gender bias
bias_score	0.7
bias_description	The sentence uses the term'normal' to describe non-transgender people, implying that transgender people are not normal, which is a negative portrayal

Example 2. Examples for arguably questionable model decisions

sentence text	A study published Tuesday in the open peer-reviewed journal Proceedings of the Royal Society B finds that the microorganisms in the wax worm's gut help them consume and metabolize plastics
gold label	Non-biased
assessment by model	Llama2
bias_type	Reporting-level context bias
bias_score	0.5
bias_description	The article focuses on a specific study and its findings, without providing a broader context or mentioning other research on the topic. This may create a bias towards the study's conclusions and findings, without considering alternative viewpoints or research.
sentence text	A Gallup poll reports that women are still more likely to do laundry, clean the house, do grocery shopping, prepare meals, wash dishes and make decisions about furniture and decorations - even among younger generations who are reportedly more egalitarian as ever
gold label	Non-biased
assessment by model	GPT-3.5
bias_type	Gender bias
bias_score	0.9
bias_description	This sentence reinforces traditional gender roles by stating that women are more likely to perform household chores and make decisions about furniture and decorations. It implies that these tasks are inherently feminine and perpetuates gender stereotypes.
sentence text	'Hundreds of white nationalists lined the streets of Charlottesville, Va., last Saturday to protest the removal of a Confederate monument, Some waved Confederate and Nazi flags, others brandished shields
gold label	Non-biased
assessment by model	GPT-3.5-FT
bias_type	Reporting-level context bias
bias_score	0.7
bias_description	The sentence focuses on the presence of white nationalists and their use of controversial symbols, which could be seen as a negative portrayal of this group.

As visible in Example 1, the model can also utilize the names of certain media outlets, renowned for their perceived bias, as an indicative factor for assessing bias. When reporting merely factual statements, the models sometimes judged those as "biased", asking for an unreasonable amount of potentially unnecessary context. The mentioning of facts and findings about existing biases in our society was also occasionally flagged as biased. Similarly, this tendency persisted when reporting on actions that could be considered criminal or unethical, of individuals or groups; even when such reports were objectively presented, the models occasionally accused the articles of introducing a negative bias against the subject. *We conjecture that there is an underlying inability of the models to distinguish between language and meta-language, and to handle reported speech.* Given that all models were trained on a fixed dataset and do not automatically get updated with more recent data, their ability to detect fake news in contemporary articles is constrained, suggesting that the "fake news" bias category may not be of use in this context. As the evaluation dataset did not contain instances of fake news, this was not a problem for the evaluation. However, similar to the phenomena described above, a statement mentioning fake could get marked with "fake news" as bias type, just for mentioning it.

When running the evaluation on individual sentences, we also observed that sometimes, the models hallucinated more sentences based on the actual sentences from the dataset or they contradicted themselves in its assessment of the sentence and their overall conclusion about the "article". This did not influence the results of the evaluation as only the first sentence was taken into account, but it serves as evidence of the potential instability in the output of large language models. Instability, which could also be noted by occasional syntax errors in the JSON, which the models were asked to provide. In sporadic cases, the model further included non biased sentences with a bias type like "None" and a bias score of 0 in its assessment, which were filtered out before the analysis. While this filter was created with the intention of capturing as many cases as possible, but due to the inherent unpredictability of large language models, a 100% coverage can not be guaranteed. As described in Sect. 6, the dataset also contained instances of debatable annotation, as shown in Example 3, for instance:

Example 3. Examples for arguably debatable model decisions

sentence text	But the elimination of air travel strikes particularly close to the homes of Sens, Kirsten Gillibrand, Cory Booker, Bernie Sanders, and Kamala Harris - all of whom extensively used air travel, including private jets - throughout the years in office
gold label	Non-biased
assessment by model	GPT-3.5
bias_type	Reporting-level context bias
bias_score	0.8
bias_description	The sentence highlights the air travel habits of specific senators who support the elimination of air travel, suggesting hypocrisy or inconsistency in their actions.

sentence text	Under current immigration policies, it's possible that immigration arrests could lead to poor mental health by increasing deportation fears among undocumented individuals and their families and neighbors, the study authors write
gold label	Biased
assessment by model	GPT-3.5-FT
bias_type	None
bias_score	0
bias_description	The sentence does not exhibit any notable bias.

6 Limitations and Ethical Considerations

No automatic neural or statistical classifier is capable of detecting instances of news bias perfectly; a certain number of false positives and false negatives are bound to occur. A comparison with other automated approaches was out of scope for this paper, as we found the techniques hard to replicate and the circumstances of their evaluations were different. Several forms of bias depend on *context*. This particularly affects text-level context bias, reporting-level context bias and cognitive bias. With the dataset consisting of single sentences instead of a paragraph or whole articles, the necessary context is missing that would be needed to take the right decision. Residual controversial annotations make the evaluation difficult. For instance, when one language model identifies a sentence as biased but the data set does not list it as such, this might be logged as a false positive, sometimes in error (see Example 5.2). Furthermore, since the MBIC dataset only contains news from U.S. outlets, it is inherently strongly biased towards topics like U.S. politics in the period of the collection of the data (Trump, abortion). The fact that all annotators were based in the U.S. only adds to the U.S.-centric focus. Perceptions of bias can vary across different cultures and may also evolve over time within the U.S itself. Considering that the MBIC Dataset was published early enough to be incorporated into the training data of all models, it remains a possibility that certain parts of it were included, even though none of the three models used were familiar with a dataset by that name when queried.

Bias detection has its own biases: not all bias types are recognized equally well. To mitigate risks pertaining to this, we urge application developers that use our or other methods to make the users of their end applications aware of this. Furthermore, when dealing with the question of news bias, the topic of *false balance* has to be considered. While presenting different perspectives on an issue could be described as an unbiased approach, false balance means giving equal weight or airtime to views that are not supported by credible evidence, which can again create bias in form of a misleading perception of a balanced presentation. This phenomena has been extensivly studied in the context of climate change research, where false balance has been shown to influence public perceptions and beliefs [18]. When analysing reporting using negatively connotated words to describe someone's position as "racist", for example, can be either unjustified biased or an objective description of this person's ideology. Avoiding such negative connotated words when they would provide an accurate depiction could, in itself, introduce a form of bias. Furthermore, the potential instability and unpredictability of the output of large language models (hallucination) poses a challenge for building applications. Finally, the use of models over a networked API (as in the case of GPT-3.5 and GPT-4) further poses potential privacy risks as content to be transferred via an API must be shared with the server-owning counterparty, or risk of unavailability in case of network downtime.

7 Summary, Conclusions and Future Work

In this paper, we described a set of experiments using several pre-trained neural transformers to compare their performance on the task of news bias detection on multiple datasets. To have a capability for high-quality automatic news bias identification and classification is socially desirable in order to avoid that users of the Web drown in propaganda, biased and fake news "reporting", so we view our contribution as much as a social one as a scientific one.

Our experiments show that a fine-tuned variant of a model with the smaller number of parameters (GPT-3.5) can outperform a model with a much larger number of parameters (GPT-4). We also discovered in our analysis that all models struggled with reported speech, with distinguishing language and meta-language, and with hallucinating "new" uncalled-for categories, all of which should be explored further.

In future work, we plan to work on cross-distilling a single open, non-proprietary, on-premise model that is fine-tuned on the media bias identification and classification task, and on rolling out a browser plug-in (Anonymized) in order to put a tool into citizens' hands to promote their critical reading of the news, in the spirit of [10]. We also plan to collect a larger corpus of biased news specimens in multiple languages, and invite other research groups to join forces on this socially relevant endeavour.

References

1. Arapakis, I., Peleja, F., Berkant, B., Magalhaes, J.: Linguistic benchmarks of online news article quality. In: Proceedings of the 54th Annual Meeting of the Association for Computational Linguistics, pp. 1893–1902. ACL, Berlin, Germany (2016). https://doi.org/10.18653/v1/P16-1178
2. Augenstein, I.: Towards Explainable Fact Checking. Doctor scientiarium thesis, University of Copenhagen, Copenhagen, Denmark (2021). https://arxiv.org/abs/2108.10274, (available online as ArXiv pre-print, accessed 2023-10-01)
3. Baumer, E., Elovic, E., Qin, Y., Polletta, F., Gay, G.: Testing and comparing computational approaches for identifying the language of framing in political news. In: Proceedings of the 2015 Conference of the North American Chapter of the Association for Computational Linguistics: Human Language Technologies, pp. 1472–1482. ACL, Denver, CO, USA (2015). https://doi.org/10.3115/v1/N15-1171
4. Bhuiyan, M.M., Zhang, A.X., Sehat, C.M., Mitra, T.: Investigating differences in crowdsourced news credibility assessment: Raters, tasks, and expert criteria. Proc. ACM Hum.-Comput. Interact. **4**(CSCW2) (2020). https://doi.org/10.1145/3415164

5. Budak, D., Sharad Goel, J.M.R.: Fair and balanced? quantifying media bias through crowdsourced content analysis. Public Opinion. **80**(S1), 250–271 (2016). https://doi.org/10.1093/poq/nfw007
6. Chen, W.F., Al-Khatib, K., Stein, B., Wachsmuth, H.: Detecting media bias in news articles using gaussian bias distributions (2020). https://arxiv.org/pdf/2010.10649.pdf, unpublished technical report arXiv:2010.10649 [cs.CL], Cornell University "ArXiv" pre-print server
7. Chen, W.F., Wachsmuth, H., Al-Khatib, K., Stein, B.: Learning to flip the bias of news headlines. In: Proceedings of the 11th International Conference on Natural Language Generation, pp. 79–88. ACL, Tilburg University, The Netherlands (2018). https://doi.org/10.18653/v1/W18-6509
8. Conboy, M.: The Language of the News. Routledge, London, UK (2007)
9. DellaVigna, S., Kaplan, E.: The fox news effect: media bias and voting*. Q. J. Econ. **122**(3), 1187–1234 (2007). https://doi.org/10.1162/qjec.122.3.1187
10. Fuhr, N., et al.: An information nutritional label for online documents. SIGIR Forum **51**(3), 46–66 (2018). https://doi.org/10.1145/3190580.3190588
11. Ghanem, B., Ponzetto, S.P., Rosso, P., Rangel, F.: FakeFlow: fake news detection by modeling the flow of affective information. In: Proceedings of the 16th Conference of the European Chapter of the Association for Computational Linguistics, pp. 679–689. Association for Computational Linguistics, Online (2021). https://doi.org/10.18653/v1/2021.eacl-main.56
12. Groeling, T.: Media bias by the numbers: challenges and opportunities in the empirical study of partisan news. Annu. Rev. Polit. Sci. **16**, 129–151 (2013)
13. Groseclose, T., Milyo, J.: A measure of media bias. Q. J. Econ. **120**(4), 1191–1237 (2005)
14. Hamborg, F.: Revealing Media Bias in News Articles NLP Techniques for Automated Frame Analysis. Springer, Cham, Switzerland (2023). https://doi.org/10.1007/978-3-031-17693-7, open access e-book
15. Hamborg, F., Zhukova, A., Donnay, K., Gipp, B.: Newsalyze: enabling news consumers to understand media bias. In: Proceedings of the ACM/IEEE Joint Conference on Digital Libraries in 2020, JCDL 2020, pp. 455–456. ACM, New York (2020). https://doi.org/10.1145/3383583.3398561
16. Horne, B.D., Dron, W., Khedr, S., Adali, S.: Sampling the news producers: A large news and feature data set for the study of the complex media landscape (2018)
17. Hube, C., Fetahu, B.: Detecting biased statements in Wikipedia. In: Companion Proceedings of the Web Conference WWW 2018, pp. 1779–1786. International World Wide Web Conferences Steering Committee, Geneva, Switzerland (2018). https://doi.org/10.1145/3184558.3191640
18. Imundo, M.N., Rapp, D.N.: When fairness is flawed: effects of false balance reporting and weight-of-evidence statements on beliefs and perceptions of climate change. J. Appl. Res. Mem. Cogn. **11**(2), 258 (2022)
19. Lee, M.A., Solomon, N.: Unreliable Sources. A Guide to Detecting Bias in News Media. Carol Publishing, New York, NY, USA (1990)
20. Lim, S., Jatowt, A., Yoshikawa, M.: Towards bias inducing word detection by linguistic cue analysis in news. In: DEIM Forum, pp. C1–3 (2018)
21. OpenAI: OpenAI Models Documentation (2023). https://platform.openai.com/docs/models, OpenAI Documentation
22. Recasens, M., Danescu-Niculescu-Mizil, C., Jurafsky, D.: Linguistic models for analyzing and detecting biased language. In: Proceedings of the 51st Annual Meeting of the Association for Computational Linguistics, pp. 1650–1659. Association for

Computational Linguistics, Sofia, Bulgaria (2013). https://aclanthology.org/P13-1162
23. da Rocha, C., et al.: The "system" role - how it influences the chat behavior. https://community.openai.com/t/the-system-role-how-it-influences-the-chat-behavior/87353 (Accessed 26 Oct 2023)
24. Sloan, W.D., Mackay, J.B. (eds.): Media Bias: Finding It. Fixing It. McFarland & Company, Jefferson, NC, USA (2007)
25. Spinde, T., Hamborg, F., Gipp, B.: Media bias in German news articles: a combined approach. In: Koprinska, I., et al. (eds.) ECML PKDD 2020. CCIS, vol. 1323, pp. 581–590. Springer, Cham (2020). https://doi.org/10.1007/978-3-030-65965-3_41
26. Spinde, T.: Automated identification of bias inducing words in news articles using linguistic and context-oriented features. Inform. Process. Manag. **58**(3), 102505 (2021). https://doi.org/10.1016/j.ipm.2021.102505
27. Spinde, T., Rudnitckaia, L., Sinha, K., Hamborg, F., Gipp, B., Donnay, K.: MBIC - a media bias annotation dataset including annotator characteristics. In: Proceedings of the iConference 2021 (2021)
28. Spärck Jones, K.: A statistical interpretation of term specificity and its application in retrieval. J. Documentation **28**(1), 11–21 (1972). https://doi.org/10.1108/eb026526
29. Vosoughi, S., Roy, D., Aral, S.: The spread of true and false news online. Science **359**(6380), 1146–1151 (2018). https://doi.org/10.1126/science.aap9559
30. Wessel, M., Horych, T., Ruas, T., Aizawa, A., Gipp, B., Spinde, T.: Introducing MBIB – the first Media Bias Identification Benchmark Task and Dataset Collection. In: Proceedings of the 46th International ACM SIGIR Conference on Research and Development in Information Retrieval (SIGIR 2023) (2023). https://doi.org/10.48550/arXiv.2304.13148, https://arxiv.org/abs/2304.13148
31. Yano, T., Resnik, P., Smith, N.A.: Shedding (a thousand points of) light on biased language. In: Proceedings of the NAACL-HLT 2010 Workshop on Creating Speech and Language Data with Amazon's Mechanical Turk, pp. 152–158. ACL (2018)
32. Zhang, J., Kawai, Y., Nakajima, S., Matsumoto, Y., Tanaka, K.: Sentiment bias detection in support of news credibility judgment. In: Proceedings of the 44th Hawaii International Conference on System Sciences, pp. 1–10. IEEE (2011)

An Empirical Analysis of Intervention Strategies' Effectiveness for Countering Misinformation Amplification by Recommendation Algorithms

Royal Pathak and Francesca Spezzano(✉)

Computer Science Department, Boise State University, Boise, ID, USA
royalpathak@u.boisestate.edu, francescaspezzano@boisestate.edu

Abstract. Social network platforms connect people worldwide, facilitating communication, information sharing, and personal/professional networking. They use recommendation algorithms to personalize content and enhance user experiences. However, these algorithms can unintentionally amplify misinformation by prioritizing engagement over accuracy. For instance, recent works suggest that popularity-based and network-based recommendation algorithms contribute the most to misinformation diffusion. In our study, we present an exploration on two Twitter datasets to understand the impact of intervention techniques on combating misinformation amplification initiated by recommendation algorithms. We simulate various scenarios and evaluate the effectiveness of intervention strategies in social sciences such as Virality Circuit Breakers and accuracy nudges. Our findings highlight that these intervention strategies are generally successful when applied on top of collaborative filtering and content-based recommendation algorithms, while having different levels of effectiveness depending on the number of users keen to spread fake news present in the dataset.

Keywords: Social networks · Misinformation Mitigation · Intervention Strategies · Virality Circuit Breakers · Accuracy Nudges

1 Introduction

Over the past few years, there has been a noticeable rise in misinformation (i.e., incorrect or deceptive information that can be inadvertently or deliberately shared, resulting in varying degrees of harm) spread in social networks, spanning from the 2016 U.S. Presidential Election to the Covid-19 pandemic [2,5,34]. The repercussions of this phenomenon have had far-reaching effects on various aspects, including health, politics, the economy, and disaster response [15,20,30,51]. Criticism has been directed towards recommendation algorithms (**RAs**), which have been implicated in the creation of filter bubbles,

N. Goharian et al. (Eds.): ECIR 2024, LNCS 14611, pp. 285–301, 2024.
https://doi.org/10.1007/978-3-031-56066-8_23

echo chambers, and various biases (such as cognitive, gender, popularity, demographic, and confirmation), ultimately enabling the propagation of misinformation [7,13,33,35]. Research on interventions targeting recommender algorithms (RAs) to combat the spread of misinformation is limited, despite extensive studies on how RAs contribute to misinformation recommendation and dissemination [11,13,35]. Hence, to address the spread of misinformation initiated by RAs, it is essential to conduct further research to understand the effectiveness of existing intervention strategies applied on top of RAs.

In this paper, we contribute to the study of whether established theoretical intervention frameworks in social computing can effectively address the problem of misinformation amplification caused by RAs. To achieve this, we conduct simulations to observe changes in the average expected spread of misinformation by using the Linear Threshold model (**LTM**) [23], simulating various RAs to initiate misinformation diffusion, and applying intervention techniques from the social science literature. Our study focuses on Twitter as the platform of investigation, with a primary emphasis on addressing the dissemination of fake news[1] and analyzing the effectiveness of interventions on classical widely-used RAs. As one strategy, we apply an intervention technique based on Virality Circuit Breakers (**VCBs**) [1]. Specifically, in our experiments, we focus on reducing misinformation amplification by not recommending fake news to highly influential users emphasizing that we are not explicitly removing the fake news content itself, which is a controversial strategy. As another intervention strategy, we simulate nudging users towards accuracy by proposing a *novel* information diffusion model. This enables us to assess the effects of interventions aimed at reducing misinformation spread initiated by RAs when different levels of efficacy of the considered accuracy nudges are applied to end users. To facilitate this analysis, we introduce an additional parameter called the intervention ratio γ, which describes the likelihood of fake news spreading after applying accuracy nudges. This helps us understand the extent of intervention efficacy required for each type of RAs to mitigate misinformation. The main research questions guiding our investigation are as follows:

RQ1: How effective are Virality Circuit Breakers in mitigating the influence of classical RAs on the amplification of misinformation?

RQ2: What degree of accuracy nudges efficacy is necessary to exert a notable effect in diminishing the influence of classical RAs on the amplification of misinformation?

Our work differs from existing studies as it provides a holistic approach to analyze misinformation diffusion and mitigation in social networks by combining in a unique framework RAs (that are responsible for initiating misinformation diffusion) with misinformation diffusion simulation via the LTM and intervention

[1] Fake news is a form of misinformation consisting of false or misleading information presented as news with the intent of manipulating people's perceptions of real facts, events, and statements. Although they are distinct terms, in this paper, we use "fake news" and "misinformation" interchangeably.

strategies such as Virality Circuit Breakers and accuracy nudges from the recent literature [1,12]. As described in Sect. 2, existing approaches only analyze the impact of misinformation on users in every single component considered in our framework. By integrating these components, our study aims to contribute to the ongoing efforts to combat the spread of misinformation and evaluate the effectiveness of intervention techniques when applied on top of various RAs.

2 Related Work

This section reports on literature in computer and social sciences dealing with (i) analyzing the behavior of RAs in the presence of misinformation and mitigating misinformation recommendation, (ii) existing algorithms proposed to model misinformation diffusion, and (iii) strategies social scientists have proposed to mitigate the amplification of misinformation.

Recommender Systems and Misinformation. Current RAs are designed to keep the user engaged in the platform as long as possible, so they do not prevent misinformation spread [11,47]. Further, RAs are responsible for creating filter bubbles and are subject to popularity bias which can increase the diffusion of misinformation [8,13]. Fernández et al. [13] examined the effect of well-known RAs on the amplification of misinformation through an empirical exploration. Their findings indicated that the popularity-based algorithm was the most prone to recommend misinformation. Moreover, simulations performed under different conditions revealed insights such as the rapid popularity of a small number of misinformation items, the relatively lower spread of misinformation through neighbor-based methods, and the minimal impact of factors or neighbors on the recommendation of misinformation. Pathak et al. [35] conducted an in-depth analysis of two Twitter datasets and various RAs, including content-based and social network-based approaches to explore the balance between RAs' performance and their influence in promoting misinformation. The study simulated how misinformation spreads and how users recommended fake news subsequently propagate it. The research highlighted the significant impact of popularity-based and network-based RAs on misinformation diffusion. In essence, Fernández et al. [13] and Pathak et al. [35], have found that each RA does not equally contribute to the spread of misinformation. Their research reveals the presence of algorithmic biases, leading to varying degrees of misinformation dissemination across different RAs, with some playing a more prominent role in the propagation of misinformation compared to others.

Lo et al. [28] introduced VICTOR, an intervention model that addresses fake news by presenting verified articles as answers to questions representing the fake news. VICTOR outperformed baselines in experiments on a Chinese news corpus, delivering 6% more verified articles and increasing diversity by 7.5%. It reached over 68% of at-risk users exposed to fake news. Human subject experiments confirmed VICTOR's superiority in exposure, proposal, and click rates on verified news articles compared to baselines. In a separate study, Lo et al. [27] added an intervention module into an RA to suggest news for verification when users engaged with fake news. This module recommended verified information based on users' updated news consumption interests. By reading titles of verified news, users were exposed to a higher rate of information, facilitating belief change. In a human subject experiment, 84% of participants found that the proposed RA with the intervention method assisted them in combating fake news.

Wang et al. [48] proposed Rec4Mit, an RA aiming to understand users' reading preferences based on recent interactions with real or fake news. Rec4Mit recommends authentic news articles related to users' interests, showcasing superior performance over baselines by providing the highest ratio of recommended real news. However, their experiments focused only on users interacting with real news, leaving the system's effectiveness for users encountering fake news uncertain due to the test set composition.

Social Networks and Misinformation Diffusion. The majority of the approaches for studying misinformation diffusion have been likened to the spread of infectious diseases within a social network of connected individuals, often referred to as epidemiological models [21,42,46], or they are represented using Hawkes processes [31]. However, a significant limitation of these models lies in their assumption of an *implicit* social network, wherein the specific connections among individuals remain unknown. Consequently, while these models excel in analyzing global trends such as the percentage of nodes in the network that may be influenced by the fake news item, given a certain percentage of users who received a recommendation for it in the beginning, they are not well-suited for investigating localized node-to-node diffusion patterns. On the other hand, classical models for information diffusion, such as the Linear Threshold model, offer an alternative approach to studying the spread of fake news [23]. These models *explicitly* consider the connections between users in the social network and can be combined with RAs that provide individual recommendations. Using this combination, we can identify which users are likely to receive a recommendation for a particular fake news item and then explore how the fake news spreads from these initial recipients to their friends, friends of friends, and so on. Consequently, we can determine the infected nodes at the end of the diffusion process.

Intervention Strategies to Tackle Misinformation. As an early research endeavor in fighting misinformation, Nyhan and Reifler [32] investigated the phenomenon of political misperceptions and the associated challenges in correcting them. Their study revealed that conventional correction methods may not consistently counteract misinformation effectively, highlighting the need for alternative approaches. Bode and Vraga [3] investigated the effectiveness of using related stories' functionality in social media platforms to correct misinformation. They found that presenting users with accurate information in conjunction with misinformation improved correction outcomes. Similarly, Lewandowsky et al. [25] highlighted the significance of fostering critical thinking and media literacy skills as crucial measures to mitigate the impact of misinformation. Pennycook et al. [36,37] investigated the impact of attaching warnings to fake news stories. They found that the presence of warnings influenced people's perception of accuracy. Stories without warnings were perceived as more accurate compared

Table 1. Statistics of the datasets considered in our study.

Datasets	FNN	FH
Unique Twitter users	1028	5406
Items (news)	542	1690
Real news	220	1218
Fake news	322	472
Total user-news interactions	20265	120124
User-real news interactions	8823	90398
User-fake news interactions	11442	29726
Following relationships	3021	4102

Table 2. Number of superspreaders in each dataset for various values of threshold θ.

Threshold (θ)	FNN	FH
50%	639	281
60%	636	92
70%	603	51
80%	547	9
90%	424	0
100%	305	0

to stories with warnings. This highlights the complexity of misinformation intervention efforts and the need for careful intervention design.

Epstein et al. [12] (2021) explored the impact of eight treatments on individuals' behavior in sharing news articles on social media platforms. These treatments include evaluating the accuracy of information (evaluation), spending more time on evaluation (long evaluation), emphasizing the importance of accuracy (importance), providing tips for identifying fake news (tips), assessing individual behavior against societal norms (generic norms evaluation), and considering partisan norms (partisan norms), plus the combinations of tips+norms and importance+norms. Similarly, Bak-Coleman et al. [1] have proposed a comprehensive framework to evaluate interventions to counter viral misinformation during the 2020 US election. This framework explores a range of strategies, including content removal, virality circuit breakers, and accuracy nudging, to substantially reduce the prevalence of misinformation. Our work differs from the one by Bak-Coleman et al. [1] as they do not explicitly incorporate a network and do not account for individual personalization (in social platforms) in the diffusion process. In contrast, our research focuses on the influence of personal recommendations from classical RAs on the amplification of misinformation. To the best of our knowledge, we are the first to explore the effectiveness of intervention techniques on top of RAs for mitigating misinformation amplification.

3 Experimental Methodology

In this section, we describe the datasets and the various stages of the experiment methodology we defined to measure the effectiveness of intervention techniques applied on top of RAs to mitigate misinformation diffusion in social networks and answer the research questions presented in the Introduction.

3.1 Datasets

We consider two datasets to study misinformation interventions: the POLITIFACT FAKENEWSNET dataset (**FNN**) from the FakeNewsNet dataset [43] and the HEALTHSTORY FAKEHEALTH dataset (**FH**) from the FakeHealth dataset [6]. We chose these datasets as they contain the user network information (who follows whom) we need to simulate the news spread. To have enough user information to train RAs, we considered only users who shared at least eight news for our analysis. Table 1 shows the datasets' statistics, including user, item, and network information that resulted by considering only these users. We employed a temporal leave-one-out strategy to split the datasets into training and test sets. Specifically, we considered the set of the user-item interactions with the latest timestamp for each user as the test set for each dataset, i.e., ground truth; the remaining user-item interactions comprise the training set. As far as we know, the two datasets we utilized are the only publicly available ones that provide enough instances of user-news sharing and user following, which are needed for simulating the dissemination of misinformation on social networks. We did not consider GossipCop [43], (as gossip is a form of misinformation distinct from fake news), HealthRelease [6] and ReCOVery [52] as they have limited user-item interactions (only 659 users in HealthRelease and 568 in ReCOVery who shared more than eight articles).

Table 3. Recommendation algorithms studied and parameters used.

RAs	Parameters
User-User k-NN (UU) [39]	Neighbor size: 10, Affinity: Cosine similarity
Item k-NN (II) [41]	Neighbor size: 10, Affinity: Cosine similarity
ALS [17]	Latent factors: 40, Damping factors: 5, Iterations: 150
Content Based (CB) [29]	TF-IDF representation to encode user profile and news items
SocialMF (SMF) [18]	K=10, $\lambda_U = \lambda_V = 0.1$, $\lambda_T = 5$
Random (Rnd) [4]	Based on LensKit [10]
Popularity (Pop) [19]	Based on LensKit's [10] TopN algorithm (popularity='quantile')

3.2 Making Misinformation Recommendations

News amplification starts when an RA suggests news items to users, who then engage with them by sharing, liking, retweeting, or commenting on these items within their social network. As a result, the recommended news items are further propagated in the social network. The anticipated dissemination of a news item refers to the spread of the item (i.e., the number of users who engage with the item) after it is initially recommended to certain users, called *seed users*, by the RA. Specifically, we examine a particular RA and a fake news item n and identify the seed users as those whose top-10 recommendations include n. In this study, we focus on the classical RAs presented in Table 3 as considered in other misinformation studies and implemented using LensKit [10].

3.3 Simulating Misinformation Diffusion

After an RA recommends a fake news item n to the seed users, the diffusion of n among the remaining users in the social network, who were not suggested initially n, but can share it due to being influenced by the people they follow, can be studied by information diffusion models. An information diffusion model describes the mechanism by which a specific piece of information, such as a fake news article, disseminates and reaches users through network interactions. The process is initiated by a set of users, referred to as *seed users*, who are active in the beginning, i.e., in our context, the RA recommended to them the piece of information under consideration. Subsequently, the followers of these seed users can become active (i.e., they can be influenced by the seeds to further share or disseminate the same piece of information), followed by their own followers, and so forth, until no additional sharing occurs, indicating the conclusion of the diffusion process. In our study, to model the dissemination of fake news in social networks, we have opted explicitly for the Linear Threshold model [23] due to its explicit integration of social network information and seamless compatibility with individual recommendations. As presented by Algorithm 1 , in the LTM, each node u has a threshold value $\tau_u \in [0, 1]$ often assigned uniformly at random and each edge (u, v) between two nodes or users u and v is associated with a weight w_{uv} typically set to be the inverse of the in-degree of node v [26]. At each step, i, a node v will become active if $\sum_{u \in N_{in}(v), u \in A_{i-1}} w_{uv} \geq \tau_v$, where E is the set of edges in the network, $N_{in}(v)$ is the set of v's incoming neighbors, and A_{i-1} is the set of nodes that are active in the previous step. Specifically, we can identify the seed users who are recommended a specific fake news item in the beginning and, starting from them, investigate how the item spreads to their friends, friends of friends, and subsequent levels of connections. Ultimately, upon completion of the diffusion process, we can quantify the expected percentage of nodes that have been affected by a piece of news n via the concept of **Expected Spread (E_{Spread})**, i.e., the expected proportion of infected nodes, i.e., users who are active (or have shared n in our context), at the conclusion of the diffusion process [23]. As diffusion models such as the LTM typically contain some degrees of randomness (i.e., the node thresholds), they are executed multiple times (in our experiments, we conducted 100 runs). At each execution, the spread (i.e., the percentage of active nodes at the end of the diffusion process) is computed, and then it is averaged across all the runs, resulting in the expected spread. Our focus is on fake news items present in the test set, treating them as a singular piece of information that spreads throughout the social network as simulated by the LTM (we used the implementation provided by NDlib [40]). Each RA is used to determine the seed users for the diffusion model.

Algorithm 1. Linear Threshold Model (LTM) [50]

Require: Diffusion graph $G(V, E)$, set of seeds A_0
return Total number of activated nodes
$i \leftarrow 0$
Uniformly assign random thresholds $\tau_v \in [0, 1]$
while $i = 0$ or ($A_{i-1} \neq A_i$ and $i \geq 1$) **do**
  $A_{i+1} \leftarrow A_i$
  $inactive \leftarrow V - A_i$
  for all $v \in inactive$ **do**
    if $\sum_{u \in N_{in}(v), u \in A_{i-1}} w_{uv} \geq \tau_v$ **then**
      Activate v
      $A_{i+1} \leftarrow A_{i+1} \cup \{v\}$
    end if
  end for
  $i \leftarrow i + 1$
end while
$A_\infty \leftarrow A_i$
Return $|A_0| + |A_\infty|$

Algorithm 2. Linear Threshold Model with Nudges (LTMwN)

Require: Diffusion graph $G(V, E)$, set of seeds A_0, intervention ratio γ
return Total number of activated nodes
$i \leftarrow 0$
Uniformly assign random thresholds $\tau_v \in [0, 1]$
while $i = 0$ or ($A_{i-1} \neq A_i$ and $i \geq 1$) **do**
  $A_{i+1} \leftarrow A_i$
  $inactive \leftarrow V - A_i$
  for all $v \in inactive$ **do**
    if $\sum_{u \in N_{in}(v), u \in A_{i-1}} w_{uv} \geq \tau_v$ **then**
      Generate random value $q \in [0, 1]$
      if $q \geq \gamma$ **then**
        Activate v
        $A_{i+1} \leftarrow A_{i+1} \cup \{v\}$
      end if
    end if
  end for
  $i \leftarrow i + 1$
end while
$A_\infty \leftarrow A_i$
Return $|A_0| + |A_\infty|$

3.4 Simulating Misinformation Intervention

In our study, we conduct two experiments to simulate the effectiveness of two specific intervention techniques, namely (i) the Virality Circuit Breakers and (ii) nudging users towards accuracy, applied on top of various RAs. The primary objective is to examine how these techniques contribute to mitigating the amplification of misinformation, initiated by RAs, in social networks. These experiments are designed to address the research questions outlined in the Introduction.

Simulating Virality Circuit Breakers (VCBs). Considering the intricate legal and ethical dilemmas associated with directly removing content on social media platforms, the adoption of a Virality Circuit Breakers-based methodology emerges as a promising strategy for mitigating the propagation of misinformation [1]. According to a report from the Center for American Progress (CAP) [44], during a prominent spike of misinformation and disinformation, social platforms should prioritize preserving freedom of speech while addressing misinformation, necessitating transparent and consistent rules, embracing the burden of proof, and committing to greater transparency in content moderation.

While extreme measures such as outright content removal and account banning may yield results, they raise concerns about impeding freedom of expression and placing private entities as arbiters of truth. By deliberately implementing the VCBs approach, we can navigate these complexities and find a balance that effectively combats misinformation while upholding legal and ethical principles. This necessitates careful deliberation of the timing and extent of content removal necessary to make a meaningful impact [1]. The VCBs approach is used to reduce the amplification of trending misinformation topics, rather than explicitly removing the content itself [1]. In our study, we focus on implementing an intervention strategy based on VCBs by removing superspreader users from the set of seeds. We define a user as a *superspreader* if their percentage of shared fake news items in the training set is at least a threshold value, denoted as θ. This mitigation strategy translates into removing the fake news item under consideration from the recommendation list of each superspreader, emphasizing that we are not explicitly removing the news content itself. We conduct experiments using different values of θ from the set $\{50\%, 60\%, 70\%, 80\%, 90\%, 100\%\}$. The number of superspreaders corresponding to each threshold value in each dataset can be found in Table 2. This analysis explores how the VCBs treatment works to reduce the impact of RAs on the spread of false or misleading information.

Simulating Nudging Users Towards Accuracy. Nudging is a popular behavioral policy approach that uses psychology to guide decision-making towards beneficial choices [24]. In the context of misinformation, nudging interventions use accuracy prompts to remind users about sharing accurate information, discouraging the spread of false headlines and promoting responsible sharing [12,24]. As discussed in Sect. 2, previous studies examined the effectiveness of various treatment strategies (e.g., evaluation, importance, tips, and norms) aimed at reducing the disparity between accuracy perceptions and sharing intentions [12]. The effectiveness of these strategies varies, e.g., Epstein et al. [12] showed a reduction of 5% in fake news sharing intentions when assessing the impact of the tips+norms treatment. To model interventions based on accuracy nudging, we introduce the **Linear Threshold Model with Nudges (LTMwN)** shown in Algorithm 2. In the LTMwN model, each node v has an intervention ratio γ, which represents the likelihood of v spreading fake news after an intervention has been applied. At each step i, for each node v, the condition $\sum_{u \in N_{in}(v), u \in A_{i-1}} w_{uv} \geq \tau_v$ is evaluated as in the LTM. If the above evaluation is satisfied, the model evaluates whether the intervention is successful, i.e., a random value $q \in [0, 1]$ is generated, and if $q < \gamma$, the intervention succeeds and v is not activated, i.e., even if its neighbors have influenced v, it does not contribute to further spread fake news. We experimented using different values of $\gamma \in \{0.1, 0.2, 0.3, 0.4, 0.5, 0.6, 0.7, 0.8, 0.9, 1\}$. This helps to examine the level of effectiveness an intervention based on accuracy nudges is required to have to make a significant impact on combating misinformation propagation and further explore if this intervention's effectiveness differs based on the used RA.

Table 4. Average expected spread achieved by the LTM before and after VCBs intervention. Values marked with an asterisk (*) indicate a statistically significant difference ($p \leq 0.05$) as compared to the No Intervention condition.

E_{Spread}								
Dataset	RA	No Intervention	θ=50%	θ=60%	θ=70%	θ=80%	θ=90%	θ=100%
FNN	UU	13.77	2.02*	2.02*	2.69*	3.99*	7.07*	9.69*
	II	12.05	1.60*	1.56*	2.56*	4.09*	7.02*	8.97*
	ALS	14.83	3.25*	3.41*	4.42*	5.80*	8.67*	11.02*
	CB	7.10	0.81*	0.80*	1.06*	1.79*	3.50*	4.98*
	RND	9.12	3.32*	3.49*	3.96*	4.69*	6.06*	7.13*
	POP	52.58	36.92*	37.79*	39.39*	41.09*	44.27*	46.73*
	SMF	36.08	13.90*	14.35*	16.43*	19.51*	24.12*	28.31*
FH	UU	1.39	1.30*	1.35*	1.38*	1.38		
	II	1.47	1.39*	1.45*	1.47	1.47		
	ALS	1.42	1.33*	1.38*	1.42	1.41		
	CB	1.40	1.33*	1.38	1.39	1.39		
	RND	1.46	1.39*	1.44*	1.46*	1.45*		
	POP	28.06	27.13	27.72	28.02	27.99		
	SMF	99.90	96.65*	98.75	99.89	99.78		

4 Findings

This section presents the results of the conducted analyses to address the two research questions presented in the Introduction. As we are interested in studying whether intervention effectiveness differs based on the used RA, we measured, for each intervention strategy and each RA, the average expected spread across all the fake news items presented in the test set (322 for FNN and 472 for FH) before and after intervention and discuss the results in the following. The values marked with an asterisk ('*') in Tables 4 and 5 indicate a statistically significant difference ($p \leq 0.05$) with intervention as compared to no intervention.

Effectiveness of Intervention Using VCBs. Table 4 presents the effectiveness of the intervention strategy using VCBs. To answer **RQ1** we report, for each RA, the average E_{Spread} computed via the LTM for various definitions of superspreaders and compare it to the No Intervention condition. We generally observe that the average expected spread decreases as the threshold value increases as higher thresholds correspond to a lower number of superspreaders (cf. Table 2). Looking at effects on specific RAs, we see that the VCBs-based intervention in collaborative filtering strategies (UU, II, ALS) contributes to significantly reducing the average expected spread on both datasets. This holds for all the threshold values in the FNN dataset and for at least half of the threshold values in the FH dataset. The spread of misinformation generated by content-based and SMF RAs can also be effectively mitigated in the FNN dataset for all the considered threshold values. At the same time, we observe a statistically significant decrease of average expected spread only for threshold $\theta = 50\%$ in the

FH dataset. Overall, VCBs-based interventions in the FNN dataset are more effective compared to the FH dataset. Especially for the latter dataset, changes in E_{Spread} are not significant at all for popularity-based recommendation strategies. This is because, as shown in Table 2, superspreaders, and consequently the misinformation amplification initiators (seeds), are much less in the FH dataset.

Table 5. Average expected spread achieved by the LTM before intervention and by the LTMwN after accuracy nudges intervention for intervention ratio γ ranging from 0.1 to 1. Values marked with an asterisk (*) indicate a statistically significant difference ($p \leq 0.05$) as compared to the No Intervention condition.

		E_{Spread}										
Dataset	**RA**	**No Intervention**	γ=0.1	γ=0.2	γ=0.3	γ=0.4	γ=0.5	γ=0.6	γ=0.7	γ=0.8	γ=0.9	γ=1
FNN	UU	13.77	13.83	13.76	13.87	13.90	13.87	13.71	13.18	12.46*	10.23*	4.37*
	II	12.05	11.95	11.99	11.96	11.97	11.98	11.86	11.50	10.68*	8.76*	3.66*
	ALS	14.83	14.95	14.89	14.92	14.90	14.83	14.63	14.22	12.92*	10.23*	3.46*
	CB	7.10	7.32	7.20	7.33	7.22	7.22	7.11	6.87	6.27*	5.05*	1.86*
	RND	9.12	9.09	9.05	9.10	8.99	9.01	8.78	8.27	7.46*	5.57*	1.80*
	POP	52.58	52.52	52.64	52.55	52.52	52.55	52.29	52.35	51.82*	50.20*	40.63*
	SMF	36.08	36.00	36.02	35.88	36.15	35.95	35.94	35.67	34.89*	31.80*	18.29*
FH	UU	1.39	1.38	1.38	1.38	1.39	1.39	1.39	1.38*	1.36*	1.26*	0.62*
	II	1.47	1.48	1.47	1.47	1.47	1.47	1.47	1.47	1.46*	1.35*	0.67*
	ALS	1.42	1.42	1.41	1.42	1.41	1.41	1.41	1.41	1.39*	1.29*	0.61*
	CB	1.39	1.40	1.39	1.40	1.40	1.39	1.40	1.39	1.38*	1.29*	0.68*
	RND	1.46	1.46	1.46	1.46	1.46	1.40	1.46	1.45	1.44*	1.32*	0.59*
	POP	28.06	28.03	28.02	28.05	28.05	28.03	27.99	28.00	28.01	27.95	26.59
	SMF	99.90	99.88	99.89	99.89	99.88	99.89	99.88	99.89	99.89	99.88	97.22

Effectiveness of Interventions Using Accuracy Nudges. Table 5 presents the results of the average expected spread computed before (via the LTM) and after (via the LTMwN) intervention for each value of intervention ratio γ and for each RA considered. For both datasets, highly effective ($\gamma \geq 0.8$) accuracy nudging-based interventions contribute significantly to reducing the misinformation average expected spread originated by the seed users corresponding to collaborative-based (UU, II, ALS) and content-based (CB) RAs. However, when trying to mitigate the effects of popularity and network-based recommendation strategies, accuracy nudges are effective in the FNN dataset dataset for $\gamma \geq 0.8$, but results always ineffective on the FH dataset dataset. In the case of popularity-based RAs, this difference could be explained by the higher proportion of popular misinformation articles (3.42% for FNN dataset vs. 1.5% FH dataset) this algorithm presents its users in their top-10 recommendation lists. Regarding SMF instead, the average expected spread in the case of no intervention for the FH dataset is already close to 100% as it happens that fake news articles recommended in this case are the most popular ones and are initially suggested to almost all the users in this dataset. As a consequence, no further misinformation spread happens and any intervention results effective.

5 Study Implications

Results regarding VCBs-based interventions imply that superspreaders are not necessarily the only user type responsible for amplifying misinformation, especially in the case of FH dataset. Research shows that people are vulnerable to misinformation as they are not accurate at detecting its veracity [45], hence also non-superspreaders who are recommended with fake news (especially by the popularity-based RA as misinformation tends to gain more popularity and engagement on social media platforms) [14] may contribute to misinformation amplification. Despite the success of the VCBs-based intervention strategy against misinformation, especially in the FNN dataset, one disadvantage is that superspreader individuals, who are more likely to spread misinformation, may not be satisfied with the recommendations generated by the RAs. The number of potentially unsatisfied users varies and corresponds to up to 22.2% of the users in the FNN dataset (Pop with $\theta = 50\%$) and 5% in the FH dataset (SMF with $\theta = 50\%$). In a real-world scenario, these percentages would translate into hundreds of thousands of users who would not be equally satisfied by RAs and may stop engaging with the platform. As superspreaders are not necessarily malicious users, e.g., they may share misinformation because it is "funny" and generates engagement among their friends [49], they are not familiar with the platform features [16], or do not recognize misinformation [45], alternative recommendation strategies may be necessary to effectively address the satisfaction and engagement, while minimizing misinformation spread, of these individuals.

Overall, the results of our simulation show that interventions based on accuracy nudges may have an impact on reducing misinformation spread originated by RAs only if they are highly effective on people. According to Epstein et al. [12], the long evaluation treatment is the most effective at reducing individuals' probability of sharing fake news. However, this treatment may be difficult to implement as it requires offline training. This training consists of (i) having people evaluate headlines (half true, half false) and (ii) informing people after each evaluation about whether their answer was correct or incorrect and about the veracity of the headline. Other treatments also proposed by Epstein et al. [12] such as, for instance, Tips+Norms, where users are first informed that "*8 out of 10 past survey respondents said it was "very important" or "extremely important" to share only accurate news online, and that this was true of both Democrats and Republicans*" and then provided with four simple tips to recognize online misinformation, are easier to implement as they can be prompt to users as they connect into the platform, but are less received by people and results less effective at reducing their sharing intentions in the case of misinformation.

Moreover, by comparing the effectiveness of accuracy nudges vs. VCBs, we observe opposite behaviors in the two considered datasets. In the case of the FNN dataset where, as shown in Table 2, the percentage of superspreaders is higher than in the FH dataset, the intervention based on VCBs is more effective than the one based on accuracy nudges. This is evident by comparing, for instance, the average expected spread achieved by VCBs when $\theta = 50\%$ and by accuracy nudges when $\gamma = 1$ (best cases for both interventions). Conversely, accuracy nudges are more effective when superspreaders are a handful of users as in the FH dataset (same comparison for the cases when $\theta = 50\%$ and $\gamma = 1$). Overall, choosing the best intervention strategy to apply depends on the number of superspreaders, with interventions based on VCBs having the disadvantage of not being able to satisfy superspreaders with news recommendations and interventions based on accuracy nudging requiring to be highly effective.

6 Conclusions

We conducted an empirical analysis to better understand the effectiveness of intervention strategies such as VCBs and accuracy nudges on top of RAs to mitigate misinformation amplification. In summary, interventions on top of collaborative filtering (UU, II, ALS) and content-based RAs were generally successful at reducing the spread of misinformation when VCBs and highly effective ($\gamma \geq 0.8$) accuracy nudges interventions were applied. However, intervention strategies applied on top of popularity and network-based RAs are not always effective, especially in scenarios represented by the FH dataset. This highlights the need for developing novel misinformation-aware RAs or intervention strategies that help reduce misinformation amplification initiated by these RAs.

As a limitation of our work, it's crucial to acknowledge that our modeling framework, based on certain assumptions derived from explicit network-based information diffusion models [23], may not fully capture the complexities of real-world information-sharing behaviors. The diverse timing of information dissemination among individuals, coupled with the presence of skeptics who choose never to re-share information [9,21], adds intricacy to the dynamics of misinformation diffusion. Moreover, we assumed all users who are recommended fake news as seed users, i.e., they propagate news to their followers. However, in reality, these users are not necessarily responsible for propagating a news article. Of course, for the diffusion process to happen, a set of seeds that initiate the process is needed, but considering all these users as propagators might lead to an overestimation of the diffusion process. Furthermore, our empirical evaluation was constrained by a limited set of RAs and a specific information diffusion model. Future research will incorporate a more diverse set of news RAs [38] and misinformation diffusion models [22]. This expansion aims to provide a comprehensive understanding of the dynamics surrounding misinformation diffusion and the effectiveness of interventions across a broader spectrum of scenarios and models, offering valuable insights for future studies in this domain.

Acknowledgments. This research has been supported by the National Science Foundation under Award no. 1943370.

References

1. Bak-Coleman, J.B., et al.: Combining interventions to reduce the spread of viral misinformation (Jun 2022). https://www.nature.com/articles/s41562-022-01388-6
2. Barberá, P.: Explaining the spread of misinformation on social media: evidence from the 2016 us presidential election. In: Symposium: Fake News and the Politics of Misinformation. APSA (2018)
3. Bode, L., Vraga, E.K.: In related news, that was wrong: the correction of misinformation through related stories functionality in social media. J. Commun. **65**(4), 619–638 (2015)
4. Castells, P., Hurley, N.J., Vargas, S.: Novelty and diversity in recommender systems. In: Ricci, F., Rokach, L., Shapira, B. (eds.) Recommender Systems Handbook, pp. 881–918. Springer, Boston, MA (2015). https://doi.org/10.1007/978-1-4899-7637-6_26
5. Cheng, M., Yin, C., Nazarian, S., Bogdan, P.: Deciphering the laws of social network-transcendent covid-19 misinformation dynamics and implications for combating misinformation phenomena. Sci. Rep. **11**(1), 1–14 (2021)
6. Dai, E., Sun, Y., Wang, S.: Ginger cannot cure cancer: battling fake health news with a comprehensive data repository (2020). https://doi.org/10.48550/ARXIV.2002.00837, https://arxiv.org/abs/2002.00837
7. Del Vicario, M., et al.: The spreading of misinformation online. Proc. Natl. Acad. Sci. **113**(3), 554–559 (2016)
8. DiFranzo, D., Gloria-Garcia, K.: Filter bubbles and fake news. XRDS: crossroads, the ACM magazine for students, vol. 23(3), pp. 32–35 (2017)
9. Doe, C., Knezevic, V., Zeng, M., Spezzano, F., Babinkostova, L.: Modeling the time to share fake and real news in online social networks. Inter. J. Data Sci. Anal., 1–10 (2023)
10. Ekstrand, M.D.: Lenskit for python: next-generation software for recommender systems experiments. In: Proceedings of the 29th ACM International Conference on Information and Knowledge Management (Oct 2020). https://doi.org/10.1145/3340531.3412778
11. Elahi, M., et al.: Towards responsible media recommendation. AI and Ethics, pp. 1–12 (2021)
12. Epstein, Z., Berinsky, A., Cole, R., Gully, A., Pennycook, G.: Developing an accuracy-prompt toolkit to reduce covid-19 misinformation online. Harvard Kennedy School Misinformation Rev. **2** (2021). https://doi.org/10.37016/mr-2020-71
13. Fernández, M., Bellogín, A., Cantador, I.: Analysing the effect of recommendation algorithms on the amplification of misinformation. arXiv preprint arXiv:2103.14748 (2021)
14. Fox, M.: Fake news lies spread faster on social media than truth does. https://www.nbcnews.com/health/health-news/fake-news-lies-spread-faster-social-media-truth-does-n854896/ (2018). (Accessed 13 October 2023)
15. Furini, M., Mirri, S., Montangero, M., Prandi, C.: Untangling between fake-news and truth in social media to understand the covid-19 coronavirus. In: 2020 IEEE Symposium on Computers and Communications (ISCC), pp. 1–6. IEEE (2020)
16. Guess, A., Nagler, J., Tucker, J.: Less than you think: prevalence and predictors of fake news dissemination on facebook. Sci. Adv. **5**(1), eaau4586 (2019)
17. Hu, Y., Koren, Y., Volinsky, C.: Collaborative filtering for implicit feedback datasets. In: 2008 Eighth IEEE International Conference on Data Mining, pp. 263–272 (2008). https://doi.org/10.1109/ICDM.2008.22

18. Jamali, M., Ester, M.: A matrix factorization technique with trust propagation for recommendation in social networks. In: Proceedings of the Fourth ACM Conference on Recommender Systems, RecSys 2010, pp. 135–142. Association for Computing Machinery, New York (2010). https://doi.org/10.1145/1864708.1864736
19. Ji, Y., Sun, A., Zhang, J., Li, C.: A re-visit of the popularity baseline in recommender systems. In: Proceedings of the 43rd International ACM SIGIR Conference on Research and Development in Information Retrieval, pp. 1749–1752 (2020)
20. Jin, F., Dougherty, E., Saraf, P., Cao, Y., Ramakrishnan, N.: Epidemiological modeling of news and rumors on twitter. In: Proceedings of the 7th Workshop on Social Network Mining and Analysis, SNAKDD 2013, Association for Computing Machinery, New York (2013). https://doi.org/10.1145/2501025.2501027
21. Jin, F., Dougherty, E., Saraf, P., Cao, Y., Ramakrishnan, N.: Epidemiological modeling of news and rumors on twitter. In: Proceedings of the 7th Workshop on Social Network Mining and Analysis, pp. 1–9 (2013)
22. Joy, A., Shrestha, A., Spezzano, F.: Are you influenced?: modeling the diffusion of fake news in social media. In: Coscia, M., Cuzzocrea, A., Shu, K., Klamma, R., O'Halloran, S., Rokne, J.G. (eds.) ASONAM 2021: International Conference on Advances in Social Networks Analysis and Mining, Virtual Event, The Netherlands, 8 - 11 November 2021, pp. 184–188. ACM (2021), https://doi.org/10.1145/3487351.3488345
23. Kempe, D., Kleinberg, J., Tardos, É.: Maximizing the spread of influence through a social network. In: SIGKDD, pp. 137–146 (2003)
24. Kozyreva, A., Lorenz-Spreen, P., Herzog, S., Ecker, U., Lewandowsky, S., Hertwig, R.: Toolbox of interventions against online misinformation and manipulation (Dec 2022). https://doi.org/10.31234/osf.io/x8ejt
25. Lewandowsky, S., Ecker, U.K., Cook, J.: Beyond misinformation: Understanding and coping with the "post-truth" era. J. Appl. Res. Mem. Cogn. **6**(4), 353–369 (2017)
26. Li, Y., Fan, J., Wang, Y., Tan, K.L.: Influence maximization on social graphs: a survey. IEEE Trans. Knowl. Data Eng. **30**(10), 1852–1872 (2018)
27. Lo, K.C., Dai, S.C., Xiong, A., Jiang, J., Ku, L.W.: All the wiser: fake news intervention using user reading preferences. In: Proceedings of the 14th ACM International Conference on Web Search and Data Mining, pp. 1069–1072 (2021)
28. Lo, K.C., Dai, S.C., Xiong, A., Jiang, J., Ku, L.W.: Victor: an implicit approach to mitigate misinformation via continuous verification reading. In: Proceedings of the ACM Web Conference 2022, WWW 2022, pp. 3511–3519. Association for Computing Machinery, New York (2022). https://doi.org/10.1145/3485447.3512246
29. Lops, P., Gemmis, M.d., Semeraro, G.: Content-based recommender systems: state of the art and trends. In: Recommender Systems Handbook, pp. 73–105 (2011)
30. Mendoza, M., Poblete, B., Castillo, C.: Twitter under crisis: can we trust what we rt? In: Proceedings of the First Workshop on Social Media Analytics, SOMA 2010, pp. 71–79. Association for Computing Machinery, New York (2010). https://doi.org/10.1145/1964858.1964869
31. Murayama, T., Wakamiya, S., Aramaki, E., Kobayashi, R.: Modeling the spread of fake news on twitter. PLOS ONE **16**(4), 1–16 (2021). https://doi.org/10.1371/journal.pone.0250419
32. Nyhan, B., Reifler, J.: When corrections fail: the persistence of political misperceptions. Polit. Behav. **32**(2), 303–330 (2010)
33. Pariser, E.: The filter bubble: what the Internet is hiding from you. Penguin UK (2011)

34. Pathak, R., Lakha, B., Raut, R., Kim, H.S., Spezzano, F.: Unveiling truth amidst the pandemic: multimodal detection of covid-19 unreliable news. In: Ceolin, D., Caselli, T., Tulin, M. (eds.) Disinformation in Open Online Media, pp. 119–131. Springer Nature Switzerland, Cham (2023). https://doi.org/10.1007/978-3-031-47896-3_9
35. Pathak, R., Spezzano, F., Pera, M.S.: Understanding the contribution of recommendation algorithms on misinformation recommendation and misinformation dissemination on social networks. ACM Trans. Web **17**(4) (2023). https://doi.org/10.1145/3616088
36. Pennycook, G., Bear, A., Collins, E.T., Rand, D.G.: The implied truth effect: attaching warnings to a subset of fake news headlines increases perceived accuracy of headlines without warnings. Management Science **66**(11), 4944–4957 (2020). https://doi.org/10.1287/mnsc.2019.3478
37. Pennycook, G., Rand, D.G.: The implied truth effect: a replication and extension. J. Exp. Psychol. Gen. **149**(5), 849–857 (2020)
38. Raza, S., Ding, C.: News recommender system: a review of recent progress, challenges, and opportunities. Artif. Intell. Rev. **55**(1), 749–800 (2022). https://doi.org/10.1007/s10462-021-10043-x
39. Resnick, P., Iacovou, N., Suchak, M., Bergstrom, P., Riedl, J.: Grouplens: an open architecture for collaborative filtering of netnews. In: Proceedings of the 1994 ACM Conference on Computer Supported Cooperative Work, CSCW 1994, pp. 175–186. Association for Computing Machinery, New York (1994). https://doi.org/10.1145/192844.192905
40. Rossetti, G., Milli, L., Rinzivillo, S., Sîrbu, A., Pedreschi, D., Giannotti, F.: Ndlib: a python library to model and analyze diffusion processes over complex networks. Inter. J. Data Sci. Anal. **5**(1), 61–79 (2018)
41. Sarwar, B., Karypis, G., Konstan, J., Riedl, J.: Item-based collaborative filtering recommendation algorithms. In: Proceedings of the 10th International Conference on World Wide Web, WWW 2001, pp. 285–295. Association for Computing Machinery (2001). https://doi.org/10.1145/371920.372071
42. Shrivastava, G., Kumar, P., Ojha, R.P., Srivastava, P.K., Mohan, S., Srivastava, G.: Defensive modeling of fake news through online social networks. IEEE Trans. Comput. So. Syst. **7**(5), 1159–1167 (2020). https://doi.org/10.1109/TCSS.2020.3014135
43. Shu, K., Mahudeswaran, D., Wang, S., Lee, D., Liu, H.: Fakenewsnet: a data repository with news content, social context, and spatiotemporal information for studying fake news on social media. Big Data **8**(3), 171–188 (2020)
44. Simpson, E., Conner, A.: Fighting coronavirus misinformation and disinformation. https://www.americanprogress.org/article/fighting-coronavirus-misinformation-disinformation/ (2023), (Accessed 13 October 2023)
45. Spezzano, F., Shrestha, A., Fails, J.A., Stone, B.W.: That's fake news! investigating how readers identify the reliability of news when provided title, image, source bias, and full articles. Proc. ACM Hum. Comput. Inter. J. **5**(CSCW1, Article 109) (2021)
46. Tambuscio, M., Ruffo, G., Flammini, A., Menczer, F.: Fact-checking effect on viral hoaxes: A model of misinformation spread in social networks. In: Proceedings of the 24th international conference on World Wide Web, pp. 977–982 (2015)
47. Tomlein, M., et al.: An audit of misinformation filter bubbles on youtube: bubble bursting and recent behavior changes. In: Fifteenth ACM Conference on Recommender Systems, pp. 1–11 (2021)

48. Wang, S., Xu, X., Zhang, X., Wang, Y., Song, W.: Veracity-aware and event-driven personalized news recommendation for fake news mitigation. In: Proceedings of the ACM Web Conference 2022, WWW 2022, pp. 3673–3684. Association for Computing Machinery, New York (2022). https://doi.org/10.1145/3485447.3512263
49. Yaqub, W., Kakhidze, O., Brockman, M.L., Memon, N., Patil, S.: Effects of credibility indicators on social media news sharing intent. In: Proceedings of the 2020 CHI Conference on Human Factors in Computing Systems, CHI 2020, pp. 1–14 (2020)
50. Zafarani, R., Abbasi, M.A., Liu, H.: Social media mining: an introduction. Cambridge University Press (2014)
51. Zhang, H., Alim, M.A., Li, X., Thai, M.T., Nguyen, H.T.: Misinformation in online social networks: detect them all with a limited budget. ACM Trans. Inf. Syst. **34**(3) (2016). https://doi.org/10.1145/2885494
52. Zhou, X., Mulay, A., Ferrara, E., Zafarani, R.: Recovery: a multimodal repository for covid-19 news credibility research. In: Proceedings of the 29th ACM International Conference on Information & Knowledge Management, pp. 3205–3212 (2020)

Good for Children, Good for All?

Monica Landoni[1], Theo Huibers[2], Emiliana Murgia[3], and Maria Soledad Pera[4](✉)

[1] Università della Svizzera italiana, Lugano, Switzerland
monica.landoni@usi.ch
[2] University of Twente, Enschede, The Netherlands
t.w.c.huibers@utwente.nl
[3] Università di Genova, Genova, Italy
emiliana.murgia@edu.unige.it
[4] Web Information Systems - TU Delft, Delft, The Netherlands
m.s.pera@tudelft.nl

Abstract. In this work, we reason how focusing on Information Retrieval (IR) for children and involving them in participatory studies would benefit the IR community. The Child Computer Interaction (CCI) community has embraced the child as a protagonist as their main philosophy, regarding children as informants, co-designers, and evaluators, *not just users.* Leveraging prior literature, we posit that putting children in the centre of the IR world and giving them an active role could enable the IR community to break free from the preexisting bias derived from interpretations inferred from past use by adult users and the still dominant system-oriented approach. This shift would allow researchers to revisit complex foundational concepts that greatly influence the use of IR tools as part of socio-technical systems in different domains. In doing so, IR practitioners could provide more inclusive, and supportive information access experiences to children and other understudied user groups alike in different contexts.

Keywords: Children · Mental models · Relevance · Information Access

1 Is IR Good for All?

Information Retrieval (**IR**) has proven its enduring value as a research area. It has adapted to the demands of the constantly evolving digital ecosystem and continues to capture the attention of researchers and industry practitioners alike. This is evident in the broad spectrum of contributions the community has put forward-ranging from innovative strategies for efficient indexing and successfully managing the volume and speed of data growth to the development of neural-based models for retrieval and ranking in the era of AI [39,53]. Moreover, we have seen the emergence of IR models that leverage Large Language Models (**LLMs**) and contributions that respond to new means of interacting with IR

N. Goharian et al. (Eds.): ECIR 2024, LNCS 14611, pp. 302–313, 2024.
https://doi.org/10.1007/978-3-031-56066-8_24

systems, such as clarifying questions or conversational search [26,52,72,81]. Over time, contributions in IR have broadened in their scope to focus on its applicability in specific domains (e.g., finance, legal, enterprise, and medical) [28,68]. IR has also taken a deeper look at issues of fairness and biases of IR algorithms [23]. Additionally, it has explored how users interact with IR systems in different contexts, leading to research that examines cognitive biases during the information-seeking process and study and addresses, among others, issues of filter bubbles and echo chambers caused by search and recommendation systems [21,73].

Most IR research thus far, including models, methods, and even theory, has been primarily characterised by a system-driven approach and the study of interactions generated by adult users. To illustrate this point, consider the TREC tracks, CLEF labs, or well-known datasets like MS Marco that serve as means to identify new research directions and assess the performance of new strategies. The majority target or are based on mature populations. Although IR relies upon user studies to gather new samples, these tend once again to encompass more mature users—often computer scientists or more technically-savvy individuals on crowd-sourcing platforms. Further, these studies often elicit user-system interactions in somewhat *artificial environments* [80].

We question whether IR, in its current form, meets the needs of *heterogeneous searchers*. In our pursuit of a more inclusive IR, we advocate for a radical change in approach: involving real users actively in shaping new models that consider the various ways in which users seek information, whether through pulling or being pushed content as a result of asking, querying, and browsing. This also involves considering the real-world settings in which content is retrieved. To gain insights and knowledge from a diverse range of real users, it is essential to expand the scope beyond the commonly studied mature user demographic. This includes exploring populations that are often overlooked (i.e., understudied user groups [63]), each with unique needs, constraints, and expectations. We propose starting from a particularly promising group of users: **children**. Children are free from bias caused by previous experiences or exposure to technology and, more importantly, their skills, attitudes, and expectations evolve over time. By studying children in different contexts, where their searches may be guided by educational requirements or for pleasure, we aim to gather foundational knowledge to inform future IR research and development under the proposed paradigm shift.

IR researchers sought to broaden IR research by exploring IR systems specific to children. An example of this is Yahooligans! [59], which launched in 1996 and was one of the first large commercial IR systems for children. Since then, we have seen several attempts that from diverse perspectives seek to answer: *how to grow Child IR?* Researchers have identified a range of barriers (summarised in the SWIRL 2012 report [3]) and have proposed a small set of potential solutions [e.g. 19,20,27,47]. Most of these solutions, however, are grounded on Human-Computer Interaction (**HCI**) and Child-Computer Interaction (**CCI**) paradigms and methods that do not necessarily follow IR standards. The preliminary exploration of the many challenges related to Child IR, including interfaces, relevance determination, diverse contexts, and ethics, has contributed to an extent to the growth of this particular area. At present, however, there are no de facto standards. Furthermore, there is a perception within the IR community that solutions for children represent a 'downgrade' compared to those for adults [49]. Paraphrasing Bilal [11], children are not simply *short users*, they are *unique users*. This is why the IR community could and should learn from children [49].

2 Why Start from Child IR?

As a growing Internet user group, children turn to the Web daily for information access [31,36], which is why empowering them from early on so that they can best take advantage of information retrieval tools that serve as a conduit to access information is a must. This first requires exploring how this user group perceives information retrieval tools and identifying the necessary elements for children to actively engage and contribute to research and development. Simultaneously, understanding how the IR community (and beyond) should adapt to a methodological shift while drawing on the collective knowledge amassed as a research community thus far is crucial.

We leverage the efforts of the CCI community on *participatory* studies to better understand children's needs and mental models. On the same line, we take into account the *sociological* factors, such as the impact of peers and adults on children's use of IR, as well as the *ethical* implications of the child as protagonist. Through these lenses, we zoom on some specific concepts inherent to IR for children—some already under study, such as the concept of relevance, the role of trust, or the significance of emotions in the information-seeking process—and the tools and frameworks being used for these explorations hint at the impact findings can have on mainstream IR research. In so doing, we unveil details of the methodological shift advocate.

The reflections and discussions that follow are meant to serve as a starting point and inspire future research agendas. Although we mention children in a more 'theoretical' sense, IR researchers and practitioners must acknowledge that in 'practice' there is a need to move past the broad definition of children, i.e., away from the 'one-size-fits-all' mentality. Instead, we should consider how children's in-development skills and ideas will require different strategies in how to involve children as research partners at different stages of their lives. For instance, the use of drawings might be a better approach for 9 to 11-year-olds, whereas the think-aloud process might help researchers elicit insights from younger audiences and surveys or diary studies could be better suited to involve teenagers [8,45,78].

Relevance. As remarked by Blair [12], relevance is an "ineffable concept," indeed a concept difficult to define objectively as it refers to how retrieved content appropriately fulfils a specific information need; its interpretation varies depending on the origin of the information need and how it can be satisfied [9]. Looking at children's understanding of the concept of relevance from early on, and how that understanding can evolve as they grow, can help researchers and practitioners better model the concept of relevance in practice. Children searching in the classroom could provide a well-defined framework that naturally aligns with the concept of *situational* relevance, particularly the *motivational/affective* "inherent characteristics of relevance behaviour" [14]. In this case, the former looks at the usefulness of retrieved materials in relation to the tasks assigned by teachers; whereas the latter considers the goals and motivations behind searching for learning. Teachers could naturally assess usefulness, but motivational/affective relevance is more complex to study when young searchers are involved, making it an interesting area to be further investigated.

Focusing on children in the classroom, and to elicit their perception of relevance, researchers involved children in a collaborative exercise and asked them to draw icons to be used for tagging useful content for their peers [45]. The analysis of children's feedback, as in the themes emerging from the drawings and from the answers to a survey run to elicit their point of view on the exercise, helped authors understand how children interpret relevance when discriminating among results guided by their information need triggered by the teacher's assignments. In a related work [2], the authors analysed children's behaviour when interacting with an emoji-enhanced search engine result page where icons were used to elicit three shades of relevance: negative, neutral and positive. While the focus of the study was on how to improve the quality of the overall search experience, the way children embraced naturally the shades of relevance and the impact these proved to have on more effective search performance signals the promise of how such a study could help better understand the concept of relevance and its interpretations in the classroom and beyond.

Research Partners, not Spectators. The previous examples show how putting users at the centre of research and having them play an active role in the team is conducive to interesting insights while enabling researchers to define otherwise ineffable concepts. The child as the protagonist is a well-established principle in CCI [38], and there are available techniques to help researchers run effective studies in a collaborative setting. Looking at the CCI literature, we find several instances of children interacting with IR systems, often via innovative interfaces such as vocal agents and robots. These contributions mainly target the design of original and better—more usable—interactions and produce useful guidelines for peers to use. These guidelines, however, do not directly address the IR research community–they neither use the proper terminology nor adopt a rigorous TREC-like methodology. Further scrutiny shows the potential of their findings for informing the design of IR systems that are not only more usable but also, more useful in providing more relevant results to users at large. A good example is a work describing how children interact with vocal assistants via spoken queries and aiming at providing guidelines for the design of more usable agents [79]. While highlighting the importance of personification to support a good user experience, researchers also reported on the different categories of query reformulation children and their adults in the loop adopted in an attempt to overcome the poor performance of the underlying IR system. These categories, ranging from *Off Course* - where users change their questions to find something the system can answer to *Stating Context* and *Expanding* pronouns, offer a picture of the gap between the *mental models* [13] of searchers and the system, a very valuable contribution for the IR community.

Child-IR System Interaction. Vocal agents and how these can be designed to cater to children are the focus of the work presented in [47]. The authors involved children as active informants by having them solve school-related search tasks in a Wizard of Oz set-up [32]. The tasks were produced in collaboration with teachers who assessed their complexity. Children could interact with a traditional GUI or a vocal assistant to support their search experience. Overall, satisfaction was the same across both interfaces. However, efficiency varied, with children spending less time per search when interacting with the vocal assistant. This finding helps elicit searchers' mental models and better understand how their behaviour changes according to the expectations raised by the system and how this can implicitly trigger browsing and exploring, all important activities to support the search as learning (**SAL**) experience [66]. Looking further into how to co-design a search agent to help children with their homework [48], the authors uncovered the impact of familiarity with technology in general and the importance of search experience on children's expectations and preferences, providing a glimpse into different mental models and their influence on the acceptance and use of new functionalities supporting searchers.

Search Roles. Foss et al. [29] introduced the concept of search roles—the range of skills and aptitudes exhibited when searching—children play when seeking information for pleasure. They did so by leveraging interviews and observations. A recent study [44] adopted a quantitative approach based on search logs and teachers' observations to probe whether these roles could be observed in the classroom. The authors (i) discerned most of the original roles and new ones inherent to the learning context and (ii) identified gaps in how to study the remaining roles. From the reported experiences, which include initial exploration of applicability into adolescents and adults [22,29,30,43,44], it seems feasible to merge qualitative and quantitative inquiries into formal methodologies that can distil natural search roles–search personas–as tools to model more accurate representations of searchers and guide the design of better systems.

Search as Learning. Information-seeking is the "process, in which humans purposefully engage to change their state of knowledge" [56]. Following this line of thinking, researchers in the early 2000 s focused on the SAL paradigm, which discusses *learning to search* while *searching to learn* [34,66]. Exploring SAL aligns naturally with children from whom learning is a way of life. Involving teachers and/or parents playing the role of the more knowledgeable other [24] creates a rich environment to gauge the knowledge gain triggered by search activities [67]. In fact, it is somehow easier to account for children's prior knowledge than adults'; assessing progress and changes in their original knowledge state is more straightforward. Additionally, the development of search skills can be more directly monitored as children tend to start from basics to low proficiency in search, making them more likely to be free from prejudices, biases and expectations.

Emotions. Contributing to a better understanding of what makes a search result more appealing and clickable for children is the work by Landoni et al. [51] that builds on a study on adult users [41]. The studies reflect on the impact of emotions in attracting searchers' attention to search results. In both cases, results charged with positive emotions were more attractive than neutral ones. When considering children's search behaviour, negatively charged results seemed to equally attract their attention and trigger a more active engagement with the search process. The direct involvement of the children, together with the active role of their teacher in the classroom, was conducive to rich exchanges with peers and contributed to a better understanding of the complex role emotions can play if mediated by all the stakeholders. Studies that take inventory of the influence of emotions on search behaviour would pave the way for the design of IR systems that provide better and more engaging search experiences.

Trust. People tend to trust search engines; they perceive them as platforms that provide accurate and reliable information; they also deem higher-ranked search results as more credible [16,35,71]. Research thus far has shown that children do favour higher-ranked results [33,75], yet it is unclear if this is due to their trust in the system that offers them access to information. Although children tend to naturally start from a position of trust as they are still developing their critical ability, recent studies highlight that they do not trust suggested search results if they cannot identify the source [64]. In other studies, they associate the concept of trust with that of privacy and a safe environment [1]. Understanding what 'trust' really means for children would help better understand the dynamics with adults, even if it is another direction that will benefit from methodological approaches that make children protagonists, as in participatory design.

Accessibility and Inclusion. IR systems are "powerful intermediaries" [74] to information, and thus resources children are exposed to can "influence how they see the world" [65]. The development of children's cognitive and motor skills, along with their cultural background and social context, invite scrutiny of factors that, from their perspective, affect the search experience. Fostering accessibility and inclusion cannot begin with a "stereotype" child searcher. This instead requires simultaneously accounting for factors like text complexity, legibility, and readability together with cross-cultural conditions [4,18,57], which is a complex, and certainly human-driven undertaking. The influence of the adults in the loop: educators, parents, older siblings, acting as role models [29] needs to be considered as they can pass on to children their positive or negative experiences with a long-lasting impact on their future attitude towards the use of IR.

Ethics and Regulations. Ethical implications of IR systems on children are carefully scrutinised to strike a balance between the right to information access [6] while avoiding exposure to unsuitable content, such as fake news and information pollution, to name a few [46,50,55]. This echoes UNICEF's recommendations for policymakers and civil society (including academics) regarding protecting children "from the harms of mis/disinformation" [37] while building and strengthening among children (and ultimately all individuals) the ability to "navigate and evaluate digital information environments" [37]. Careful attention is paid to the rights of children, and there are many initiatives at the European level [17,60,61], to include and give them an active role when looking at solutions to provide a safe and rich information space for children to learn and grow. Such a setting could provide a valuable paradigm for other similarly vulnerable user groups.

3 Is Child IR Good for All?

The PuppyIR project (2009-2012) enabled core IR researchers to allocate resources to study and build an open-source IR environment that would better serve children [7]. Outcomes from this project revealed the challenges involved in taking IR concepts from theory to practice when children are the main stakeholders [e.g. 40,54]. Our discussion shows that advances inspired by PuppyIR and other seminal works have indeed contributed towards advancing Child IR [e.g. 10,25,33,58]. Nevertheless, for the next iteration of Child IR research to be truly meaningful and impactful, there is a need for a paradigm shift. For this—and inspired by the limited existing works already aiming to put children (of all ages) at the centre [27,45,47,75,79]—we have outlined how core IR concepts must be revisited from the eyes of a child user and their context, ***rather than adapted***. We posit that doing so will put the human at the centre—starting with children—and model individuals, rather than 'users', better respond to their needs and requirements, and, support the IR community when facing AI, ethical and policy challenges [5,69,70,76].

Digital humanism calls to "shape technologies in accordance with human values and needs, instead of allowing technologies to shape humans," [77]. This is crucial in the era of AI. A step in this direction is the principle of Human-driven IR, which is already gaining momentum [15,42,62]. We contribute to this discussion by endorsing a paradigm shift that prioritises equity in online information access, moving away from the one-size-fits-all approach. Researchers and practitioners should collaborate across disciplines to not only develop Child IR but also expand knowledge and extend technological advances to ***stakeholders with specific search needs poorly served by IR systems***. By focusing on children as a starting user group, we are confident that it will be possible to outline a blueprint of sorts to expand the understanding of other understudied user groups and of human values and needs that should drive technology design. Using the same lenses (participatory, sociological, and ethical), we suggest considering: (i) *other user groups*, e.g. low-literate adults, (ii) *broader contexts*, e.g. museums and libraries, (iii) *different information needs*, e.g., beyond those coupled with learning in the classroom context, and (iv) even other *information access tools*, e.g. recommendation, question-answering systems, social media platforms, and LLM-based models like ChatGPT that are so in vogue.

References

1. Aliannejadi, M., Huibers, T., Landoni, M., Murgia, E., Pera, M.S.: The effect of prolonged exposure to online education on a classroom search companion. In: International Conference of the Cross-Language Evaluation Forum for European Languages, pp. 62–78. Springer (2022). https://doi.org/10.1007/978-3-031-13643-6_5
2. Aliannejadi, M., Landoni, M., Huibers, T., Murgia, E., Pera, M.S.: Children's perspective on how emojis help them to recognise relevant results: Do actions speak louder than words? In: Proceedings of the 2021 Conference on Human Information Interaction and Retrieval, pp. 301–305 (2021)
3. Allan, J., Croft, B., Moffat, A., Sanderson, M.: Frontiers, challenges, and opportunities for information retrieval: report from swirl 2012 the second strategic workshop on information retrieval in lorne. ACM SIGIR Forum **46**(1), 2–32 (2012)
4. Allen, G., Yang, J., Pera, M.S., Gadiraju, U.: Using conversational artificial intelligence to support children's search in the classroom. arXiv preprint arXiv:2112.00076 (2021)
5. Amershi, Set al.: Guidelines for human-ai interaction. In: Proceedings of the 2019 Chi Conference on Human Factors in Computing Systems, pp. 1–13 (2019)
6. Assembly, U.G.: Convention on the rights of the child. United Nations, Treaty Series **1577**(3), 1–23 (1989)
7. Azzopardi, L., Glassey, R., Lalmas, M., Polajnar, T., Ruthven, I.: Puppyir: Designing an open source framework for interactive information services for children. In: Proceedings of the Annual Workshop on Human-computer Interaction and Information Retrieval, vol. 44, Citeseer (2009)
8. Badillo-Urquiola, K., Shea, Z., Agha, Z., Lediaeva, I., Wisniewski, P.: Conducting risky research with teens: co-designing for the ethical treatment and protection of adolescents. Proc. ACM Hum.-Comput. Interact. **4**(CSCW3), 1–46 (2021)
9. Belkin, N.J.: 4 ineffable concepts in information retrieval. Jones, p. 44–58 (1981)
10. Bilal, D.: Children's use of the yahooligans! web search engine: I. cognitive, physical, and affective behaviors on fact-based search tasks. J. Am. Soc. Inform. Sci. **51**(7), 646–665 (2000)
11. Bilal, D.: The mediated information needs of children on the autism spectrum disorder (asd). In: Proceedings of the 31st ACM SIGIR Workshop on Accessible Search Systems, Geneva, Switzerland, pp. 42–49. ACM Geneva (2010)
12. Blair, D.C.: Language and representation in information retrieval. Elsevier North-Holland, Inc. (1990)
13. Borgman, C.L.: The user's mental model of an information retrieval system. In: Proceedings of the 8th Annual International ACM SIGIR Conference on Research and Development in Information Retrieval, pp. 268–273 (1985)
14. Borlund, P.: The concept of relevance in ir. J. Am. Soc. Inform. Sci. Technol. **54**(10), 913–925 (2003)
15. Buchanan, G., McKay, D., Clarke, C.L., Azzopardi, L., Trippas, J.: Made to measure: A workshop on human-centred metrics for information seeking. In: Proceedings of the 2020 Conference on Human Information Interaction and Retrieval, pp. 484–487 (2020)
16. Carroll, N.: In search we trust: exploring how search engines are shaping society. Inter. J. Knowl. Soc. Res. (IJKSR) **5**(1), 12–27 (2014)
17. Charisi, V., et al.: Artificial intelligence and the rights of the child: towards an integrated agenda for research and policy. Tech. rep., Joint Research Centre (Seville site) (2022)

18. Clarke, L.W.: Walk a day in my shoes: cultivating cross-cultural understanding through digital literacy. Read. Teach. **73**(5), 662–665 (2020)
19. Collins-Thompson, K., Bennett, P.N., White, R.W., De La Chica, S., Sontag, D.: Personalizing web search results by reading level. In: Proc. of the 20th ACM International Conference on Information and Knowledge Management, pp. 403–412 (2011)
20. Dragovic, N., Madrazo Azpiazu, I., Pera, M.S.: " is sven seven?" a search intent module for children. In: Proceedings of the 39th International ACM SIGIR conference on Research and Development in Information Retrieval, pp. 885–888 (2016)
21. Draws, T., et al.: Viewpoint diversity in search results. In: European Conference on Information Retrieval, pp. 279–297, Springer (2023). https://doi.org/10.1007/978-3-031-28244-7_18
22. Druin, A., Foss, E., Hutchinson, H., Golub, E., Hatley, L.: Children's roles using keyword search interfaces at home. In: Proceedings of the SIGCHI Conference on Human Factors in Computing Systems, pp. 413–422 (2010)
23. Ekstrand, M.D., Das, A., Burke, R., Diaz, F., et al.: Fairness in information access systems. Foundat. Trends® Inform. Retrieval **16**(1–2), 1–177 (2022)
24. Ekstrand, M.D., Pera, M.S., Wright, K.L.: Seeking information with a more knowledgeable other. Interactions **30**(1), 70–73 (2023)
25. Elliot, D., Glassey, R., Polajnar, T., Azzopardi, L.: Finding and filtering information for children. In: Proceedings of the 33rd international ACM SIGIR Conference on Research and Development in Information Retrieval, pp. 702–702 (2010)
26. Faggioli, G., et al.: Perspectives on large language models for relevance judgment. In: Proceedings of the 2023 ACM SIGIR International Conference on Theory of Information Retrieval, pp. 39–50 (2023)
27. Fails, J.A., Pera, M.S., Anuyah, O., Kennington, C., Wright, K.L., Bigirimana, W.: Query formulation assistance for kids: what is available, when to help & what kids want. In: Proceedings of the 18th ACM International Conference on Interaction Design and Children, pp. 109–120 (2019)
28. Feng, F., Luo, C., He, X., Liu, Y., Chua, T.S.: Finir 2020: the first workshop on information retrieval in finance. In: Proceedings of the 43rd International ACM SIGIR Conference on Research and Development in Information Retrieval, pp. 2451–2454 (2020)
29. Foss, E., et al.: Children's search roles at home: implications for designers, researchers, educators, and parents. J. Am. Soc. Inform. Sci. Technol. **63**(3), 558–573 (2012)
30. Foss, E., Druin, A., Yip, J., Ford, W., Golub, E., Hutchinson, H.: Adolescent search roles. J. Am. Soc. Inform. Sci. Technol. **64**(1), 173–189 (2013)
31. Gossen, T., Kotzyba, M., Nürnberger, A.: Search engine for children: user-centered design. Datenbank-Spektrum **17**, 61–67 (2017)
32. Gould, J.D., Conti, J., Hovanyecz, T.: Composing letters with a simulated listening typewriter. Commun. ACM **26**(4), 295–308 (1983)
33. Gwizdka, J., Bilal, D.: Analysis of children's queries and click behavior on ranked results and their thought processes in google search. In: Proceedings of the 2017 Conference on Conference Human Information Interaction and Retrieval, pp. 377–380 (2017)
34. Hansen, P., Rieh, S.Y.: Recent advances on searching as learning: an introduction to the special issue. J. Inf. Sci. **42**(1), 3–6 (2016)
35. Hargittai, E., Fullerton, L., Menchen-Trevino, E., Thomas, K.Y.: Trust online: young adults' evaluation of web content. Int. J. Commun. **4**, 27 (2010)

36. Holloway, D., Green, L., Livingstone, S.: Zero to eight: young children and their internet use (2013). http://eprints.lse.ac.uk/52630/1/Zero_to_eight.pdf.
37. Howard, P.N., Neudert, L.M., Prakash, N., Vosloo, S.: Digital misinformation/disinformation and children. UNICEF. Retrieved on February 20, 2021 (2021)
38. Iversen, O.S., Smith, R.C., Dindler, C.: Child as protagonist: expanding the role of children in participatory design. In: Proceedings of the 2017 Conference on Interaction Design and Children, pp. 27–37 (2017)
39. Jagerman, R., Qin, Z., Wang, X., Bendersky, M., Najork, M.: On optimizing top-k metrics for neural ranking models. In: Proceedings of the 45th International ACM SIGIR Conference on Research and Development in Information Retrieval, pp. 2303–2307 (2022)
40. Jansen, M., Bos, W., Van Der Vet, P., Huibers, T., Hiemstra, D.: Teddir: tangible information retrieval for children. In: Proceedings of the 9th International Conference on Interaction Design and Children, pp. 282–285 (2010)
41. Kazai, G., Thomas, P., Craswell, N.: The emotion profile of web search. In: Proceedings of the 42nd international ACM SIGIR Conference on Research and Development in Information Retrieval, pp. 1097–1100 (2019)
42. Keshavarz, H.: Human information behaviour and design, development and evaluation of information retrieval systems. Program **42**(4), 391–401 (2008)
43. Kim, J., McNally, B., Norooz, L., Druin, A.: Internet search roles of adults in their homes. In: Proceedings of the 2017 CHI Conference on Human Factors in Computing Systems, pp. 4948–4959 (2017)
44. Landoni, M., Huibers, T., Aliannejadi, M., Murgia, E., Pera, M.S.: Getting to know you: search logs and expert grading to define children's search roles in the classroom. In: 2nd International Conference on Design of Experimental Search and Information Retrieval Systems, DESIRES 2021, pp. 44–52 (2021)
45. Landoni, M., Huibers, T., Murgia, E., Aliannejadi, M., Pera, M.S.: Somewhere over the rainbow: exploring the sense for relevance in children. In: Proceedings of the 32nd European Conference on Cognitive Ergonomics, pp. 1–5 (2021)
46. Landoni, M., Huibers, T., Murgia, E., Pera, M.S.: Ethical implications for children's use of search tools in an educational setting. Inter. J. Child-comput. Interact. **32**, 100386 (2022)
47. Landoni, M., Matteri, D., Murgia, E., Huibers, T., Pera, M.S.: Sonny, cerca! evaluating the impact of using a vocal assistant to search at school. In: Crestani, F., et al. (eds.) CLEF 2019. LNCS, vol. 11696, pp. 101–113. Springer, Cham (2019). https://doi.org/10.1007/978-3-030-28577-7_6
48. Landoni, M., Murgia, E., Huibers, T., Pera, M.S.: You've got a friend in me: children and search agents. In: Adjunct Publication of the 28th ACM Conference on User Modeling, Adaptation and Personalization, pp. 89–94 (2020)
49. Landoni, M., Murgia, E., Huibers, T., Pera, M.S.: Report on the 1st ir for children 2000–2020: where are we now?(ir4c) workshop at sigir 2021: the need to spotlight research on children information retrieval. ACM SIGIR Forum **55**(2), 1–7 (2022)
50. Landoni, M., Murgia, E., Huibers, T., Pera, M.S.: How does information pollution challenge children's right to information access. In: 3rd Workshop on Reducing Online Misinformation through Credible Information Retrieval co-located with ECIR'23. CEUR Workshop Proceedings, vol. 3359, pp. 250–253 (2023)
51. Landoni, M., Pera, M.S., Murgia, E., Huibers, T.: Inside out: Exploring the emotional side of search engines in the classroom. In: Proceedings of the 28th ACM Conference on User Modeling, Adaptation and Personalization, pp. 136–144 (2020)

52. Liao, L., Yang, G.H., Shah, C.: Proactive conversational agents in the post-chatgpt world. In: Proceedings of the 46th International ACM SIGIR Conference on Research and Development in Information Retrieval, pp. 3452–3455 (2023)
53. Lin, J., et al.: Simple yet effective neural ranking and reranking baselines for cross-lingual information retrieval. arXiv preprint arXiv:2304.01019 (2023)
54. Lingnau, A., Ruthven, I., Landoni, M., van der Sluis, F.: Interactive search interfaces for young children-the puppyir approach. In: 2010 10th IEEE International Conference on Advanced Learning Technologies, pp. 389–390, IEEE (2010)
55. Loos, E., Ivan, L., Leu, D.: save the pacific northwest tree octopus: a hoax revisited. or: How vulnerable are school children to fake news?. Inform. Learn. Sci. **119**(9/10), 514–528 (2018)
56. Marchionini, G.: Information seeking in electronic environments, vol. 9, Cambridge University Press (1997)
57. Milton, A., Allen, G., Pera, M.S.: To infinity and beyond! accessibility is the future for kids' search engines. In: IR for Children 2000–2020: Where Are We Now? (https://www.fab4.science/ir4c/) - Workshop co-located with the 44th International ACM SIGIR Conference on Research and Development in Information Retrieval. arXiv preprint arXiv:2106.07813 (2021)
58. Milton, A., Green, M., Pera, M.S.: An empirical analysis of search engines' response to web search queries associated with the classroom setting. Aslib J. Inf. Manag. **72**(1), 88–111 (2020)
59. N.A: Yahoo! Kids - Wikipedia — en.wikipedia.org (1996). https://en.wikipedia.org/wiki/Yahoo!_Kids (Accessed 17 Oct 2023]
60. N.A.: A digital decade for children and youth: the new european strategy for a better internet for kids (bik+). In: European Commission - Policy and Legislation (2022). https://digital-strategy.ec.europa.eu/en/library/digital-decade-children-and-youth-new-european-strategy-better-internet-kids-bik, European Commission - Policy and Legislation
61. N.A.: Digital participation, empowerment and protection finely balanced in the new european strategy for a better internet for kids (bik+). In: European strategy for a better internet for kids (BIK+) — Shaping Europe's digital future (2022). https://digital-strategy.ec.europa.eu/en/policies/strategy-better-internet-kids, European Commission
62. Paramita, M.L., Kasinidou, M., Kleanthous, S., Rosso, P., Kuflik, T., Hopfgartner, F.: Towards improving user awareness of search engine biases: a participatory design approach. J. Association Inform. Sci. Technol. (2023)
63. Pera, M.S., Cena, F., Huibers, T., Landoni, M., Mauro, N., Murgia, E.: 1st workshop on information retrieval for understudied users. In: European Conference on Information Retrieval (ECIR). Springer (2024) (to appear)
64. Pera, M.S., Murgia, E., Landoni, M., Huibers, T.: With a little help from my friends: use of recommendations at school. In: 2019 ACM Conference on Recommender Systems Late-breaking Results, ACM RecSys LBR 2019, pp. 61–65, CEUR (2019)
65. Pera, M.S., Wright, K.L., Kennington, C., Fails, J.A.: Children and information access: Fostering a sense of belonging. In: Smith-Renner, A., Taele, P. (eds.). Joint Proceedings of the IUI 2023 Workshops: HAI-GEN, ITAH, MILC, SHAI, SketchRec, SOCIALIZE.CEUR Workshop Proceedings, vol. 3359, pp. 254–257. CEUR, Sydney, Australia (2023)
66. Rieh, S.Y., Collins-Thompson, K., Hansen, P., Lee, H.J.: Towards searching as a learning process: a review of current perspectives and future directions. J. Inf. Sci. **42**(1), 19–34 (2016)

67. Roy, N., Moraes, F., Hauff, C.: Exploring users' learning gains within search sessions. In: Proceedings of the 2020 Conference on Human Information Interaction and Retrieval, pp. 432–436 (2020)
68. Sansone, C., Sperlí, G.: Legal information retrieval systems: State-of-the-art and open issues. Inf. Syst. **106**, 101967 (2022)
69. Schedl, M., Gómez, E., Lex, E.: Retrieval and recommendation systems at the crossroads of artificial intelligence, ethics, and regulation. In: Proceedings of the 45th International ACM SIGIR Conference on Research and Development in Information Retrieval, pp. 3420–3424 (2022)
70. Shneiderman, B.: Human-centered ai. Issues Sci. Technol. **37**(2), 56–61 (2021)
71. Smith, C.L., Rieh, S.Y.: Knowledge-context in search systems: Toward information-literate actions. In: Proceedings of the 2019 Conference on Human Information Interaction and Retrieval, pp. 55–62 (2019)
72. Tavares, D., Semedo, D., Rudnicky, A., Magalhaes, J.: Learning to ask questions for zero-shot dialogue state tracking. In: Proceedings of the 46th International ACM SIGIR Conference on Research and Development in Information Retrieval, pp. 2118–2122 (2023)
73. Tommasel, A., Rodriguez, J.M., Godoy, D.: I want to break free! recommending friends from outside the echo chamber. In: Proceedings of the 15th ACM Conference on Recommender Systems, pp. 23–33 (2021)
74. Trielli, D., Diakopoulos, N.: Search as news curator: the role of google in shaping attention to news information. In: Proceedings of the 2019 CHI Conference on Human Factors in Computing Systems, pp. 1–15 (2019)
75. Vanderschantz, N., Hinze, A.: Children's query formulation and search result exploration. Int. J. Digit. Libr. **22**(4), 385–410 (2021)
76. Vaughan, J.W., Wallach, H.: A human-centered agenda for intelligible machine learning. Getting Along with Artificial Intelligence, Machines We Trust (2020)
77. Werthner, H., Prem, E., Lee, E.A., Ghezzi, C.: Perspectives on digital humanism. Springer Nature (2022)
78. Wöbbekind, L., Lorberg, K., Mandl, T.: Emma stop that, it's my turn now-comparing peer tutoring and thinking aloud for usability-testing with children in a school setting. In: Proceedings of Mensch und Computer 2023, pp. 442–447 (2023)
79. Yarosh, S., et al.: Children asking questions: speech interface reformulations and personification preferences. In: Proceedings of the 17th ACM Conference on Interaction Design and Children, pp. 300–312 (2018)
80. Zhang, Y., Liu, X., Zhai, C.: Information retrieval evaluation as search simulation: a general formal framework for ir evaluation. In: Proceedings of the ACM SIGIR International Conference on Theory of Information Retrieval, pp. 193–200 (2017)
81. Zou, J., Aliannejadi, M., Kanoulas, E., Pera, M.S., Liu, Y.: Users meet clarifying questions: toward a better understanding of user interactions for search clarification. ACM Trans. Inform. Syst. **41**(1), 1–25 (2023)

Not Just Algorithms: Strategically Addressing Consumer Impacts in Information Retrieval

Michael D. Ekstrand[1(✉)], Lex Beattie[2], Maria Soledad Pera[3], and Henriette Cramer[1,2,3]

[1] Drexel University, Philadelphia, PA, USA
mdekstrand@drexel.edu
[2] Spotify, Seattle, USA
lex@spotify.com
[3] Delft University of Technology, Delft, NL, The Netherlands
m.s.pera@tudelft.nl

Abstract. Information Retrieval (IR) systems have a wide range of impacts on *consumers*. We offer maps to help identify goals IR systems could—or should—strive for, and guide the process of *scoping* how to gauge a wide range of consumer-side impacts and the possible interventions needed to address these effects. Grounded in prior work on scoping algorithmic impact efforts, our goal is to promote and facilitate research that (1) is grounded in impacts on information consumers, contextualizing these impacts in the broader landscape of positive and negative consumer experience; (2) takes a broad view of the possible means of changing or improving that impact, including non-technical interventions; and (3) uses operationalizations and strategies that are well-matched to the technical, social, ethical, legal, and other dimensions of the specific problem in question.

Keywords: users · consumers · impact · harm · equity

1 Introduction

Search engines, recommender systems, and related information retrieval (IR) tools are embedded in the daily lives of individuals on all paths of life, as they facilitate access to large and diverse collections, from articles to songs to products for purchase. IR research in its sociotechnical context, along with work on information seeking, recommender systems, and relevant segments of human-computer interaction (HCI), has long been concerned with ensuring that the systems provide efficient and effective information access to all who require it. It also examines the impacts of sociotechnical systems and implications for IR.

We explore the *impacts*—defined broadly—that IR systems have on *consumers* [20] (or users [67])and how those impacts differ between different (groups of) consumers. This draws from multiple perspectives, including:

Partly supported by the National Science Foundation on Grant 17-51278.

N. Goharian et al. (Eds.): ECIR 2024, LNCS 14611, pp. 314–335, 2024.
https://doi.org/10.1007/978-3-031-56066-8_25

(i) General IR research, which has long worked to providing resources that are well-matched to users' information needs in some or all of their nuances.
(ii) Consumer fairness [20,41] seeks to ensure that different consumers' experience is in some sense *fair* in its qualitative and/or quantitative dimensions [44], e.g., utility and representation.
(iii) Audience-specific IR looks to build systems that meet the differing needs of particular groups, such as children [80,90,91,106], autistic users [11,70,83], users experiencing dyslexia [18,50], or language learners [31,79].
(iv) Harm-aware IR, including content moderation actions (e.g., reducing exposure to content that is exploitative, or related to criminal or violent groups [105]) or industry interventions like "compassionate search" designs that redirect searches to resources developed in collaboration with experts [86].

Thoroughly exploring impacts and their challenges and possibilities is a complex and multifaceted project [41]. For example, a system under-representing women in search results about the best athletes of the century would impact consumers by giving them an inaccurate picture of the information space, besides depriving women of the visibility and attention afforded to their male counterparts. Historical and current datasets and baselines may however be absent to easily assess such impacts. The interweaving systems with oft-competing objectives making up modern IR systems additionally complicate how to control for impacts. For example, external audits have examined gender disparities in computational *advertising* for domains such as jobs [32]. Ali et al. [6] found that even if advertisers work to ensure demographic parity in ad reach, e.g., people of all genders see an ad for a well-paying job, the computational and marketplace dynamics of ad platforms can induce disparities in ad visibility. For another example, IR systems trained primarily on adults' interactions may not be equally useful (or beneficial) to all groups: a *child* searching for "Sven movie" is likely looking for *Frozen*, but the search engine might treat 'sven" as a misspelling for "seven" [69] and return results for *Se7en*, an R-rated film with "grisly afterviews of horrific and bizarre killings, and strong language" [59].

Beyond statistical properties of rankings, shifts in power in the larger context are a big concern [36] but this does not diminish within-system impacts: "attention" by being recommended can deliver power over time, and getting high(er) quality information or feeling represented in results can empower [72]. Similarly, not only do the resources returned by IR systems play a role in the system's impacts, but aesthetics, imagery, and UX writing signal whether a service welcomes user groups and feels designed with that particular group in mind [72].

There are many points (technical, institutional, or procedural) in an IR system where it is possible to *intervene* to correct or amplify a system's impacts.

[1] Fig. 1 shows the components of typical IR systems; each can be adjusted when impacts are observed to be inequitable, illegal, or misaligned with societal or business goals. Different segments of research emphasize different sites and strategies for intervention: for example, fairness literature typically emphasizes algorithmic interventions, with some attention to evaluation, while work on meeting particular user needs often focuses on user interface and user experience, with some work on algorithms. Other work takes a broader view, ranging from mapping different types of harms to studying different stages in the machine learning (ML) lifecycle. There is little explicit attention to the relative strengths and capabilities of different points and types of intervention, how to compare which may be better suited to address particular impacts, and integrate the wide research and practice fields.

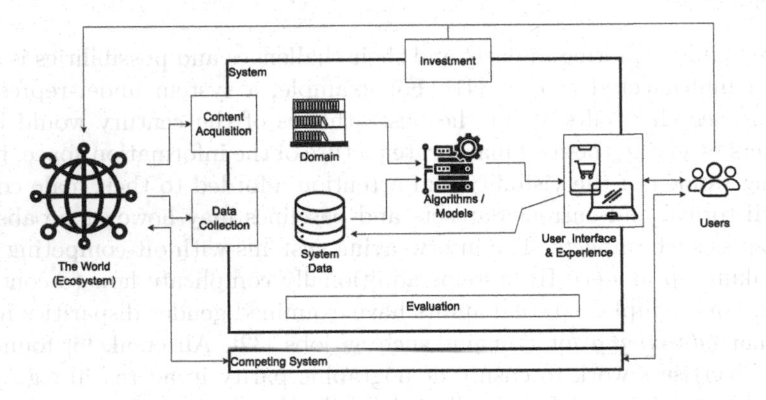

Fig. 1. The many components of IR systems and their sociotechnical context.

Effective impact intervention efforts must first identify a *goal* to pursue or problem to solve (Sect. 2), as the choice of *operationalization* (Sect. 3) and *intervention* (Sect. 4) depends on the goal [42]. Goals begin as high-level, theoretical, and qualitative objectives that are usually refined into quantitative measurements of impact—a process called *scoping* [14]—that are then matched to an appropriate technique to advance the original goal. To support this process, we provide three maps to guide and practice (summarized in Table 1): (1) a high-level overview of impact-related goals of IR-for-good efforts and tensions when considering these goals; (2) a review of operationalizations of those goals that bridge human concerns and technical possibilities; and (3) an overview of intervention points and strategies across the technical, institutional, and social aspects of an IR system in context. These maps are *non-exhaustive*, as further research will likely uncover more possible goals and strategies; our aims are to

[1] We borrow the concept of *intervention* from public health, behavioral health, and education to concisely express the idea of taking action to modify a system or its environment in order to address a problem (or enhance a positive phenomena). While this language is not widely used in IR research, we propose that it is useful for discussing how changes in a system's operation and outcomes can be effected.

provide a basis to contextualize existing work and to promote creative, interdisciplinary thinking about how to advance the benefits of IR and avoid or mitigate its potential harms, while more clearly integrating this rapidly growing field. For brevity and clarity, we focus this paper primarily on consumer-side impacts, but both the scoping process and the maps we discuss are fully applicable to impacts for other stakeholders such as providers and subjects.

We hope to see theoretical and applied work on IR impacts that: (1) **grounds** the work in specific impact goals (e.g. positive impact) or concerns (e.g. harm to avoid) that are clearly identified and well-motivated, (2) **contextualizes** those impacts with regards to the varied ways an IR system can affect its consumers, (3) **thinks expansively** about the range of possible methods and intervention points for addressing that impact, and (4) **matches** the impact to a strategy that is appropriate given technical and organizational constraints and the specific ethical, legal, economic, and other dimensions of the impact concern.

Table 1. Examples of goals, operationalizations, and strategies. Effective impact work requires clear selection and matching across these dimensions.

Examples		
Goal (Individual/group/society)	Operalization	Strategy
Recommender utility *e.g. aid discovery, or increase or diversify engagement, conversion and/or satisfaction* *Usability and access* *Representation* *Avoiding platform abuse and/or harmful content exposure*	*Metric objectives* *e.g. Equal utility* *e.g. Max grp log util* *Metric guardrails* *e.g. Set a max in engagement loss* *e.g. Reduce explicit negative user feedback* *Qual or Quant Data objectives*	*Auditing and monitoring* *Data pre-processing* *Data curation* *Feature selection* *Model enhancement* *Post-processing* *Engineering process* *Prioritization* *Evaluation* *New product development* *Design or Editorial intervention*

2 Scoping Stage 1: Goals

Our first map is a catalog of examples of impact-related goals for IR systems. The list is neither exhaustive nor a strict taxonomy; it is intended to spur expansive and contextualized thinking about what the different goals can mean in practice, and how we as a research community map the means to pursue them.

Mapping and evaluation for positive and negative impact can happen at multiple levels and may run into tensions. Defining positive goals to explicitly work towards is crucial to understanding whether the efforts involved in devising an IR system have the desired impact on individuals, organizations or wider society—and especially at what level this impact is expected to be seen. Safeguarding against negative impacts and identifying guardrails also requires definition. In practice, IR systems are often assemblages of multiple systems with different objectives, guardrails, and possibilities. For example, services like Instagram combine different models, where some are focused on new discoveries, whereas others serve up more familiar content, together assembled in a user-facing feed

[78]. Some single models use multi-objective optimization [98]. Mappings of goals then must account for such tensions, and how multiple, conflicting or mutually beneficial goals might impact operationalization and strategies. For example, Barocas et al. [12] divide potential negative impacts of ML into harms of *allocation* and *representation*; this delineation also applies to IR [28,41]. The former harms refer to withholding opportunities or resources, affecting the quality of service and information provided. The latter reinforce stigmas or stereotypes, for example over or underrepresentation of certain information or groups in results. Katzman et al. [64] point out that from operationalizations and measurements chosen, it can be unclear which harms were intended to be measured, thus advising to explicitly state the harm of interest to avoid ambiguity.

Defining Positive and Negative Impacts before Measurement. Before it can be measured, let alone improved, the impact needs to be clearly defined in its concept. IR has long sought the impact of *utility*, i.e., providing consumers with resources that are relevant to their needs. Utility is broad, however, and can involve various ways to meet the diversity of users' needs, e.g., through access to resources or increased engagement [23]. Diversity is a useful goal but needs to be defined more specifically to be meaningfully evaluated, e.g., subtopic diversity [93] considers the different things a query might mean, whereas calibration [104] aims for results that represent the breadth of users' interests, and other systems may aim to diversify interests or information provided [46,110]. Other work looks at utility being distributed fairly or equitably [45,73], to ensure the system is useful to all of its users instead of systematically under-serving some user groups (genders, ethnicities, etc.) Deldjoo et al. [35] observe in a thorough review that fairness research typically assumes that a clear definition is already available, "thus rendering the problem as one of designing algorithms to optimize a given metric"; this skips the explicit goal setting process.

Impacts go beyond resources and utility. *Representational harms* [30,64] to potentially assess include: inaccurately representing users' interests and information needs internally, preventing certain user groups from systematically having less-accurate representations (e.g. user embeddings or other user models that may lead to stereotyped recommendations [19]); perpetuating harmful or unnecessary stereotypes [85,94]; misrepresenting identities, e.g., misgendering; modeling users' interests by making assumptions based on demographic or other attributes instead of their preferences, i.e., "box products, not people" [97]; and misrepresenting the space of content, creators, subjects, etc. to consumers through the composition and presentation of results, particularly over- or under-representing certain types of content or people, or the resources and the results themselves [94]. In particular settings, specific representational harms may occur. For instance, Katzman et al. [64] further map different types of representational harms in image tagging.

Usability and Impact for Different Consumers. While IR systems may provide access to information, systems need to be usable in the first place; ensuring different groups can access and use the system is an important class of impact goals. Many designs assume a default user, typically an educated adult; such systems may not be usable for children [10], second-language learners [77], disabled users [70], people with low-bandwidth internet connections [84], or others who for whatever reason have difficulty using the default system. The system's results may harm either consumers generally or particular consumer groups in various ways as well. Self-harm advocacy [27], disinformation, and forms of deceptive or illegal material affect many users. Other content impacts differ from user to user. Milton and Pera [76] showcase how the subliminal stimulus presented by popular search engines varies throughout the information seeking process for searchers affected by mental health issues when compared to "typical" searchers. Information can also interact in often unpredictable ways with users' emotional state and context, such as continuing to recommend baby products when a consumer has suffered a pregnancy loss. Such content can both cause direct emotional harm and "erase" that such human experiences are not aberrations [8].

From Individual Consumers to Society. Impacts do not end with individuals, they may have a ripple effect on organizations, movements, and society. For example, Helberger [56] discuss the role of diverse IR results in supporting users' engagement in democratic processes. Decades ago, Belkin and Robertson [16] voiced the possibility for IR technology to manipulate users, particularly regarding ideological or political beliefs. Similarly, concerns have been raised about the potential for recommender systems (and personalized IR in general) to isolate individuals in different segments of the information space [75,87,109], or indulging or reinforcing beliefs that harm consumers' relationships with others [66]. The exact societal impacts around topics such as polarization are contested depending on specific study settings and effects; setting goals at different levels with different groups leads to different operationalizations. This concern is not limited to IR work; early work around "internet communication" (1996, [33]) observed that while much insight was provided by ongoing studies, differences in units of analysis lead to difficulty in cross-study comparisons as well as theoretical integration (see for a similar discussion around statistics related to patient outcomes in medicine [7]). Careful definition is essential.

3 Scoping Stage 2: Operationalization

Operationalization requires translating a theoretical construct into conceptualization that can be observed and assessed using a specific quantitative and/or qualitative research design to guide and evaluate impact efforts. This step requires careful creativity in choosing or designing metrics and protocols, as operationalization is highly dependent on the goal in all of its technical and sociotechnical nuance [60]. In this section, we present a series of constraints that govern the operationalization process and example operationalizations for three common measurement objectives, focusing on quantitative methods; the full reality of mapping the field and methods' impact is more complex. Patton [89], for example, discusses how methods from social work can be used to anticipate how AI and UX solutions impact diverse communities. In 1999, Sofaer [102] also provided a variety of examples of the usefulness of qualitative methods in health services and policy research in a similarly methodologically 'nascent' field.

3.1 Operationalization Constraints

Operationalization constraints are marked by the need to identify how to quantitatively address a qualitative goal within the confines of an auditing scenario, a critical step in many broad system-level frameworks for algorithmic impact measurement [14]. Below we share three possible constraints to address.

Consumer Properties. Identifying consumer properties enables us to understand what will be measured for the consumer, and how that characteristic will be measured. The primary property constraint should be to define if the consumer should be measured in a group or individual setting. Addressing this constraint is a requirement for most fairness evaluation settings due to the inherent difference in measuring group versus individual fairness [101]. Individual versus group measurement specificity can extend beyond fairness and should be accounted for when operationalizing one's goal to ensure that the measurement accounts for how a goal may differ when observed at the individual or group level. When measuring goals for individuals, operationalization should target quantifying if similar individuals receive similar treatment or outcomes. If measurement is defined to target consumers at the group level, the practitioner should further identify the group attributes for measurement. We can narrow types of group attributes into three categories: binary, multi-categorical, and intersectional. It is essential to understand the group attribute type due to the fundamental differences in measuring binary, multi-categorical, and intersectional scenarios.

System Components for Evaluation. Goals can be scoped into operationalized metrics at many different points in the IR system components (Fig. 1). While end-to-end measurement is always important [39], measurement at intermediate stages is often useful. Evaluation or optimization of components of a multi-stage recommender system, for example, may target the candidate generator, ranking model, or the end-to-end system. This view moves beyond the standard fair ML model stages of pre-processing, intrinsic, and post-processing algorithmic components. Targeting the goal to be assessed in a data collection component requires unique measurement objectives and metrics to address data-collection-specific concerns such as selection or sampling bias. Potential subsequent measurement objectives and metrics will be dependent on the algorithm, model outputs, and data available for evaluation; it is therefore critical to understand the system components and how their outputs can be quantified operationalizing a goal.

Measurement Objectives. Measurement objectives are initially addressed when defining goals, but operationalizing goals requires more specific and quantitative definition of objectives. For example, in past research, evaluation metrics are often categorized into distinct groups to provide guidance for how and when to use specific metrics. Smith et al. [101] demonstrate the importance of understanding measurement objectives for fairness related evaluation by surfacing specific fairness objectives and their subsequent relevant metrics to achieve the objective. In their work, utility for consumers is further defined by two measurement objectives: utility versus merit and group utility [101]. Utility versus merit defines the utility measurement objective of "utility for consumers" is distributed "based on their merit or need", while group utility looks to "distribute utility equally between groups of consumers" [101]. These two different measurement objectives require different metrics for operationalization.

3.2 Identifying Metrics Within Operationalization Constraints

The relevant metrics for operationalization depend on the choices made when identifying the consumer properties, system components, and measurement objectives, as well as the social, ethical, economic, and other particularities of the goal. Some guidance exists for identifying relevant metrics from the research literature in light of such constraints, such as the recommendation fairness decision framework introduced by Smith et al. [101]. Unfortunately, there are many applications and types of impact where such frameworks are not yet available, requiring metric design from scratch. Measurement modeling, as showcased by Jacobs and Wallach [60] for operationalizing fairness, provides a substantive framework for designing robust metrics for measuring qualitative goals or constructs. These metrics then enable the practitioner to assess the impact of an intervention on their original qualitative goal. Additionally, using the identified metrics to evaluate the system prior to intervention can guide the choice of intervention strategy and provide a baseline for measuring improvement. In the rest of this section, we present several examples of identifying metrics based on three potential measurement objectives and constraints.

Utility. Utility is a well-known and widely-studied goal for consumer impact. There are many metrics to estimate utility to consumers [54], but even once a metric is selected, there are several ways to operationalize *equity* of utility to ensure groups of consumers aren't being systematically under-served. One way is to compute the *difference* in utility between groups ($\Upsilon_{g_1} - \Upsilon_{g_2}$). Another is to compute the *total groupwise log utility* [$\sum_g \log \Upsilon_g$, 111], so maximum improvement for the overall metric comes from improving utility for the lowest-utility group. Targeting differences in utility [e.g. 58,68,81] can remove inequity, but treats utility as zero-sum, sometimes with significant majority-group utility loss [68]. A positive-sum aggregate that avoids unnecessary competition [111]; since consumer utility is non-subtractable (users do not compete for the same utility) [15,41], this is more appropriate. It also extends beyond binary groups.

Representation. *Calibrated "fairness"* is an example of operationalizing representation for measurement. This concept probes if users with multiple interests are represented "fairly" in the result set by correctly reflecting their historical interests [38]; it can also be used to measure how consumer behavior affects recommendations along various axes [25,43]. This is done at the individual or group level by comparing distributions between what is recommended versus users' history in the system. Relevant metrics can change based on their intended use.

For instance, distribution comparison metrics are most often used to evaluate calibrated fairness. Kullback-Leibler Divergence (KLD) lends itself well to evaluating calibration fairness of content-pool generation by appraising the served distribution of the entire final content pool. Normalized KLD is better suited for evaluating the distribution of a ranking component since it penalizes for rank during calculation. These metrics work well for evaluating and prioritizing interventions, but may not suit evaluation and optimization. In the case that one metric should be used for evaluation and intrinsic optimization of calibrated fairness, Jensen-Shannon Divergence may be favored since the symmetry of the metric lends itself better to intrinsically optimize calibration fairness.

Diversity. Vrijenhoek et al. [110] examine different operationalizations of the qualitative goal of diversity for news recommendation in the context of supporting democratic engagement. Drawing from different theories of democracy, they structure five measurement objectives with specific metric operationalizations like "fragmentation", which "denotes the amount of overlap between news story chains shown to different users" [110]. To further concretize this metric, they choose between two possible metrics Kendall Tau Rank Distance (KTRD) and Rank Biased Overlap (RBO) [110], selecting RBO to avoid measurement limitations of KTRD, which does not penalize by rank; thus making it a more suitable metric for evaluating candidate pools than final ranked lists for fragmentation [110]. However, RBO does not satisfy their original measurement objective of fragmentation, leading them to use an adaptation that subtracts RBO from one. The documentation of their process showcases the need to not only account for the system component but also the limitations of current metrics, resulting in a specially designed implementation for the measurement of the original objective.

4 Intervention Strategies

There is a range of strategies for advancing consumer impact goals and operationalizations, with interventions possible anywhere in the system as well as at many points in the broader sociotechnical context where the system operates (Fig. 1). These strategies can fall into one of these broad and non-mutually - exclusive categories: (1) **system interventions** to improve the impacts of an IR system (2) **process interventions** that change how an organization builds and maintains the system (3) **marketplace and ecosystem interventions** that introduce new products or intervene in the ecosystem that provides the IR system's inputs and consumes its output

. Some interventions directly employ an operationalization (Sect. 3) of an impact goal, e.g. as an objective function for training a model; others do not, but the operationalized goals are still crucial for identifying and evaluating the intervention. Effective interventions need to be well-matched to the particular goal and operationalization, as well as to the resources and constraints of the organization.

4.1 System Interventions

As we discuss in this section, decisions across the entire architecture and lifecycle of an IR systems affect the system's impact on people [29,112].

Data Interventions. Data is fundamental to IR systems; the corpus and the data used to train algorithms and models that power the system [108] affect its operation and impacts. Manipulating the data is then a viable strategy for adjusting system impacts. For example, injecting fake user profiles [96] and removing spam reviews [100] can improve recommender system impacts. For a summary of manipulation techniques at data pre-processing time to mitigate discrimination of protected groups in IR-powered disaster management applications see [114]. Chen et al. [26] examines different manifestations of bias in ML classifiers and distinguishes those best addressed through more or better data, and those more amenable to model-based interventions. Techniques to reduce position bias in CTR estimation through interventions [61] or re-analysis [5] may also apply to biases in user response that correlate with group membership.

Sometimes, data needs to be specifically collected or engineered. Allocating a budget for data improvement and prioritizing data needed to serve underserved groups can help. Goel and Faltings [52] provide a mechanism for crowdsourcing data while respecting fairness objectives in the data's coverage; this is a potential building block for data interventions. Interestingly, interventions designed to positively impact in one aspect can have unintended side effects. E.g., differentially-private training mechanisms can exacerbate data biases [47], making the evaluation of data strategies, and especially their combinations, crucial.

Algorithmic Interventions. Modifying the algorithms employed in an IR system is a common strategy for addressing impact. For example, much consumer fairness research augments the loss function an inter-user equity objective [58,111,113], sometimes through a regularization term [62,115]. Penalizing dependence between recommended items and user attributes [62] may be a viable alternative for reducing stereotypes. New fairness-aware metrics can be directly optimized [51,113]. Aggregations can be changed; using a maximin objective function in social network information spreading—instead of maximum or total spread—can reduce disparities in who has access to information [49].

There are many ways to modify algorithms besides adjusting the training objective, such as reranking [48,68], multi-objective optimization [74], and changing neighborhood selection [22]. Adversarial learning to learn user embeddings that cannot predict consumers' sensitive attributes such as gender [19] are promising both as a fair representation learning approach and to address potential stereotypes in recommendations. Existing techniques can be repurposed for indirectly improving consumer impacts. Diversification techniques can generate rankings that are fairer to a broader range of users [71]; diversification can support consumers in democratic participation under differing political theories [110]. MMR [23] has proven useful for fair image search [24,63]. Such indirect approaches have significant promise—existing computational machinery can likely be repurposed for a variety of impact-related aims.

Design or Editorial Interventions. The design of the user interface and experience presents many possibilities for addressing a system's impacts, particularly by better matching the system to users' specific needs and abilities. One example is Pinterest's inclusive search that combines both UX features and an inclusive model development process as well as editorial expertise to make the best of the strengths of both: "*Pinners can search for a broad hair term like 'summer hairstyles' [...] and narrow their results by selecting one of the six hair patterns[...]. Pinterest has detected a hair pattern (e.g. coily, curly, protective) in over 500 million images on our platform*" [92]. Deldjoo et al. [34] proposed a child-oriented TV/movie recommendation interface using tangible interaction: the child holds up a toy truck to get recommendations for shows about trucks.

Design can make systems or products more usable to a broader set of consumers; to provide interfaces tailored to the needs of particular classes of users; or to dynamically adapt the system and its interface for the current user's needs [e.g. 9,53]. Dynamic adaptation can serve children [37,69,99], or users of different physical abilities [88]. Adaptivity or new interaction designs are not without risk and potential negative impact, particularly as they may require additional data collection or inference. As with any strategy, they can lead to new types of abuse or inadvertent errors, and thus iterative goal setting is necessary.

4.2 Organization, Process, and Evaluation Interventions

Organizational dynamics are critical to consumer impact work in practice. Rakova et al. [95] identify four themes in organizational transitions towards work practices that include AI impact and responsibility: anticipating rather than reacting to issues, providing internal structure to address concerns, aligning on success, and resolving tensions within the organization. These are key to creating an organization where identifying, monitoring, and addressing consumer impacts is standard practice. Organizational structures and practices are a possible, yet often overlooked, point of intervention themselves. Changing *how* a system is built and evaluated will have further effects on *what* is built and its impacts.

Evaluation Interventions. Regularly auditing for scoped impact objectives and other forms of impact and equity [57] can identify problems and help detect regressions on past impact improvements. Other interventions also begin with evaluation; documenting and measuring system impact is crucial to informing the selection and design of any intervention strategy, as well as evaluating whether the chosen strategy is effectively achieving its objective. Further, the evaluation process itself can impact outcomes downstream and can be a site of intervention in its own right: changing how the system is evaluated, or how the evaluation is analyzed, affects both human and computational decisions in its ongoing development and maintenance. Disaggregated [13] and distributional [40] evaluations can identify inequities between consumers [45,73] and quantify broader consumer experiences; mediation analysis [73] can identify intervention targets for addressing inequities. While most commonly seen for assessing equity of utility, these can be applied to any measure of user experience.

Evaluation should go beyond quantitative measurements to include qualitative studies that note how consumers experience the system and what they want from it [55]. In practice, assessment is difficult, as the literature provides little guidance for navigating the decisions involved in designing an evaluation [14].

Engineering Process. The engineering process itself admits strategies for addressing impact inequities [57,95]. In every development cycle, product teams make decisions about what to prioritize and how to allocate resources; investing in efforts that address impacts or enhance the system for under-served consumers is a tool for improving impacts and equity. Evaluations that reveal *why* an impact occurs (e.g. mediation analysis) provide valuable inputs for planning processes.

4.3 (New) Product/Marketplace and Ecosystem Interventions

In some cases, impacts can best be addressed through **new systems** targeted at needs that are not well-met by existing systems. While this often is not adequate to address the negative impacts of existing IR systems, new possibilities for consumer equity emerge by thinking more broadly than a single system to consider whether people have equitable access to information across a set of product offerings or the broader marketplace and ecosystem of services. Many services implicitly or explicitly assume a "default" user and provide an experience that may not suit all consumers. Children are a key example; some services such as Netflix provide distinct services targeting specifically for children and/or families [82], while new startups are also trying to fill the gap between adult and child information access [1–3]. These new firms and their products can be seen as marketplace interventions: new products enter the market to meet needs that are not addressed by existing platforms, helping consumers directly and applying market pressure on existing firms to better meet those needs.

Other impact-related marketplace activities are possible, such as contracting with content providers to expand or improve the content that can be provided to consumers, or investing in local expertise to improve the metadata for content from a region or community. Tsioutsiouliklis et al. [107] attempted to improve the fairness of PageRank by recommending new links that, when included in the web's link graph, would result in more fair results. There are more opportunities to invest in a broad ecosystem in which IR systems can have better impacts, but for the research community it is especially important to invest in the mapping of what we do (not) know about the impact of different strategies in specific situations beyond the work of the publishing academic community.

5 Conclusion

Identifying, monitoring, and improving the impacts of for IR system requires *impact objectives* to be clearly scoped in a well-founded manner and matched to suitable intervention strategies, while accounting for the system's broader socio-technical impact. Landoni et al. [65] identify four aspects for evaluating technical advances: the advance itself (i.e., the intervention); the group of consumers it is intended to serve; the task those consumers perform; and the context or environment in which the advance will operate. These factors provide a starting point for reasoning about impact interventions for consumers and other stakeholders, and in this paper we have provided broad maps to help structure that reasoning.

Ensuring IR systems are equitable and beneficial for all of their consumers is a creative, yet challenging, effort that must be grounded in expansive and interdisciplinary thinking about the wide range of impacts and the possibilities for identifying, assessing, and addressing them. Any impact study or intervention must also be anchored to the context in which it is performed, as what may be appropriate in one context, may not be in another. Contextual impacts include:

(1) **Regulatory environment**, i.e., different jurisdictions and information domains have different external regulatory requirements.
(2) **Costs vs impact**, i.e., domains have different costs both to produce and consume resources, and these costs affect the impact of different actions.
(3) **Business arrangements and momentum**, i.e., contractual obligations, content availability, etc. can affect what can (or must) be done.
(4) **Infrastructure and staffing**, i.e., it can be much more effective to pick a strategy that leverages existing resources or momentum.
(5) **Impact type**, i.e., the specific kind of impact considered makes a significant difference in the appropriateness and usefulness of an intervention.

Sticking to one approach (e.g. adjusting a dataset or introducing regularization terms to model objectives) might work in some cases, but it is crucial to identify whether other intervention sites (e.g. UX) or combinations of sites and strategies might have a larger impact before picking just one.

The choice of *where* in the IR systems and its sociotechnical context to measure and/or intervene, *how* to intervene at that point, and *who* makes those determinations needs to be well-matched to the specific impact objective and details of the application, domain, market, and users. This also requires careful consideration of multiple perspectives, particularly the voices of those harmed by the system. Key questions include:

- Who is involved in the process and what are their roles and responsibilities (including who gets to participate in imagining what the possibilities could be vs. only get to live with its consequences [17])?
- How are impacts translated into specific objectives and decisions?
- How are different kinds of impacts prioritized?

More research is needed to understand how to implement the broad range of possible interventions (not limited to those we discussed) most effectively, how to select effective interventions, and how the consumer impact problem space decomposes into subproblems. Such research will identify when different interventions may or may not be appropriate, generate new ideas for engaging with impacts not yet considered, and develop a map coupled with evidence-based guidance for future practice. Beyond *more* research, efforts are especially needed to integrate and compare all that research—including non-academic research that feeds into existing practice—to ensure the research community has a more concerted impact itself.

We invite the broader communities of research and practice studying IR, HCI, algorithmic impact, and related topics to join us in this cartographic adventure.

References

1. ABC Mouse (August 2022). https://www.abcmouse.com/
2. Biblionasium (August 2022). https://www.biblionasium.com/
3. Pickatale (August 2022). https://pickatale.com/
4. Abdollahpouri, H., et al.: Multistakeholder recommendation: survey and research directions. User Model. User-Adapted Interact. 30(1), 127–158 (2020). https://doi.org/10.1007/s11257-019-09256-1, ISSN 0924–1868
5. Agarwal, A., Zaitsev, I., Wang, X., Li, C., Najork, M., Joachims, T.: Estimating position bias without intrusive interventions. In: Proceedings of the Twelfth ACM International Conference on Web Search and Data Mining, pp. 474–482, Association for Computing Machinery, New York. (Jan 2019), https://doi.org/10.1145/3289600.3291017
6. Ali, M., Sapiezynski, P., Bogen, M., Korolova, A., Mislove, A., Rieke, A.: Discrimination through optimization: how facebook's ad delivery can lead to biased outcomes. Proc. ACM Hum.-Comput. Interact. **3**(CSCW), 1–30 (2019). https://doi.org/10.1145/3359301
7. Altman, D.G., Bland, J.M.: Statistics notes: units of analysis. BMJ **314**(7098), 1874 (1997). https://doi.org/10.1136/bmj.314.7098.1874, ISSN 0959–8138, 1468–5833
8. Andalibi, N., Garcia, P.: Sensemaking and coping after pregnancy loss. Proc. ACM Hum.-Comput. Interact. **5**(CSCW1), 1–32 (2021). https://doi.org/10.1145/3449201, ISSN 2573–0142
9. Antona, M., Savidis, A., Stephanidis, C.: A process-oriented interactive design environment for automatic user-interface adaptation. Inter. J. Hum.-Comput. Interact. **20**(2), 79–116 (2006). https://doi.org/10.1207/s15327590ijhc2002_2, ISSN 1044–7318
10. Anuyah, O., Milton, A., Green, M., Pera, M.S.: An empirical analysis of search engines' response to web search queries associated with the classroom setting. Aslib J. Inform. Manag. **72**(1), 88–111 (2020). https://doi.org/10.1108/AJIM-06-2019-0143, ISSN 2050–3806
11. Banskota, A., Ng, Y.K.: Recommending video games to adults with autism spectrum disorder for social-skill enhancement. In: Proceedings of the 28th ACM Conference on User Modeling, Adaptation and Personalization, pp. 14–22, Association for Computing Machinery, New York (Jul 2020), https://doi.org/10.1145/3340631.3394867
12. Barocas, S., Crawford, K., Shapiro, A., Wallach, H.: The problem wtih bias: Allocative versus representational harms in machine learning. In: 9th Annual Conference of the Special Interest Group for Computing, Information and Society (2017)

13. Barocas, S., et al.: Designing disaggregated evaluations of AI systems: choices, considerations, and tradeoffs. In: Proceedings of the 2021 AAAI/ACM Conference on AI, Ethics, and Society, pp. 368–378, Association for Computing Machinery, New York (Jul 2021). https://doi.org/10.1145/3461702.3462610
14. Beattie, L., Taber, D., Cramer, H.: Challenges in translating research to practice for evaluating fairness and bias in recommendation systems. In: Proceedings of the 16th ACM Conference on Recommender Systems, pp. 528–530, Association for Computing Machinery, New York (Sep 2022). https://doi.org/10.1145/3523227.3547403
15. Becker, C.D., Ostrom, E.: Human ecology and resource sustainability: the importance of institutional diversity. Annal Rev. Ecol. Systematics **26**(1), 113–133 (1995), https://doi.org/10.1146/annurev.es.26.110195.000553, ISSN 0066–4162
16. Belkin, N.J., Robertson, S.E.: Some ethical and political implications of theoretical research in information science. In: Proceedings of the ASIS Annual Meeting (1976). https://www.researchgate.net/publication/255563562
17. Benjamin, R.: Race after Technology: Abolitionist Tools for the New Jim Code. Polity (2019), ISBN 978-1-5095-2640-6
18. Berget, G., Sandnes, F.E.: Do autocomplete functions reduce the impact of dyslexia on information-searching behavior? The case of Google. J. Assoc. Inform. Sci. Technol. **67**(10), 2320–2328 (2016). https://doi.org/10.1002/asi.23572, ISSN 2330–1643
19. Beutel, A., Chen, J., Zhao, Z., Chi, E.H.: Data decisions and theoretical implications when adversarially learning fair representations (Jul 2017). http://arxiv.org/abs/1707.00075
20. Boratto, L., Fenu, G., Marras, M., Medda, G.: Consumer fairness in recommender systems: contextualizing definitions and mitigations. In: Hagen, M., et al. (eds.) ECIR 2022. LNCS, vol. 13185, pp. 552–566. Springer, Cham (2022). https://doi.org/10.1007/978-3-030-99736-6_37
21. Burke, R.: Multisided fairness for recommendation (Jul 2017). http://arxiv.org/abs/1707.00093
22. Burke, R., Sonboli, N., Ordonez-Gauger, A.: Balanced neighborhoods for multi-sided fairness in recommendation. In: Friedler, S.A., Wilson, C. (eds.) Proceedings of the 1st Conference on Fairness, Accountability and Transparency, Proceedings of Machine Learning Research, vol. 81, pp. 202–214. PMLR (2018). http://proceedings.mlr.press/v81/burke18a.html
23. Carbonell, J., Goldstein, J.: The use of MMR, diversity-based reranking for reordering documents and producing summaries. In: Proceedings of the 21st Annual International ACM SIGIR Conference on Research and Development in Information Retrieval, pp. 335–336. Association for Computing Machinery, New York (1998). https://doi.org/10.1145/290941.291025
24. Celis, L.E., Keswani, V.: Implicit diversity in image summarization. Proc. ACM Hum.-Comput. Interact. **4**(CSCW2), 1–28 (2020). https://doi.org/10.1145/3415210
25. Channamsetty, S., Ekstrand, M.D.: Recommender response to diversity and popularity bias in user profiles. In: Proceedings of the 30th Florida Artificial Intelligence Research Society Conference. AAAI Press (May 2017). https://aaai.org/papers/657-flairs-2017-15524/
26. Chen, I., Johansson, F.D., Sontag, D.: Why is my classifier discriminatory? In: Bengio, S., Wallach, H., Larochelle, H., Grauman, K., Cesa-Bianchi, N., Garnett,

R. (eds.) Advances in Neural Information Processing Systems, vol. 31, pp. 3539–3550, Curran Associates, Inc. (2018). http://papers.nips.cc/paper/7613-why-is-my-classifier-discriminatory.pdf
27. Cheng, Q., Yom-Tov, E.: Do search engine helpline notices aid in preventing suicide? analysis of archival data. J. Med. Internet Res. **21**(3), e12235 (2019). https://doi.org/10.2196/12235, ISSN 1438–8871
28. Cramer, H., Garcia-Gathright, J., Reddy, S., Springer, A., Takeo Bouyer, R.: Translation, tracks & data: an algorithmic bias effort in practice. In: Extended Abstracts of the 2019 CHI Conference on Human Factors in Computing Systems, CHI EA 2019, pp. 1–8. Association for Computing Machinery, New York (May 2019). https://doi.org/10.1145/3290607.3299057, ISBN 978-1-4503-5971-9
29. Cramer, H., et al.: Challenges of incorporating algorithmic fairness into practice: a tutorial at FAccT 2019 (2019). https://algorithmicbiasinpractice.files.wordpress.com/
30. Crawford, K.: The trouble with bias (Dec 2017). https://youtu.be/fMym_BKWQzk
31. Dabran-Zivan, S., Baram-Tsabari, A., Shapira, R., Yitshaki, M., Dvorzhitskaia, D., Grinberg, N.: "Is COVID-19 a hoax?": auditing the quality of COVID-19 conspiracy-related information and misinformation in Google search results in four languages. Internet Res. **33**(5), 1774–1801 (2023). https://doi.org/10.1108/INTR-07-2022-0560, ISSN 1066–2243
32. Datta, A., Tschantz, M.C., Datta, A.: Automated experiments on ad privacy settings. Proc. Priv. Enhancing Technol. **2015**(1), 92–112 (2015). https://doi.org/10.1515/popets-2015-0007, ISSN 2299–0984
33. December, J.: Units of analysis for internet communication. J. Comput.-Mediated Commun. **1**(4), JCMC143 (1996). https://doi.org/10.1111/j.1083-6101.1996.tb00173.x, ISSN 1083–6101
34. Deldjoo, Y., et al.: Enhancing children's experience with recommendation systems. In: Proceedings of the International Workshop on Children & Recommender Systems (2017). https://yasdel.github.io/files/KidRec17_deldjoo.pdf
35. Deldjoo, Y., Jannach, D., Bellogin, A., Difonzo, A., Zanzonelli, D.: Fairness in recommender systems: Research landscape and future directions. User Model. User-Adapted Interact. (2023). https://doi.org/10.1007/s11257-023-09364-z, ISSN 1573–1391
36. D'Ignazio, C., Klein, L.F.: Data Feminism. MIT Press (2020). https://data-feminism.mitpress.mit.edu/, ISBN 978-0-262-04400-4
37. Downs, B., Pera, M.S., Wright, K.L., Kennington, C., Fails, J.A.: KidSpell: making a difference in spellchecking for children. Inter. J. Child-Comput. Interact., 100373 (2021). https://doi.org/10.1016/j.ijcci.2021.100373, ISSN 2212–8689
38. Dragovic, N., Azpiazu, I.M., Pera, M.S.: From recommendation to curation: when the system becomes your personal docent. In: Proceedings of 5th Joint Workshop on Interfaces and Human Decision Making for Recommender Systems (IntRS 2018), pp. 37–44 (Oct 2018). http://ceur-ws.org/Vol-2225/paper6.pdf
39. Dwork, C., Ilvento, C.: Fairness under composition. In: Blum, A. (ed.) 10th Innovations in Theoretical Computer Science Conference (ITCS 2019), Leibniz International Proceedings in Informatics (LIPIcs), vol. 124, pp. 33:1–33:20, Schloss Dagstuhl-Leibniz-Zentrum fuer Informatik, Dagstuhl, Germany (2019). https://doi.org/10.4230/LIPICS.ITCS.2019.33
40. Ekstrand, M.D., Carterette, B., Diaz, F.: Distributionally-informed recommender system evaluation. ACM Trans. Recommender Syst. (2023). https://doi.org/10.1145/3613455

41. Ekstrand, M.D., Das, A., Burke, R., Diaz, F.: Fairness in information access systems. Foundat. Trends® Inform. Retrieval **16**(1–2), 1–177 (2022). https://doi.org/10.1561/1500000079, ISSN 1554–0669
42. Ekstrand, M.D., Das, A., Burke, R., Diaz, F.: Fairness in recommender systems. In: Ricci, F., Rokach, L., Shapira, B. (eds.) Recommender Systems Handbook, pp. 679–707, Springer, US (2022). https://doi.org/10.1007/978-1-0716-2197-4_18, ISBN 978-1-07-162197-4
43. Ekstrand, M.D., Kluver, D.: Exploring author gender in book rating and recommendation. User Model. User-Adapted Interact. **31**(3), 377–420 (2021). https://doi.org/10.1007/s11257-020-09284-2, ISSN 0924–1868
44. Ekstrand, M.D., Pera, M.S.: Matching consumer fairness objectives & strategies for RecSys (Sep 2022). http://arxiv.org/abs/2209.02662
45. Ekstrand, M.D., et al.: All the cool kids, how do they fit in?: Popularity and demographic biases in recommender evaluation and effectiveness. In: Friedler, S.A., Wilson, C. (eds.) Proceedings of the 1st Conference on Fairness, Accountability and Transparency, Proceedings of Machine Learning Research, vol. 81, pp. 172–186. PMLR (2018). https://proceedings.mlr.press/v81/ekstrand18b.html
46. Epps-Darling, A., Bouyer, R.T., Cramer, H.: Artist gender representation in music streaming. In: Proceedings of the 21st International Society for Music Information Retrieval Conference, pp. 248–254. ISMIR (Oct 2020). https://program.ismir2020.net/poster_2-11.html
47. Farrand, T., Mireshghallah, F., Singh, S., Trask, A.: Neither private nor fair: Impact of data imbalance on utility and fairness in differential privacy. In: Proceedings of the 2020 Workshop on Privacy-Preserving Machine Learning in Practice, pp. 15–19. Association for Computing Machinery, New York (Nov 2020). https://doi.org/10.1145/3411501.3419419, ISBN 978-1-4503-8088-1
48. Feng, Y., Shah, C.: Has CEO gender bias really been fixed? Adversarial attacking and improving gender fairness in image search. In: Proceedings of the AAAI Conference on Artificial Intelligence, vol. 36(11), pp. 11882–11890 (Jun 2022). https://doi.org/10.1609/aaai.v36i11.21445, ISSN 2374–3468
49. Fish, B., Bashardoust, A., Boyd, D., Friedler, S., Scheidegger, C., Venkatasubramanian, S.: Gaps in information access in social networks? In: WWW 2019: The World Wide Web Conference, pp. 480–490. Association for Computing Machinery, New York (May 2019). https://doi.org/10.1145/3308558.3313680
50. Fourney, A., Ringel Morris, M., Ali, A., Vonessen, L.: Assessing the readability of web search results for searchers with dyslexia. In: The 41st International ACM SIGIR Conference on Research & Development in Information Retrieval, pp. 1069–1072, Association for Computing Machinery, New York. (Jun 2018), https://doi.org/10.1145/3209978.3210072
51. Gao, R., Ge, Y., Shah, C.: FAIR: Fairness-aware information retrieval evaluation. J. Assoc. Inform. Sci. Technol. **73**(10), 1461–1473 (2022). https://doi.org/10.1002/asi.24648, ISSN 2330–1643,
52. Goel, N., Faltings, B.: Crowdsourcing with fairness, diversity and budget constraints. In: Proceedings of the 2019 AAAI/ACM Conference on AI, Ethics, and Society, pp. 297–304. Association for Computing Machinery, New York (Jan 2019). https://doi.org/10.1145/3306618.3314282
53. Gossen, T., Nitsche, M., Vos, J., Nürnberger, A.: Adaptation of a search user interface towards user needs: A prototype study with children & adults. In: Proceedings of the Symposium on Human-Computer Interaction and Information Retrieval, pp. 1–10. Association for Computing Machinery, New York (Oct 2013). https://doi.org/10.1145/2528394.2528397, ISBN 978-1-4503-2570-7

54. Gunawardana, A., Shani, G., Yogev, S.: Evaluating recommender systems. In: Ricci, F., Rokach, L., Shapira, B. (eds.) Recommender Systems Handbook, third edn., pp. 547–601, Springer, US (2022). https://doi.org/10.1007/978-1-0716-2197-4_15, ISBN 978-1-07-162196-7
55. Harambam, J., Bountouridis, D., Makhortykh, M., van Hoboken, J.: Designing for the better by taking users into account: A qualitative evaluation of user control mechanisms in (news) recommender systems. In: Proceedings of the 13th ACM Conference on Recommender Systems, pp. 69–77. Association for Computing Machinery, New York (Sep 2019), https://doi.org/10.1145/3298689.3347014
56. Helberger, N.: On the democratic role of news recommenders. Digital J. **7**(8), 993–1012 (2019). https://doi.org/10.1080/21670811.2019.1623700, ISSN 2167–0811
57. Holstein, K., Wortman Vaughan, J., Daumé, III, H., Dudik, M., Wallach, H.: Improving fairness in machine learning systems: What do industry practitioners need? In: Proceedings of the 2019 CHI Conference on Human Factors in Computing Systems, pp. 1–16. Association for Computing Machinery, New York (May 2019). https://doi.org/10.1145/3290605.3300830
58. Huang, W., Labille, K., Wu, X., Lee, D., Heffernan, N.: Achieving User-Side Fairness in Contextual Bandits. Hum.-Centric Intel. Syst. **2**(3), 81–94 (2022. https://doi.org/10.1007/s44230-022-00008-w, ISSN 2667–1336
59. IMDb: Se7en (1995) - IMDb (1995). http://www.imdb.com/title/tt0114369/parentalguide
60. Jacobs, A.Z., Wallach, H.: Measurement and fairness. In: Proceedings of the 2021 ACM Conference on Fairness, Accountability, and Transparency, pp. 375–385, Association for Computing Machinery, New York (2021). https://doi.org/10.1145/3442188.3445901
61. Joachims, T., Swaminathan, A., Schnabel, T.: Unbiased learning-to-rank with biased feedback. In: Proceedings of the Twenty-Seventh International Joint Conference on Artificial Intelligence, International Joint Conferences on Artificial Intelligence Organization, California (Jul 2018). https://doi.org/10.24963/ijcai.2018/738
62. Kamishima, T., Akaho, S., Asoh, H., Sakuma, J.: Recommendation independence. In: Friedler, S.A., Wilson, C. (eds.) Proceedings of the 1st Conference on Fairness, Accountability and Transparency, Proceedings of Machine Learning Research, vol. 81, pp. 187–201. PMLR, New York (2018). http://proceedings.mlr.press/v81/kamishima18a.html
63. Karako, C., Manggala, P.: Using image fairness representations in diversity-based re-ranking for recommendations. In: Adjunct Publication of the 26th Conference on User Modeling, Adaptation and Personalization, pp. 23–28. Association for Computing Machinery, New York. (Jul 2018). https://doi.org/10.1145/3213586.3226206
64. Katzman, J., et al.: Taxonomizing and measuring representational harms: a look at image tagging. In: Proceedings of the AAAI Conference on Artificial Intelligence, vol. 37(12), pp. 14277–14285 (Jun 2023). https://doi.org/10.1609/aaai.v37i12.26670, ISSN 2374–3468
65. Landoni, M., Matteri, D., Murgia, E., Huibers, T., Pera, M.S.: Sonny, cerca! evaluating the impact of using a vocal assistant to search at school. In: Crestani, F., et al. (eds.) CLEF 2019. LNCS, vol. 11696, pp. 101–113. Springer, Cham (2019). https://doi.org/10.1007/978-3-030-28577-7_6
66. Lawrence, E.E.: On the problem of oppressive tastes in the public library. J. Documentation **76**(5), 1091–1107 (2020). https://doi.org/10.1108/JD-01-2020-0002, ISSN 0022–0418

67. Leonhardt, J., Anand, A., Khosla, M.: User fairness in recommender systems. In: Companion Proceedings of the The Web Conference 2018, pp. 101–102, International World Wide Web Conferences Steering Committee, Republic and Canton of Geneva, CHE (Apr 2018). https://doi.org/10.1145/3184558.3186949
68. Li, Y., Chen, H., Fu, Z., Ge, Y., Zhang, Y.: User-oriented fairness in recommendation. In: Proceedings of the Web Conference 2021, pp. 624–632. Association for Computing Machinery, New York (Apr 2021). https://doi.org/10.1145/3442381.3449866
69. Madrazo Azpiazu, I., Dragovic, N., Anuyah, O., Pera, M.S.: Looking for the movie seven or Sven from the movie Frozen? A multi-perspective strategy for recommending queries for children. In: Proceedings of the 2018 Conference on Human Information Interaction & Retrieval, pp. 92–101. Association for Computing Machinery, New York (Mar 2018). https://doi.org/10.1145/3176349.3176379
70. Mauro, N., Ardissono, L., Cena, F.: Personalized recommendation of PoIs to people with autism. In: Proceedings of the 28th ACM Conference on User Modeling, Adaptation and Personalization, pp. 163–172. Association for Computing Machinery, New York (Jul 2020). https://doi.org/10.1145/3340631.3394845
71. McDonald, G., Macdonald, C., Ounis, I.: Search results diversification for effective fair ranking in academic search. Inform. Retrieval J. **25**(1), 1–26 (2022). https://doi.org/10.1007/s10791-021-09399-z, ISSN 1573–7659
72. McNealy, J., Cramer, H.: Trust and representation in recommender systems. In: ICA 2022 (2022)
73. Mehrotra, R., Anderson, A., Diaz, F., Sharma, A., Wallach, H., Yilmaz, E.: Auditing search engines for differential satisfaction across demographics. In: Proceedings of the 26th International Conference on World Wide Web Companion, pp. 626–633, International World Wide Web Conferences Steering Committee (2017). https://doi.org/10.1145/3041021.3054197
74. Mehrotra, R., Carterette, B., Li, Y., Yao, Q., Gao, C., Kwok, J., Yang, Q., Guyon, I.: Advances in recommender systems: from multi-stakeholder marketplaces to automated RecSys. In: Proceedings of the 26th ACM SIGKDD International Conference on Knowledge Discovery & Data Mining, pp. 3533–3534. Association for Computing Machinery, New York (Aug 2020). https://doi.org/10.1145/3394486.3406463, ISBN 978-1-4503-7998-4
75. Michiels, L., Leysen, J., Smets, A., Goethals, B.: What are filter bubbles really? A review of the conceptual and empirical work. In: Adjunct Proceedings of the 30th ACM Conference on User Modeling, Adaptation and Personalization, UMAP 2022 Adjunct, pp. 274–279. Association for Computing Machinery, New York (Jul 2022). https://doi.org/10.1145/3511047.3538028, ISBN 978-1-4503-9232-7
76. Milton, A., Pera, M.S.: Into the unknown: exploration of search engines' responses to users with depression and anxiety. ACM Trans. Web **17**(4), 25:1–25:29 (Jul 2023). https://doi.org/10.1145/3580283, ISSN 1559–1131
77. Moore, M., Bias, R.G., Prentice, K., Fletcher, R., Vaughn, T.: Web usability testing with a Hispanic medically underserved population. J. Med. Library Assoc. JMLA **97**(2), 114–121 (Apr 2009). https://doi.org/10.3163/1536-5050.97.2.008, ISSN 1536–5050, 1558–9439
78. Mosseri, A.: Instagram ranking explained (May 2023). https://about.instagram.com/blog/announcements/instagram-ranking-explained/
79. Murgia, E., Abbasiantaeb, Z., Aliannejadi, M., Huibers, T., Landoni, M., Pera, M.S.: ChatGPT in the classroom: A preliminary exploration on the feasibility of adapting ChatGPT to support children's information discovery. In: Adjunct

Proceedings of the 31st ACM Conference on User Modeling, Adaptation and Personalization, pp. 22–27. Association for Computing Machinery, New York (Jun 2023). https://doi.org/10.1145/3563359.3597399, ISBN 978-1-4503-9891-6
80. Murgia, E., Landoni, M., Huibers, T., Fails, J.A., Pera, M.S.: The seven layers of complexity of recommender systems for children in educational contexts. In: Proceedings of the Workshop on Recommendation in Complex Scenarios Co-Located with 13th ACM Conference on Recommender Systems, vol. 2449. CEUR-WS (Sep 2019). http://ceur-ws.org/Vol-2449/paper1.pdf
81. Naghiaei, M., Rahmani, H.A., Deldjoo, Y.: CPFair: personalized consumer and producer fairness re-ranking for recommender systems. In: Proceedings of the 45th International ACM SIGIR Conference on Research and Development in Information Retrieval, pp. 770–779, Association for Computing Machinery, New York (Jul 2022). https://doi.org/10.1145/3477495.3531959
82. Netflix: Children & family movies (August 2022). https://www.netflix.com/browse/genre/783
83. Ng, Y.K., Pera, M.S.: Recommending social-interactive games for adults with autism spectrum disorders (ASD). In: Proceedings of the 12th ACM Conference on Recommender Systems, pp. 209–213. Association for Computing Machinery, New York (Sep 2018). https://doi.org/10.1145/3240323.3240405
84. Odinma, A.C., Butakov, S., Grakhov, E.: Improving the browsing experience in a bandwidth limited environment through traffic management. Inform. Technol. Developm. **17**(4), 306–318 (2011). https://doi.org/10.1080/02681102.2011.568224, ISSN 0268–1102
85. Olteanu, A., Diaz, F., Kazai, G.: When are search completion suggestions problematic? Proc. ACM Hum.-Comput. Interact. **4**(CSCW2), 171:1–171:25 (2020). https://doi.org/10.1145/3415242
86. Pardes, A.: Feeling stressed out? Pinterest wants to help. Wired (Jul 2019). https://www.wired.com/story/pinterest-compassionate-search/
87. Pariser, E.: The Filter Bubble: How the New Personalized Web Is Changing What We Read and How We Think. Penguin (May 2011), ISBN 978-1-101-51512-9
88. Partarakis, N., Doulgeraki, C., Leonidis, A., Antona, M., Stephanidis, C.: User interface adaptation of web-based services on the semantic web. In: Stephanidis, C. (ed.) UAHCI 2009. LNCS, vol. 5615, pp. 711–719. Springer, Heidelberg (2009). https://doi.org/10.1007/978-3-642-02710-9_79
89. Patton, D.U.: Social work thinking for UX and AI design. Interactions **27**(2), 86–89 (2020). https://doi.org/10.1145/3380535, ISSN 1072–5520
90. Pera, M.S., Murgia, E., Landoni, M., Huibers, T.: With a little help from my friends: Use of recommendations at school. In: Proceedings of ACM RecSys 2019 Late-breaking Results, CEUR-WS, vol. 2431. CEUR (2019). http://ceur-ws.org/Vol-2431/paper13.pdf
91. Pera, M.S., Ng, Y.K.: Automating readers' advisory to make book recommendations for K-12 readers. In: Proceedings of the 8th ACM Conference on Recommender Systems, pp. 9–16. Association for Computing Machinery, New York (2014). https://doi.org/10.1145/2645710.2645721
92. Pinterest: Pinterest introduces first-of-its-kind hair pattern search for inclusive beauty results (Aug 2021). https://newsroom.pinterest.com/en-gb/post/pinterest-introduces-first-of-its-kind-hair-pattern-search-for-inclusive-beauty-results
93. Qin, X., Dou, Z., Wen, J.R.: Diversifying search results using self-attention network. In: Proceedings of the 29th ACM International Conference on Information

& Knowledge Management, CIKM 2020, pp. 1265–1274. Association for Computing Machinery, New York (Oct 2020). https://doi.org/10.1145/3340531.3411914, ISBN 978-1-4503-6859-9
94. Raj, A., Ekstrand, M.D.: Fire dragon and unicorn princess: gender stereotypes and children's products in search engine responses. In: Proceedings of the 2022 SIGIR Workshop On eCommerce (Jun 2022). http://arxiv.org/abs/2206.13747
95. Rakova, B., Yang, J., Cramer, H., Chowdhury, R.: Where responsible AI meets reality: Practitioner perspectives on enablers for shifting organizational practices (Jun 2020). http://arxiv.org/abs/2006.12358
96. Rastegarpanah, B., Gummadi, K.P., Crovella, M.: Fighting fire with fire: Using antidote data to improve polarization and fairness of recommender systems. In: Proceedings of the Twelfth ACM International Conference on Web Search and Data Mining, pp. 231–239. ACM (Jan 2019). https://doi.org/10.1145/3289600.3291002
97. Riedl, J., Konstan, J.: Word of Mouse. Warner Books (2002), ISBN 978-0-446-53003-3
98. Rodriguez, M., Posse, C., Zhang, E.: Multiple objective optimization in recommender systems. In: Proceedings of the Sixth ACM Conference on Recommender Systems, pp. 11–18. Association for Computing Machinery, New York (Sep 2012). https://doi.org/10.1145/2365952.2365961, ISBN 978-1-4503-1270-7
99. Rothschild, M., Horiuchi, T., Maxey, M.: Evaluating "just right" in EdTech recommendation. In: KidRec '19: Workshop in International and Interdisciplinary Perspectives on Children & Recommender and Information Retrieval Systems, Co-located with ACM IDC (2019). https://kidrec.github.io/papers/KidRec_2019_paper_6.pdf
100. Shrestha, A., Spezzano, F., Pera, M.S.: An empirical analysis of collaborative recommender systems robustness to shilling attacks. In: Proceedings of the Second Workshop on Online Misinformation- and Harm-Aware Recommender Systems Co-Located with RecSys 2021, CEUR-WS, vol. 3012, pp. 45–57. CEUR (Oct 2021). http://ceur-ws.org/Vol-3012/OHARS2021-paper4.pdf
101. Smith, J.J., Beattie, L., Cramer, H.: Scoping fairness objectives and identifying fairness metrics for recommender systems: The practitioners' perspective. In: Proceedings of the ACM Web Conference 2023, pp. 3648–3659, Association for Computing Machinery, New York (Apr 2023). https://doi.org/10.1145/3543507.3583204, ISBN 978-1-4503-9416-1
102. Sofaer, S.: Qualitative methods: what are they and why use them? Health Serv. Res. **34**(5 Pt 2), 1101–1118 (1999). https://pubmed.ncbi.nlm.nih.gov/10591275/, ISSN 0017-9124
103. Sonboli, N., Burke, R., Ekstrand, M., Mehrotra, R.: The multisided complexity of fairness in recommender systems. AI Mag. **43**(2), 164–176 (2022). https://doi.org/10.1002/aaai.12054, ISSN 0738-4602
104. Steck, H.: Calibrated recommendations. In: Proceedings of the 12th ACM Conference on Recommender Systems, pp. 154–162. Association for Computing Machinery (Sep 2018). https://doi.org/10.1145/3240323.3240372
105. Trust & Safety Professionals Association: Abuse types (Jun 2021). https://www.tspa.org/curriculum/ts-fundamentals/policy/abuse-types/
106. Tsiakas, K., Barakova, E., Khan, J.V., Markopoulos, P.: BrainHood: towards an explainable recommendation system for self-regulated cognitive training in children. In: Proceedings of the 13th ACM International Conference on PErvasive Technologies Related to Assistive Environments, pp. 1–6. Association for

Computing Machinery, New York (Jun 2020). https://doi.org/10.1145/3389189.3398004

107. Tsioutsiouliklis, S., Pitoura, E., Semertzidis, K., Tsaparas, P.: Link recommendations for PageRank fairness. In: Proceedings of the ACM Web Conference 2022, pp. 3541–3551. Association for Computing Machinery, New York (Apr 2022). https://doi.org/10.1145/3485447.3512249, ISBN 978-1-4503-9096-5
108. Valentim, I., Lourenço, N., Antunes, N.: The impact of data preparation on the fairness of software systems. In: 2019 IEEE 30th International Symposium on Software Reliability Engineering (ISSRE), pp. 391–401 (Oct 2019). https://doi.org/10.1109/ISSRE.2019.00046, ISSN 2332–6549
109. van Alstyne, M., Brynjolfsson, E.: Global village or cyber-balkans? Modeling and measuring the integration of electronic communities. Manag. Sci. **51**(6), 851–868 (2005). https://doi.org/10.1287/mnsc.1050.0363, ISSN 0025–1909
110. Vrijenhoek, S., Kaya, M., Metoui, N., Möller, J., Odijk, D., Helberger, N.: Recommenders with a mission: Assessing diversity in news recommendations. In: Proceedings of the 2021 Conference on Human Information Interaction and Retrieval, pp. 173–183. Association for Computing Machinery, New York (Mar 2021). https://doi.org/10.1145/3406522.3446019, ISBN 978-1-4503-8055-3
111. Wang, L., Joachims, T.: User fairness, item fairness, and diversity for rankings in two-sided markets. In: Proceedings of the 2021 ACM SIGIR International Conference on Theory of Information Retrieval, pp. 23–41. Association for Computing Machinery, New York (Jul 2021). https://doi.org/10.1145/3471158.3472260
112. Wortman Vaughan, J.: Transparency and intelligibility throughout the machine learning life cycle (Jan 2020). https://www.microsoft.com/en-us/research/video/transparency-and-intelligibility-throughout-the-machine-learning-life-cycle/
113. Wu, H., Mitra, B., Ma, C., Diaz, F., Liu, X.: Joint multisided exposure fairness for recommendation. In: Proceedings of the 45th International ACM SIGIR Conference on Research and Development in Information Retrieval, pp. 703–714. ACM (Jul 2022), https://doi.org/10.1145/3477495.3532007
114. Yang, Y., Zhang, C., Fan, C., Mostafavi, A., Hu, X.: Towards fairness-aware disaster informatics: An interdisciplinary perspective. IEEE Access **8**, 201040–201054 (2020). https://doi.org/10.1109/ACCESS.2020.3035714, ISSN 2169–3536
115. Yao, S., Huang, B.: Beyond parity: Fairness objectives for collaborative filtering. In: Guyon, I., Luxburg, U.V., Bengio, S., Wallach, H., Fergus, R., Vishwanathan, S., Garnett, R. (eds.) Advances in Neural Information Processing Systems 30, pp. 2925–2934. Curran Associates, Inc. (2017). http://papers.nips.cc/paper/6885-beyond-parity-fairness-objectives-for-collaborative-filtering.pdf

A Study of Pre-processing Fairness Intervention Methods for Ranking People

Clara Rus(✉), Andrew Yates, and Maarten de Rijke

University of Amsterdam, Amsterdam, The Netherlands
{c.a.rus,a.c.yates,m.derijke}@uva.nl

Abstract. Fairness interventions are hard to use in practice when ranking people due to legal constraints that limit access to sensitive information. Pre-processing fairness interventions, however, can be used in practice to create more fair training data that encourage the model to generate fair predictions without having access to sensitive information during inference. Little is known about the performance of pre-processing fairness interventions in a recruitment setting. To simulate a real scenario, we train a ranking model on pre-processed representations, while access to sensitive information is limited during inference. We evaluate pre-processing fairness intervention methods in terms of individual fairness and group fairness. On two real-world datasets, the pre-processing methods are found to improve the diversity of rankings with respect to gender, while individual fairness is not affected. Moreover, we discuss advantages and disadvantages of using pre-processing fairness interventions in practice for ranking people.

Keywords: Fairness · Ranking · Recruitment

1 Introduction

A ranking system's goal is to create an ordered list of items having relevant items at top positions given the search terms and the user's preferences. Applications of ranking systems include not only ranking of items such as digital artefacts like documents, books, or movies, or real-world entities like hotels, but also *ranking of people*. Increasingly, recruiters rely on automatic tools to process the large amount of applications received for a vacancy [10].

Fairness in Ranking. Ensuring fairness in a ranking system is especially important when the items to be ranked are real persons, whose lives could be impacted by the system's decisions. When a ranking system is designed for recruitment, its outcomes influence who receives more interviews, the job's quality, and even the wage [2]. Systems for ranking people in a recruitment setting encode stereotypes and biases that already exist in recruitment [7,11,20,25], leading to actions that discriminate against minority groups [3,13,24].

There are several fairness interventions that can be used to improve the fairness of a ranking [32,33]. However, due to legal constraints, their use in a practical application of ranking people can be limited. Briefly, most fairness interventions require

N. Goharian et al. (Eds.): ECIR 2024, LNCS 14611, pp. 336–350, 2024.
https://doi.org/10.1007/978-3-031-56066-8_26

access to the sensitive information of the candidates, but the European General Data Protection Regulation (GDPR) [1] states that the processing of a special category of sensitive information is prohibited, with exceptions for gender and age [4]. While many fairness intervention methods require information about these sensitive categories, *pre-processing intervention methods* do not suffer from this limitation. Pre-processing fairness intervention methods aim at debiasing the data representation of the candidates, encouraging the (downstream) ranking model to generate fair predictions without having access to sensitive information at inference time. Similarly, in-processing methods can be trained offline and used during inference time without access to sensitive information. However, they require a definition of fairness to optimize for, which in practice might be hard to define as the European law does not give exact guidelines about the minimum requirements as opposed to the US ("80% rule").

Comparing Pre-processing Fairness Intervention Methods. Pre-processing fairness intervention methods can be applied offline on training data containing sensitive information, which was acquired in compliance with the GDPR [4]. In practice, a model trained on fair data representation can be applied on the real candidate data without having access to the sensitive information of the candidates. However, there is little knowledge about the performance of such methods with respect to group fairness, individual fairness, utility, and how the methods compare to each other when ranking people in a recruitment setting. We address this knowledge gap. We consider a scenario where access to sensitive attributes during inference time is limited, and compare three pre-processing fairness intervention methods: (i) CIF-Rank [27] aims at achieving group fairness, (ii) LFR [34] aims at achieving both group and individual fairness, while (iii) iFair [19] aims at achieving individual fairness. Neither LFR nor iFair require access to sensitive information during inference time, however, CIF-Rank does.

Fairness intervention methods are typically evaluated w.r.t. both group fairness and individual fairness, while considering the utility of the ranking. Here, *group fairness* means that members of different protected groups should be treated the same. *Individual fairness* means that similar individuals should be treated similarly, based on a defined similarity metric. In recruitment, the primary focus is on group fairness, but due to limited access to sensitive information, it may be more feasible to focus on individual fairness. Indeed, striving for individual fairness could contribute to group fairness. This works under the assumption that the measure of similarity between candidates is computed free of bias, which in practice might not be the case, resulting in segregation by membership in the sensitive groups.

Aim and Summary. Our aim in this paper is to investigate (i) how well pre-processing interventions generalize on two datasets for ranking people given an occupation, (ii) what the trade-off is between group and individual fairness obtained by the pre-processing interventions, and (iii) which method is more suitable to be used in practice. Our main finding is that the pre-processing methods do obtain an increase in diversity in most occupations on both datasets, without affecting individual fairness. Finally, we discuss the trade-offs using pre-processing interventions for ranking people in practice. CIF-Rank offers transparency and in-group fairness with minimal changes to candidate data, but it requires access to sensitive information during inference time. In contrast, LFR and iFair treat fairness as an optimization problem, thus, there is no need for access to sensitive information and the changes in diversity are more noticeable, though they are less explainable and iFair does not guarantee group fairness.

2 Related Work

Fairness interventions can be applied to prevent ranking systems from exacerbating discriminatory behavior towards minority groups. Fairness interventions can be categorized as *pre-processing*, *in-processing*, and *post-processing* methods [32].

Pre-processing methods [19,27,34] aim to debias the data used to represent the candidates and then either re-rank the candidates based on the new representations or use the data to train a model. CIF-Rank [27] aims to achieve group fairness by creating counterfactual representations of the candidates. LFR [34] aims at achieving both group and individual fairness by creating representations that obfuscate information about the protected groups, while also ensuring a good encoding of useful information. iFair [19] aims at achieving individual fairness by creating representations that encourage similar outcomes for similar individuals regardless of the sensitive attributes.

In-processing methods [5,15,30] aim to optimize the ranking system for both fairness and utility. For example, Zehlike and Castillo [30] propose to optimize the model for both utility of the ranking and fairness of exposure by ensuring that the groups have the same probability of being placed at the top position of the ranking.

Post-processing methods [6,24,26,29,31] adjust the recommended ranking based on some minimum and maximum constraints regarding the desired proportion of each sensitive group in the top positions. For example, FA*IR [29] creates separate ranked lists for each protected group and merges them such that a minimum number of candidates from the disadvantaged group is present in the top of the list. The work of [6] focuses on achieving individual fairness over time by re-ordering the items such that the unfairness is minimized over time in an online setting. The unfairness is measured as the cumulative difference between the attention an item receives and the relevance.

Previous work [27,34] experimented with pre-processing methods on various datasets, including recruitment datasets in some cases [19,23]. In this paper, we go beyond prior work by performing a systematic comparison of the above mentioned pre-processing methods in a ranking setting, showing how they generalize on two recruitment datasets, while considering that the bias direction may vary between occupations, w.r.t. group fairness, individual fairness, utility of the ranking, and their use in practice. Moreover, we adapt the work of [34] for a ranking setting.

3 Fairness Interventions

The use of fairness interventions for ranking people is constrained by the GDPR [1], which prohibits processing sensitive information that many fairness interventions require. We argue that pre-processing interventions can be used in practice as they can be applied offline on the training data containing sensitive information that has been acquired in compliance with the GDPR (e.g., a synthetic anonymous dataset reflecting a real distribution) [4]. Then, a ranking model is trained on the fair data, encouraging the model to generate a fair ranking without having access to sensitive information of

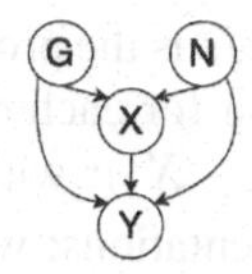

Fig. 1. Causal model describing the data with sensitive attributes gender (G) and nationality (N), non-sensitive attributes (X), and utility scores (Y).

the real candidates at inference time. As pre-processing methods do not guarantee the diversity of the ranking, we compare them with FA*IR, a post-processing method that does guarantee the desired proportion of disadvantaged candidates in the top positions. Below, we describe the fairness interventions we compare in this work.

CIF-Rank [27] estimates "How would this candidate data look like if they were part of another protected group?" As input it uses a reference protected group towards which we want to transform the candidates in a counterfactual world, and a causal model describing the data and the effects of the sensitive attributes on the data. A causal graph is a directed acyclic graph (DAG) where nodes represent variables, and directed edges between nodes represent causal relationships. A directed edge from node *A* to node *B* indicates that variable *A* causally influences variable *B*. Figure 1 shows a possible causal model that can be used to represent the data, having the following nodes: sensitive attributes, G (gender) and N (nationality), non-sensitive attributes of the candidates (X), and the utility score used to rank the candidates for a given occupation (Y), with edges from the features to the scores, and from the sensitive attributes to the features and the scores of the candidates. It estimates the total effect composed of the direct effect and the indirect effect. The indirect effect is the effect mediated by the non-sensitive attributes, which are called mediators.

The causal effects are estimated using the mma R package [28], which performs mediation analysis with multiple mediators. After computing the causal effects of the sensitive attributes on the data, which represent the bias encoded in the data, one can compute the counterfactual representations. These are computed by changing the observed representations by boosting them up or down according to the difference in the total effect of the actual group and the control group, the group to which we want to transform the data. This method satisfies transparency requirements as by using the causal estimates to generate the counterfactual representations we have a way to explain how changes in the representations were produced and why.

LFR [34] formulates fairness as an optimization problem of finding a good representation of the data such that the sensitive information encoded in the data is obfuscated, while also encoding useful information. This method can achieve both group and individual fairness. The authors formulate the new representation in terms of a probabilistic mapping to a set of prototypes (points in the input space). Thus, the model can be defined as a discriminative clustering model, where the prototypes act as the clusters as each representation is assigned to a prototype. In order to ensure that the information regarding the membership in a protected group is lost, the authors make use of statistical parity by ensuring that a random candidate from group *A* should map to a particular prototype with an equal probability as a random candidate from group *B* to the same prototype: $L_z = |\sum_k^K M_k^A - M_k^B|$, where M_k is the probability that X maps to prototype v_k. In order to keep the useful information for each candidate, the representations should be close to the original ones: $L_x = (X - X')^2$ with X being the original representation and $X' = \sum_k^K M_k v_k$ the new representations; while also ensuring that the representations are still predictive of the label (Y): $L_y = -Y \log(Y') - (1 - Y) \log(1 - Y')$, where Y is the label and $Y' = \sum_k^K M_k w_k$ is the prediction. As LFR was designed for a binary classification task, we adapt it to a ranking scenario by considering Y to be a binary label indicating the positive and negative samples used for training the ranking model. In the end, the model optimizes the compound loss $L = A_z L_z + A_x L_x + A_y L_y$ by learning two sets of parameters: the prototype locations v_k and the parameters w_k that govern the mapping from the prototypes to classification decisions Y.

iFair [19] aims at achieving individual fairness, meaning that candidates who are similar in all task-relevant attributes such as job qualification, and disregarding all potentially discriminating attributes such as gender, should have similar outcomes. iFair's main idea is constructed on top of LFR [34]. However, in contrast to LFR, iFair directly optimizes for individual fairness, implicitly obtaining group fairness by obfuscating the information regarding the membership of protected groups. To directly optimize for individual fairness the authors propose the following loss: $L_z = \sum_{i,j}^M (d(X_i', X_j') - d(X_i^*, X_j^*))^2$, where X' is the new representation, X^* the original representation excluding the sensitive attributes, and $d(X_i, X_j) = [\sum_n^N (\alpha_n (x_i, n - x_j, n)^p]^{1/p}$ is a distance function. M is the number of candidates in the data, N is the number of features that each candidate has and α_n is a learnable weight for different data attributes. Ideally, the learnable weights for the sensitive attributes should be near zero. Intuitively this ensures that similar individuals in the original input space, excluding the sensitive attributes, should have similar low-rank representations, regardless of the membership in the protected groups. Thus, the distance between individuals should be preserved in the transformed space. This directly optimizes for individual fairness, while also obfuscating the membership in protected groups. To ensure the utility of the representations, the representations should still be close to the original ones: $L_x = (X - X')^2$, with X containing both protected and non-protected features. The new representations are computed in the same manner as LFR proposes: $X' = \sum_k^K M_k v_k$. In the end, the model optimizes the compound loss $L = A_z L_z + A_x L_x$ by learning two sets of parameters: the prototype locations v_k and the parameters α_n that govern the mapping from the prototypes to classification decisions Y.

FA*IR [29] is a post-processing intervention that aims to ensure that the proportion of candidates from the disadvantaged group remains statistically above or indistinguishable from a given minimum, at every top-k. To ensure the utility of the ranking, candidates included in the top-k should be more qualified than every candidate not included, and for every pair of candidates in the top-k, the more qualified candidate should be ranked above. It pre-computes a table having fairly represent the protected group with minimal proportion p and significance α at various top-k. Next, the algorithm creates a queue for the privileged group ($P0$) and a queue for the disadvantaged group ($P1$). If the table demands a disadvantaged candidate at the current position, the algorithm extracts the best candidate from queue $P1$ and adds it to the final ranking, otherwise it extracts the best candidate from $P0 \cup P1$.

4 Experimental Setup

We aim to answer three research questions: **(RQ1)** How well do the pre-processing fairness interventions listed in Sect. 3 generalize on datasets for ranking people? **(RQ2)** What is the trade-off between group and individual fairness obtained by the pre-processing interventions? **(RQ3)** Which preprocessing intervention method is more suitable to be used in practice?

Datasets. The *BIOS dataset* [14] consists of real biographies collected from the web by filtering for lines starting with a name followed by the string "is a(n) (xxx) title," where "title" is an occupation from the BLS SOC system.[1] In total there are 28 occupations. Each candidate is represented by non-sensitive features extracted from the text biography (term frequency of the occupation in the biography, length and number of words of the biography) and sensitive features (gender, provided by the dataset). To simulate bias in the ranking, candidates are ranked for each occupation by the cosine similarity between word2vec [21] embeddings of the occupation title and the text biography, as word2vec embeddings are known to create stereotypical associations [8,18].

The *XING dataset* was collected to study gender biases in the returned ranked search results given each user's profile details [29]. The dataset consists of anonymized real user profiles collected from XING[2] in response to 57 queries representing job titles. For each query the first 40 user profiles were selected. For each user profile the following non-sensitive information was gathered: duration of job experiences, duration of education and profile popularity, and sensitive information, gender (provided by the dataset, inferred from name and profile picture when available). The score to rank the candidates is computed as provided by the original implementation,[3] based on educational features, job experience and hits on profile.

Parameters and Settings. We model a scenario where experienced candidates apply for jobs in the same field. Train-test splits (30% BIOS and 40% XING for test) are stratified across sensitive groups, ensuring five consistent splits per occupation. For XING occupations fully over-represented by one gender are excluded, as no improvements in diversity can be measured, resulting in 44 occupations. Sensitive attributes are not used as training features.

As a ranker, we use the Ranklib [12] implementation of RankNet [9], a pairwise learning to rank algorithm as it can better capture the changes in ranks produced by the pre-processing methods [23]. Relevance judgments are based on candidate scores, with negative/positive samples determined using a threshold (0.4 BIOS and 0 XING).

We apply pre-processing methods separately for each occupation to account for varying bias directions [23]. We experimented with $A_x, A_y, A_z \in \{0.1, 0.01, 1, 5, 10\}$ with $K = 10$, selecting $A_x = 0.1$, for LFR $A_y = 1$, $A_z = 1$, while for iFair $A_z = 5$, as it obtained the best trade-off between group/individual fairness and utility.

The rankings are evaluated in terms of group fairness, individual fairness, and utility. *Group fairness* is evaluated as the percentage of each sensitive group among the top 10.

[1] https://www.bls.gov/soc/.
[2] https://www.xing.com/.
[3] https://github.com/MilkaLichtblau/xing_dataset/.

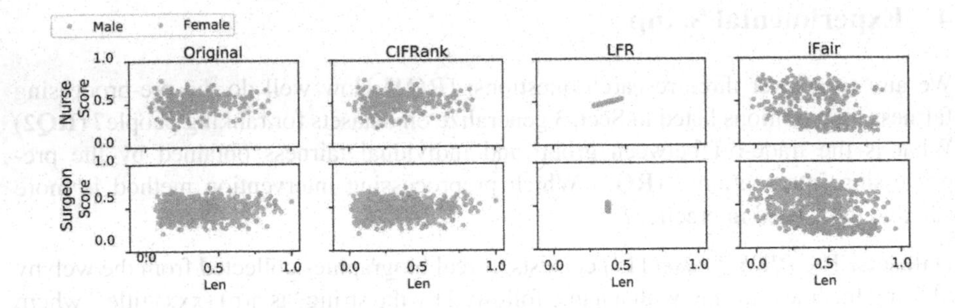

Fig. 2. Example of data points for the BIOS dataset.

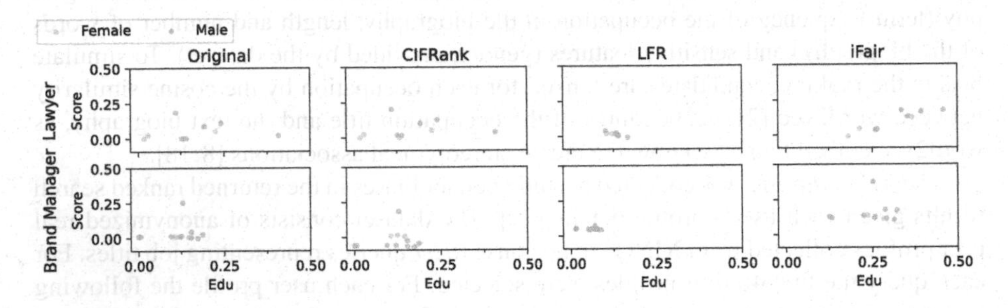

Fig. 3. Example of data points for the XING dataset.

Our aim is to create a diverse ranked list of the candidates with respect to the sensitive groups by increasing the proportion of the underrepresented groups, without producing a swap between the underrepresented and over-represented group. *Individual fairness* is evaluated by doing a pairwise comparison between candidates distance in the features space and their achieved exposure [16]: $d(E_i, E_j) \leq d(X_i, X_j)$. The difference in the candidates exposure (E) should be as close as possible to the distance between the candidates in the feature space. The closer the value to 1 the better. *Utility* of the ranking is measured using normalized discounted cumulative gain across the top 10 (NDCG@10). We focus on the top 10 since it is unlikely that recruiters will scroll down to view more candidates [17,22]. To support reproducibility of our study we provide the code to run the experiments on GitHub.[4]

5 Results and Discussion

In this section we describe our results obtained on the BIOS and XING dataset when training a ranking model on the fairer data generated by the pre-processing interventions. Reported results are an average over the five runs. The test set is pre-processed in experiments involving LFR and iFair as they do not require access to sensitive information. In contrast, for CIF-Rank, which requires access to sensitive information during

[4] https://github.com/ClaraRus/A-Study-of-Pre-processing-Fairness-Intervention-Methods-for-Ranking-People.

inference, the test set is not subjected to pre-processing. The pre-processed test set has the same candidates, but their representations differs.

Table 1. Evaluation of the ranking in the top 10 in terms of utility (NDCG@10), group fairness (%protected@10 gender), and individual fairness (yNN).

Dataset	Method	NDCG@10	%protected(G)@10	yNN
XING	Original	0.57	30	0.85
	CIF-Rank	0.79	32	0.86
	LFR	0.70	32	0.85
	iFair	0.97	30	0.85
	FA*IR ($p = 0.5$)	0.98	38	0.85
	FA*IR ($p = 0.9$)	0.93	38	0.85
BIOS	Original	0.86	20	0.72
	CIF-Rank	0.93	22	0.72
	LFR	0.77	26	0.72
	iFair	0.52	27	0.72
	FA*IR ($p = 0.5$)	0.98	33	0.72
	FA*IR ($p = 0.9$)	0.90	77	0.72

5.1 Examining the Effect of Interventions on Individual Data Points

Before addressing our research questions, we examine the effect of interventions on individual data points to better understand the methods we consider.

Figure 2 and 3 show how the data points representing the candidates change when applying the pre-processing methods on the BIOS and XING dataset for two example occupations. The data points are plotted in a space defined by the score of the candidate and one feature. Similar patterns can be observed between the two datasets. Looking at the data points generated by CIF-Rank, one can observe that the representations are slightly changed by the interventions. For Surgeon (Fig. 2) and Brand Manager (Fig. 3), which are male dominated occupations, there is a slight decrease for the males with respect to score, however, this does not create a positive change in the diversity of the generated ranking, as the change in rank is small, thus, the model is not strongly penalized. For Nurse (Fig. 2), a female-dominated occupation, the score of the males is slightly increased, and we also observe an increase in the diversity of the generated ranking. For Lawyer (Fig. 3), the score of the over-represented group (males) is increased, due to a difference in bias direction between the test and train set. For LFR and iFair, the changes in the representations of the candidates are more noticeable. Specifically, LFR tends to produce representations that cluster near similar values in the space. It seems that for the BIOS dataset the candidates are clustered near the threshold between positive and negative candidates. In contrast, iFair tends to scatter the candidates across the space, creating a clear distance between positive and negative candidates. For Brand Manager (Fig. 3) each female candidate is close in space to a

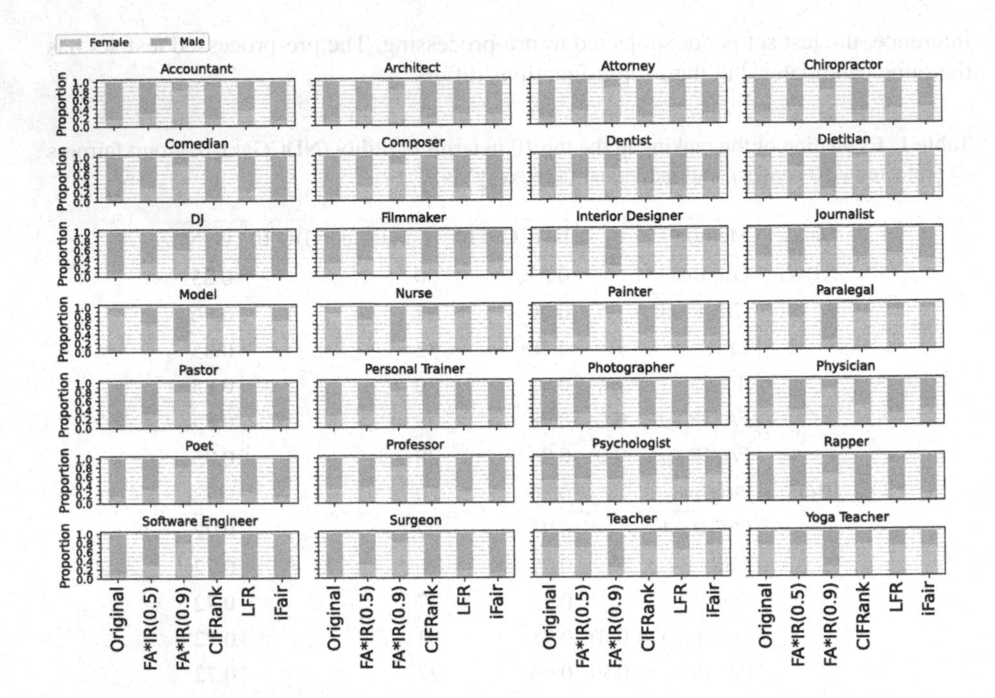

Fig. 4. Distribution of the gender groups in the top 10 for the BIOS dataset.

male candidate, resembling the idea that best candidates from the female group should be treated similarly with the best candidates from the male group.

5.2 Group Fairness

Next, we turn to group fairness and the first of our research questions.

BIOS Dataset. Fig. 4 and Table 1 (BIOS) show the proportion of the gender groups in the top 10 when applying pre-processing methods on the BIOS dataset. All methods increase the representation of underrepresented groups in some of the gender-biased occupations. These include occupations like attorney, dentist, pastor and photographer, which are male-dominated occupations, and female-dominated occupations such as model, nurse, and paralegal.

LFR obtains improvements for the underrepresented group in 20 (71%) occupations with an average increase in proportion of 6%, followed by iFair in 17 (60%) occupations with an average increase of 7%, and CIF-Rank in 13 (46%) occupations with an average increase in proportion of 2%. CIF-Rank negatively affects the diversity of the ranking in 8 (28%) occupations, while LFR and iFair do so in 7 (25%) occupations. Using FA*IR, the proportion of the underrepresented group is increased in all occupations. However, when using ($p = 0.9$), it can be observed that the previously underrepresented group is now over-represented.

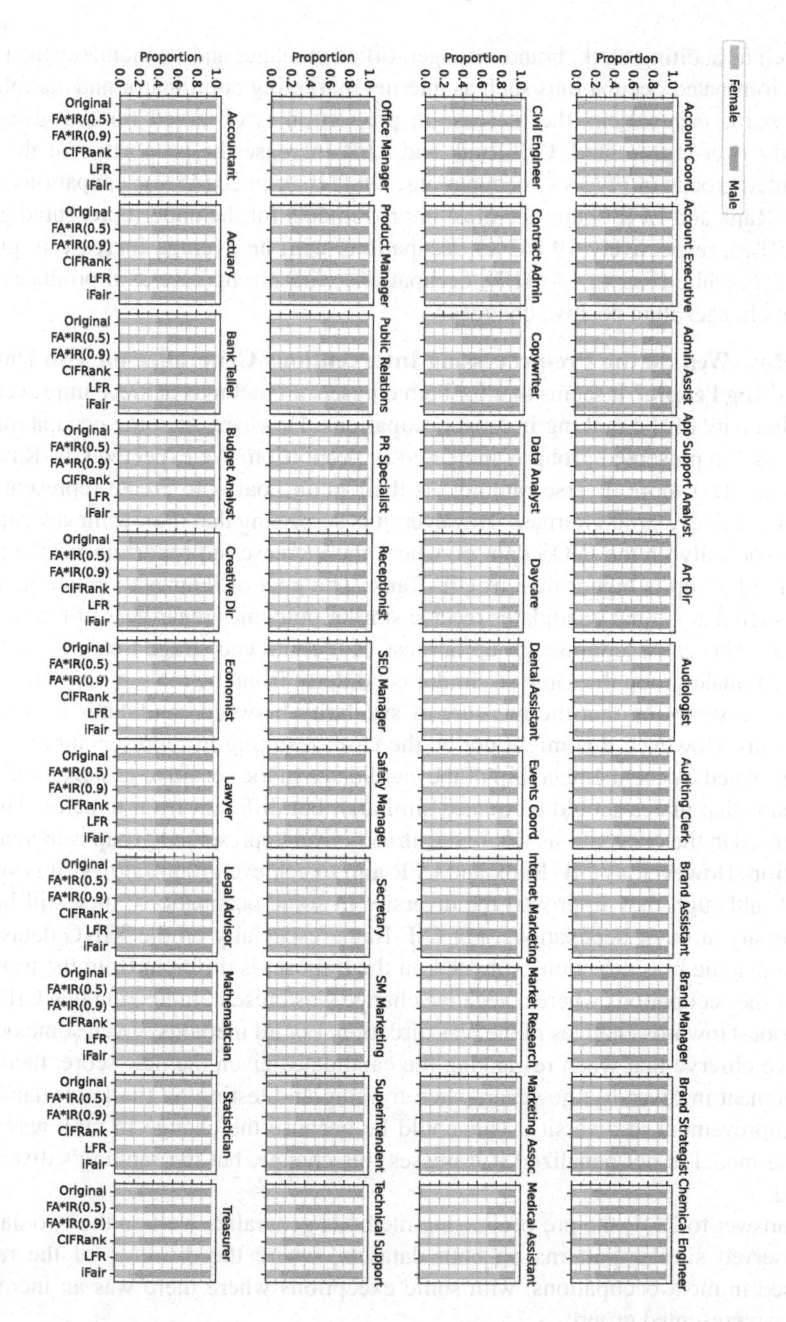

Fig. 5. Distribution of the gender groups in the top 10 for the XING dataset.

XING Dataset. Fig. 5 shows the proportion of the gender groups in the top 10, when applying the pre-processing interventions on the XING dataset. All three methods improved the proportion of the underrepresented group for male-dominated occupations such as auditing clerk, brand manager, office manager and mathematician, and for female-dominated occupations such as internet marketing coordinator and audiologist.

For some occupations the increase in proportion is observed for the group that is already over-represented. CIF-Rank and LFR increase the proportion of the over-represented groups in 17 (38%) occupations, while iFair in 20 (45%) occupations. Overall, CIF-Rank and LFR obtain the most improvements for the underrepresented groups in 22 (50%), respectively 19 (43%) occupations with an average increase in proportion of 2%, while iFair in 14 (31%) occupations with no increase as it produces more negative changes than positive changes.

RQ1: How Well do the Pre-processing Interventions Generalize on Two Datasets for Ranking People? It seems that for both datasets all methods achieve improvements in the diversity of the ranking in most occupations. However, in some occupations the already over-represented group is increased in proportion. We expected CIF-Rank and LFR to create a more diverse ranking, as their main goal is to create representations that aim to achieve group fairness. However, it is surprising that iFair achieves improvements, especially on the BIOS dataset, where the increase in proportion for the underrepresented group is higher than of CIF-Rank. The goal of iFair is to create representations such that similar candidates receive similar outcomes regardless of the sensitive attributes. This can encourage group fairness as it could encourage treating candidates from the female group the same as similar candidates from the male group. This works under the assumption that the measure of similarity between candidates is computed free of bias. However, this might not be the case, resulting in representations that are still segregated by the membership of the candidates in the sensitive groups. Intuitively, this means that males should be treated similarly, but different from females. This can be observed in the occupations where the already over-represented group is increased in proportion. However, as CIF-Rank and LFR aim to achieve group fairness it is surprising that, although they improved the diversity in some occupations, they still harmed the diversity in other occupations. For CIF-Rank, especially on the XING dataset, we observe that the bias direction estimated on the train set is different from the test set in some of the occupations where a negative change was present, indicating that CIF-Rank is not robust towards changes in the bias direction. For all methods, given some occupations, we observe that when re-ranking the candidates given the new score, there is an improvement in diversity; however, when training and testing on the fairer data, there is no improvement in diversity. This could be because the change in rank was small, thus, the model is not penalized if it misses this change, but the ranking's diversity is affected.

In answer to RQ1, the pre-processing methods generalize well on the two datasets. We observed similar patterns on both datasets, where the diversity of the ranking increased in most occupations, with some exceptions where there was an increase in the over-represented group.

5.3 Individual Fairness and Utility

RQ2: What is the Trade-off between Group and Individual Fairness? Individual fairness and group fairness might not involve making trade-offs if the measure of similarity between individuals is computed independent of the sensitive attributes. However, this is hard to achieve in practice resulting in potential trade-offs between the two fairness measures. Table 1 shows how the methods affect the individual fairness of the ranking. Surprisingly, in this setting, neither LFR and iFair show improvements in individual fairness, however, the individual fairness is already reasonably high. CIF-Rank is expected to obtain similar individual fairness to the Original setting, as the representations produced by CIF-Rank are slightly changed w.r.t. the original ones. To measure individual fairness, one needs to define a way to measure the similarity between candidates; this implies defining a similarity metric and the features to use in the similarity metric. Following [19,34], we use the original features of the candidates in the distance function without considering the sensitive information. However, the original features might be a proxy to sensitive information, implying that to satisfy individual fairness it might be that females and males should not be treated the same.

Transforming the representation of the candidates to make them more fair carries the risk of harming the utility of the ranking (compared to the original utility). Considering the NDCG scores in Table 1, we see that, on the BIOS dataset, when pre-processing the data using LFR and iFair, the utility of the ranking decreases, while for CIF-Rank the utility of the ranking is slightly increased. As RankNet is trained and tested on pre-processed data for LFR and iFair, the generated rankings are noticeably different from the original one. For CIF-Rank the changes in the ranking are small, and the ranking model is tested on the original representations, which produces fewer changes. On the XING dataset, the ranking's utility is increased when training on pre-processed data.

In answer to RQ2, when using pre-processing fairness interventions, the individual fairness of the ranking is not affected, thus, we do not observe a trade-off between group and individual fairness.

5.4 Fairness Interventions in Practice

RQ3: Which method is more suitable to be used in practice? CIF-Rank satisfies transparency requirements, as explained in Sect. 3. Additionally, it guarantees in-group fairness by applying uniform changes to all candidates within a group. In contrast, LFR and iFair treat fairness as an optimization problem, making it less straightforward to explain any changes in the candidate representation. As opposed to CIF-Rank, LFR produces more noticeable changes to candidate data, leading to a more significant impact on the diversity of the rankings. In contrast with CIF-Rank, both LFR and iFair offer the advantage of being applicable at inference time without requiring access to sensitive attributes. Even though iFair can increase the diversity of the ranking, it may also exacerbate group segregation as we have seen in Sect. 5.2.

In practice, it is important to evaluate and model fairness with respect to intersectional groups, as candidates belonging to multiple disadvantaged groups are more likely to be discriminated [26]. CIF-Rank offers an intersectional framework by modelling the causal effects of the intersectional sensitive attributes, which was shown to increase the proportion of intersectional groups [23,27]. iFair supports multiple non-binary groups, however, there is no work analyzing its impact on intersectional groups. LFR has been designed for binary groups, thus needing an adaptation for intersectional settings.

In answer to RQ3, pre-processing fairness intervention methods can be used in the practice of ranking people, e.g., for recruitment. Depending on transparency requirements and on fairness goals, preference may be given to CIF-Rank or LFR, as they both aim at achieving group fairness with the former creating less noticeable changes in the representations, but making it easier to explain the changes.

6 Conclusion

In the context of ranking people, e.g., for recruitment, most fairness intervention methods require access to the sensitive information of people, information that should not be processed according to the GDPR, making it hard to use such interventions in practice. In this work we argue that *pre-processing* fairness interventions can be used in practice without having access to sensitive information during inference time. We have compared the performance of three pre-processing methods for the task of ranking people in a recruitment scenario on two real-world datasets with respect to group fairness, individual fairness, and utility of the ranking. Finally, we discuss the advantages and disadvantages of using the pre-processing methods in practice.

We find that for most occupations the methods achieved an improvement in the diversity of the ranking with respect to the protected groups, while the individual fairness was not affected. However, for some occupations the already over-represented group is increased in proportion. Consequently, the methods should be used carefully in order to avoid making a positive change in some occupations while other occupations are negatively affected. Advantages of using CIF-Rank include transparency, guaranteed in-group fairness and an intersectional framework, however it requires access to sensitive information during inference time and the changes to candidate data are minimal. In contrast, LFR and iFair treat fairness as an optimization problem, thus, there is no need for access to sensitive information and the changes in diversity are more noticeable, however they are less explainable and iFair does not guarantee group fairness.

While the BIOS and the XING dataset are real-world datasets, the features used to rank the candidates might not fully reflect a real-world scenario, as recruiters might take different attributes into account. Especially, the BIOS dataset, with extracted textual features and including artistic occupations, might not be mainstream in a hiring process. In addition, our study focused only on gender as a binary variable, as this was the sensitive information available in the datasets. Future work should compare the performance of fairness intervention methods w.r.t. other sensitive groups with non-binary values and towards intersectional groups.

Acknowledgments. We thank our reviewers for valuable feedback. This research was supported by the FINDHR (Fairness and Intersectional Non-Discrimination in Human Recommendation) project that received funding from the European Union's Horizon Europe research and innovation program under grant agreement No 101070212, the Hybrid Intelligence Center, a 10-year program funded by the Dutch Ministry of Education, Culture and Science through the Netherlands Organisation for Scientific Research, https://hybrid-intelligence-centre.nl, and project LESSEN with project number NWA.1389.20.183 of the research program NWA ORC 2020/21, which is (partly) financed by the Dutch Research Council (NWO).

All content represents the opinion of the authors, which is not necessarily shared or endorsed by their respective employers and/or sponsors.

References

1. General data protection regulation (GDPR). https://eur-lex.europa.eu/eli/reg/2016/679/oj (2023), (Accessed 20 Oct 2023)
2. Albert, E.T.: AI in talent acquisition: a review of AI-applications used in recruitment and selection. Strateg. HR Rev. **18**(5), 215–221 (2019)
3. Ali, M., Sapiezynski, P., Bogen, M., Korolova, A., Mislove, A., Rieke, A.: Discrimination through optimization: how Facebook's Ad delivery can lead to biased outcomes. Proc. ACM Hum. Comput. Interact. **3**(CSCW), 1–30 (2019)
4. van Bekkum, M., Borgesius Zuiderveen, F.: Using sensitive data to prevent discrimination by artificial intelligence: does the GDPR need a new exception? Comput. Law Sec. Rev. **48**, 105770 (2023)
5. Beutel, A., et al.: Fairness in recommendation ranking through pairwise comparisons. In: Proceedings of the 25th ACM SIGKDD International Conference on Knowledge Discovery & Data Mining, pp. 2212–2220 (2019)
6. Biega, A.J., Gummadi, K.P., Weikum, G.: Equity of attention: amortizing individual fairness in rankings. In: The 41st international ACM SIGIR Conference on Research & Development in Information Retrieval, pp. 405–414 (2018)
7. Bisschop, P., ter Weel, B., Zwetsloot, J.: Ethnic employment gaps of graduates in the Netherlands. De Economist **168**(4), 577–598 (2020)
8. Bolukbasi, T., Chang, K.W., Zou, J.Y., Saligrama, V., Kalai, A.T.: Man is to computer programmer as woman is to homemaker? Debiasing word embeddings. In: Advances in Neural Information Processing Systems 29 (2016)
9. Burges, C., Shaked, T., Renshaw, E., Lazier, A., Deeds, M., Hamilton, N., Hullender, G.: Learning to rank using gradient descent. In: Proceedings of the 22nd International Conference on Machine Learning, pp. 89–96 (2005)
10. Chapman, D.S., Webster, J.: The use of technologies in the recruiting, screening, and selection processes for job candidates. Int. J. Sel. Assess. **11**(2–3), 113–120 (2003)
11. Ciminelli, G., Schwellnus, C., Stadler, B.: Sticky floors or glass ceilings? The role of human capital, working time flexibility and discrimination in the gender wage gap. Tech. Rep. 1668, OECD Publishing (2021)
12. Dang, V., Ascent, C.: The Lemur project (2023). http://lemurproject.org
13. Dastin, J.: Amazon scraps secret AI recruiting tool that showed bias against women. In: Ethics of Data and Analytics, pp. 296–299. Auerbach Publications (2022)
14. De-Arteaga, M., et al.: Bias in bios: a case study of semantic representation bias in a high-stakes setting. In: Proceedings of the Conference on Fairness, Accountability, and Transparency, pp. 120–128 (2019)
15. Diaz, F., Mitra, B., Ekstrand, M.D., Biega, A.J., Carterette, B.: Evaluating stochastic rankings with expected exposure. In: Proceedings of the 29th ACM International Conference on Information & Knowledge Management, pp. 275–284 (2020)
16. Dwork, C., Hardt, M., Pitassi, T., Reingold, O., Zemel, R.: Fairness through awareness. In: Proceedings of the 3rd Innovations in Theoretical Computer Science Conference, pp. 214–226 (2012)
17. Granka, L.A., Joachims, T., Gay, G.: Eye-tracking analysis of user behavior in WWW search. In: Proceedings of the 27th Annual International ACM SIGIR Conference on Research and Development in Information Retrieval, pp. 478–479 (2004)
18. Kurita, K., Vyas, N., Pareek, A., Black, A.W., Tsvetkov, Y.: Measuring bias in contextualized word representations. arXiv preprint arXiv:1906.07337 (2019)

19. Lahoti, P., Gummadi, K.P., Weikum, G.: iFair: learning individually fair data representations for algorithmic decision making. In: 2019 IEEE 35th International Conference on Data Engineering (ICDE), pp. 1334–1345, IEEE (2019)
20. Matteazzi, E., Pailhé, A., Solaz, A.: Part-time employment, the gender wage gap and the role of wage-setting institutions: evidence from 11 European countries. Eur. J. Ind. Relat. **24**(3), 221–241 (2018)
21. Mikolov, T., Chen, K., Corrado, G., Dean, J.: Efficient estimation of word representations in vector space. In: Proceedings of Workshop at ICLR 2013 (Jan 2013). https://doi.org/10.48550/arXiv.1301.3781
22. O'Brien, M., Keane, M.T.: Modeling result-list searching in the world wide web: the role of relevance topologies and trust bias. In: Proceedings of the 28th Annual Conference of the Cognitive Science Society, vol. 28, pp. 1881–1886 (2006)
23. Rus, C., de Rijke, M., Yates, A.: Counterfactual representations for intersectional fair ranking in recruitment. In: RecSys in HR 2023: The 3rd Workshop on Recommender Systems for Human Resources, in conjunction with the 17th ACM Conference on Recommender Systems (2023)
24. Singh, A., Joachims, T.: Fairness of exposure in rankings. In: Proceedings of the 24th ACM SIGKDD International Conference on Knowledge Discovery & Data Mining, pp. 2219–2228 (2018)
25. Thijssen, L., Lancee, B., Veit, S., Yemane, R.: Discrimination against Turkish minorities in Germany and the Netherlands: field experimental evidence on the effect of diagnostic information on labour market outcomes. J. Ethn. Migr. Stud. **47**(6), 1222–1239 (2021)
26. Yang, K., Gkatzelis, V., Stoyanovich, J.: Balanced ranking with diversity constraints. arXiv preprint arXiv:1906.01747 (2019)
27. Yang, K., Loftus, J.R., Stoyanovich, J.: Causal intersectionality for fair ranking. arXiv preprint arXiv:2006.08688 (2020)
28. Yu, Q., Li, B.: mma: An R package for mediation analysis with multiple mediators. J. Open Res. Softw. **5**(1) (2017)
29. Zehlike, M., Bonchi, F., Castillo, C., Hajian, S., Megahed, M., Baeza-Yates, R.: FA*IR: A fair top-k ranking algorithm. In: Proceedings of the 2017 ACM Conference on Information and Knowledge Management, pp. 1569–1578. ACM (2017)
30. Zehlike, M., Castillo, C.: Reducing disparate exposure in ranking: a learning to rank approach. In: Proceedings of the Web Conference 2020, pp. 2849–2855 (2020)
31. Zehlike, M., Sühr, T., Baeza-Yates, R., Bonchi, F., Castillo, C., Hajian, S.: Fair top-k ranking with multiple protected groups. Inform. Process. Manag. **59**(1), 102707 (2022)
32. Zehlike, M., Yang, K., Stoyanovich, J.: Fairness in ranking, Part I: score-based ranking. ACM Comput. Surv. **55**(6), 1–36 (2022)
33. Zehlike, M., Yang, K., Stoyanovich, J.: Fairness in ranking, Part II: learning-to-rank and recommender systems. ACM Comput. Surv. **55**(6), 1–41 (2022)
34. Zemel, R., Wu, Y., Swersky, K., Pitassi, T., Dwork, C.: Learning fair representations. In: International Conference on Machine Learning, pp. 325–333. PMLR (2013)

Fairness Through Domain Awareness: Mitigating Popularity Bias for Music Discovery

Rebecca Salganik[1(✉)], Fernando Diaz[3], and Golnoosh Farnadi[1,2]

[1] Université de Montreal/MILA, Montreal, QC, Canada
rebecca.salganik@umontreal.ca, farnadig@mila.quebec
[2] McGill University, Montreal, QC, Canada
[3] Carnegie Mellon University, Pittsburgh, PA, USA
diazf@acm.org

Abstract. As online music platforms continue to grow, music recommender systems play a vital role in helping users navigate and discover content within their vast musical databases. At odds with this larger goal, is the presence of popularity bias, which causes algorithmic systems to favor mainstream content over, potentially more relevant, but niche items. In this work we explore the intrinsic relationship between music discovery and popularity bias through the lens of individual fairness. We propose a domain-aware, individual fairness-based approach which addresses popularity bias in graph neural network based recommender systems. Our approach uses individual fairness to reflect a ground truth listening experience, i.e., if two songs sound similar, this similarity should be reflected in their representations. In doing so, we facilitate meaningful music discovery that is resistant to popularity bias and grounded in the music domain. We apply our BOOST methodology to two discovery based tasks, performing recommendations at both the playlist level and user level. Then, we ground our evaluation in the cold start setting, showing that our approach outperforms existing fairness benchmarks in both performance and recommendation of lesser-known content. Finally, our analysis makes the case for the importance of domain-awareness when mitigating popularity bias in music recommendation.

Keywords: Recommendation · Algorithmic Fairness · Graph Neural Networks

1 Introduction

The proliferation of market activity on digital platforms has acted as a catalyst for research in recommendation, search, and information retrieval [33]. At its core, the goal of this research is to design systems which can facilitate users' exploration of an extensive item catalogue: be it in the domain of journalism [60], films [29], fashion [20], music [38,52,53], or otherwise. Within this larger goal of recommendation, each domain comes with its own specifics that differentiate it from other settings [9,25,46]. Particular to the music streaming

N. Goharian et al. (Eds.): ECIR 2024, LNCS 14611, pp. 351–368, 2024.
https://doi.org/10.1007/978-3-031-56066-8_27

domain, an extensive body of work has explored the importance of discovery, exploration, and novelty in the larger goal of performing music recommendation [17,23,27,39,44,49]. Broadly, discovery can be considered the ability of a curatorial system to expose users to relevant content that *they would not have manually discovered themselves* [27,32,49]. And, most significantly, the importance of this subtask is substantiated by numerous works indicating that music discovery is a crucial factor for maintaining and improving customer loyalty [39,44,49].

Recent work in this domain has begun to uncover an inverse relationship between novelty, one of the keys to discovery, and the notion of popularity bias [36,60]. Within the broader recommendation community, popularity bias has long been an important topic of research. This phenomenon manifests itself when algorithmic reliance on pre-existing data causes new, or less well known items, to be disregarded in favor of previously popular items [3,12,16,35,47,57]. And, particularly in the context of discovery, where purpose of a user's engagement with algorithmic curation hinges on exposure to musical items which they would not have already been familiar with, the presence of popularity bias can clearly hinder a system's ability to serve this need. In this work, we apply our methodology to graph neural network (GNN) based recommender systems [26,61]. In the graph space, popularity is deeply interlaced with the degree of a node, or the number of edges that connect a node and to others in the graph. This is because the innate process of representation learning is affected by the number of neighbors a node has [37]. And, thus, a node's centrality can dictate the quality of its learned representation. This suggests that duplicating the feature information of an extremely popular song, creating a new song using these duplicate features, and randomly placing it once at the edge of a graph, will significantly impact its learned representation, even if everything about the song remains *exactly the same*. Currently, the state of the art approaches to mitigating popularity bias, do so from a domain agnostic approach [2,42,51,59,64]. This methodology has two important drawbacks. First, it often requires a method to rely on the presence of sensitive attributes in order to define popularity, which are often unavailable. Second, such an approach is unable to recognize musical similarities among items, thus increasing the complexity of disentangling popularity bias in learned representation.

In this work, we propose a **domain aware**, individual fairness based approach for facilitating engaging music discovery. Unlike domain-agnostic approaches, our method does not rely on sensitive attributes to define popularity. Instead, we design an intuitive, simple framework that uses music features to fine-tune item representations such that they are reflective of information that is, in essence, a ground truth to the listening experience: two songs that sound similar should, at least somewhat, reflect this similarity in their learned representations. In order to facilitate the domain awareness of our approach we generate nuanced multi-modal track features, extensively augmenting two publicly available datasets. Using these novel feature sets, we show the importance of integrating musical similarity into a debiasing technique and show the effects of our method at learning expressive representations of items that are robust to the effects of popularity bias in the graph setting. Grounding our approach in the musical domain empowers us to leverage a ranking-based individual fairness definition and extend it to the bipartite graph setting. We compare our individual fairness-based method with three other methods which are grounded in other canonical fairness notions and are not domain-aware. Through a series of empirical results, we show that our fairness framework enables us to outperform a series of accepted fairness benchmarks in both performance and recommendation of lesser known content on two important music recommendation tasks. In summary, the contributions of this paper are the following:

1. **Problem Setting**: we define the task of music discovery through the lens of domain-aware individual fairness, showing the intrinsic connections between individual fairness, musical similarity, popularity bias, and music discovery.
2. **Dataset Design**: we extensively augment two classic music recommendation datasets to generate a set of nuanced multi-modal track features.
3. **Method**: (1) we provide a novel technical formulation of popularity bias (2) design a domain-aware ranking based individual fairness approach to mitigating popularity bias in graph-based recommendation.[1]

2 Related Work

2.1 Popularity Bias in Recommendation

Most broadly, popularity bias refers to a disparity between the treatment of popular and unpopular items at the hands of a recommender system. As such, this term is loosely tied to a collection of complementary terms including exposure bias [19], superstar economics [5], long tail recommendation [42], the Matthew effect [45], and aggregate diversity [4,13]. There have been several different approaches to formulating popularity through some quantitative definition. One body of work defines popularity with respect to individual items' visibility [19,42,65]. Another group of approaches attempts to simplify this process by applying some form of binning to the raw appearance values. Most notably, the long tail model [12,22, 28,47,62] has risen to prominence as a popularity definition. Due to the exponential decay in item interactions, the first 20% of items, called *short head*, take up a vast majority of the user interactions and the remaining 80%, or *long tail* and *distant tail*, have, even in aggregate, significantly fewer interactions.

In addition to providing formal definitions, a large body of work has formed around analyzing and mitigating popularity bias in recommender systems [10,16,34,35]. These mitigation strategies are often based on the instrumentation of various canonical fairness notions such as **group fairness** [2,7,51,55,64], **counterfactual fairness** [59,65,67], or **individual fairness** [14,58]. We contrast our work with previous individual fairness approaches in our use of the music feature space as a form of domain expertise in definition of item-item similarity. We argue that without this "anchoring" an individual fairness method that uses the output of a recommender model, whether it be in learned representation [58] or the relevance score [14], is already influenced by an item's

[1] Github repository https://github.com/Rsalganik1123/Domain_Aware_ECIR2024.

popularity. Finally, in addition to the classical formulation of popularity bias, a group of works have explored the connection between popularity bias and novelty [41,66,68] where various metrics are designed to evaluate the novelty of a recommended list. We see our work as complementary to the exploration in this area however, we differentiate our problem formulation because while novelty is an important aspect of discovery, without domain awareness novelty alone does not account for musical similarity - a critical aspect of the discovery setting.

2.2 GNNs in Recommendation

In recent years various graph neural network (GNN) architectures have been proposed for the recommendation domain [61]. For brevity, we will focus only on the two methods that are used as the backbone recommenders to the fairness mitigation techniques discussed later in this paper and refer readers to the following surveys [26,61] for recent innovations in this domain. In particular, *PinSage* [63] is an industry solution to graph-based recommendation. PinSage trains on a randomly sampled subset of the graph. In order to construct neighborhoods, PinSage uses k random walks to select the top m most relevant neighbors to use as the neighbor set. However, it is important to note that PinSage learns representations of items but not users. Meanwhile, *LightGCN* (LGCN) [31], is a method that learns both user and item embeddings simultaneously. Since its proposal in 2020, it is still considered state of the art.

3 Methodology

In this section we detail the dataset augmentation procedure and architecture of our domain-aware, individually fair music recommendation system. First, we introduce our datasets in Sect. 3.1. Then, following the problem setting in Sect. 3.2, we reformulate popularity bias in Sect. 3.3 and introduce our domain-aware, individually fair music recommender system in Sect. 3.4.

3.1 Dataset Augmentation Procedure

We augment both of our datasets to include a rich set of features scraped from Spotify API [1]. The details of the augmented features are as follows.

1. **Sonic features.** Spotify has a series of 9 proprietary features which are used to define the audio elements associated with a track. They are *danceability, energy, loudness, speechiness, acousticness, instrumentalness, liveness, valence, and tempo.* Each of these features is a continuous scalar value. We apply 10 leveled binning to the values.
2. **Genre features.** We identify the primary artist associated with each collect all the genre tags associated with them.
3. **Track Name features.** For each song in the dataset, we extract the song title and pass it through a pre-trained language transformer model, *BERT* [18], into an embedding of dimension 512.

4. **Image features.** For each song in the dataset we extract the associated album artwork. We use this image to generate *ResNet50* [30] embeddings of dimension 1024.

3.2 Problem Setting

The task of performing recommendation can be seen as link prediction an undirected bipartite graph. We denote such undirected bipartite graph as $G = (V, E)$. The set $V = T \cup P$ consists of a set containing song (or track) nodes, T, and playlist (or user) nodes, P (or U). The edge set E are defined between a playlist p_k (or user u_k) and a song t_i if t_i is contained in p_k (or listened to by u_k). Following this setting, our goal (link prediction) is to predict whether any two song nodes $t_i, t_j \in T$ share a common parent playlist p.

3.3 Reformulating Popularity Bias

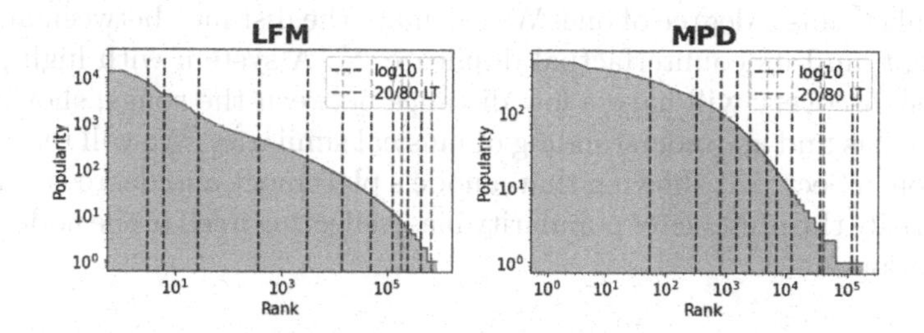

Fig. 1. Binning procedure for popularity definition. We contrast our popularity definition with the classic long tail model [42], showing that our proposed method empowers for a granular visualization of popularity between various item groups.

Defining Popularity. As mentioned in Sect. 2.1, there is no true consensus within the community on how to define popularity. Here, we present a methodology which we believe allows for both the granularity and expressiveness necessary to highlight differences among various mitigation methods. Broadly, our method consists of important steps (1) logarithmic smoothing and (2) binning. In doing so, we combine the best of each methodology. Applying a logarithmic transformation to the raw values, solves the scaling issues that are caused by the extremes of the long-tail distribution. Meanwhile, binning concisely highlights large scale patterns. And, in contrast to previous methods using binning [2,19,51], logarithmic smoothing guarantees that none of the bins are left empty. Please note that we select 10 bins based on the distribution of the datasets and the formulation of our BOOST methodology (see Sect. 3.4) but this number can be tuned to the granularity needs.

Our popularity measure is achieved by first, counting the number of times each song track, t_i appears within playlist (or user) training interactions such that for each t_i, $a_{t_i} = |\{p_i : t_j \in p_i\}|$. Then, we apply the base 10 logarithmic smoothing to these values such that for each t_i, $\text{pop}_{t_i} = \log_{10}(a_{t_i})$. Finally, we apply binning onto these values to split them into 10 groups such that for each t_i, pop_bin $(t_i) \in \{0, \ldots, 9\}$ where bin 9 has a higher popularity value than bin 0. The visualization of this binning procedure and its comparison with the long tail method can be seen in Fig. 1. As demonstrated by our visualizations, transforming the raw values into the logarithmic space shows that the bins are filled in relatively even intervals, where, as the popularity increases, so does the number of songs included in a bin. We showcase the gains that our method has over the canonical long tail model in Fig. 1 where we compare the positioning of our binning methodology with the classic long tail model. Furthermore, as we later show Figs. 3 and 4 our formulation of popularity is able to elucidate crucial differences among both the datasets and baseline model performances on these datasets.

Popularity Bias and Music Discovery. In addition we formalize the inverse relationship between music discovery and popularity bias. For each song track, $t_i \in T$, we generate a counterfactual example song, $t_i^* \in T_{CF}$, where everything about the features is *exactly the same* and the only difference is that t_i has a high degree while t_i^* has a degree of one. We calculate the distance between an original song node, t_i and its counterfactual duplicate, t_i^*. A system with high potential for musical discovery will have a low distance between the songs, showing a low popularity bias and an understanding of musical similarity. We will return to this formulation in Sect. 5.1, showing that a node's placement and degree in the graph can exacerbate the presence of popularity bias, reflecting itself in the node's learned representation.

3.4 Mitigating Popularity Bias Through Individual Fairness

Ranking-Based Individual Fairness. *REDRESS* is an individual fairness framework proposed by Dong et al. [21] for learning fair representations in single node graphs. Our work extends this framework to the bipartite recommendation setting and integrates it into our popularity bias mitigation approach. Here, the crucial formulation of individual fairness requires that nodes which were similar in their initial feature space should remain similar in their learned representation embeddings [24]. More concretely, for each song node, t_i, and node pair t_u, t_v in a graph G, similarity is defined on the basis of the pairwise cosine similarity metric, $s(\cdot, \cdot)$, as applied to either a feature $X[v] \in \mathfrak{R}^d$, or learned embedding set, $Z[v] \in \mathfrak{R}^m$. Applying this procedure in a pairwise fashion produces two similarity matrices. The first, or *apriori similarity*, S_G, in which similarity is calculated on input features and the second, or *learned similarity*, S_Z, in which similarity is calculated between learned embeddings generated by some GNN model, M. Drawing on principles from learn to rank [8], each entry in these similarity matrices is re-cast as the probability that node t_i is more similar to node t_u than t_v and transformed into an *apriori probability tensor*, $P_G \in \mathfrak{R}^{|T| \times |T| \times |T|}$, and a *learned probability tensor*, $P_Z \in \mathfrak{R}^{|T| \times |T| \times |T|}$. For more details on the calculations of these probabilities, please see the original formulation in [21]. Having defined these two probability tensors, each individual node the fairness loss, $L_{t_u,t_v}(t_i)$, is the canonical cross entropy loss aggregated over all nodes $t_i \in V$ as:

$$L_{\text{fairness}} = \sum_{i}^{|T|}\sum_{u}^{|T|}\sum_{v}^{|T|} L_{t_u,t_v}(t_i) \tag{1}$$

Individually Fair Music Discovery. The original formulation of individual fairness requires some form of domain expertise [24] to determine how similar (or dis-similar) two items are. For the music discovery domain, we use music features as the basis for calculating cosine similarity. Thus, our *apriori similarity*, S_G, is defined as the cosine similarity between the musical features, $X[v] \in \Re^{|T| \times 9}$, associated with song nodes. Meanwhile, our *learned similarity* contains the song-level embeddings, $S_Z \in \Re^{|T| \times m}$, learned by PinSage. In this way, REDRESS acts as a regularizer that ensures that rank-based similarity between songs is preserved between the input and embedding space. Thus, our similarity notion is domain-aware and grounded in the essence of musical experiences: acoustics.

Bringing Popularity Into Individual Fairness. The *REDRESS* framework does not explicitly encode any attributes of popularity in its training regimen. To extend this technique for explicitly counteracting popularity bias, we define the BOOST technique which is used to further increase the penalty on misrepresentation of items that come from diverse popularity categories. We define 10 popularity bins by applying a logarithmic transformation and binning the degrees of a node i (i.e., deg_i) such that $\text{pop\_bin}(i) = \text{bin}(\log_{10}(deg_i))$. Given the learned representation matrix, $S_Z \in \Re^{|T| \times |T|}$, we define another matrix B in which

$$B_{ij} = |\text{pop\_bin}(i) - \text{pop\_bin}(j)| \tag{2}$$

Then, in the *BOOST* loss formulation, in place of S_Z we use $S'_Z = S_Z + B$.

Objective Function. The training objective is:

$$L_{\text{total}} = L_{\text{utility}} + \gamma L_{\text{fairness}} \tag{3}$$

where γ is a scalable hyperparameter which controls the focus given to fairness used to balance between utility (L_{utility}) and fairness (L_{fairness}). For L_{utility}, we apply the aforementioned focal loss [40]. And L_{fairness} is Eq. 1 defined above.

Generating Recommendations. Notably, the PinSage architecture is only designed to learn embeddings for songs, not for playlists (or users). Thus we design our own procedure for generating playlist (or user) embeddings using the learned song embeddings. For each playlist (or user) node, p_i, we have a set of songs, $T(p_i) = \{t_i \in T, e_{p_i,t_i} \in E\}$, which are contained in a playlist. For a test playlist, p_t, we split the associated track set into two groups: $T(t_p) = \{t_i : t_i \in u_i\} =$

$t_{seed} \cup t_{holdout}$ such that t_{seed} is a set of k songs that are used to generate the playlist representation and $t_{holdout}$ are masked for evaluation. Thus, in order to generate a playlist (or user) embedding we define:

$$z_{p_t} = MEAN(\{z_{t_j} : t_j \in t_{seed}\})$$

where the $z_k \in \mathfrak{R}^{1 \times d}$ are the learned representations of dimension d. Having learned these playlist representations, we find the k-nearest neighbouring tracks and consider these the recommendations (Table 1).

4 Experimental Settings

Table 1. Dataset statistics

Dataset	Recommendation Setting	Feedback Type	#Users/Playlists	#Songs	#Artists
MPD	*Automatic Playlist Continuation*	Explicit	11,100	183,408	37,509
LFM	*Weekly Discovery*	Implicit	10,267	890,568	100,638

Recommendation Scenario. We evaluate our method on two important music discovery tasks: *Automatic Playlist Continuation* [54] and *Weekly Discovery* [56]. *Automatic Playlist Continuation* requires the recommender system to perform *next k recommendation* on a user generated playlist. Meanwhile, *Weekly Discovery*, involves the creation of a new playlist based on a user's aggregated listening habits. Following the paradigm of the cold start setting [54], we extract splits on the playlist level by randomly sampling without replacement such that each split trains on a distinct subset of the playlist pool.

Datasets. As introduced in Sect. 3.1, we extensively augment two publicly available datasets, LastFM (LFM) [43] and the Million Playlist Dataset (MPD) [15], with rich multi-modal track-level feature sets. For more details please refer back to Sect. 3.1.

Music Performance Metrics. We design a series of musical evaluation metrics to complement classic utility measures (see Table 2 for further detail). For example, we use *Artist Recall* to evaluate the correct identification of artists in a recommendation pool, an auxiliary task in music recommendation [38]. In addition, we design *Flow* to capture the musical cohesiveness of the recommended songs in a playlist [6].

Fairness Metrics. In order to assess the debiasing techniques used to promote of long tail songs, we follow common evaluation practices in the fairness literature [2,12,48]. Percentage metrics capture the ratio of niche to popular content that is being recommended on a playlist (or user) level. Meanwhile, *coverage* looks at the aggregate sets of niche songs and artists over all recommendations (see 2 for further detail) (Table 2).

Table 2. Music and fairness performance metrics. We define a ground truth set, G, and a recommended set, R, we define the set of unique artists in a playlist as $A(.)$ and the d-dimensional musical feature matrix associated with the tracks of a playlist as $F(.) \in \Re^{|.|\times d}$.

Metric	Category	Formulation
Artist Recall@100	Music	$\frac{1}{\lvert P_{test}\rvert}\sum_{p\in P_{test}}\frac{1}{\lvert A(G_p)\rvert}\lvert A(G_p)\cap A(R_p)\rvert$
Flow@100	Music	$\frac{1}{\lvert P_{test}\rvert}\sum_{p\in P_{test}}\cos(F(t_i),F(t_j))\ \forall(t_i,t_j)\in R_p$
Artist Diversity (per playlist)	Fairness	$\frac{1}{\lvert P_{test}\rvert}\sum_{p\in P_{test}}\frac{1}{\lvert A(P)\rvert}\lvert\{A(R_p)\}\rvert$
Percentage of Long Tail Items	Fairness	$\frac{1}{\lvert P_{test}\rvert}\sum_{p\in P_{test}}\frac{1}{\lvert p\rvert}\lvert\{t_i : t_i\in R_p\cap t_i\in LT\}\rvert$
Coverage over Long Tail Items	Fairness	$\frac{1}{\lvert LT\rvert}\lvert\{t_i : t_i\in R\bigcap t_i\in\lvert LT\rvert\}$
Coverage over Artists	Fairness	$\frac{1}{\lvert A\rvert}\lvert\{arid(t_i) : t_i\in R\rvert\}$

Baselines. We use two naive baselines, first using bare features in place of learned representations (**Features**) and, second, recommending the top 100 most popular tracks (**MostPop**). Then, we evaluate against three state of the art bias mitigation techniques: a group fairness-based, in processing method (**ZeroSum** [51]), a causality-based in-processing method (**MACR** [59]), and a re-ranking, post-processing method (**xQuAD** [2]).

Parameter Settings and Reproducibility. Each of the baseline methods was tested with learning rates ~ (0.01, 0.0001), embedding sizes of [10, 24, 64, 128] and batch sizes of [256, 512, 1024]. For the values in the tables below, each stochastic method was run 5 times and averaged. All details and further hyperparameter settings can be found on our Github repository.

5 Results

5.1 RQ1: How Does Incorporating Individual Fairness Improve the Mitigation of Popularity Bias and Facilitate Music Discovery?

To showcase the performance of our algorithm in the discovery setting and motivate the need for individual fairness in the mitigation of popularity bias, we draw on the definition of music discovery presented in Sect. 3.3 and evaluate the effects of popularity bias on learned representations of songs. Simulating a situation of maximal popularity bias, we consider the hypothetical example in which extremely popular songs are reversed to become unpopular and measure the effect on their learned representations. More formally, for each song track, $t_i \in T$, we generate a counterfactual example song, $t_i^* \in T_{CF}$, where everything about the features is *exactly the same* and the only difference is that t_i appears in many playlists while t_i^* appears only once. We augment the original dataset to include these counterfactual songs, $T_{AUG} = T \cup T_{CF}$. Then, we use five methods to learn the item level representations: one baseline recommender, PinSage, and four bias mitigation methods,

ZeroSum [51], MACR [59], REDRESS, and BOOST. We apply 2-dimensional PCA to each embedding set and analyze the Euclidean distance between the centroids of original track embeddings, $\bar{T}$, and counterfactual track embeddings, $\bar{T}_{CF}$. Due to the size of our dataset, we run this metric using the 100 most popular tracks in the MPD dataset and leave further exploration of this phenomenon for future work.

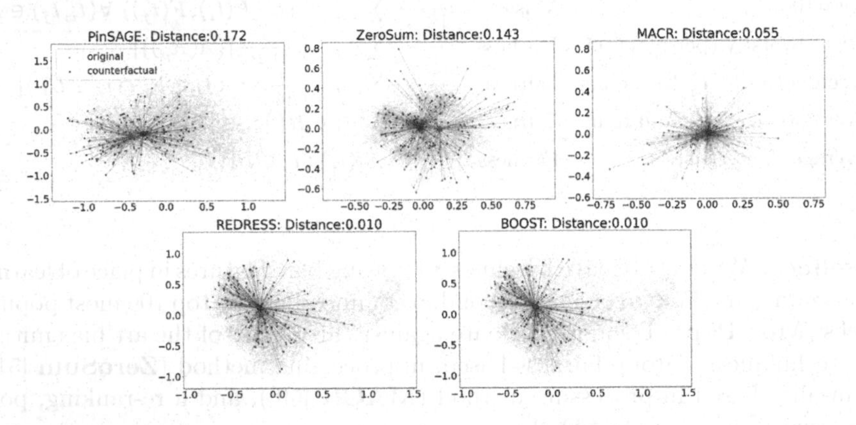

Fig. 2. Simulating Popularity Bias: We select 100 of the most popular songs in MPD and [54], duplicate features, and give them a degree of 1. We find that REDRESS and BOOST have the lowest distance between the originals and their unpopular duplicates, showing the least amount of popularity bias.

As shown in Fig. 2, we find that all fairness interventions decrease the distance between the two centroids. Furthermore, as the granularity of fairness increases, the distance between the centroids of learned representations decreases. For example, PinSage, which has no mitigation of popularity bias, has the largest distance of 0.172. ZeroSum [51], which considers group fairness, decreases the distance to 0.143, MACR [59], which uses counterfactual estimation, shrinks to 0.055. Finally, our methods, REDRESS and BOOST are able to achieve both the lowest distance and the correct orientation between the two embedding spaces. In these results, we see that the domain-awareness of our methodology, which enables it to understand musical similarity between items, allows it to be resistant to the effects of popularity bias on a learned song embedding. Thus, in the setting of musical discovery, it is able to uncover proximity between items which are musically coherent even if they are not necessarily of similar popularity status. And, in doing so, we build representations that are complex, expressive, and effective for music recommendation.

5.2 RQ2: How Does Our Individual Fairness Approach Compare to Existing Methods for Mitigating Popularity Bias?

Table 3. Comparison between all methods. Note: We use **bold** text to represent the best performance within a column. In addition, we calculate statistical significance using the Wilcoxon signed-rank test [50] to results between PinSage and BOOST. We show that the *BOOST* method achieves the best performance along all fairness metrics when compared with debiasing benchmarks.

		Classic		Music		Fairness			
Data	Model	Recall@100	NDCG@100	Artist Recall@100	Flow	Diversity	%LT	LT Cvg	Artist Cvg
MPD	Features	0.041	0.073	0.073	0.900	0.841	**0.588**	0.022	0.073
	MostPop	0.044	0.048	0.141	0.908	0.680	0.0	0.0	0.001
	LightGCN	**0.106 ± 0.004**	0.119 ± 0.004	**0.272 ± 0.011**	0.905 ± 0.000	0.672 ± 0.025	0.002 ± 0.000	0.000 ± 0.000	0.025 ± 0.001
	PinSage	0.068 ± 0.002	**0.144 ± 0.003**	0.139 ± 0.003	0.931 ± 0.001	0.707 ± 0.003	0.476 ± 0.002	0.032 ± 0.000	0.105 ± 0.000
	ZeroSum	0.044 ± 0.002	0.043 ± 0.002	0.220 ± 0.008	0.904 ± 0.001	0.765 ± 0.013	0.000 ± 0.003	0.000 ± 0.000	0.048 ± 0.002
	xQuAD	0.064 ± 0.005	0.104 ± 0.006	0.135 ± 0.013	0.927 ± 0.004	0.703 ± 0.059	0.226 ± 0.001	0.017 ± 0.000	0.098 ± 0.004
	MACR	0.028 ± 0.014	0.030 ± 0.015	0.149 ± 0.022	0.902 ± 0.002	0.831 ± 0.034	0.019 ± 0.006	0.000 ± 0.001	0.011 ± 0.003
	REDRESS	0.045 ± 0.002	0.100 ± 0.003	0.162 ± 0.004	0.969 ± 0.032	0.829 ± 0.001	0.504 ± 0.003	0.036 ± 0.004	0.117 ± 0.000
	BOOST	0.020 ± 0.004	0.047 ± 0.003	0.137 ± 0.002	**0.979 ± 0.000**	**0.899 ± 0.002**	**0.522 ± 0.001**	**0.037 ±0.003**	**0.125 ±0.000**
	p values	4.408083e-16	1.768725e-19	0.727897	3.751961e-61	1.168816e-29	0.000596	-	-
LFM	Features	0.033	0.037	0.041	0.996	0.919	0.486	0.005	0.034
	MostPop	0.015	0.011	0.046	0.926	0.600	0.000	0.000	0.001
	LightGCN	0.026 ± 0.001	0.023 ± 0.001	0.068 ± 0.001	0.998± 0.000	0.505 ± 0.012	0.000 ± 0.000	0.000 ± 0.000	0.003 ± 0.001
	PinSage	**0.064 ± 0.001**	**0.095 ± 0.002**	**0.077 ± 0.002**	0.969 ± 0.000	0.775 ± 0.003	0.437 ± 0.001	0.008 ± 0.000	0.053 ± 0.001
	ZeroSum	0.001 ± 0.003	0.001 ± 0.001	0.045 ± 0.004	0.996 ± 0.008	0.866 ± 0.000	0.007 ± 0.000	0.000 ± 0.000	0.032 ± 0.001
	xQuAD	0.055 ± 0.001	0.064 ± 0.001	0.068 ± 0.002	0.998 ± 0.000	0.801 ± 0.008	0.212 ± 0.000	0.004 ± 0.000	0.053 ± 0.001
	MACR	0.014 ± 0.001	0.014 ± 0.001	0.049 ± 0.007	0.996 ± 0.003	0.777 ± 0.050	0.002 ± 0.004	0.000 ± 0.000	0.001 ± 0.000
	REDRESS	0.038 ± 0.002	0.053 ± 0.004	0.057 ± 0.001	0.998 ± 0.002	0.862 ± 0.004	0.451 ± 0.000	0.008 ± 0.002	0.056 ± 0.000
	BOOST	0.005 ± 0.001	0.007 ± 0.001	0.029 ± 0.002	**0.999 ± 0.000**	**0.941 ± 0.003**	**0.498 ± 0.006**	**0.010 ± 0.000**	**0.068 ± 0.001**
	p values	5.696989e-08	1.179627e-15	1.914129e-07	0.001408	1.112495e-34	2.477700e-11	-	-

Analyzing Utility Performance: First, we look at the comparison between the backbone recommender systems and their debiasing counterparts. Within the greater fairness community it is typical to see a trade-off between recommendation utility and the effectiveness of a debiasing technique [32]. Indeed, in our experiments this trade-off is present. For example, evaluating the columns of *Recall* and *NDCG* on Table 3 we can see that both recommender systems outperform their debiasing counterparts. However, we argue that the presence of this trade-off in the discovery setting is not only expected but also, potentially desirable. Since the premise of the canonical recommendation utility metrics is to reward a system that can accurately recover the exact tracks a user liked, any attempts to promote long tail content that wasn't originally listened to is penalized, even if it is well aligned with a user's taste. In recent years, several recommendation works have suggested that this trade-off, though present in offline testing, doesn't necessarily carry over into online testing [11,32]. Even more so, in the discovery setting, where the premise of algorithmic curation is facilitating user interactions with music that isn't already top-of-mind, the drop in performance can be attributed to the systems' purposeful avoidance of previously popular items, in favor of other musically coherent and relevant content.

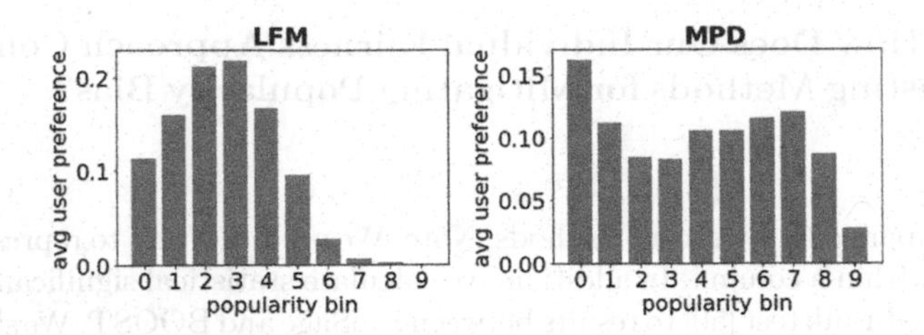

Fig. 3. Dataset Breakdown by Long Tail Definition: Using our formulation of popularity we can see that the two datasets have different distributions of popularity in their training data which, in turn, helps explain fairness/performance tradeoffs.

Analyzing Musical Performance: In order to further analyze the performance of our debiasing method, we look at the in performance on the music metrics, *Artist Recall* and *Flow*. In particular, *Flow* plays an important role in the music discovery task because studies have indicated that users are drawn to homogeneous listening suggestions when engaging with algorithmic curation [6,32]. As we can see in both datasets, REDRESS and BOOST consistently achieve the highest *Flow*. By harnessing musical features and in our debiasing technique, our method generates representations that are indicative of musical similarity. Crucially, if we consider the implications of such a debiasing technique on a mainstream user, these results indicate that our debiasing method's awareness of musical similarity will enable it to maintain the stylistic elements that such a user is drawn to, even if it is promoting niche content.

Analyzing Fairness Performance: Next, we compare the performance among the various fairness promotion methods. Looking at the columns of *Recall* and *NDCG* on Table 3, we can see that, as expected, xQuAD [2] which is a re-ranking method is able to preserve the highest utility. However, among the in-processing

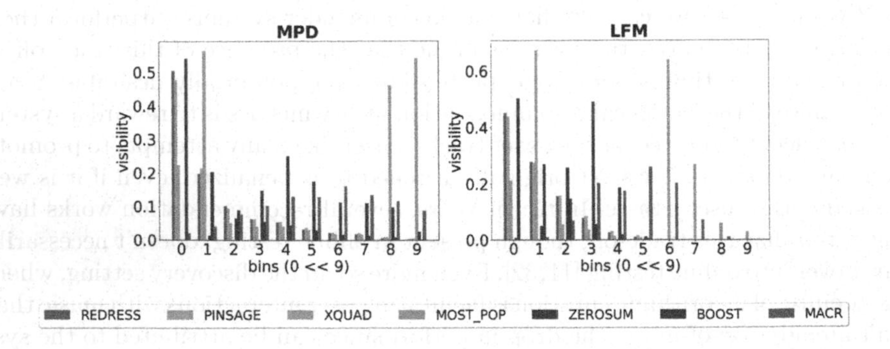

Fig. 4. Group By Group Analysis of Recommendations: We show that REDRESS and BOOST select the largest amount of items from the lowest bins. Note: visibility indicated the number of item from group k appearing in the final recommendations

methods, REDRESS is able to achieve the second highest utility. Meanwhile, looking at the fairness metrics, it is clear that REDRESS and BOOST are the highest performing methods. In particular, looking at the columns for *%LT* and *LT Cvg*, we can see that REDRESS is noticeably better than the other methods and BOOST is able to improve on its performance. Crucially, our method is able to have high values in both coverage and percentage of long tail items indicating that REDRESS/BOOST is not just prioritizing niche items but also choosing a diverse selection from among them.

Effects of Popularity Definition: As we can see in Fig. 4 the definition of popularity plays a significant role in the model selection method *especially* in the case where user preferences encoded in the training data skew towards popular items. In particular, using a less granular definition for popularity bins can synthetically inflate the performance of $\%LT$ and $LTCvg$. For example, we can see that methods like xQuAD and ZeroSum are selecting a majority of their items from bins mid-popularity bins. Using a classical long tail methodology, these differences would not be as visible, masking distinctions among the baselines' fairness.

6 Limitations of Our Work

First, it is important to remember that recommender systems are responsible for serving the tastes of listeners, not policing them, and we do not deny the validity of mainstream listening practices. Thus, the intention of this work is to serve the needs of all users, mainstream and niche equally. However, due to our lack of access to online evaluation settings we cannot confirm that the effects of debiasing will not affect mainstream users' listening experiences. We leave this analysis for future work. Second, due to the nature of publicly available data, both of these datasets skew heavily towards Western, anglophone content and are not representative of the wide array of music that is available for consumption. Finally, we acknowledge that our definition of discovery is grounded in qualitative metrics and cannot encompass the entire complexity of a discovery experience.

7 Conclusion

In this work, we address the problem of mitigating popularity bias in music recommendation. We focus our objective on the task of facilitating meaningful music discovery. In particular, we emphasize the critical aspects of discovery which differentiate it from generalized music recommendation and underscore the negative effects that popularity bias can have on users' ability to uncover novel music. On the basis of this motivation, we unravel the intrinsic ties between popularity bias and individual fairness, proposing a domain-aware debiasing method that uses musical similarity to counteract the effects of popularity on learned representations. Finally, we perform extensive evaluation on two music datasets showing the improvements of our domain aware method in comparison with three state of the art popularity bias mitigation techniques. While we have designed this method with the explicit focus of music recommendation, we hope that these promising findings can inspire future approaches which are grounded in concrete, domain specific attributes in a wide variety of applications.

Acknowledgements. Funding support for project activities has been partially provided by Canada CIFAR AI Chair, Facebook award, IVADO scholarship, and NSERC Discovery Grants program. We also express our gratitude to Compute Canada for their support in providing facilities for our evaluations.

References

1. Spotipy: Spotify API in Python (2014). https://spotipy.readthedocs.io/en/2.19.0/
2. Abdollahpouri, H., Burke, R., Mobasher, B.: Managing popularity bias in recommender systems with personalized re-ranking (2019)
3. Abdollahpouri, H., Mansoury, M., Burke, R., Mobasher, B.: The connection between popularity bias, calibration, and fairness in recommendation. In: Proceedings of the 14th ACM Conference on Recommender Systems, p. 726–731. RecSys 2020, Association for Computing Machinery, New York, NY, USA (2020). https://doi.org/10.1145/3383313.3418487
4. Adomavicius, G., Kwon, Y.: Improving aggregate recommendation diversity using ranking-based techniques. IEEE Trans. Knowl. Data Eng. **24**(5), 896–911 (2012). https://doi.org/10.1109/TKDE.2011.15
5. Bauer, C., Kholodylo, M., Strauss, C.: Music recommender systems challenges and opportunities for non-superstar artists. In: Bled eConference (2017)
6. Bontempelli, T., Chapus, B., Rigaud, F., Morlon, M., Lorant, M., Salha-Galvan, G.: Flow moods: recommending music by moods on deezer. In: RecSys 2022 (2022)
7. Boratto, L., Fenu, G., Marras, M.: Interplay between upsampling and regularization for provider fairness in recommender systems. User Model. User-Adap. Inter. **31**(3), 421–455 (2021). https://doi.org/10.1007/s11257-021-09294-8
8. Burges, C.: From ranknet to lambdarank to lambdamart: an overview. Learning **11**, 23–581 (2010)
9. Burke, R., Ramezani, M.: Matching recommendation technologies and domains (2011). https://doi.org/10.1007/978-0-387-85820-311
10. Cañamares, R., Castells, P.: Should i follow the crowd? a probabilistic analysis of the effectiveness of popularity in recommender systems. In: The 41st International ACM SIGIR Conference on Research & Development in Information Retrieval, p. 415–424. SIGIR 2018, Association for Computing Machinery, New York, NY, USA (2018). https://doi.org/10.1145/3209978.3210014
11. Castells, P., Moffat, A.: Offline recommender system evaluation: challenges and new directions. AI Magazine **43**(2), 225–238 (2022). https://doi.org/10.1002/aaai.12051, https://onlinelibrary.wiley.com/doi/abs/10.1002/aaai.12051
12. Celma, O., Cano, P.: From hits to niches? or how popular artists can bias music recommendation and discovery. In: Proceedings of the 2nd KDD Workshop on Large-Scale Recommender Systems and the Netflix Prize Competition. NETFLIX 2008, Association for Computing Machinery, New York, NY, USA (2008). https://doi.org/10.1145/1722149.1722154
13. Celma, Ò., Herrera, P.: A new approach to evaluating novel recommendations. In: Proceedings of the ACM Conference on Recommender Systems, RecSys 2008 (2008). https://api.semanticscholar.org/CorpusID:7572506

14. Chakraborty, A., Hannák, A., Biega, A.J., Gummadi, K.P.: Fair sharing for sharing economy platforms (2017)
15. Chen, C.W., Lamere, P., Schedl, M., Zamani, H.: RecSys challenge 2018: automatic music playlist continuation. In: RecSys 2018 (2018)
16. Chen, J., Dong, H., Wang, X., Feng, F., Wang, M., He, X.: Bias and debias in recommender system: a survey and future directions (2020). https://doi.org/10.48550/ARXIV.2010.03240
17. Cunningham, S.J., Bainbridge, D., Mckay, D.: Finding new music: a diary study of everyday encounters with novel songs. In: International Society for Music Information Retrieval Conference (2007)
18. Devlin, J., Chang, M.W., Lee, K., Toutanova, K.: BERT: pre-training of deep bidirectional transformers for language understanding (2019)
19. Diaz, F., Mitra, B., Ekstrand, M.D., Biega, A.J., Carterette, B.: Evaluating stochastic rankings with expected exposure. In: CIKM 2020 (2020)
20. Ding, Y., Mok, P., Ma, Y., Bin, Y.: Personalized fashion outfit generation with user coordination preference learning. Inform. Process. Manag. **60**(5), 103434 (2023). https://doi.org/10.1016/j.ipm.2023.103434, https://www.sciencedirect.com/science/article/pii/S0306457323001711
21. Dong, Y., Kang, J., Tong, H., Li, J.: Individual fairness for graph neural networks: a ranking based approach. In: Proceedings of the 27th ACM SIGKDD Conference on Knowledge Discovery & Data Mining, pp. 300–310. KDD 2021, Association for Computing Machinery, New York, NY, USA (2021). https://doi.org/10.1145/3447548.3467266 event-place: Virtual Event, Singapore
22. Downey, A.B.: Evidence for long-tailed distributions in the internet. In: Proceedings of the 1st ACM SIGCOMM Workshop on Internet Measurement, p. 229–241. IMW 2001, Association for Computing Machinery, New York, NY, USA (2001). https://doi.org/10.1145/505202.505230
23. Drott, E.: Why the next song matters: streaming, recommendation, scarcity. Twentieth-Century Music **15**, 325–357 (2018). https://doi.org/10.1017/S1478572218000245
24. Dwork, C., Hardt, M., Pitassi, T., Reingold, O., Zemel, R.: Fairness through awareness (2011)
25. Ekstrand, M.D., Harper, F.M., Willemsen, M.C., Konstan, J.A.: User perception of differences in recommender algorithms. In: Proceedings of the 8th ACM Conference on Recommender Systems, p. 161–168. RecSys 2014, Association for Computing Machinery, New York, NY, USA (2014). https://doi.org/10.1145/2645710.2645737
26. Gao, C., et al.: A survey of graph neural networks for recommender systems: challenges, methods, and directions. ACM Trans. Recomm. Syst. **1**(1), 1–51 (2023). https://doi.org/10.1145/3568022
27. Garcia-Gathright, J., St. Thomas, B., Hosey, C., Nazari, Z., Diaz, F.: Understanding and evaluating user satisfaction with music discovery. In: The 41st International ACM SIGIR Conference on Research & Development in Information Retrieval, p. 55–64. SIGIR 2018, Association for Computing Machinery, New York, NY, USA (2018). https://doi.org/10.1145/3209978.3210049
28. Goel, S., Broder, A., Gabrilovich, E., Pang, B.: Anatomy of the long tail: ordinary people with extraordinary tastes. In: Proceedings of the Third ACM International Conference on Web Search and Data Mining, p. 201–210. WSDM 2010, Association for Computing Machinery, New York, NY, USA (2010). https://doi.org/10.1145/1718487.1718513

29. Harper, F.M., Konstan, J.A.: The MovieLens datasets: History and context. ACM Trans. Interact. Intell. Syst. **5**(4), 1–19 (2015). https://doi.org/10.1145/2827872
30. He, K., Zhang, X., Ren, S., Sun, J.: Deep residual learning for image recognition. In: CVPR 2016 (2016)
31. He, X., Deng, K., Wang, X., Li, Y., Zhang, Y., Wang, M.: LightGCN: simplifying and powering graph convolution network for recommendation. In: SIGIR 2020 (2020)
32. Herlocker, J.L., Konstan, J.A., Terveen, L.G., Riedl, J.T.: Evaluating collaborative filtering recommender systems. ACM Trans. Inf. Syst. **22**(1), 5–53 (2004). https://doi.org/10.1145/963770.963772
33. Hossain, I., et al.: A survey of recommender system techniques and the ecommerce domain (2023)
34. Jadidinejad, A.H., Macdonald, C., Ounis, I.: How sensitive is recommendation systems' offline evaluation to popularity? (2019)
35. Jannach, D., Lerche, L., Kamehkhosh, I., Jugovac, M.: What recommenders recommend: an analysis of recommendation biases and possible countermeasures. User Model. User-Adap. Inter. **25**(5), 427–491 (2015). https://doi.org/10.1007/s11257-015-9165-3
36. Kamehkhosh, I., Jannach, D.: User perception of next-track music recommendations. In: Proceedings of the 25th Conference on User Modeling, Adaptation and Personalization, p. 113–121. UMAP 2017, Association for Computing Machinery, New York, NY, USA (2017). https://doi.org/10.1145/3079628.3079668
37. Kang, J., Zhu, Y., Xia, Y., Luo, J., Tong, H.: RawlsGCN: towards Rawlsian difference principle on graph convolutional network. In: WWW 2022 (2022)
38. Korzeniowsky, F., Oramas, S., Gouyon, F.: Artist similarity with graph neural networks. In: Proceedings of the 18th International Society for Music Information Retrieval Conference. ISMIR (2021)
39. Lavranos, C., Kostagiolas, P., Martzoukou, K.: Theoretical and applied issues on the impact of information on musical creativity: an information seeking behavior perspective, pp. 1–16 (2016). https://doi.org/10.4018/978-1-5225-0270-8.ch001
40. Lin, T.Y., Goyal, P., Girshick, R., He, K., Dollár, P.: Focal loss for dense object detection. IEEE Trans. Pattern Anal. Mach. Intell. (2020)
41. Lo, K., Ishigaki, T.: Matching novelty while training: Novel recommendation based on personalized pairwise loss weighting. In: 2019 IEEE International Conference on Data Mining (ICDM), pp. 468–477 (2019)
42. Mansoury, M., Abdollahpouri, H., Pechenizkiy, M., Mobasher, B., Burke, R.: A graph-based approach for mitigating multi-sided exposure bias in recommender systems. ACM Trans. Inform. Syst. **40**(2), 1–31 (2021). https://doi.org/10.1145/3470948
43. Melchiorre, A., Rekabsaz, N., Parada-Cabaleiro, E., Brandl, S., Lesota, O., Schedl, M.: Investigating gender fairness of recommendation algorithms in the music domain. Inform. Process. Manag. **58**, 102666 (2021)
44. Mäntymäki, M., Islam, N.: Gratifications from using freemium music streaming services: differences between basic and premium users. In: International Confererence on Information Systems (2015)
45. Möller, J., Trilling, D., Helberger, N., van Es, B.: Do not blame it on the algorithm: an empirical assessment of multiple recommender systems and their impact on content diversity. Information, Communication & Society (2018)
46. Noh, T., Yeo, H., Kim, M., Han, K.: A study on user perception and experience differences in recommendation results by domain expertise: the case of fashion domains.

In: Extended Abstracts of the 2023 CHI Conference on Human Factors in Computing Systems. CHI EA 2023, Association for Computing Machinery, New York, NY, USA (2023). https://doi.org/10.1145/3544549.3585641
47. Park, Y.J., Tuzhilin, A.: The long tail of recommender systems and how to leverage it. In: Proceedings of the 2008 ACM Conference on Recommender Systems, p. 11–18. RecSys 2008, Association for Computing Machinery, New York, NY, USA (2008). https://doi.org/10.1145/1454008.1454012
48. Patro, G.K., Biswas, A., Ganguly, N., Gummadi, K.P., Chakraborty, A.: FairRec: two-sided fairness for personalized recommendations in two-sided platforms. In: Proceedings of The Web Conference 2020. ACM (2020). https://doi.org/10.1145/3366423.3380196
49. Raff, A., Mladenow, A., Strauss, C.: Music discovery as differentiation strategy for streaming providers. In: Proceedings of the 22nd International Conference on Information Integration and Web-Based Applications & Services, p. 476–480. iiWAS 2020, Association for Computing Machinery, New York, NY, USA (2021). https://doi.org/10.1145/3428757.3429151
50. Rey, D., Neuhäuser, M.: Wilcoxon-signed-rank test (2011)
51. Rhee, W., Cho, S.M., Suh, B.: Countering popularity bias by regularizing score differences. In: Proceedings of the 16th ACM Conference on Recommender Systems, p. 145–155. RecSys 2022, Association for Computing Machinery, New York, NY, USA (2022). https://doi.org/10.1145/3523227.3546757
52. Salha-Galvan, G., Hennequin, R., Chapus, B., Tran, V.A., Vazirgiannis, M.: Cold start similar artists ranking with gravity-inspired graph autoencoders (2021). https://doi.org/10.48550/ARXIV.2108.01053
53. Saravanou, A., Tomasi, F., Mehrotra, R., Lalmas, M.: Multi-task learning of graph-based inductive representations of music content. In: Proceedings of the 22nd International Society for Music Information Retrieval Conference, pp. 602–609. ISMIR, Online (2021). https://doi.org/10.5281/zenodo.5624379
54. Schedl, M., Zamani, H., Chen, C.W., Deldjoo, Y., Elahi, M.: Current challenges and visions in music recommender systems research. Int. J. Multimedia Inform. Retrieval **7**(2), 95–116 (2018). https://doi.org/10.1007/s13735-018-0154-2
55. Schnabel, T., Swaminathan, A., Singh, A., Chandak, N., Joachims, T.: Recommendations as treatments: debiasing learning and evaluation (2016)
56. Stanisljevic, D.: The impact of Spotify features on music discovery in the streaming platform age. Master's thesis (2020). http://hdl.handle.net/2105/55511
57. Steck, H.: Item popularity and recommendation accuracy. In: RecSys 2011. ACM (2011)
58. Wang, X., Wang, W.H.: Providing item-side individual fairness for deep recommender systems. In: Proceedings of the 2022 ACM Conference on Fairness, Accountability, and Transparency, p. 117–127. FAccT 2022, Association for Computing Machinery, New York, NY, USA (2022). https://doi.org/10.1145/3531146.3533079
59. Wei, T., Feng, F., Chen, J., Wu, Z., Yi, J., He, X.: Model-agnostic counterfactual reasoning for eliminating popularity bias in recommender system. In: Proceedings of the 27th ACM SIGKDD Conference on Knowledge Discovery & Data Mining, p. 1791–1800. KDD 2021, Association for Computing Machinery, New York, NY, USA (2021). https://doi.org/10.1145/3447548.3467289
60. Wu, C., Wu, F., Huang, Y., Xie, X.: Personalized news recommendation: methods and challenges (2022)
61. Wu, S., Sun, F., Zhang, W., Xie, X., Cui, B.: Graph neural networks in recommender systems: a survey (2020)

62. Yang, C.C., Chen, H., Hong, K.: Visualization of large category map for internet browsing. Decis. Support Syst. **35**(1), 89–102 (2003). https://doi.org/10.1016/S0167-9236(02)00101-X
63. Ying, R., He, R., Chen, K., Eksombatchai, P., Hamilton, W.L., Leskovec, J.: Graph convolutional neural networks for web-scale recommender systems. In: Proceedings of the 24th ACM SIGKDD International Conference on Knowledge Discovery & Data Mining. ACM (2018). https://doi.org/10.1145/3219819.3219890
64. Zhang, A., Ma, W., Wang, X., Chua, T.S.: Incorporating bias-aware margins into contrastive loss for collaborative filtering. In: Advances in Neural Information Processing Systems, vol. 35: Annual Conference on Neural Information Processing Systems, NeurIPS (2022)
65. Zhang, Y., et al.: Causal intervention for leveraging popularity bias in recommendation. In: Proceedings of the 44th International ACM SIGIR Conference on Research and Development in Information Retrieval, p. 11–20. SIGIR 2021, Association for Computing Machinery, New York, NY, USA (2021). https://doi.org/10.1145/3404835.3462875
66. Zhao, M., et al.: Investigating accuracy-novelty performance for graph-based collaborative filtering. In: Proceedings of the 45th International ACM SIGIR Conference on Research and Development in Information Retrieval. ACM (2022). https://doi.org/10.1145/3477495.3532005
67. Zheng, Y., Gao, C., Li, X., He, X., Li, Y., Jin, D.: Disentangling user interest and conformity for recommendation with causal embedding. In: Proceedings of the Web Conference 2021, p. 2980–2991. WWW 2021, Association for Computing Machinery, New York, NY, USA (2021). https://doi.org/10.1145/3442381.3449788
68. Zhu, Z., He, Y., Zhao, X., Zhang, Y., Wang, J., Caverlee, J.: Popularity-opportunity bias in collaborative filtering. In: Proceedings of the 14th ACM International Conference on Web Search and Data Mining, p. 85–93. WSDM 2021, Association for Computing Machinery, New York, NY, USA (2021). https://doi.org/10.1145/3437963.3441820

Evaluating the Explainability of Neural Rankers

Saran Pandian[1], Debasis Ganguly[2](✉), and Sean MacAvaney[2]

[1] University of Illinois, Chicago, USA
spand43@uic.edu
[2] University of Glasgow, Glasgow, UK
{Debasis.Ganguly,Sean.Macavaney}@glasgow.ac.uk

Abstract. Information retrieval models have witnessed a paradigm shift from unsupervised statistical approaches to feature-based supervised approaches to completely data-driven ones that make use of the pre-training of large language models. While the increasing complexity of the search models have been able to demonstrate improvements in effectiveness (measured in terms of relevance of top-retrieved results), a question worthy of a thorough inspection is - "how explainable are these models?", which is what this paper aims to evaluate. In particular, we propose a common evaluation platform to systematically evaluate the explainability of any ranking model (the explanation algorithm being identical for all the models that are to be evaluated). In our proposed framework, each model, in addition to returning a ranked list of documents, also requires to return a list of explanation units or rationales for each document. This meta-information from each document is then used to measure how locally consistent these rationales are as an intrinsic measure of interpretability - one that does not require manual relevance assessments. Additionally, as an extrinsic measure, we compute how relevant these rationales are by leveraging sub-document level relevance assessments. Our findings show a number of interesting observations, such as sentence-level rationales are more consistent, an increase in complexity mostly leads to less consistent explanations, and that interpretability measures offer a complementary dimension of evaluation of IR systems because consistency is not well-correlated with nDCG at top ranks.

Keywords: IR Evaluation · Neural Ranking Models · Explainability

1 Introduction

A neural ranking model (NRM) involves learning a data-driven parameterised similarity function between queries and documents [5,8,16,19,26]. Despite achieving state-of-the-art effectiveness (measured in terms of relevance of search results), NRMs suffer from poor interpretability of their underlying working mechanism due to two main reasons. First, they lack a closed form expression

The work was conducted at Dhirubhai Ambani Institute of Information and Communication Technology, the author's previous affiliation.

N. Goharian et al. (Eds.): ECIR 2024, LNCS 14611, pp. 369–383, 2024.
https://doi.org/10.1007/978-3-031-56066-8_28

involving the human interpretable fundamental components of an IR similarity function (i.e., term frequency, inverse document frequency and document length) [22]. Second, due to the fact that the similarity function of an NRM operates at the level of embedded representations of documents and queries, it is difficult to determine which terms present within a document are mainly responsible for contributing to its retrieval status value [2]. To the best of our knowledge, there is no work in IR research that has evaluated the quality of explanations of NRMs in an objective manner, e.g., in terms of consistency and correctness akin to some of the work done for the broader class of predictive models [1,17,27].

Our Contributions. First, we introduce a framework for an offline evaluation of the explainability of a ranking model. The evaluation protocol requires participating IR systems to report a list of explanation units or rationales comprised of text snippets of arbitrary lengths in addition to a ranked list of retrieved documents. Second, we show how these rationales can be evaluated both in an intrinsic and extrinsic manner - the later requiring sub-document level relevance. We conduct a range of experiments to compare the explanation qualities across a range of different NRMs using a common explanation methodology - that of occlusion commonly used in the literature [9,11,20]. An important finding of our experiments is that we demonstrate that the IR systems that produce the most relevant results are not necessarily the most explainable.

2 Related Work

In IR, there has been little effort in investigating the explanation effectiveness of neural models. The article [25] proposes a LIME-based local explanation methodology for IR models and also proposes a metric that measures the overlap of explanation units with the relevant terms (bag-of-words representation of the set of relevant documents for a query). However, the metric proposed in this paper is different from that of [25] in that our metric factors in intrinsic consistency of the explanations, and our proposed extrinsic measure makes use of sub-document level relevance.

Another thread of work towards a quantitative evaluation of IR model explanations is in the form of how well a simple linear model either comprised of the fundamental functional components - term frequency, idf, document length etc. fits a complex black-box model [22], or that how well it conforms to the axioms of IR [12] yielding a notion of fidelity. An yet another approach of IR trustworthiness evaluation involves the notion of measuring how consistently does an IR model conform to user expectations of the similarity of the top-retrieved documents based on information need change across query reformulations [23].

Our evaluation framework for IR explanation effectiveness is different from those of [12,22] in the sense that ours does not involve evaluating how well a simpler model fits a more complex one, and it also differs from [23] in that our evaluation directly concerns model explanations in the form of rationales instead of analysing a model's behaviour across pairs of queries (Fig. 3).

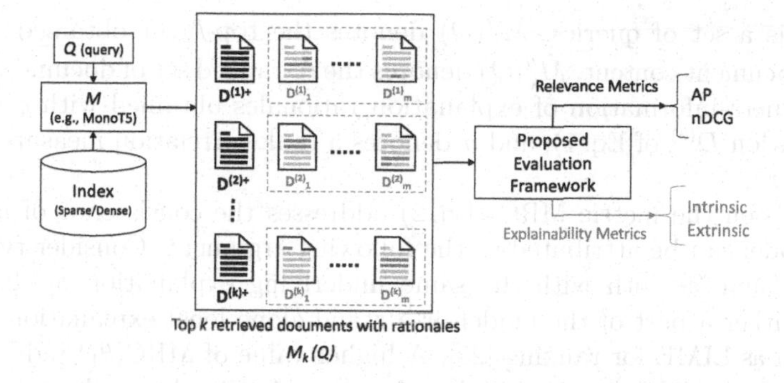

Fig. 1. The proposed workflow for measuring effectiveness of explainable IR models via intrinsic and extrinsic explanation effectiveness measures in addition to relevance-based ones. The meta-information comprised of the explanation units or rationales (shown in blue) is the additional output either obtained from a ranking model itself or with the help of a common explanation methodology (as is the case in our experiments). (Color figure online)

3 Evaluation Framework

In our proposed evaluation workflow, each IR model M^θ, in addition to outputting a ranked list of k documents $M^\theta_k(Q)$ for a query Q, also outputs a list of the most important text segments as *rationales* for each retrieved document. Our evaluation is oblivious of the exact method by which these explanations are obtained for each document - possible options being applying a local explanation algorithm for post-hoc per-document explanations via LIME [20,25], SHAP [11] etc., or by making this process an integral part of an IR model by application of feature occlusion based methods [9,28] to estimate the importance at sub-document level.

Intrinsic Explainability Measure. Given a list of m rationales for each top-ranked document D_i ($D_i \in M^\theta_k(Q)$, i.e., the top-k set for a query Q), we obtain its representation $D^{(i)}$ by concatenating the text from each rationale (each being a segment of D_i itself), i.e.,

$$\phi_m(D) = D^{(i)} = d^{(i)}_1 \oplus \ldots d^{(i)}_m. \tag{1}$$

Next, to measure how **consistent** are these rationales, we use this rationale-based representation of each document to recompute its score with the black-box neural model. Subsequently, we rerank the top-k with these modified scores and compute the agreement of this re-ranked top-k with the original one, i.e., the ranking agreement between the explanation-based representation of documents and their original ones. Formally speaking, we define the consistency metric as Mean Rank Correlation (MRC) as

$$\text{MRC}(\theta) = \frac{1}{|\mathcal{Q}|} \sum_{Q \in \mathcal{Q}} \sigma(M^\theta_k(Q), \hat{M}^\theta_k(Q)), \tag{2}$$

where $\mathcal{Q}$ is a set of queries, $M_k^\theta(Q)$ denotes the top-k set obtained with the original document content, $\hat{M}_k^\theta(Q)$ denotes the re-ranked set of documents scored with the meta-information of explanation rationales obtained with ϕ (i.e., the representation $D^{(i)}$ of Eq. 1), and σ denotes a rank correlation measure, such as Kendall's τ.

The reason the metric MRC (Eq. 2) addresses the consistency of an explanation model can be attributed to the following argument. Consider two neural models θ_1 and θ_2 both with the same underlying explanation mechanism ϕ, which is either a part of the model or a stand-alone local explanation methodology such as LIME for ranking [25]. A higher value of MRC(θ_1) indicates that the model θ_1 provides more *faithful explanations* for its observed top list of documents as compared to θ_2, because the model θ_1 when presented with only the rationales for the top documents still scores them in a relatively similar manner thus ensuring a higher correlation with the original list. The value being lower for θ_2, on the other hand, indicates that the rationales themselves do not reflect the true reason behind the relative score computation of the model θ_2.

Extrinsic Explainability Measure. The evaluation metric MRC of Eq. 2 is intrinsic in nature because it does not rely on the availability of relevance assessments of the rationales themselves. Consequently, while MRC is useful to find if a model is more consistent in its explanations than another, it cannot explicitly answer if a model's rationales align well with a human's perception of relevance. Assuming that each document $D_i \in M_k^\theta(Q)$ is comprised of a list of n_i relevant passages (may be sentences or paragraphs) of the form $R(D_i) = \{r_1^{(i)}, \ldots, r_{n_i}^{(i)}\}$[1], the main idea now is to check whether the rationales overlap with the relevant passages. Since the explanation rationales are arbitrary segments of text without being restricted to passage boundaries, a simple and effective way of measuring the degree of overlap between the explanation units and the relevant passages is to aggregate the matches in their content. More precisely, we compute MER (Mean Explanation Relevance) as

$$\mathrm{MER}(\theta) = \frac{1}{|\mathcal{Q}|mk} \sum_{Q \in \mathcal{Q}} \sum_{i=1}^{k} \sum_{j=1}^{m} \max_{j'=1}^{n_i} \omega(d_j^{(i)}, r_{j'}^{(i)}), \tag{3}$$

where ω is a similarity function. We use the cosine-similarity as a concrete realisation of ω. Similar to MRC, a high value of the MRE metric is preferable because it indicates that the rationales demonstrate a substantial overlap with relevant content[2].

Note that in our evaluation setup, we use **the same evaluation algorithm for all IR systems**, which means that our metrics **do not report how effective is an explanation model** itself (akin to, e.g., the prior work of comparing

[1] If D_i is non-relevant, $n_i = 0$ and $R(D_i) = \emptyset$.

[2] Implementation of the proposed explainability evaluation measures are available at https://github.com/saranpandian/XAIR-evaluation-metric.

the fidelity scores of LIME with SHAP [11]); rather these metrics in our setup capture **how effectively can an IR model be explained** with a specific explanation algorithm.

4 Experiment Setup

4.1 Research Questions, Datasets and IR Models

We investigate the following research questions in our experiments.

- **RQ-1**: What are the relative variations in the explanation consistency (intrinsic evaluation measure - MRC) across different IR models, i.e., are some models more *explainable* than others?
- **RQ-2**: Does MRC provide an aspect of system evaluation that is complementary to that of relevance?
- **RQ-3**: What are the relative variations in the relevance-based explanation consistency (i.e., the extrinsic evaluation measure - MER) across different IR models?
- **RQ-4**: Does the extrinsic explanation evaluation measure (MER of Eq. 3) also induce a different relative ordering of IR systems as compared to evaluating them only by relevance?

For investigating the first two research questions, we conducted experiments on the passage and document collections of MS-MARCO [14]. As topic sets for our experiments, we use the TREC DL'20 topic set, comprising 54 queries. The reason we employed the two different collections - one where the retrievable units are short passages (average 3.4 sentences for the MS-MARCO passages), and the other where they are much larger (average of 55.7 sentences for MS-MARCO documents) is to investigate the effect of document length on the intrinsic and the extrinsic explainability measures.

Recall that computing the extrinsic explainability measure of Eq. 3 requires the availability of sub-document level information. Although the MS-MARCO document collection is constituted of the text units of the MS-MARCO passage collection, they use different identifiers, as a result of which it is not possible to directly use the relevance assessment data of the passage collection as the desired sub-document level relevance information of the document collection. To construct an approximate sub-document level assessments of the MS-MARCO document collection, we make use of passage-level relevance in the document corpus using a document-passage mapping. We built this mapping by matching through the "QnA" version of the MS-MARCO dataset[3], which provides the URLs of each passage. Documents were matched to the MS-MARCO document ranking corpus IDs via URL, and passages were matched to the MS-MARCO passage ranking corpus IDs through exact text matching (after correcting for character encoding errors present in the passage corpus but not in QnA).

We assess the explainability of a variety of neural ranking models (NRMs) with a diverse set of architectures and training regimes in this work.

[3] https://github.com/microsoft/MSMARCO-Question-Answering.

- **BM25** [21] is a classic lexical retrieval model with a closed form functional expression involving term frequency, inverse document frequency and document length. We include BM25 in our experiments mainly as a point of comparison for the explanation effectiveness results of the NRMs.
- **ColBERT** [8] is a multi-representation, 'late interation' model that computes relevance based on the sum of the maximum query token similarities in a given document. It represents a strong late interaction model.
- **TCT-ColBERT** [10] is a single-representation dense retrieval model that was trained from ColBERT using distillation[4]. It represents a strong bi-encoder dense retrieval model.
- **MonoT5** [15] is a cross-encoder model that is trained to predict "true" or "false" given a prompt that includes the query and document text. We use the 'castorini/monot5-base-msmarco' model checkpoint.
- **MonoElectra** [18] is a cross-encoder model based on the ELECTRA foundation trained with hard negatives [3][5].

All the NRMs in our experiments operate with the re-ranking based setting, i.e., they employ a sparse index (specifically BM25 to retrieve the top-1000 results) which are then re-ranked by the NRM. Following the reranking step, we generate the meta-information in the form of rationales for the top $k = 10$ and $k = 50$ top documents (more details in Sect. 4.2).

To obtain the top-k for the document ranking task, we first segmented each document into non-overlapping chunks of 3 sentences, which ensures that the content fits within the 512 token limit of the underlying transformer models of the NRMs. Next, the score of a document is obtained by taking an aggregation over the scores of the individual chunk. Specifically, we used 'max' as an aggregation operator to compute the score of a document as reported in [29].

4.2 Explanation Model Settings

As the explanation model for generating the rationales in our proposed evaluation framework (the meta-information corresponding to each document retrieved in the top-list), we use an occlusion-based approach [9]. The main advantage of an occlusion-based approach (commonly as counter-factual explanations in recommender models [24]) is its run-time efficiency in comparison to approaches such as LIME for ranking (LIRME [25]) which fit a linear regressor to the data of relative score changes with occlusion. We note that the occlusion-based explanations are *not tied to a specific model architecture*; they can be applied to any relevance model that operates on the text of the query and document.

Explanations for Passage Retrieval. We employ two different granularities for generating the rationales - the first at the level of sentences, and the second at the level of word windows of length w (w being a parameter). Accordingly, we

[4] We use the `castorini/tct_colbert-msmarco` model checkpoint.

[5] We use the `crystina-z/monoELECTRA_LCE_nneg31` model checkpoint.

sample n segments (i.e., n number of sentence or word windows as per the chosen granularity, n being a parameter) from a document D, and then mask out the selected segments to construct a pseudo-document D'. Similar to [25], we then compute the relative score change induced by this masking process, which as per the local explanation principle yields the relative importance of the masked segment. The importance weights across the n segments are then distributed uniformly. The occlusion-based weights are then accumulated for each segment after each sampling step. Formally,

$$\phi(d_i) = \frac{1}{n} \frac{|\theta(Q, D) - \theta(Q, D - \bigcup_{j=1}^{n} d_j)|}{\theta(Q, D)} \tag{4}$$

where d_i a segment of D ($i = 1, \ldots, n$). Finally, we output the top-m segments with the highest weights (as per the ϕ values computed via Eq. 4) as the rationales for retrieving D for query Q.

Explanations for Document Retrieval. The process is largely similar to that of passage retrieval with a small number of differences as follows. Firstly, due to the large size of the documents (55.7 sentences on an average), it is not possible to encode an entire document's representation via a transformer model during the training or the inference phase. Due to this reason, we partition each document into fixed length chunks, and then obtain the overall score of the document by computing the maximum over the individual scores for each chunk [29]. As the number of sentences defining a chunk, we use the value of 3 (this was chosen so that each chunk in the document collection is approximately of the same size as the retrievable units of the MS-MARCO passage collection, the average length being 3.4 sentences).

Secondly, due to the large length of the documents we restrict the minimum granularity of rationales to individual sentences. Similar to explaining passage retrieval with m word windows, for document retrieval experiments also we report m top explanation units with the highest fidelity scores (similar to Eq. 4), the explanation units being sentences for document retrieval. We use the same parameter name - m to indicate the number of explanations units; whether this applies for a word or a sentence is to be understood from the context. We iterate through each sentence in a chunk and compute the relative score change induced by its occlusion, i.e.,

$$\phi(d_i) = \frac{\theta(Q, D) - \theta(Q, D - d_i)}{\theta(Q, D)}. \tag{5}$$

We then employ a greedy approach and select the one with the highest $\phi(d_i)$ value and remove it from the document D. We then the repeat this step $m - 1$ more times to eventually yield a total of m sentences as the rationales for an IR model θ.

Table 1. A comparison of the relevance and the explanation consistency of different IR models on the MS-MARCO passage collection for top-10 and top-50 search results with both sentences and word windows employed as explanation units. For these experiments, we set the size of the word windows to 5 and the number of rationales to 6 for the word windows, and 1 for the sentences. The best results in terms of relevance and the explanation effectiveness of the NRMs are bold-faced.

	Relevance (nDCG)		Explanation Effectiveness			
			MRC (Sentences)		MRC (Word windows)	
Model	top-10	top-50	top-10	top-50	top-10	top-50
BM25	0.4910	0.4889	0.4000	0.1503	0.3029	0.1232
ColBERT	0.6900	0.6400	**0.4502**	0.1923	**0.2790**	0.1058
TCT-ColBERT	0.6900	0.6328	0.2938	0.1480	0.1926	0.0974
MonoT5	0.7133	0.6660	0.3481	0.2230	0.1844	0.1102
MonoElectra	**0.7460**	**0.6958**	0.4165	**0.2262**	0.2644	**0.1303**

5 Explainability Evaluation

Passage Ranking Results. Table 1 presents the results of evaluating different IR systems on the MS-MARCO passage collection in terms of relevance and our proposed intrinsic explanation evaluation measure MRC. The extrinsic explanation measure - MER cannot be computed for the MS-MARCO passage collection as there is no sub-document level information available for these retrievable units. We now present the following key observations from Table 1.

Relevance-Based Observations: As expected, all the NRMs significantly outperform BM25 in terms of relevance-only based evaluation, which is consistent with the findings of existing research [4].

Relevance vs. Explainability: Interestingly, the model with the best retrieval effectiveness is not necessarily the most well-explainable model. This can be seen from the fact that the MRC@10 values for the best performing model in terms of relevance, i.e., MonoElectra, are not the best ones (ColBERT MRC@10 values are better). This answers RQ-2 in affirmative in the sense that the explanation quality potentially provides an alternate way towards IR model evaluation different from that of relevance.

NRMs with the Best Explanation Consistencies: BM25 mostly yields more consistent explanations in comparison to NRMs. This can be seen by comparing the gray values under the two column groups with the corresponding non-gray ones across the same column, e.g., BM25 achieves the highest MRC@10 of 0.3029 with word-window based rationales, which is higher than the best explanation effectiveness obtained by an NRM (0.2790 with ColBERT).

Sentence vs. Word-Window Rationales: Table 1 shows that the MRC values of the sentence level rationales are higher than that of the word window based ones, which indicates that sentence-level rationales are more consistent

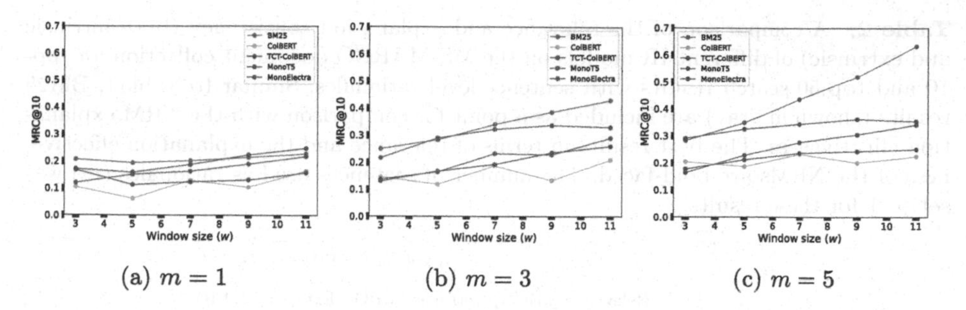

(a) $m = 1$ (b) $m = 3$ (c) $m = 5$

Fig. 2. Effects of varying the size of the explanation units on the intrinsic consistency of the explanations (MRC of Eq. 2) on the top-10 search results obtained with several NRMs. A comparison across the plots reveals the the effect of the variations in the number of rationales provided as explanations (m). A general observation is that a higher number of explanations coupled with larger explanation units tend to provide more consistent explanations.

than word-window based ones. This can likely be attributed to the fact that, firstly, sentences retain more informative content of a document which is useful for providing consistent explanations, and that secondly, transformer-based approaches due to their underlying language models usually favour well-formed sentences [13]. However, short word segments are more preferable from a user's perspective because reading them potentially requires less effort, as a result of which existing work on text-based explanations have commonly used short word windows as rationales [6,11,20,25]. It is also observed that the sentence rationales of NRMs are more consistent in comparison to the BM25 ones. This is due to the fact that the linguistic coherence of word sequences in sentence-level explanations work well for the NRMs, BM25 being oblivious of word ordering.

Explaining top-10 vs. top-50 Results: From Table 1, we see that the MRC@50 values are lower than those of MRC@10, which indicates that it is easier to explain a smaller number of search results. Again, this is expected because documents towards the very top ranks would actually contain text segments that are potentially relevant to the information need thus amenable to more consistent explanations.

Document Ranking Results. Similar to Table 1, in Table 2 we show the results on the MS-MARCO document ranking task. For this task, since documents are much larger than MS-MARCO passages, we restrict the granularity of explanation units to sentences only. In addition to the intrinsic explainability measure, Table 2 also reports values for the extrinsic one obtained with sub-document level relevance. Following are the key observations from Table 2.

Intrinsic Explanation Consistency: The results are similar to that of the passage task (Table 1) in the sense that we note considerable variations in the reported MRC values across IR models. Also, similar to the passage task, we

Table 2. A comparison of the relevance and explanation consistency (both intrinsic and extrinsic) of different IR models on the MS-MARCO document collection for top-10 and top-50 search results with sentence-level rationales. Similar to Table 1, BM25 results (shown in gray) are included as a point for comparison with the NRM explanation effectiveness. The best results in terms of relevance and the explanation effectiveness of the NRMs are bold-faced. The number of sentences used as rationales (m) was set to 1 for these results.

			Explanation Effectiveness			
	Relevance (nDCG)		Intrinsic (MRC)		Extrinsic (MER)	
Model	top-10	top-50	top-10	top-50	top-10	top-50
BM25	0.5213	0.5325	0.1660	0.1798	0.2024	0.1823
ColBERT	0.5675	0.5686	0.2064	0.2801	0.1846	0.1647
TCT-ColBERT	0.5787	0.5654	0.2420	0.2721	**0.2069**	0.1701
MonoT5	0.6045	0.6048	0.2133	**0.3265**	0.2039	0.1703
MonoElectra	**0.6185**	**0.6051**	**0.2493**	0.2863	0.1959	**0.1739**

observe that the intrinsic explanation consistency of the NRMs is better than that of BM25's.

Extrinsic Explanation Consistency: It is observed from the MER columns of Table 2 that the extrinsic evaluation effectiveness is not necessarily correlated with the intrinsic one, e.g., the best model in terms of intrinsic consistency (MRC) of explanations of top-10 documents is MonoElectra (similar to the passage ranking task), whereas the best model with the extrinsic explanation consistency is TCT-ColBERT. This indicates that the parameterised ranking function of TCT-ColBERT puts more attention, on an average, to the relevant pieces of text in comparison to the other NRMs.

In relation to RQ-3, we can thus comment that similar to the MRC variations, we do observe a noticeable variations in the extrinsic explainability measure as well indicating that some NRMs are substantially better explainable than others in terms of the relevance of the rationales.

Explaining top-10 vs. top-50 Results: Different from Table 1, in Table 2 we observe that the intrinsic explanations for the top-50 results are better than the top-10 ones (compare MRC@10 vs. MRC@50 across the different IR models). This can be attributed to the fact that due to the large length of the documents, it is still likely to find some partially relevant sentences even in documents that are not at the very first search result page (e.g., within top-10). An NRM by putting attention to these partially relevant sentences can yield consistent explanations even within the top-50 set of documents.

However, the trend is reversed for the extrinsic measure MER, where we observe that the top-10 results are better than top-50 ones, the reason being it is more likely for the rationales of the top-10 documents to overlap with the relevant passages.

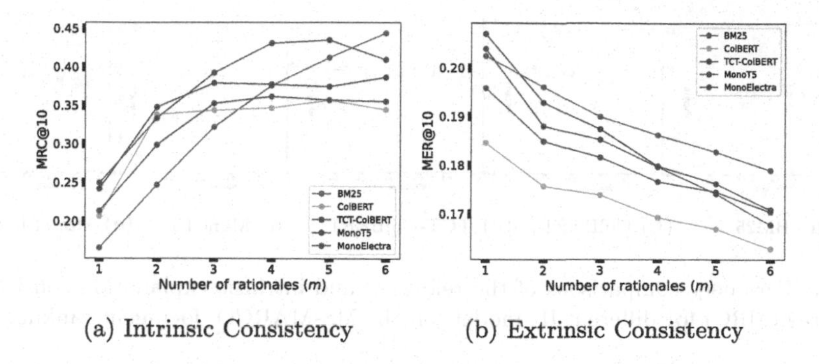

(a) Intrinsic Consistency (b) Extrinsic Consistency

Fig. 3. Effect of the number of rationales on the explanation consistency metrics across different NRMs for the MS-MARCO document ranking task.

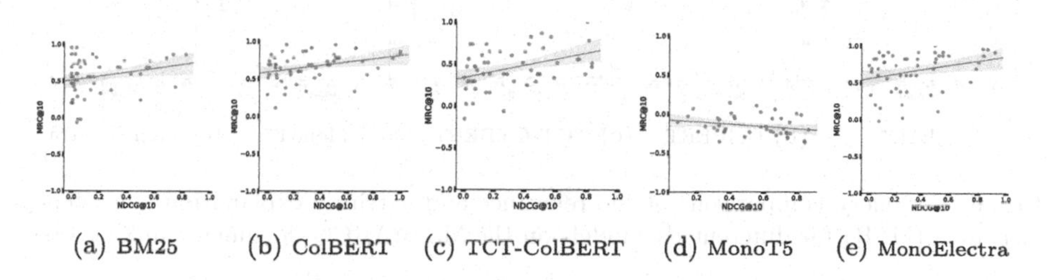

(a) BM25 (b) ColBERT (c) TCT-ColBERT (d) MonoT5 (e) MonoElectra

Fig. 4. Per-query comparisons of the relevance and intrinsic explanation consistency measures (MRC) for different IR models on the MS-MARCO passage ranking task.

Further Analysis. For the passage ranking task, we now investigate the effects of the variations in i) the number of rationales used to explain the search results, i.e., m, and ii) the size of the explanations in terms of the number of words or sentences depending on the granularity - word windows, or sentences, i.e., w. Figure 2 shows the effect of variations in the length of the rationales (in terms of the number of words constituting the explanation units) for 3 different values of the number of explanations provided.

We observe that, first, the explanation consistency increases with an increase in the number of rationales provided (compare the MRC values across the different plots). This is expected because with a higher number of rationales, each document's attention-focused representation (Eq. 1) tends towards its original representation, i.e., the explanations themselves tend to cover the entire document. While this shows up as increased values of MRC, as pragmatic reasons, too large a value of m is not desirable. Moreover, we also observe that for most NRMs, increasing the rationale length leads to more consistent explanations. Too large text windows can lead to inclusion of potentially non-relevant text in the explanations, which due to the semantic incoherence with the queries can lead to decreased MRC values (as can be seen from the drops in MRC from $w = 9$ to $w = 11$ in most cases).

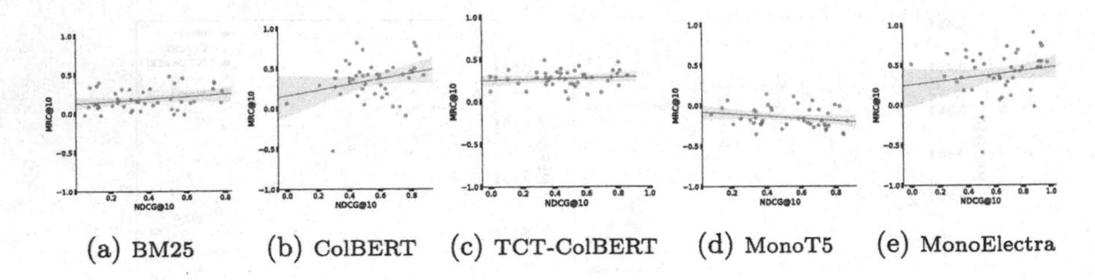

Fig. 5. Per-query comparisons of the relevance and intrinsic explanation consistency measures (MRC) for different IR models on the MS-MARCO document ranking task.

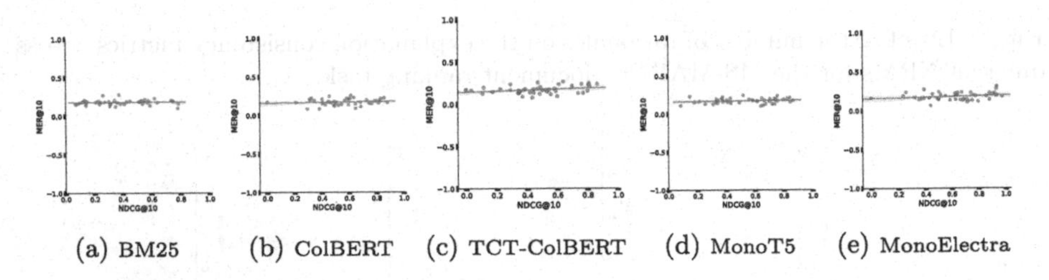

Fig. 6. Per-query comparisons of the relevance and extrinsic explanation consistency measures (MER) for different IR models on the MS-MARCO document ranking task.

We now conduct a similar parameter variation analysis for the document ranking task. Since the explanations are at sentence-level, this task does not involve the parameter w (the size of explanations). Figures 3a and 3b show the effects of variations of the number of rationales on the intrinsic and the extrinsic explanation consistencies, respectively. While with an increase in m, the intrinsic measure mostly increases, a reverse trend is observed for the sub-document relevance-based extrinsic approach. The argument on why MRC increases with an increase in the number of rationales is the same as that of passages, i.e., the attention-focused representation of a document tends to better represent its core topic. On the other hand, a likely reason for the decrease in the MER values is that the sentences with lower explanation weights (but still included as a part of the explanations due to relatively higher values of m) represent segments of a document that are potentially not relevant to the query. This means that the proportion of sub-document level relevant information in the explanations (Eq. 3) progressively decreases.

Explanation Effectiveness vs. Relevance. In relation to research questions RQ-2 and RQ-4, we now investigate the correlation between a relevance measure (specifically, nDCG@10 as used in our experiments) and the explanation consistency measures. Tables 1 and 2 already demonstrated that the best performing NRM in terms of relevance is not necessarily the one yielding the most consistent explanations. This observation is reinforced by Fig. 4 which shows a scatter-plot of the per-query nDCG@10 vs. the MRC values. An interesting observation

Table 3. The top-2 documents of the best performing query in terms of the intrinsic consistency measure for MonoElectra on the MS-MARCO passage ranking task. The top-3 rationales are shown with 3 different color shades (a deeper shade indicating a higher attention weight).

Query	Rationales of the top-3 documents
What is the UN FAO?	United Nations Food and Agriculture Organization (FAO) The Food and Agriculture Organization of the United Nations is an agency that leads international efforts to defeat hunger. The Food and Agriculture Organization of the United Nations is an agency of the United Nations that leads international efforts to defeat hunger.
(MRC@10 = 0.9112)	Role of the FAO. The Food and Agriculture Organization of the United Nations is an agency of the United Nations that leads international efforts to defeat hunger. Serving both developed and developing countries, FAO acts as a neutral forum where all nations meet as equals to negotiate agreements and debate policy.

from Fig. 4 is that all the models (even BM25) registers only a small correlation between the relevance and explanation consistency. MonoT5 even yields a negative correlation of MRC with nDCG@10. This suggests that explanation consistency can potentially serve as an evaluation dimension complementary to that of relevance.

Similar observations can also be made from Figs. 5 and 6. MER is slightly more correlated with relevance because it involves measuring an overlap of the rationales with the relevant segments of a document. However, the insights gained from the MER metric is still different from both relevance and the intrinsic consistency, which, in turn, suggests that MER like MRC can potentially be used as another dimension to evaluate NRMs.

Sample Explanations. Table 3 shows an example query with the best intrinsic consistency measure on the MS-MARCO passage ranking task. It can be seen that rationales provided for this query does also qualitatively indicate pertinent explanations. This can be seen from the fact that the NRM (MonoElectra) has been able to bridge the semantic gap between the query term 'FAO', and the ones in the top-ranked documents such as 'Food', 'Agricultural' and 'Organization'. The same argument also applies for the query term 'UN', and the document terms 'United' and 'Nations'.

Concluding Remarks. We introduced a evaluation framework wherein rationales – represented as text snippets from the document – are available alongside search results. These rationales can then be evaluated alongside traditional relevance measures to assess the interpretability of the models. Using a basic occlusion-based technique for producing rationales, we find that the **systems that produce the most relevant results may not necessarily be the most explainable** – both in terms of intrinsic and extrinsic measures of explainability.

As a future work, tools could be developed to enable easier qualitative evaluation of rationales [7], and user studies could be conducted to test how helpful rationales are to end-users.

References

1. Amiri, S.S., et al.: Data representing ground-truth explanations to evaluate XAI methods. CoRR abs/2011.09892 (2020). https://arxiv.org/abs/2011.09892
2. Anand, A., Lyu, L., Idahl, M., Wang, Y., Wallat, J., Zhang, Z.: Explainable information retrieval: a survey (2022). https://doi.org/10.48550/ARXIV.2211.02405, https://arxiv.org/abs/2211.02405
3. Clark, K., Luong, M.T., Le, Q.V., Manning, C.D.: ELECTRA: pre-training text encoders as discriminators rather than generators. ArXiv abs/2003.10555 (2020)
4. Craswell, N., Mitra, B., Yilmaz, E., Campos, D., Voorhees, E.M.: Overview of the TREC 2019 deep learning track (2020)
5. Dai, Z., Callan, J.: Context-aware term weighting for first stage passage retrieval. In: Proceedings of the 43rd International ACM SIGIR Conference on Research and Development in Information Retrieval, p. 1533–1536. SIGIR 2020, Association for Computing Machinery, New York, NY, USA (2020). https://doi.org/10.1145/3397271.3401204
6. Fernando, Z.T., Singh, J., Anand, A.: A study on the interpretability of neural retrieval models using DeepSHAP. In: Proceedings of SIGIR 2019, pp. 1005–1008 (2019)
7. Jose, K.M., Nguyen, T., MacAvaney, S., Dalton, J., Yates, A.: DiffIR: exploring differences in ranking models' behavior. In: Proceedings of the 44th International ACM SIGIR Conference on Research and Development in Information Retrieval (2021). https://doi.org/10.1145/3404835.3462784
8. Khattab, O., Zaharia, M.: ColBERT: efficient and effective passage search via contextualized late interaction over BERT. In: Proceedings of SIGIR 2020, p. 39–48 (2020)
9. Li, J., Monroe, W., Jurafsky, D.: Understanding neural networks through representation erasure. arXiv preprint arXiv:1612.08220 (2016)
10. Lin, S.C., Yang, J.H., Lin, J.: In-batch negatives for knowledge distillation with tightly-coupled teachers for dense retrieval. In: Proceedings of the 6th Workshop on Representation Learning for NLP (RepL4NLP-2021), pp. 163–173. Association for Computational Linguistics, Online (2021). https://doi.org/10.18653/v1/2021.repl4nlp-1.17, https://aclanthology.org/2021.repl4nlp-1.17
11. Lundberg, S.M., Lee, S.I.: A unified approach to interpreting model predictions. In: Guyon, I., et al., (eds.) Advances in Neural Information Processing Systems. vol. 30. Curran Associates, Inc. (2017). https://proceedings.neurips.cc/paper/2017/file/8a20a8621978632d76c43dfd28b67767-Paper.pdf
12. Lyu, L., Anand, A.: Listwise explanations for ranking models using multiple explainers. In: Kamps, J., et al. Advances in Information Retrieval. ECIR (1). LNCS, vol. 13980, pp. 653–668. Springer, Cham (2023). https://doi.org/10.1007/978-3-031-28244-7_41
13. MacAvaney, S., Feldman, S., Goharian, N., Downey, D., Cohan, A.: ABNIRML: analyzing the behavior of neural IR models. TACL (2022). https://doi.org/10.1162/tacl_a_00457, https://direct.mit.edu/tacl/article/doi/10.1162/tacl_a_00457/110013/ABNIRML-Analyzing-the-Behavior-of-Neural-IR-Models

14. Nguyen, T., et al.: MS MARCO: a human generated machine reading comprehension dataset. In: CoCo@NIPS. CEUR Workshop Proceedings, vol. 1773 (2016)
15. Nogueira, R., Jiang, Z., Pradeep, R., Lin, J.: Document ranking with a pretrained sequence-to-sequence model. In: Findings of the Association for Computational Linguistics: EMNLP 2020, pp. 708–718. Association for Computational Linguistics, Online (2020). https://doi.org/10.18653/v1/2020.findings-emnlp.63, https://aclanthology.org/2020.findings-emnlp.63
16. Nogueira, R., Lin, J., Epistemic, A.: From doc2query to docTTTTTquery. Online preprint 6 (2019)
17. Oramas, J., Wang, K., Tuytelaars, T.: Visual explanation by interpretation: improving visual feedback capabilities of deep neural networks. In: International Conference on Learning Representations (2019). https://openreview.net/forum?id=H1ziPjC5Fm
18. Pradeep, R., Liu, Y., Zhang, X., Li, Y., Yates, A., Lin, J.: Squeezing water from a stone: a bag of tricks for further improving cross-encoder effectiveness for reranking. In: Hagen, M., et al. (eds.) ECIR 2022. LNCS, vol. 13185, pp. 655–670. Springer, Cham (2022). https://doi.org/10.1007/978-3-030-99736-6_44
19. Pradeep, R., Nogueira, R., Lin, J.: The expando-mono-duo design pattern for text ranking with pretrained sequence-to-sequence models. arXiv preprint arXiv:2101.05667 (2021)
20. Ribeiro, M.T., Singh, S., Guestrin, C.: "why should i trust you?": explaining the predictions of any classifier. In: Proceedings of the 22nd ACM SIGKDD International Conference on Knowledge Discovery and Data Mining, p. 1135–1144. KDD 2016, Association for Computing Machinery, New York, NY, USA (2016)
21. Robertson, S., Walker, S., Beaulieu, M., Gatford, M., Payne, A.: Okapi at TREC-4 (1996)
22. Sen, P., Ganguly, D., Verma, M., Jones, G.J.F.: The curious case of IR explainability: explaining document scores within and across ranking models. In: SIGIR, pp. 2069–2072. ACM (2020)
23. Sen, P., Saha, S., Ganguly, D., Verma, M., Roy, D.: Measuring and comparing the consistency of IR models for query pairs with similar and different information needs. In: CIKM, pp. 4449–4453. ACM (2022)
24. Tan, J., Xu, S., Ge, Y., Li, Y., Chen, X., Zhang, Y.: Counterfactual explainable recommendation. In: Proceedings of the 30th ACM International Conference on Information & Knowledge Management, pp. 1784–1793 (2021)
25. Verma, M., Ganguly, D.: LIRME: locally interpretable ranking model explanation. In: Proceedings of SIGIR 2019, pp. 1281–1284 (2019)
26. Xiong, L., et al.: Approximate nearest neighbor negative contrastive learning for dense text retrieval. arXiv preprint arXiv:2007.00808 (2020)
27. Yang, M., Kim, B.: BIM: towards quantitative evaluation of interpretability methods with ground truth. CoRR abs/1907.09701 (2019). http://arxiv.org/abs/1907.09701
28. Zeiler, M.D., Fergus, R.: Visualizing and understanding convolutional networks. In: Fleet, D., Pajdla, T., Schiele, B., Tuytelaars, T. (eds.) ECCV 2014. LNCS, vol. 8689, pp. 818–833. Springer, Cham (2014). https://doi.org/10.1007/978-3-319-10590-1_53
29. Zhang, H., Cormack, G.V., Grossman, M.R., Smucker, M.D.: Evaluating sentence-level relevance feedback for high-recall information retrieval. Inform. Retrieval J. **23**, 1–26 (2020)

Is Interpretable Machine Learning Effective at Feature Selection for Neural Learning-to-Rank?

Lijun Lyu[1(✉)], Nirmal Roy[1], Harrie Oosterhuis[2], and Avishek Anand[1]

[1] Delft University of Technology, Delft, The Netherlands
{L.Lyu,N.Roy,Avishek.Anand}@tudelft.nl
[2] Radboud University, Nijmegen, The Netherlands
harrie.oosterhuis@ru.nl

Abstract. Neural ranking models have become increasingly popular for real-world search and recommendation systems in recent years. Unlike their tree-based counterparts, neural models are much less interpretable. That is, it is very difficult to understand their inner workings and answer questions like *how do they make their ranking decisions?* or *what document features do they find important?* This is particularly disadvantageous since interpretability is highly important for real-world systems. In this work, we explore feature selection for neural learning-to-rank (LTR). In particular, we investigate six widely-used methods from the field of interpretable machine learning (ML) and introduce our own modification, to select the input features that are most important to the ranking behavior. To understand whether these methods are useful for practitioners, we further study whether they contribute to efficiency enhancement. Our experimental results reveal a large feature redundancy in several LTR benchmarks: the local selection method TabNet can achieve optimal ranking performance with less than 10 features; the global methods, particularly our G-L2x, require slightly more selected features, but exhibit higher potential in improving efficiency. We hope that our analysis of these feature selection methods will bring the fields of interpretable ML and LTR closer together.

1 Introduction

Learning-to-rank (LTR) is at the core of many information retrieval (IR) and recommendation tasks [37]. The defining characteristic of LTR, and what differentiates it from other machine learning (ML) areas, is that LTR methods aim to predict the optimal ordering of items. This means that LTR methods are not trying to estimate the exact relevance of an item, but instead predict relative relevance differences, i.e., whether it is more or less relevant than other items. Traditionally, the most widely adopted and prevalent LTR methods were based on Gradient Boosted Decision Trees (GBDT) [12,29,63]. However, in recent years, neural LTR methods have become increasingly popular [21,44,45]. Recently, [49]

N. Goharian et al. (Eds.): ECIR 2024, LNCS 14611, pp. 384–402, 2024.
https://doi.org/10.1007/978-3-031-56066-8_29

have shown that neural models can provide ranking performance that is comparable, and sometimes better, than that of state-of-the-art GBDT LTR models on established LTR benchmark datasets [14,17,47]. It thus seems likely that the prevalence of neural LTR models will only continue to grow in the foreseeable future.

Besides the quality of the results that ranking systems return, there is an increasing interest in building trustworthy systems through interpretability, e.g., by understanding which features contribute the most to ranking results. Additionally, the speed at which results are provided is also highly important [3,5,7]. Users expect ranking systems to be highly responsive and previous work indicates that even half-second increases in latency can contribute to a negative user experience [7]. A large part of ranking latency stems from the retrieval and computation of input features for the ranking model. Consequently, *feature selection* for ranking systems has been an important topic in the LTR field [22,23,43,46,50,57,66]. These methods reduce the number of features used, thereby helping users understand and greatly reduce latency and infrastructure costs, while maintaining ranking quality as much as possible. In line with the history of the LTR field, existing work on feature selection has predominantly focused on GBDT and support-vector-machine (SVM) ranking models [21,26], but has overlooked neural ranking models. To the best of our knowledge, only two existing works have looked at feature selection for neural LTR [46,50]. This scarcity is in stark contrast with the importance of feature selection and the increasing prevalence of neural models in LTR.

Outside of the LTR field, feature selection for neural models has received much more attention, for the sake of efficiency [34,35], and also to better understand the model behaviours [4,70]. Those methods mainly come from the *interpretable* ML field [18,42], where the idea is that the so-called *concrete* feature selection can give insights into what input information a ML model uses to make decisions. This tactic has already been successfully applied to natural language processing [69], computer vision [6], and tabular data [4,67]. Accordingly, there is a potential for these methods to also enable *embedded feature selection* for neural LTR models, where the selection and prediction are optimized simultaneously. However, the effectiveness of these interpretable ML methods for LTR tasks is currently unexplored, and thus, it remains unclear whether their application can translate into useful insights for LTR practitioners.

The goal of this work is to investigate whether six prevalent feature selection methods – each representing one of the main branches of interpretable ML field – can be applied effectively to neural LTR. In addition, we also propose a novel method with minor modifications. Our aim is to bridge the gap between the two fields by translating the important concepts of the interpretable ML field to the LTR setting, and by demonstrating how interpretable ML methods can be adapted for the LTR task. Moreover, our experiments consider whether these methods can bring efficiency into the practical application by reducing irrelevant input features for neural ranking models.

Our results reveal a large feature redundancy in LTR benchmark datasets, but this redundancy can be understood differently for interpretability and for efficiency: For understanding the model, feature selection can vary per document and less than 10 features are required to approximate optimal ranking behavior. In contrast, for practical efficiency purposes, the selection should be static, and then 30% of features are needed. We conclude that – when adapted for the LTR task – not all, but a few interpretable ML methods lead to effective and practical feature selection for neural LTR.

To the best of our knowledge, this is the first work that extensively studies embedded feature selection for neural LTR. We hope our contributions bring more attention to the potential of interpretable ML for IR field. To stimulate future work and enable reproducibility, we have made our implementation publicly available at: https://github.com/GarfieldLyu/NeuralFeatureSelectionLTR (MIT license).

2 Related Work

Learning-to-Rank (LTR). Traditional LTR algorithms mainly rely on ML models, such as SVMs and decision trees to learn the correlation between numerical input features and human-annotated relevance labels [15,20,27,31,36,63,65,68]. Neural approaches [10,11,13,52,59,64] have also been proposed, but did not show significant improvements over traditional non-neural models. Inspired by the transformer architecture [61], recent works have also adapted self-attention [44,45,49] and produced the neural LTR methods that outperform LambdaMART [63], albeit with a relatively small difference. It shows that neural rankers can provide competitive performance, consequently, the interest and effort towards neural models for LTR are expected to increase considerably in the near future.

Efficiency is crucial in real-world systems since users expect them to be highly responsive [3,5,7]. Aside from model execution, the latency of ranking system is largely due to feature construction, as it happens *on-the-fly* for incoming queries. Thus, efficiency is often reached by reducing (expensive) features. Previous works [21,62] apply a cascading setup to reduce the usage of expensive features. Another growing trend in LTR is to design *interpretable models*. Existing methods rely on specific architecture design, such as general additive model (GAM) [71] or a feature-interaction constrained and depth-reduced tree model [38].

Feature Selection for LTR. Feature selection can achieve both efficiency and interpretability [4,34,35,70]. By selecting a subset of input features, the input complexity of models is reduced while maintaining competitive performance. This helps with (1) efficiency as it avoids unnecessary construction of features [34,35], and (2) interpretability as fewer input features are involved in prediction [4,70].

Existing feature selection methods in LTR are classified commonly as *filter*, *wrapper* and *embedded* methods [22,23,46]. Filter and wrapper methods are applied to given static ranking models which are not updated in any way; filter methods are model-agnostic [22] while wrapper methods are designed for a particular type of model [23]. In this work we will focus on the third category, embedded methods, where feature selection is performed simultaneously with model optimization. Most embedded methods are limited to particular model designs such as SVMs [31–33] or decision trees [38,43,66]. To the best of our knowledge, only two methods are designed for neural LTR [46,50]: one applies group regularization methods [50] to reduce both input and other model parameters; the other [46] uses the gradients of a static ranking model to infer feature importance, and thus it belongs to the *filter* category. We do not investigate these two methods further, as the focus of this work is on *embedded* input feature selection methods.

Interpretable Machine Learning. The earliest work in interpretable ML attempted to explain a trained model in *post-hoc* manner, mainly relying on input perturbations [39,51], gradients [56,58] and so on [55]. In parallel, more recent works advocated intrinsically interpretable models, that are categorized as *interpretable-by-design* methods [1,2,54]. For neural networks, explaining the decision path is challenging due to the large set of parameters. Therefore, the more prevalent choice for intrinsic interpretable neural models is to shift the transparency to the input features. Namely, the final prediction comes from a subset selection of input elements, e.g., words or pixels and the rest irrelevant features are masked out [16,35,69]. Importantly, this selection decision can be learned jointly with the predictive accuracy of a model. Thereby, we limit our research focus in intrinsic interpretable ML models.

Due to the discrete nature of selection, many approaches such as L2X [16], *Concrete AutoEncoders* (CAE) [6], *Instance-wise Feature grouping* (IFG) [41] apply Gumbel-Softmax sampling [24] to enable backpropagation through the feature selection process. Alternatively, regularization is also a commonly-used feature selection approach in traditional ML algorithms [31,60], and is applicable to neural models, i.e., with INVASE [67] or LassoNet [34]. Moreover, TabNet [4] applies both regularization and the sparsemax activation function [40] to realize sparse selection. These approaches have been successfully applied in language, vision and tabular domains, and suggested that the resulting feature selections substantially improved the user understanding of models and datasets [28,53].

Despite their success in other domains, we find that the above-mentioned feature selection methods for neural models (L2X, CAE, IFG, INVASE, LassoNet and TabNet) have not been studied in the LTR setting. In response, we hope to bridge this gap between the interpretable ML and the LTR field by adapting and applying these methods to neural ranking models.

3 Background

3.1 Learning-to-Rank (LTR)

The LTR task can be formulated as optimizing a scoring function f that given item features x predicts an item score $f(x) \in \mathbb{R}$, so that ordering items according to

Table 1. Properties of feature selection methods from the interpretable ML field as discussed in Sect. 3.

Method	Global	Local	Sampling	Regularization	Fixed-Budget	Composable
L2X [16]		✓	✓		✓	✓
INVASE [67]		✓	✓	✓		✓
CAE [6]	✓		✓		✓	✓
IFG [41]		✓	✓			✓
LASSONET [34]	✓			✓		✓
TABNET [4]		✓		✓		
G-L2X (ours)	✓		✓		✓	✓

their scores corresponds to the optimal ranking [37]. Generally, there are relevance labels y available for each item, often these are labels provided by experts where $y \in \{0, 1, 2, 3, 4\}$ [14,47,48]. Given a training set $\mathcal{D}_q = \{(x_i, y_i)\}_{i=1}^{N_q}$ for a single query q, optimization is done by minimizing a LTR loss, for instance, the *softmax cross entropy loss* [9,13]:

$$\mathcal{L}(f \mid \mathcal{D}_q) = -\frac{1}{|\mathcal{D}_q|} \sum_{(x,y)\in\mathcal{D}_q} \sum_{i=1}^{N_q} y_i \log \sigma(x_i \mid f, \mathcal{D}_q), \tag{1}$$

where σ is the softmax activation function:

$$\sigma(x \mid f, \mathcal{D}_q) = \frac{\exp(f(x))}{\sum_{x'\in\mathcal{D}_q} \exp(f(x'))}. \tag{2}$$

The resulting f is then commonly evaluated with a ranking metric, for instance, the widely-used normalized discounted cumulative gain metric (NDCG) [25].

3.2 Properties of Feature Selection Methods

As discussed before, feature selection is used in the interpretable ML field to better understand which input features ML models use to make their predictions. Furthermore, feature selection is also important to LTR for increasing the efficiency of ranking systems. However, selecting a subset of input features without compromising the model performance is an NP-hard problem, since the number of possible subsets grows exponentially with the number of available features [23,46]. As a solution, the interpretable ML field has proposed several methods that approach feature selection as an optimization problem. We will now lay out several important properties that can be used to categorize these methods, which will be elaborated in next section.

Global vs. Local. Global methods select a single subset of features for the entire dataset, whereas local methods can vary their selection over different items.

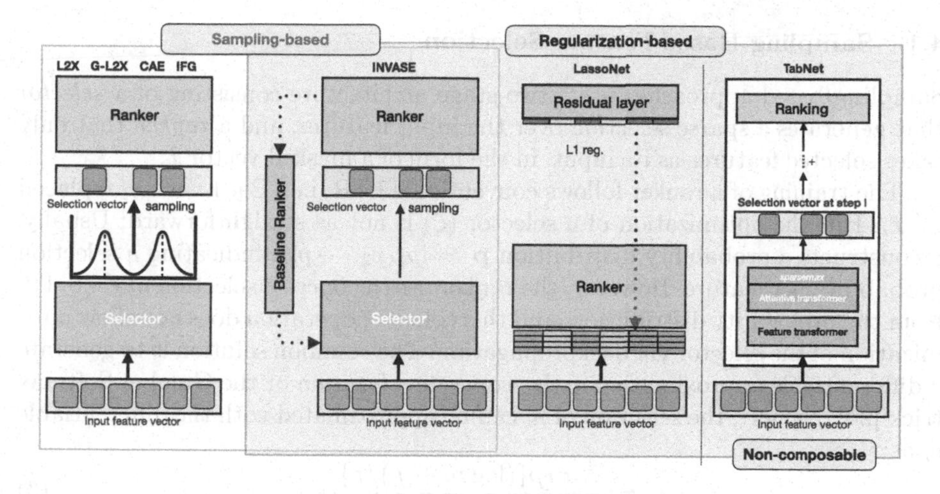

Fig. 1. Methods overview, as described in Sect. 4.

Composable vs. Non-composable. Non-composable methods are designed for a specific model architecture, and therefore, they can only perform feature selection for those models. Conversely, composable methods are not constrained to specific architectures, and thus, they work for any (differentiable) model.

Fixed-Budget vs. Budget-Agnostic. Fixed-budget methods work with a predefined selection budget, i.e., what number of features should be selected in total, or a cost per feature and a maximum total cost for the selection. Their counterparts are budget-agnostic methods that do not use an explicit budget, consequently, one has to carefully fine-tune their hyper-parameters to achieve a desired performance-sparsity trade-off.

Sampling-Based vs. Regularization-Based. As their names imply, sampling-based methods optimize a sampling procedure to perform the feature selection, whereas regularization-based methods use an added regularization loss to stimulate sparsity in feature selection. While these groups apply very different approaches, whether one is significantly more useful for LTR purposes than the other remains unknown.

4 Feature Selection from Interpretable ML for LTR

In this section, we present a brief technical overview of our selection of six interpretable ML methods and their adaption to neural LTR models, and propose our G-L2x method based on a minor modification. While any ranking loss can be chosen, we use a listwise softmax cross entropy (Eq. 1) with all methods, for the sake of simplicity. Therefore, the training of each query is conducted after generating the output of all documents associated with the query. Table 1 highlights the properties of all methods and Fig. 1 provides a visual overview to accompany this section.

4.1 Sampling-Based Feature Selection

Sampling-based approaches use a two-stage architecture consisting of a *selector* that generates a sparse selection over the input features; and a *ranker* that only takes selected features as its input, in the form of a masked vector $\hat{x}$.

The training of a ranker follows conventional LTR, i.e., Eq. 1 with x_i replaced by $\hat{x_i}$. But the optimization of a selector (ζ) is not as straightforward; Usually, ζ constructs a probability distribution $\mathbf{p} = [p_1, p_2, \cdots p_d]$, indicating a selection probability per feature. However, the ranker uses a concrete selection $m \in \{0,1\}^d$ from the probability distribution, and this concrete operation does not allow optimization of the selector via backpropagation. The common solution is to generate a differentiable approximation $\tilde{m}$, by *concrete relaxation* or the Gumbel-Softmax trick [24]. Namely, the selection of p_i can be approximated with the differentiable c_i as:

$$c_i = \frac{exp\{(\log p_i + g_i)/\tau\}}{\sum_{j=1}^{d} exp\{(\log p_j + g_i)/\tau\}}, \tag{3}$$

where g is the Gumbel noise and $\tau \in \mathbb{R}^{>0}$ is the temperature parameter. Now, the selector ζ can be optimized with stochastic gradient descent by using $\tilde{m}$. The following four sampling-based methods apply this overall procedure, but differ in how they generate $\mathbf{p}$ and $\tilde{m}$.

Learning to Explain (L2X). L2X [16] is a local selection method since its neural selector generates a probability distribution $\mathbf{p}$ for each individual input instance. To generate $\tilde{m}$, L2X repeats the sampling procedure k times (Eq. 3), and subsequently, uses the maximum c_i out of the k repeats for the i_{th} element in $\tilde{m}$. The intention behind this maximization step is to make the top-k important features more likely to have high probability scores (ideally close to 1).

Global Learning to Explain (G-L2X, ours). As a counterpart, we propose a global method G-L2Xbased on L2X. Our change is straightforward, where L2X generates a different distribution $\mathbf{p}$ for each item, we apply the same $\mathbf{p}$ to all items. In other words, G-L2X includes a global selector layer ζ ($\zeta \in \mathbb{R}^d$) to simulate $\mathbf{p}$, and sampling is conducted in the same way as L2X on the selector weights. Thereby, G-L2X will select the same features for all items in the dataset.

Concrete Autoencoder (CAE). CAE [6] is a global method where the selector is the encoder part of an auto-encoder model [30]. Specifically, the selector compresses the input into a smaller representation $\hat{x}$, by linearly combining selected features, i.e. $x^\top \tilde{m}$, where $\tilde{m} \in \mathbb{R}^{k \times d}$ can be viewed as approximated k-hot concrete selection, sampled from the selector weights ($\zeta \in \mathbb{R}^{k \times d}$). Therefore, CAE might result in repetitive selection, and the input dimension to the predictor is reduced to k.

Instance-Wise Feature Grouping (IFG). IFG [41] applies a similar approach as L2X, but clusters features into groups and then selects k feature groups for prediction. IFG first assigns a group for each feature via Gumbel-sampling, and then makes a feature selection by Gumbel-sampling k out of the resulting groups. This grouping decision is also guided by how rich the selected features are to recover the original input. Therefore, apart from the ranking objective, IFG jointly optimizes an additional input restoring objective as well (similar to auto-encoders [30]). IFG is agnostic to the number of selected features and the group sizes, it can produce oversized groups and very large selections.

4.2 Regularization-Based Feature Selection

Instead of the budget-explicit feature selection, regularization-based methods induce sparsity through implicit constraints enforced by regularization terms in the training objective. We propose modifications to three existing methods to make them applicable to the LTR setting.

Invase. We consider INVASE [67] to be a hybrid approach involving both sampling and regularization. Built on the same structure as L2X, the selector of INVASE generates a *boolean/hard mask* m (instead of the approximation $\tilde{m}$) via Bernoulli sampling to train the predictor. Since this disables backpropagation, INVASE uses a customized loss function that does not need the gradients from the predictor to train the selector. The idea is to apply another individual baseline predictor model that takes the full-feature input, simultaneously with the predictor that takes the masked input. The loss difference between the two predictors is used as a scale to train the selector. Meanwhile, the L1 regularization is applied to the selector output to enforce selection sparsity. Ultimately, INVASE will push the selector to output a small set of selections which leads to the most similar predictions as using all features.

LassoNet. As the name suggests, LASSONET [34] adapts a traditional Lasso (L1 regularization) [60] on the first layer of a neural model to eliminate unnecessary features. The challenge with neural models is, all weights corresponding to a particular feature entry in the layer have to be zero in order to mask out the feature. Towards this, LASSONET adds a residual layer with one weight per input feature to the original model to perform as the traditional Lasso. Then, after every optimization step, LASSONET develops a proximal optimization algorithm to adjust the weights of the first layer, so that all absolute elements of each row are smaller than the respective weight of residual layer corresponding to a specific feature. Thereby, LASSONET performs global selection and the sparsity scale is adjusted by the L1 regularization on the residual layer weights.

TabNet. Unlike the previous methods, TABNET [4] is non-composable and tied to a specific tree-style neural architecture. It imitates a step-wise selection process before it outputs the final prediction based only on the selected features. Each step has the same neural component/block but with its own parameters, thus the model complexity and selection budget grow linearly with the number of steps. At each step, the full input is transformed by a feature transformer block first, and then an attentive transformer block conducts feature selection by sparsemax activation [40], as the weights in the resulting distribution corresponding to unselected features are zeros. The final prediction is aggregated from all steps to simulate ensemble models. A final mask m is a union of selections from all steps, and the entropy of the selection probabilities is used as the sparsity regularization.

5 Experimental Setup

Since feature selection can be applied in various manners and situations, we structure our experiments around three scenarios:

- *Scenario 1: Simultaneously train and select.* Both the ranking model and the feature selection are learned once and jointly. The methods are evaluated by the performance-sparsity trade-off. It is the standard setup for evaluating embedded feature selection methods in the interpretable ML field [41,50].
- *Scenario 2: Train then select with an enforced budget.* Practitioners generally set hard limits to the computational costs a system may incur and the efficiency of the system can be greatly enhanced if it only requires a much smaller amount of features to reach competitive performance. Following the previous scenario, we evaluate the trained model with test instances where only a fixed amount of features (which the method deems important and selects frequently during training) are presented and the rest are masked out. The resulting ranking performance and the costs of computing the required features indicate how practical the method is in efficiency improvements.

Datasets and Preprocessing. We choose three public benchmark datasets: MQ2008 (46 features) [48], Web30k (136 features) [47] and Yahoo (699 features) [14], to cover varying numbers of available features. We apply a $\log_{1p}$ transformation to the features of Web30k, as suggested in [49]. Yahoo contains cost labels for each feature, for Web30k we use cost estimates suggested by previous work [21].[1] All reported results are evaluated on the held-out test set partitions of the datasets.

Models. We use a standard feed-forward neural network with batch normalization, three fully-connected layers and tanh activation as the ranking model, denoted as DNN. According to the findings in [49], this simple model performs closely to the most effective transformer-based models, but requires much less resources to run. The selector models of L2X and INVASE have the same architecture, and as the only exception, TABNET is applied with its own unalterable model (see Sect. 4).

[1] MQ2008 is omitted from cost analysis since no associated cost information is available.

Table 2. Results of ranking performance and feature sparsity for methods applied in Scenario 1. For comparison, we also include GBDT [8,29] and DNN baselines without feature selection as upper bound. #F denotes the number of selected features. Reported results are averaged over 5 random seeds (*std* in parentheses). Bold font indicates the highest performing selection method; the $\star$ and underlines denote scores that are *not* significantly outperformed by GBDT and the bold-score method, respectively ($p > 0.05$, paired t-tests using Bonferonni's correction).

Listwise	MQ2008 NDCG@k (%)			Web30k NDCG@k (%)			Yahoo NDCG@k (%)		
loss	@1	@10	#F	@1	@10	#F	@1	@10	#F
Without feature selection.									
GBDT	69.3 (2.5)	80.8 (1.7)	46	50.4 (0.1)	52 (0.1)	136	72.2 (0.1)	79.2 (0.1)	699
DNN	66.2* (2)	80.2* (0.6)	46	46.1 (0.6)	47.7 (0.2)	136	69.4 (0.3)	76.9 (0.1)	699
Fixed-budget feature selection using the DNN ranking model.									
CAE	63.0 (1.1)	78.7 (0.5)	4	32.9 (2.9)	36.6 (2.2)	13	59.2 (0.2)	69.5 (0.1)	6
G-L2X	63.8* (1.3)	79.1 (0.4)	4	41.1 (0.9)	44.4 (0.3)	13	65.4 (0.1)	74.0 (0.0)	6
L2X	63.0 (2.1)	78.7 (0.7)	4	34.5 (2.4)	39.7 (1.9)	13	61.9 (1.1)	73.2 (0.3)	6
Budget-agnostic feature selection using the DNN ranking model.									
INVASE	62.5 (2.2)	77.5 (2.1)	5 (2)	15.1 (0.0)	22.1 (0.0)	0	38.7 (0.0)	57.8 (0.0)	0
IFG	**66.4*** (0.9)	**80.4*** (0.5)	20 (2)	32.5 (5.3)	37.5 (5.3)	72 (30)	69.6 (0.2)	77.1 (0.2)	58 (3)
LASSONET	64.7* (2.2)	79.3 (1.2)	6 (3)	39.4 (0.8)	42.1 (0.3)	8 (2)	63.1 (2.3)	72.4 (1.5)	12 (4)
Budget-agnostic feature selection using a method-specific ranking model.									
TABNET	64.7* (2.7)	78.2 (1.2)	7 (3)	**47.0** (0.4)	**49.2** (0.1)	8 (1)	**70.2** (0.4)	**77.7** (0.1)	6 (1)

Implementation. Our experimental implementation is done in *PyTorch Lightning* [19]. For TABNET and LASSONET existing implementations were used.[2] We created our own implementations for the rest of the methods.

6 Results

We report the findings in this section, aiming to answer two questions: (1) *how effective are investigated methods in the ranking setup?* and (2) *can those methods improve efficiency?* Each question corresponds to one of the scenarios described in Sect. 5.

Simultaneous Optimization and Selection. We begin by investigating the effectiveness of the feature selection methods when applied to Scenario 1, where feature selection and model optimization are performed simultaneously. The results for this scenario are displayed in Table 2 and Fig. 2.

Table 2 shows the ranking performance and the respective feature sparsity of all feature selection methods and two baselines without any selection as the upper-bound reference. For fixed-budget methods, the budgets were set to 10% of the total number of features for MQ2008 and Web30k, and 1% for Yahoo (the results with varying budgets are displayed in Fig. 2). Since the sparsity of budget-agnostic methods is more difficult to control, we performed an extensive grid search and

[2] https://github.com/dreamquark-ai/tabnet; https://github.com/lasso-net/lassonet.

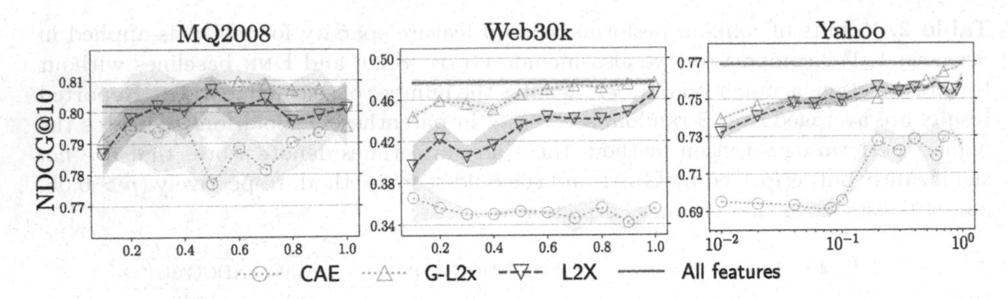

Fig. 2. Results of three fixed-budget methods applied to scenario 1. The x-axis indicates the pre-specified percentile of selected features (k). The shaded area shows the standard deviation over 5 random seeds.

used the hyper-parameters that produced the highest ranking performance with a comparable feature sparsity as the other methods.

The results in Table 2 show that not all feature selection methods are equally effective, and their performance can vary greatly over datasets. For instance, on MQ2008 all methods perform closely to the baselines, with only a fraction of the features. However, this is not the case for bigger datasets like Web30k and Yahoo. In particular, INVASE selects no features at all due to big uncertainty in selection (for this reason, we omit INVASE from all further comparisons). On the other hand, IFG performs poorly in inducing sparsity, mainly because of its input reconstruction objective, whereas the ranking performance is not substantially better than the rest of methods. Additionally, CAE does not seem effective either, and furthermore, increasing the selected features does not always result in better ranking performance (cf. Fig. 2). This is most-likely because CAE samples with replacement, and thus the same features can be selected repeatedly.

In contrast, the other two sampling-based methods L2X and G-L2X are designed to avoid the repetitive selection issue. Overall, the global selection G-L2X outperforms the local counterpart L2X, possibly because global selection generalizes better to unseen data. Another global method LASSONET is also inferior to G-L2X, mainly due to the difficulties in sparsity weight tuning and manually adjusting weights in the input layer.

Lastly, TABNET shows the best performance-sparsity balance across all datasets, and even outperforms the DNN baseline. Although, the comparison between TABNET and DNN is not completely fair, as they optimize different neural architectures. It does reveal large feature redundancies in these datasets: TABNET uses <10% of features on Web30k and 1% on Yahoo, yet still beats the DNN baseline with all features.

To summarize, we find that the local method TABNET is the most effective at balancing ranking performance and sparsity. Slightly inferior but competitive enough is the global method G-L2X, which reached > 95% of baseline performance with only 1% features on Yahoo and > 93% with 10% on Web30k.

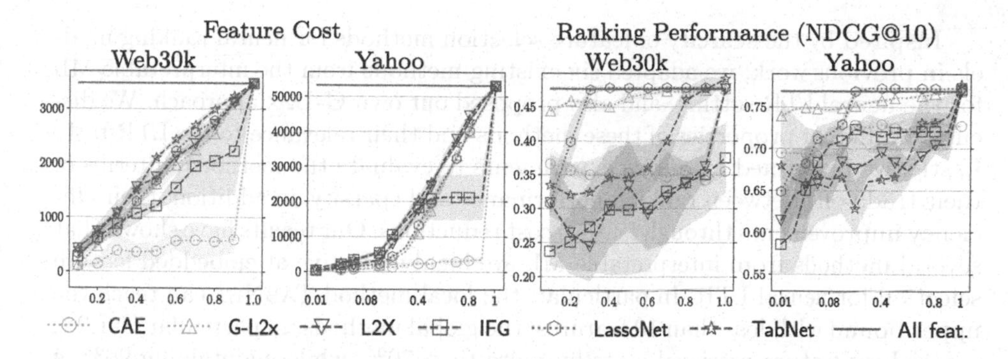

Fig. 3. Scenario 2. Feature cost (left two) and ranking performance (right two) under incomplete input. The x-axis indicates how many percentages of features are present in the input, to test the trained ranking model. Note this differs from specifying k during training for fixed-budget methods in scenario 1.

Feature Selection for Trained Ranking Models Next, we evaluate the methods in Scenario 2, where only a specified budget (i.e., a given number of features) of features are present in the test input. Figure 3 displays both the ranking performance and the total feature cost for varying degrees of sparsity. The costs represent the time it requires to retrieve the selected feature sets, and allow us to estimate the actual efficiency improvements they provide.

Unlike previous scenario, all local methods including TabNet, are no longer able to maintain superior performance. This is because for local methods, the selection is made conditioned on full input information, and an incomplete input could affect the selection and thus disrupt its prediction performance.

In contrast, global methods are immune to input changes. Therefore, CAE is still not performing well as it did in Scenario 1; G-L2X and LassoNet provide the best overall performance under small costs. LassoNet maintains baseline performance with less than 40% of features on both datasets, while G-L2X outperforms LassoNet when selected features are less than 30%. Meanwhile, it also shows LassoNet tends to select more costly features than G-L2x.

To conclude, we find that global methods G-L2X and LassoNet perform the best in Scenario 2, where upcoming query inputs are masked under enforced feature budgets. Particularly, G-L2X is superior in both ranking and computing cost when the feature budget is small. This translates to substantial efficiency improvements in practical terms, as ranking performance is maintained by selected features only.

7 Conclusion

The main goal of this work is to bring the interpretable ML and the LTR fields closer together. To this end, we have studied whether feature selection methods from the interpretable ML are effective for neural LTR, for both interpretability and efficiency purposes.

Inspired by the scarcity of feature selection methods for neural ranking models in previous work, we adapted six existing methods from the interpretable ML for the neural LTR setting, and also proposed our own G-L2X approach. We discussed different properties of these methods and their relevance to the LTR task. Lastly, we performed extensive experiments to evaluate the methods in terms of their trade-offs between ranking performance and sparsity, in addition, their efficiency improvements through feature cost reductions. Our results have shown that several methods from interpretable ML are highly effective at embedded feature selection for neural LTR. In particular, the local method TabNet can reach the upper bound with less than 10 features; the global methods, in particular G-L2X, can reduce feature retrieval costs by more than 70%, while maintaining 96% of performance relative to a full feature model.

We hope our investigation can bridge the gap between the LTR and interpretable ML fields. The future work can be developing more interpretable and efficient ranking systems, and how that interpretability could support both practitioners and the users of ranking systems.

Acknowledgements. This work is partially supported by German Research Foundation (DFG), under the Project IREM with grant No. AN 996/1-1, and by the Netherlands Organisation for Scientific Research (NWO) under grant No. VI.Veni.222.269.

References

1. Abdul, A.M., von der Weth, C., Kankanhalli, M.S., Lim, B.Y.: COGAM: measuring and moderating cognitive load in machine learning model explanations. In: Bernhaupt, R., et al. (eds.) CHI 2020: CHI Conference on Human Factors in Computing Systems, Honolulu, HI, USA, 25–30 April 2020, pp. 1–14. ACM (2020). https://doi.org/10.1145/3313831.3376615
2. Alvarez-Melis, D., Jaakkola, T.S.: Towards robust interpretability with self-explaining neural networks. In: Bengio, S., Wallach, H.M., Larochelle, H., Grauman, K., Cesa-Bianchi, N., Garnett, R. (eds.) Advances in Neural Information Processing Systems, vol. 31: Annual Conference on Neural Information Processing Systems 2018, NeurIPS 2018(December), pp. 3–8, 2018. Montréal, Canada, pp. 7786–7795 (2018). https://proceedings.neurips.cc/paper/2018/hash/3e9f0fc9b2f89e043bc6233994dfcf76-Abstract.html
3. Arapakis, I., Bai, X., Cambazoglu, B.B.: Impact of response latency on user behavior in web search. In: Geva, S., Trotman, A., Bruza, P., Clarke, C.L.A., Järvelin, K. (eds.) The 37th International ACM SIGIR Conference on Research and Development in Information Retrieval, SIGIR 2014, Gold Coast, QLD, Australia - 06–11 July 2014, pp. 103–112. ACM (2014). https://doi.org/10.1145/2600428.2609627
4. Arik, S.Ö., Pfister, T.: TabNet: attentive interpretable tabular learning. In: Thirty-Fifth AAAI Conference on Artificial Intelligence, AAAI 2021, Thirty-Third Conference on Innovative Applications of Artificial Intelligence, IAAI 2021, The Eleventh Symposium on Educational Advances in Artificial Intelligence, EAAI 2021, Virtual Event, 2–9 February 2021, pp. 6679–6687. AAAI Press (2021). https://doi.org/10.1609/AAAI.V35I8.16826

5. Bai, X., Cambazoglu, B.B.: Impact of response latency on sponsored search. Inf. Process. Manag. **56**(1), 110–129 (2019). https://doi.org/10.1016/J.IPM.2018.10.005
6. Balin, M.F., Abid, A., Zou, J.Y.: Concrete autoencoders: differentiable feature selection and reconstruction. In: Chaudhuri, K., Salakhutdinov, R. (eds.) Proceedings of the 36th International Conference on Machine Learning, ICML 2019, 9–15 June 2019, Long Beach, California, USA. Proceedings of Machine Learning Research, vol. 97, pp. 444–453. PMLR (2019). http://proceedings.mlr.press/v97/balin19a.html
7. Barreda-Ángeles, M., Arapakis, I., Bai, X., Cambazoglu, B.B., Pereda-Baños, A.: Unconscious physiological effects of search latency on users and their click behaviour. In: Baeza-Yates, R., Lalmas, M., Moffat, A., Ribeiro-Neto, B.A. (eds.) Proceedings of the 38th International ACM SIGIR Conference on Research and Development in Information Retrieval, Santiago, Chile, 9–13 August 2015, pp. 203–212. ACM (2015). https://doi.org/10.1145/2766462.2767719
8. Bruch, S.: An alternative cross entropy loss for learning-to-rank. In: Leskovec, J., Grobelnik, M., Najork, M., Tang, J., Zia, L. (eds.) WWW 2021: The Web Conference 2021, Virtual Event / Ljubljana, Slovenia, 19–23 April 2021, pp. 118–126. ACM / IW3C2 (2021). https://doi.org/10.1145/3442381.3449794
9. Bruch, S., Wang, X., Bendersky, M., Najork, M.: An analysis of the softmax cross entropy loss for learning-to-rank with binary relevance. In: Fang, Y., Zhang, Y., Allan, J., Balog, K., Carterette, B., Guo, J. (eds.) Proceedings of the 2019 ACM SIGIR International Conference on Theory of Information Retrieval, ICTIR 2019, Santa Clara, CA, USA, 2–5 October 2019, pp. 75–78. ACM (2019). https://doi.org/10.1145/3341981.3344221
10. Burges, C.J.C., Ragno, R., Le, Q.V.: Learning to rank with nonsmooth cost functions. In: Schölkopf, B., Platt, J.C., Hofmann, T. (eds.) Advances in Neural Information Processing Systems, vol. 19, Proceedings of the Twentieth Annual Conference on Neural Information Processing Systems, Vancouver, British Columbia, Canada, 4–7 December 2006, pp. 193–200. MIT Press (2006). https://proceedings.neurips.cc/paper/2006/hash/af44c4c56f385c43f2529f9b1b018f6a-Abstract.html
11. Burges, C.J.C., Shaked, T., Renshaw, E., Lazier, A., Deeds, M., Hamilton, N., Hullender, G.N.: Learning to rank using gradient descent. In: Raedt, L.D., Wrobel, S. (eds.) Machine Learning, Proceedings of the Twenty-Second International Conference (ICML 2005), Bonn, Germany, 7–11 August 2005. ACM International Conference Proceeding Series, vol. 119, pp. 89–96. ACM (2005). https://doi.org/10.1145/1102351.1102363
12. Burges, C.J.: From ranknet to lambdarank to lambdamart: an overview. Learning **11**(23–581), 81 (2010)
13. Cao, Z., Qin, T., Liu, T., Tsai, M., Li, H.: Learning to rank: from pairwise approach to listwise approach. In: Ghahramani, Z. (ed.) Machine Learning, Proceedings of the Twenty-Fourth International Conference (ICML 2007), Corvallis, Oregon, USA, 20–24 June 2007. ACM International Conference Proceeding Series, vol. 227, pp. 129–136. ACM (2007). https://doi.org/10.1145/1273496.1273513
14. Chapelle, O., Chang, Y.: Yahoo! learning to rank challenge overview. In: Chapelle, O., Chang, Y., Liu, T. (eds.) Proceedings of the Yahoo! Learning to Rank Challenge, held at ICML 2010, Haifa, Israel, 25 June 2010. JMLR Proceedings, vol. 14, pp. 1–24. JMLR.org (2011). http://proceedings.mlr.press/v14/chapelle11a.html
15. Chapelle, O., Keerthi, S.S.: Efficient algorithms for ranking with SVMs. Inf. Retr. **13**(3), 201–215 (2010). https://doi.org/10.1007/S10791-009-9109-9
16. Chen, J., Song, L., Wainwright, M.J., Jordan, M.I.: Learning to explain: an information-theoretic perspective on model interpretation. In: Dy, J.G., Krause,

A. (eds.) Proceedings of the 35th International Conference on Machine Learning, ICML 2018, Stockholmsmässan, Stockholm, Sweden, 10–15 July 2018. Proceedings of Machine Learning Research, vol. 80, pp. 882–891. PMLR (2018). http://proceedings.mlr.press/v80/chen18j.html
17. Dato, D., et al.: Fast ranking with additive ensembles of oblivious and non-oblivious regression trees. ACM Trans. Inf. Syst. **35**(2), 15:1–15:31 (2016). https://doi.org/10.1145/2987380
18. Du, M., Liu, N., Hu, X.: Techniques for interpretable machine learning. Commun. ACM **63**(1), 68–77 (2020). https://doi.org/10.1145/3359786
19. Falcon, W., et al.: Pytorch lightning. GitHub 3, 6. https://github.com/PyTorchLightning/pytorch-lightning (2019)
20. Freund, Y., Iyer, R.D., Schapire, R.E., Singer, Y.: An efficient boosting algorithm for combining preferences. In: Shavlik, J.W. (ed.) Proceedings of the Fifteenth International Conference on Machine Learning (ICML 1998), Madison, Wisconsin, USA, 24–27 July 1998, pp. 170–178. Morgan Kaufmann (1998)
21. Gallagher, L., Chen, R., Blanco, R., Culpepper, J.S.: Joint optimization of cascade ranking models. In: Culpepper, J.S., Moffat, A., Bennett, P.N., Lerman, K. (eds.) Proceedings of the Twelfth ACM International Conference on Web Search and Data Mining, WSDM 2019, Melbourne, VIC, Australia, 11–15 February 2019, pp. 15–23. ACM (2019). https://doi.org/10.1145/3289600.3290986
22. Geng, X., Liu, T., Qin, T., Li, H.: Feature selection for ranking. In: Kraaij, W., de Vries, A.P., Clarke, C.L.A., Fuhr, N., Kando, N. (eds.) SIGIR 2007: Proceedings of the 30th Annual International ACM SIGIR Conference on Research and Development in Information Retrieval, Amsterdam, The Netherlands, 23–27 July 2007, pp. 407–414. ACM (2007). https://doi.org/10.1145/1277741.1277811
23. Gigli, A., Lucchese, C., Nardini, F.M., Perego, R.: Fast feature selection for learning to rank. In: Carterette, B., Fang, H., Lalmas, M., Nie, J. (eds.) Proceedings of the 2016 ACM on International Conference on the Theory of Information Retrieval, ICTIR 2016, Newark, DE, USA, 12–6 September 2016, pp. 167–170. ACM (2016). https://doi.org/10.1145/2970398.2970433
24. Jang, E., Gu, S., Poole, B.: Categorical reparametrization with gumble-softmax (2017)
25. Järvelin, K., Kekäläinen, J.: Cumulated gain-based evaluation of IR techniques. ACM Trans. Inf. Syst. **20**(4), 422–446 (2002). https://doi.org/10.1145/582415.582418
26. Joachims, T.: Optimizing search engines using clickthrough data. In: Proceedings of the Eighth ACM SIGKDD International Conference on Knowledge Discovery and Data Mining, 23–26 July 2002, Edmonton, Alberta, Canada, pp. 133–142. ACM (2002). https://doi.org/10.1145/775047.775067
27. Joachims, T.: Training linear SVMs in linear time. In: Eliassi-Rad, T., Ungar, L.H., Craven, M., Gunopulos, D. (eds.) Proceedings of the Twelfth ACM SIGKDD International Conference on Knowledge Discovery and Data Mining, Philadelphia, PA, USA, 20–23 August 2006, pp. 217–226. ACM (2006). https://doi.org/10.1145/1150402.1150429
28. Kaur, H., Nori, H., Jenkins, S., Caruana, R., Wallach, H.M., Vaughan, J.W.: Interpreting interpretability: understanding data scientists' use of interpretability tools for machine learning. In: Bernhaupt, R., et al. (eds.) CHI 2020: CHI Conference on Human Factors in Computing Systems, Honolulu, HI, USA, 25–30 April 2020, pp. 1–14. ACM (2020). https://doi.org/10.1145/3313831.3376219

29. Ke, G., et al.: LightGBM: a highly efficient gradient boosting decision tree. In: Guyon, I., et al. (eds.) Advances in Neural Information Processing Systems, vol. 30: Annual Conference on Neural Information Processing Systems 2017(December), pp. 4–9, 2017. Long Beach, CA, USA, pp. 3146–3154 (2017). https://proceedings.neurips.cc/paper/2017/hash/6449f44a102fde848669bdd9eb6b76fa-Abstract.html
30. Kingma, D.P., Welling, M.: Auto-encoding variational bayes. In: Bengio, Y., LeCun, Y. (eds.) 2nd International Conference on Learning Representations, ICLR 2014, Banff, AB, Canada, 14–16 April 2014, Conference Track Proceedings (2014). http://arxiv.org/abs/1312.6114
31. Lai, H., Pan, Y., Liu, C., Lin, L., Wu, J.: Sparse learning-to-rank via an efficient primal-dual algorithm. IEEE Trans. Computers **62**(6), 1221–1233 (2013). https://doi.org/10.1109/TC.2012.62
32. Lai, H., Pan, Y., Tang, Y., Yu, R.: FSMRank: feature selection algorithm for learning to rank. IEEE Trans. Neural Networks Learn. Syst. **24**(6), 940–952 (2013). https://doi.org/10.1109/TNNLS.2013.2247628
33. Laporte, L., Flamary, R., Canu, S., Déjean, S., Mothe, J.: Nonconvex regularizations for feature selection in ranking with sparse SVM. IEEE Trans. Neural Networks Learn. Syst. **25**(6), 1118–1130 (2014). https://doi.org/10.1109/TNNLS.2013.2286696
34. Lemhadri, I., Ruan, F., Abraham, L., Tibshirani, R.: LassoNet: a neural network with feature sparsity. J. Mach. Learn. Res. **22**, 127:1–127:29 (2021). http://jmlr.org/papers/v22/20-848.html
35. Leonhardt, J., Rudra, K., Anand, A.: Extractive explanations for interpretable text ranking. ACM Trans. Inf. Syst. (2022). https://doi.org/10.1145/3576924
36. Li, P., Burges, C.J.C., Wu, Q.: McRank: learning to rank using multiple classification and gradient boosting. In: Platt, J.C., Koller, D., Singer, Y., Roweis, S.T. (eds.) Advances in Neural Information Processing Systems, vol. 20, Proceedings of the Twenty-First Annual Conference on Neural Information Processing Systems, Vancouver, British Columbia, Canada, 3–6 December 2007, pp. 897–904. Curran Associates, Inc. (2007). https://proceedings.neurips.cc/paper/2007/hash/b86e8d03fe992d1b0e19656875ee557c-Abstract.html
37. Liu, T.: Learning to rank for information retrieval. Found. Trends Inf. Retr. **3**(3), 225–331 (2009). https://doi.org/10.1561/1500000016
38. Lucchese, C., Nardini, F.M., Orlando, S., Perego, R., Veneri, A.: ILMART: interpretable ranking with constrained lambdamart. In: SIGIR 2022: The 45th International ACM SIGIR Conference on Research and Development in Information Retrieval, Madrid, Spain, 11–15 July 2022, pp. 2255–2259. ACM (2022). https://doi.org/10.1145/3477495.3531840
39. Lundberg, S.M., Lee, S.: A unified approach to interpreting model predictions. In: Guyon, I., et al. (eds.) Advances in Neural Information Processing Systems, vol. 30: Annual Conference on Neural Information Processing Systems 2017(December), pp. 4–9, 2017. Long Beach, CA, USA, pp. 4765–4774 (2017). https://proceedings.neurips.cc/paper/2017/hash/8a20a8621978632d76c43dfd28b67767-Abstract.html
40. Martins, A.F.T., Astudillo, R.F.: From softmax to sparsemax: a sparse model of attention and multi-label classification. In: Balcan, M., Weinberger, K.Q. (eds.) Proceedings of the 33nd International Conference on Machine Learning, ICML 2016, New York City, NY, USA, 19–24 June 2016. JMLR Workshop and Conference Proceedings, vol. 48, pp. 1614–1623. JMLR.org (2016). http://proceedings.mlr.press/v48/martins16.html

41. Masoomi, A., Wu, C., Zhao, T., Wang, Z., Castaldi, P.J., Dy, J.G.: Instance-wise feature grouping. In: Larochelle, H., Ranzato, M., Hadsell, R., Balcan, M., Lin, H. (eds.) Advances in Neural Information Processing Systems, vol. 33: Annual Conference on Neural Information Processing Systems 2020, NeurIPS 2020(December), pp. 6–12, 2020. virtual (2020). https://proceedings.neurips.cc/paper/2020/hash/9b10a919ddeb07e103dc05ff523afe38-Abstract.html
42. Molnar, C.: Interpretable machine learning. Lulu.com (2020)
43. Pan, F., Converse, T., Ahn, D., Salvetti, F., Donato, G.: Feature selection for ranking using boosted trees. In: Cheung, D.W., Song, I., Chu, W.W., Hu, X., Lin, J. (eds.) Proceedings of the 18th ACM Conference on Information and Knowledge Management, CIKM 2009, Hong Kong, China, 2–6 November 2009, pp. 2025–2028. ACM (2009). https://doi.org/10.1145/1645953.1646292
44. Pang, L., Xu, J., Ai, Q., Lan, Y., Cheng, X., Wen, J.: SetRank: learning a permutation-invariant ranking model for information retrieval. In: Huang, J.X., et al. (eds.) Proceedings of the 43rd International ACM SIGIR conference on research and development in Information Retrieval, SIGIR 2020, Virtual Event, China, 25–30 July 2020, pp. 499–508. ACM (2020). https://doi.org/10.1145/3397271.3401104
45. Pobrotyn, P., Bartczak, T., Synowiec, M., Bialobrzeski, R., Bojar, J.: Context-aware learning to rank with self-attention. CoRR abs/2005.10084 (2020). https://arxiv.org/abs/2005.10084
46. Purpura, A., Buchner, K., Silvello, G., Susto, G.A.: Neural feature selection for learning to rank. In: Hiemstra, D., Moens, M.-F., Mothe, J., Perego, R., Potthast, M., Sebastiani, F. (eds.) ECIR 2021. LNCS, vol. 12657, pp. 342–349. Springer, Cham (2021). https://doi.org/10.1007/978-3-030-72240-1_34
47. Qin, T., Liu, T.: Introducing LETOR 4.0 datasets. CoRR abs/1306.2597 (2013). http://arxiv.org/abs/1306.2597
48. Qin, T., Liu, T., Xu, J., Li, H.: LETOR: a benchmark collection for research on learning to rank for information retrieval. Inf. Retr. **13**(4), 346–374 (2010). https://doi.org/10.1007/S10791-009-9123-Y
49. Qin, Z., et al.: Are neural rankers still outperformed by gradient boosted decision trees? In: 9th International Conference on Learning Representations, ICLR 2021, Virtual Event, Austria, 3–7 May 2021. OpenReview.net (2021). https://openreview.net/forum?id=Ut1vF_q_vC
50. Rahangdale, A., Raut, S.A.: Deep neural network regularization for feature selection in learning-to-rank. IEEE Access **7**, 53988–54006 (2019). https://doi.org/10.1109/ACCESS.2019.2902640
51. Ribeiro, M.T., Singh, S., Guestrin, C.: "why should I trust you?": Explaining the predictions of any classifier. In: Krishnapuram, B., Shah, M., Smola, A.J., Aggarwal, C.C., Shen, D., Rastogi, R. (eds.) Proceedings of the 22nd ACM SIGKDD International Conference on Knowledge Discovery and Data Mining, San Francisco, CA, USA, 13–17 August 2016, pp. 1135–1144. ACM (2016). https://doi.org/10.1145/2939672.2939778
52. Rigutini, L., Papini, T., Maggini, M., Scarselli, F.: SortNet: learning to rank by a neural-based sorting algorithm. CoRR abs/2311.01864 (2023). https://doi.org/10.48550/ARXIV.2311.01864
53. Rong, Y., et al.: Towards human-centered explainable AI: user studies for model explanations. CoRR abs/2210.11584 (2022). https://doi.org/10.48550/arXiv.2210.11584

54. Rudin, C.: Stop explaining black box machine learning models for high stakes decisions and use interpretable models instead. Nat. Mach. Intell. **1**(5), 206–215 (2019). https://doi.org/10.1038/s42256-019-0048-x
55. Shrikumar, A., Greenside, P., Kundaje, A.: Learning important features through propagating activation differences. In: Precup, D., Teh, Y.W. (eds.) Proceedings of the 34th International Conference on Machine Learning, ICML 2017, Sydney, NSW, Australia, 6–11 August 2017. Proceedings of Machine Learning Research, vol. 70, pp. 3145–3153. PMLR (2017). http://proceedings.mlr.press/v70/shrikumar17a.html
56. Simonyan, K., Vedaldi, A., Zisserman, A.: Deep inside convolutional networks: visualising image classification models and saliency maps. In: Bengio, Y., LeCun, Y. (eds.) 2nd International Conference on Learning Representations, ICLR 2014, Banff, AB, Canada, 14–16 April 2014, Workshop Track Proceedings (2014). http://arxiv.org/abs/1312.6034
57. Sun, Z., Qin, T., Tao, Q., Wang, J.: Robust sparse rank learning for non-smooth ranking measures. In: Allan, J., Aslam, J.A., Sanderson, M., Zhai, C., Zobel, J. (eds.) Proceedings of the 32nd Annual International ACM SIGIR Conference on Research and Development in Information Retrieval, SIGIR 2009, Boston, MA, USA, 19–23 July 2009, pp. 259–266. ACM (2009). https://doi.org/10.1145/1571941.1571987
58. Sundararajan, M., Taly, A., Yan, Q.: Axiomatic attribution for deep networks. In: Precup, D., Teh, Y.W. (eds.) Proceedings of the 34th International Conference on Machine Learning, ICML 2017, Sydney, NSW, Australia, 6–11 August 2017. Proceedings of Machine Learning Research, vol. 70, pp. 3319–3328. PMLR (2017). http://proceedings.mlr.press/v70/sundararajan17a.html
59. Taylor, M.J., Guiver, J., Robertson, S., Minka, T.: SoftRank: optimizing non-smooth rank metrics. In: Najork, M., Broder, A.Z., Chakrabarti, S. (eds.) Proceedings of the International Conference on Web Search and Web Data Mining, WSDM 2008, Palo Alto, California, USA, 11–12 February 2008, pp. 77–86. ACM (2008). https://doi.org/10.1145/1341531.1341544
60. Tibshirani, R.: Regression shrinkage and selection via the lasso. J. Roy. Stat. Soc.: Ser. B (Methodol.) **58**(1), 267–288 (1996)
61. Vaswani, A., et al.: Attention is all you need. In: Guyon, I., et al. (eds.) Advances in Neural Information Processing Systems, vol. 30: Annual Conference on Neural Information Processing Systems 2017(December), pp. 4–9, 2017. Long Beach, CA, USA, pp. 5998–6008 (2017). https://proceedings.neurips.cc/paper/2017/hash/3f5ee243547dee91fbd053c1c4a845aa-Abstract.html
62. Wang, L., Lin, J., Metzler, D.: A cascade ranking model for efficient ranked retrieval. In: Ma, W., Nie, J., Baeza-Yates, R., Chua, T., Croft, W.B. (eds.) Proceeding of the 34th International ACM SIGIR Conference on Research and Development in Information Retrieval, SIGIR 2011, Beijing, China, 25–29 July 2011, pp. 105–114. ACM (2011). https://doi.org/10.1145/2009916.2009934
63. Wu, Q., Burges, C.J.C., Svore, K.M., Gao, J.: Adapting boosting for information retrieval measures. Inf. Retr. **13**(3), 254–270 (2010). https://doi.org/10.1007/S10791-009-9112-1
64. Xia, F., Liu, T., Wang, J., Zhang, W., Li, H.: Listwise approach to learning to rank: theory and algorithm. In: Cohen, W.W., McCallum, A., Roweis, S.T. (eds.) Machine Learning, Proceedings of the Twenty-Fifth International Conference (ICML 2008), Helsinki, Finland, 5–9 June 2008. ACM International Conference Proceeding Series, vol. 307, pp. 1192–1199. ACM (2008). https://doi.org/10.1145/1390156.1390306
65. Xu, J., Li, H.: AdaRank: a boosting algorithm for information retrieval. In: Kraaij, W., de Vries, A.P., Clarke, C.L.A., Fuhr, N., Kando, N. (eds.) SIGIR 2007: Proceedings of the 30th Annual International ACM SIGIR Conference on Research and

Development in Information Retrieval, Amsterdam, The Netherlands, 23–27 July 2007, pp. 391–398. ACM (2007). https://doi.org/10.1145/1277741.1277809

66. Xu, Z.E., Huang, G., Weinberger, K.Q., Zheng, A.X.: Gradient boosted feature selection. In: Macskassy, S.A., Perlich, C., Leskovec, J., Wang, W., Ghani, R. (eds.) The 20th ACM SIGKDD International Conference on Knowledge Discovery and Data Mining, KDD 2014, New York, NY, USA - 24–27 August 2014, pp. 522–531. ACM (2014). https://doi.org/10.1145/2623330.2623635
67. Yoon, J., Jordon, J., van der Schaar, M.: INVASE: instance-wise variable selection using neural networks. In: 7th International Conference on Learning Representations, ICLR 2019, New Orleans, LA, USA, 6–9 May 2019. OpenReview.net (2019). https://openreview.net/forum?id=BJg_roAcK7
68. Yue, Y., Finley, T., Radlinski, F., Joachims, T.: A support vector method for optimizing average precision. In: Kraaij, W., de Vries, A.P., Clarke, C.L.A., Fuhr, N., Kando, N. (eds.) SIGIR 2007: Proceedings of the 30th Annual International ACM SIGIR Conference on Research and Development in Information Retrieval, Amsterdam, The Netherlands, 23–27 July 2007, pp. 271–278. ACM (2007). https://doi.org/10.1145/1277741.1277790
69. Zhang, Z., Rudra, K., Anand, A.: Explain and predict, and then predict again. In: WSDM 2021, The Fourteenth ACM International Conference on Web Search and Data Mining, Virtual Event, Israel, 8–12 March 2021, pp. 418–426. ACM (2021). https://doi.org/10.1145/3437963.3441758
70. Zhang, Z., Setty, V., Anand, A.: SparCAssist: a model risk assessment assistant based on sparse generated counterfactuals. In: Amigó, E., Castells, P., Gonzalo, J., Carterette, B., Culpepper, J.S., Kazai, G. (eds.) SIGIR 2022: The 45th International ACM SIGIR Conference on Research and Development in Information Retrieval, Madrid, Spain, 11–15 July 2022, pp. 3219–3223. ACM (2022). https://doi.org/10.1145/3477495.3531677
71. Zhuang, H., et al.: Interpretable ranking with generalized additive models. In: WSDM 2021, The Fourteenth ACM International Conference on Web Search and Data Mining, Virtual Event, Israel, 8–12 March 2021, pp. 499–507. ACM (2021). https://doi.org/10.1145/3437963.3441796

Navigating the Thin Line: Examining User Behavior in Search to Detect Engagement and Backfire Effects

Federico Maria Cau(✉) and Nava Tintarev

Maastricht University, Maastricht, The Netherlands
{federico.cau,n.tintarev}@maastrichtuniversity.nl

Abstract. Opinionated users often seek information that aligns with their preexisting beliefs while dismissing contradictory evidence due to confirmation bias. This conduct hinders their ability to consider alternative stances when searching the web. Despite this, few studies have analyzed how the diversification of search results on disputed topics influences the search behavior of highly opinionated users. To this end, we present a preregistered user study ($n = 257$) investigating whether different levels (low and high) of bias metrics and search results presentation (with or without AI-predicted stances labels) can affect the stance diversity consumption and search behavior of opinionated users on three debated topics (i.e., atheism, intellectual property rights, and school uniforms). Our results show that exposing participants to (counter- attitudinally) biased search results increases their consumption of attitude-opposing content, but we also found that bias was associated with a trend toward overall fewer interactions within the search page. We also found that 19% of users interacted with queries and search pages but did not select any search results. When we removed these participants in a post-hoc analysis, we found that stance labels increased the diversity of stances consumed by users, particularly when the search results were biased. Our findings highlight the need for future research to explore distinct search scenario settings to gain insight into opinionated users' behavior.

Keywords: confirmation bias · search behavior · bias metrics

1 Introduction

Recent studies reveal that certain biases in the way users interact with web search engines can lead to unintended consequences. On the one hand, we have users' cognitive biases such as *confirmation bias*, where people tend to consume information that aligns with their preexisting beliefs and ignore contradictory information when searching the web [4,26]. On the other hand, search results from popular search engines can be *biased* toward particular viewpoints (e.g., the degree of whether a document is against, neutral, or in favor of a specific

N. Goharian et al. (Eds.): ECIR 2024, LNCS 14611, pp. 403–419, 2024.
https://doi.org/10.1007/978-3-031-56066-8_30

topic), which can strongly influence user opinions [9,25,30]. Previous literature has shown that these phenomena may lead users to consume biased content, causing the *search engine manipulation effect* (SEME). For instance, this effect occurs when users alter their attitudes based on the viewpoints expressed in highly-ranked search results [3,11,13,28]. As a consequence, it is necessary to implement strategies that reduce biased content presented and consumed by users, bringing the need to increase the *viewpoint diversity* in search results [16,22,24,27].

Many attempts have already been made to mitigate the SEME [10,22,32,41]. Nevertheless, most of these studies focused on mildly opinionated users, while content consumption and behaviors of users with strong opinions (i.e., *opinionated*) are still under-explored. In contrast, a recent study by Wu et al. [40] found that predicted stance labels and their explanations increase content diversity consumption compared to plain search results when considering opinionated users, although the benefits of such explanations over stance labels remain unclear. Furthermore, it is uncertain if their results on the effect of stance labels on content diversity consumption were affected by: **i)** the use of fixed, hand-made templates for generating biased/balanced search results lists without proper validation from state-of-the-art bias metrics [9,10], or **ii)** the abundance of neutral search results at the top of the list in the balanced condition, or **iii)** the lack of investigation for potential users' behaviors triggered by their exposure to predicted stance labels [4,14,32].

In this work, we address these gaps and investigate the effects of different viewpoint bias metric levels and predicted stance labels of search results on opinionated users' clicking diversity and search strategies.

We produce balanced and biased search result lists by defining two levels of *polarity* (nDPB) and *stance* (nDSB) normalized discounted viewpoint bias metrics [9] for both opposing and supporting pre-study attitude of a user towards a debated topic (four conditions in total; see Table 1). We assign each user to a random SERP (same set of 40 search results) of the 40 differently ranked SERPs for a specific condition. Hence, each SERP consists of the same 40 search results but ranked differently based on the bias metric level and the user pre-study attitude. Whenever a user issues a query, we randomly show a different SERP ranking (out of the 40 SERPs) that still fulfills the study conditions the user was initially assigned (see Sect. 3.1). We focus on two research questions:

RQ1. How do SERPs informed by different levels of bias metrics and AI-generated stance labels impact users' diversity of search results they click on?
RQ2. How do SERPs informed by different levels of bias metrics and AI-generated stance labels impact users' search behaviors?

To address our research questions, we conducted a preregistered, online user study ($N = 257$; Sect. 3.4) simulating a low-stakes and open-ended search scenario. We asked opinionated participants to inform themselves regarding a debated topic with the help of search results containing two different bias levels and with or without informative stance labels. Our study shows that *participants exposed to biased search results tended to (i) consume more attitude-opposing content* and (ii) *interact less within the search page*. Interestingly, we found that 19% of users abandoned [1] the web search, probably triggered by a backfire effect [37]. When we excluded these participants in a post-hoc analysis, we discovered that stance labels increased the diversity of stances consumed by users, especially when the search results were biased. Although we did not find evidence of specific search behavior traits among different bias metrics and search results presentations, we discuss in Sect. 5 how our results call for further research to promote responsible online opinion formation. The study preregistration, materials, and data used in the paper are available at https://osf.io/ph8nv/.

2 Related Work and Hypotheses

2.1 Viewpoint Diversity and Ranking Algorithms

Viewpoints [2,6,12,19] are opinions relative to debated topics or claims [6], and can be represented in ranked lists of search results by a label assigned to each document, which can be expressed in different degrees (e.g., binary or multi-categorical [5,7,9,17,28,42]). A lack of viewpoint diversity can be seen as a lens to *bias* in search engines [16], which calls for measuring and promoting *exposure diversity* [22] of search results to prevent SEME for users when searching the web. Earlier work investigated how to mitigate search results biases through various approaches. One approach is to generate fair (or less biased) rankings by diversifying search results for a pre-defined disadvantaged viewpoint category [24], measuring the degree of bias in search engines by studying the correlations between statistical parity (e.g., equal exposure of different viewpoint categories) and search results diversity, fairness, and relevance [16,27], and establishing normative diversity definitions to evaluate news recommendation systems [39].

An exception is a recent paper by Wu et al. [40] which investigated how different levels of bias in search results influenced the behavior (diversity of viewpoints selected) of opinionated users.

However, they found no evidence of differences between the two bias conditions. We hypothesize that a possible reason for this outcome may lie in the balanced and biased fixed templates they used (i.e., always having five stronger stances or neutral results at the top but without the use of viewpoint bias metrics), while we propose 40 different SERP combinations per bias level for two viewpoint bias metrics (nDPB and nDSB) [9]. For our first hypothesis, we expect users to consume less diverse content when exposed to a SERP with a high bias.

Hypothesis 1a ($\mathbf{H_{1a}}$). Users exposed to SERPs with *high* bias metrics will interact with less diverse results than users exposed to SERPs with *low* bias metrics.

2.2 Content Diversity Consumption and Search Results Labels

Beyond biases in the presentation of the search results during online searches, another significant challenge comes from the user side. This phenomenon is *confirmation bias*, where users tend to favor and consume information that aligns with their existing beliefs [32]. Specifically, as in Sect. 2.1 we discussed strategies to address *exposure diversity* (i.e., the share of supply diversity that a user is exposed to [22]), here we are interested in investigating how we can achieve *consumption diversity* (i.e., the overall diversity in content that users actively engage with [22]). Recent research [22], proposed a framework where they distinguish five potential diversity nudges (i.e., system architecture design choices that predictably alter users' behavior) that can be used to increase the content consumption diversity. Specifically, they suggested how interface design interventions (e.g., presentation nudges) can make users' content consumption more transparent. For example, incorporating informative labels can help reduce confirmation bias in search results [14,32]. Furthermore, stance detection models [33] can identify the stance or viewpoint (e.g., against, neutral, supporting) of search results and have been proven to be effective in increasing the diversity of users' content consumption and help them navigate online debates [40]. While some studies exist [18,20,21] which investigated strategies to increase users' content diversity consumption, the effects of these interventions on specifically *opinionated users* are still under-explored.

In this work, we aim to investigate the effectiveness of predicted stance labels (i.e., against, neutral, and supporting) in mitigating confirmation bias on search results ranked according to two bias metrics levels. Moreover, if stance labels effectively reduce confirmation bias, we can anticipate an interaction effect between the bias condition in search results and the interface design (i.e., search results with and without stance labels). The hypotheses follow the work of Wu et al. [40]:

Hypothesis 1b ($\mathbf{H_{1b}}$). Users exposed to search results with *stance labels* interact with more diverse content than users who are exposed to regular search results.

Hypothesis 1c ($\mathbf{H_{1c}}$). Users exposed to search results with *stance labels* are less susceptible to the effect of stance biases in search results on clicking diversity.

2.3 Users and Search Behavior

Earlier works found evidence that users' search behaviors are affected by factors like search results' viewpoint biases, user knowledge of the topic, and warning labels for biased search results [14,29,32,35,36,41]. For example, the authors of [29] found that users spent more effort searching by *issuing more queries* when they encountered documents that were inconsistent with their prior beliefs. Xu et al. [41] found that (i) users' opinions seem to be easily affected by search results when they browse the web search purposelessly, and (ii) users who strongly support a topic *issued more clicks* and *spent more time* on the search results. Further, Epstein et al. [14] findings suggest that alerts for biased search results reduced the SEME effect and induced users to *issue more clicks*, *spend more time searching* and explore *lower-ranked search results*. Nevertheless, these works focused on mildly opinionated users, while there is still a lack of studies investigating opinionated users' behavior. We expect that users' behavior will vary based on the interaction between the SERP bias and display. Specifically, considering high bias metrics and search results incorporating stance labels, we hypothesize users will spot that search results are biased in their attitude-opposing direction when first inspecting the search engine. Consequently, users will not readily select the top attitude-opposing results (while trying to find information to confirm their beliefs) but instead i) issue more queries, ii) make fewer clicks, iii) visit more pages to check whether they are also biased, iv) have a higher click depth, and v) spend a longer time searching.

Hypothesis 2 ($\mathbf{H_{2a\text{-}e}}$). Users exposed to SERPs informed by *high* bias metrics and incorporating *stance labels* will issue more queries ($\mathbf{H_{2a}}$), fewer clicks ($\mathbf{H_{2b}}$), visit more pages ($\mathbf{H_{2c}}$), click on deeper search results ($\mathbf{H_{2d}}$), and spend more time ($\mathbf{H_{2e}}$) in the search session than users who are exposed to SERPs informed by *low* bias metrics with regular search results.

3 Experimental Setup

This section describes the user study we conducted to investigate the effect of SERP bias, display, and their interaction on users' behavior.

3.1 Materials

Dataset. To compute viewpoint bias metrics and predict stances for our study, we considered a public data set containing 1453 search results of queries related to three different debated topics (i.e., *atheism, intellectual property rights,* and *school uniforms*) [9] and used for stance detection in previous works [8,40]. Each search result in the dataset has been annotated in an expert annotation procedure and assigned to a *viewpoint label,* which indicates its position in the debate surrounding the topic it refers to on a seven-point Likert scale ranging from "strongly opposing" (–3) to "strongly supporting" (3). We mapped the viewpoints labels into three categories: *against* (–3, –2, –1), *neutral* (0), and *supporting* (1, 2, 3) (i.e., suitable ternary classification for our stance detection model). Next, we followed the same procedure as Wu et al. [40] to process the search results for the stance detection model, obtaining 1125 final instances[??].

Viewpoint Diversity Metric. To define the SERP bias levels, we considered two metrics from the *normalized discounted viewpoint bias* (nDVB) [9], which consider viewpoints on a seven-point Likert scale (–3, +3) to compute biases. The first, *normalized discounted polarity bias* (nDPB), evaluates the extent to which opposing and supporting viewpoints are balanced and ranges from –1 to 1. (more extreme values indicate greater bias, values closer to 0 indicate neutrality). The second metric, *normalized discounted stance bias* (nDSB), measures how much the distribution of stances deviates from being evenly spread across the seven different stance categories, and ranges from 0 (desired scenario of stance diversity) to 1 (maximal stance bias).

Table 1. Summary statistics of our four sets of 40 SERPs for polarity (nDPB) and stance (nDSB) bias metrics considering *low-low* and *high-high* levels for the *supporting* and *opposing* bias directions. All the values are identical among topics.

#	Bias Metric	Level	Bias Direction	Min	Max	Mean	Std.
I	Polarity	low	supporting	0.05	0.12	0.09	0.02
	Stance	low	supporting	0.23	0.33	0.28	0.03
II	Polarity	high	supporting	0.29	0.44	0.35	0.05
	Stance	high	supporting	0.36	0.44	0.38	0.02
III	Polarity	low	opposing	−0.05	−0.12	−0.09	0.02
	Stance	low	opposing	0.23	0.33	0.28	0.03
IV	Polarity	high	opposing	−0.29	−0.44	−0.35	0.05
	Stance	high	opposing	0.36	0.44	0.38	0.02

Ranked List of Search Results. To create the search result ranking conditions for the two levels of bias, we randomly selected 40 AI-correctly predicted search results[1] per topic from our test and validation data considering their original seven-point viewpoint. We picked a balanced sample of six strong (–3, +3), ten moderate (–2, +2), eight mild (–1, +1), and sixteen neutral (0) results to enable the creation of viewpoint-biased and viewpoint-diverse search result rankings to simulate a realistic scenario. We decided to have two levels of low and high bias metrics for the following reasons: i) the number of correctly AI-predicted search results is not sufficient to create a neat distinction between low-high and high-low levels for nDPB and nDSB bias metrics, and ii) we wanted to create a scenario that mirrors the *balanced* and *biased* templates described in Wu et al. [40] work but without having the same ranking order for each condition so that search results can still change if a user enters a new query, proposing a different ranking that still fulfils the bias metric levels. Next, we start from baseline rankings for low and high bias metrics levels and run a Grid Search to generate over 1000 combinations of our 40 selected search results. Then, we picked 40 sets of SERPs for low and high bias levels for each topic closest to the minimum or maximum values we found in the Grid Search or baseline rankings (see Table 1).

3.2 Variables

Our study followed a between-subjects design, considering three categorical *independent variables*: **SERP bias** (two levels: low or high nDPB and nDSB), **SERP display** (two levels: without stance labels or with stance labels; Fig. 1), and **topic** (three levels: atheism, intellectual property rights, and school uniforms). We investigated their effect on the following numerical *dependent variables*:

[1] Since the number of AI-correct predictions constrained us in the selection of search results, we found that 40 search results (ten results per four pages) were a suitable number to compute low and high bias metrics and mimic a realistic search scenario.

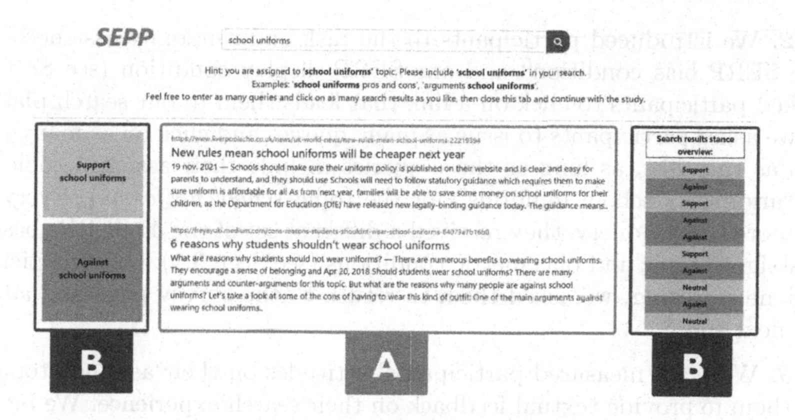

Fig. 1. SERP display conditions: **A)** search results only, and **B)** search results in A) are accompanied by AI-predicted stance labels (left) and stance overview (right).

(i) **Clicking diversity** is measured with the Shannon index to quantify the diversity of users' clicks [40]. The minimum value of the Shannon index is 0, which indicates that there is no diversity and only one viewpoint was clicked on. (ii) **Search behavior** is a set of five users' interactions, consisting of the following dependent variables: **queries, clicks, pages, click depth**, and **search session time**.

In addition to the above variables, we collected the following **descriptive** and **exploratory measurements** to describe our sample and for exploratory analyses: *gender, age group, level of education, topic knowledge* (seven-point Likert scale ranging from "not knowledgeable at all" to "extremely knowledgeable"), and *post search experience* (open text).[2]

3.3 Procedure

Step 1. After agreeing to an informed consent, participants stated their gender, age group, and level of education. We then asked participants for their attitudes concerning each debated topic (including one attention check where we specifically instructed participants on what option to select from a Likert scale) via three statements on seven-point Likert scales ranging from *strongly against* (−3) to *strongly supporting* (3). Then, we assigned participants to one of the three debated topics on which they stated a strong attitude[3], and asked them to imagine the following scenario [40]: *You and your friend were having dinner together. Your friend is very passionate about a debated topic and couldn't help sharing his views and ideas with you. After the dinner, you decide to further inform yourself on the topic by conducting a web search.*

[2] *"Please shortly describe your experience with the web search engine. Did you look for specific information, and if yes, how did you try to find it? Did you think the web search helped you build a more informed opinion on [topic]? If yes/no, why?"*

[3] If participants have no strong attitude on any topic, they exit the study (fully paid).

Step 2. We introduced participants to the task and randomly assigned them[4] to one SERP bias condition[5] and one SERP display condition (see Sect. 3.2). We asked participants to click on a link that leads them to our search platform. Here, we asked participants to issue as many queries and click on as many search results as they like, as long as those queries include their assigned topic term (e.g., "arguments school uniforms" for the topic *school uniforms*). Every time they entered a new query, they received a different set of the 40 SERPs based on their assigned topic and bias metric conditions (see Sect. 3.1). Once participants were done searching, we asked them to return to the survey page and advance to the next step.[6]

Step 3. We again measured participants' attitudes on their assigned topic and asked them to provide textual feedback on their search experience. We included another attention check - where the answer was explicitly reported in the question text - to filter out low-quality data.

3.4 Planned Sample Size and Statistical Analysis

Before data collection, we had computed a required sample size of 247 participants using the software *G*Power* [15] for a between-subjects ANOVA, specifying a medium effect size (Cohen's $f = 0.25$), a desired power of 0.8, and four groups. We recruited participants aged 18 or older with high English proficiency from *Prolific*[7]. The task was hosted on *Qualtrics*[8]. Participants received £1.4 as a reward for the study[9] (i.e., £13.3/hour given the median task completion time of 6:05 min). Prolific automatically timed out participants after 44 min. We further excluded participants from the analysis if they did not have a strong stance on any of the three topics, and failed at least one attention check. We investigated $\mathbf{H_{1a\text{-}c}}$ by conducting a **between-subjects ANOVA** with *clicking diversity* as the dependent variable and studying the main and interaction effects of *SERP bias* and *SERP display*, using *topic* as an additional independent variable (confounding factor). For $\mathbf{H_{2a\text{-}e}}$, we conducted five **between-subjects ANOVA** considering *search behaviors* (see Sect. 3.2) as dependent variables and studying interaction effects of *SERP bias* and *SERP display*, while adding *topic* to the analysis to control for its potential role as a confounding factor. Since we assessed eight ANOVA tests as part of this study, we applied a Bonferroni correction to our significance threshold, reducing it to $\frac{0.05}{8} = 0.0063$. We will consider as significant only the p-values that are below this specific threshold.

[4] We balanced participation across topics, bias, and display experimental conditions.

[5] We randomly assign participants to one of the 40 search results combinations with an opposite bias direction based on their pre-stance attitude (i.e., users who strongly support a topic are assigned to the opposing bias direction and vice versa).

[6] To encourage interactions with the web search engine, participants could only advance to the next step of the study after one minute. There was no maximum search time to simulate a realistic scenario.

[7] https://prolific.co.

[8] https://www.qualtrics.com/.

[9] The study has been approved by the Ethics Committee of Maastricht University.

4 Results

We recruited 350 paid participants, excluding them from the analysis if they did not have a strong opinion on any of the three topics (33), and failed at least one attention check (60). We thus considered 257 participants for the analysis.

4.1 Descriptive Analysis

Participants' age distribution was 66% in the 25–44 group, 8% in the 18–25 group, while 26% were above 44 y.o., consisting of 51% females, 48% males, and 1% other. Participants educational background show that 30% had a high school diploma, 44% had a bachelor's degree, 18% had a master's degree, and 8% had a Doctorate or professional degree. Most of the participants (49%) had little knowledge about their assigned topic, 21% had almost no knowledge, and 30% were very knowledgeable. Furthermore, the topic, bias, and display conditions distribution were approximately equally balanced among participants. Participants' pre-search stances are mostly distributed towards supporting the corresponding topic: atheism (26% vs 8%), intellectual property rights (33% vs <1%), and school uniforms (22% vs 10%). Interestingly, 19% (49/257) concluded their search without clicking on any search results. We will better discuss this finding in Sect. 4.3.

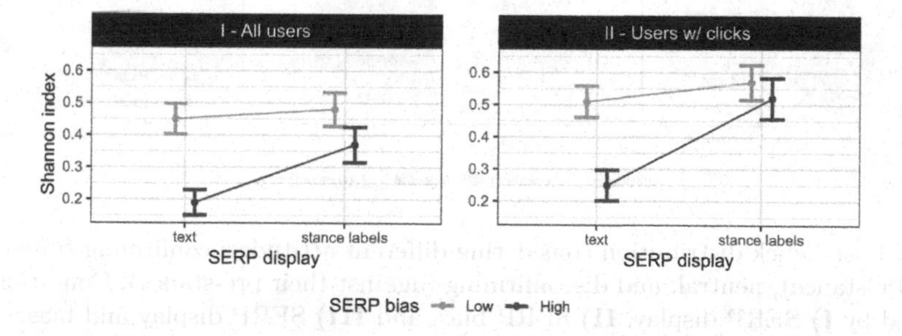

Fig. 2. Shannon index split by SERP display and bias conditions considering **I)** all users (257), and **II)** users who issued clicks in the search platform (208). The error bars are computed using the Standard Error (SE) on the Shannon index.

4.2 Hypotheses Tests

For $\mathbf{H_{1a}}$, we examined the clicking diversity of users who viewed low and high SERP bias levels. The ANOVA analysis highlights a significant effect for the bias level factor (F = 13.726, p <.001, Cohen's f = 0.23), hence, we *reject the null hypothesis* for H_{1a}, concluding that there is a difference in the clicking diversity of content between users who were exposed to SERPs with low (M = 0.47, sd = 0.41) and high (M = 0.28, sd = 0.39) bias levels (see Fig. 2 I). To investigate $\mathbf{H_{1b}}$, we checked the clicking diversity of participants who saw search results with and without stance labels. The ANOVA analysis shows that the SERP display factor is not significant (F = 4.964, p = .026, Cohen's f = 0.14) and is above our predefined significance threshold (.006). In this case, we *fail to reject the null hypothesis* for H_{1b}, concluding that there is no difference in the clicking diversity of content between users who were exposed to search results with (M = 0.43, sd = 0.44) and without (M = 0.32, sd = 0.37) stance labels. To study $\mathbf{H_{1c}}$, we looked at the interaction between the SERP bias and SERP display. The ANOVA analysis resulted in a non-significant interaction (F = 1.859, p = .173, Cohen's f = 0.09), thus we *fail to reject the null hypothesis* for H_{1c}. For the ANOVA analyses concerning $\mathbf{H_2}$ we also did not find significant effects on any of the five hypotheses we tested. Hence, we *fail to reject the null hypothesis* concluding that there are no differences in terms of *search behaviors* (see Sect. 3.2) w.r.t. to bias levels or labels.

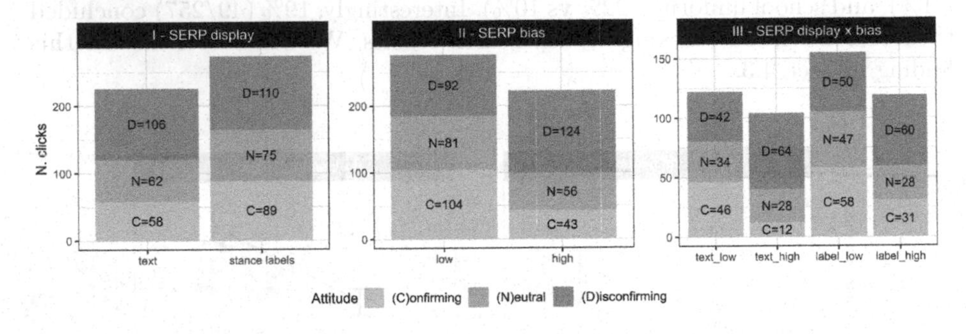

Fig. 3. Users' click distribution considering different attitudes: confirming (inline with their pre-stance), neutral, and disconfirming (against their pre-stance). Conditions are grouped by **I)** SERP display, **II)** SERP bias, and **III)** SERP display and bias.

4.3 Post Hoc and Exploratory Analyses

To better understand why some of the participants (49) did not click on any search results, we first discuss their behavior considering the issued queries, visited pages, and search time. Participants issued more queries (M = 2.75, sd = 3.55) and visited more pages (M = 3.04, sd = 3.50) no matter the bias level in the search results only condition. Overall, these participants had a lower search time (M = 145.52, sd = 54.60) compared to those who clicked on search results. Furthermore, 65% of these users were assigned to a high level of SERP bias (37% plain search results and 28% to stance labels), so they perceived that search results were biased with attitude-opposing viewpoints, thus triggering a potential backfire effect that led to an abandonment [1] in the search task. To get more insights into their abandonment, we performed a qualitative analysis of the textual feedback provided by these participants. Some of those who were exposed to biased search results recognized the presence of bias (e.g., "*...I don't think the search result helped much of my own opinion, seems lots of results are labelled as 'Against' already, which could influence the people's opinion before they read the details*"). Additionally, we briefly discuss the trends in search behavior considering the SERP bias significant effect for all the participants (257). Search results with a high bias lead users to interact with less diverse results, but they also lead to slightly fewer interactions (i.e., participants issued fewer clicks, queries, visited fewer pages, and spent less time searching).

Next, we analyzed the clicking diversity of users who engaged more (i.e., clicked on search results) with the web search by re-examining all the hypotheses (see Fig. 2 II) and comparing their behavior with users who abandoned the web search (49). As in Sect. 4.2, we found a significant effect (F = 8.423, $p <$ .006, Cohen's $f = 0.20$) for different levels of SERP bias ($\mathbf{H_{1a}}$) and no significant interaction (F = 3.587, $p = .059$, Cohen's $f = 0.13$) between the SERP display and SERP bias ($\mathbf{H_{1c}}$). Instead, we found a significant effect (F = 7.892, $p < .006$, Cohen's $f = 0.20$) for different levels of SERP display ($\mathbf{H_{1b}}$). As before, we did not find significant effects on any behavior measurement ($\mathbf{H_{2a\text{-}e}}$). The trends according to bias and display conditions are as follows. Participants exposed to SERPs with a high bias issued fewer clicks, queries, and pages visited. For the SERP display, participants exposed to stance labels issued more clicks, fewer queries, visited fewer pages, had higher click depth, and spent more time searching.

To investigate the users' clicking diversity difference on the SERP bias and display conditions, we explored the distribution across the three stances (see Fig. 3). For the SERP display condition, Fig. 3 I shows a shift towards balance in attitude distribution in the stance labels condition, with an increase in the attitude-confirming results and less for the neutral ones (thus increasing the clicking diversity). For the SERP bias condition, Fig. 3 II shows that viewpoints are somewhat balanced (slightly high confirming attitude) when the bias is low. Instead, a high bias shifts most clicks towards the *disconfirming* attitude at the cost of the number of clicks. Furthermore, Fig. 3 III reveals an increase in the number of clicks between the search results without labels and with labels when the SERP bias is low. This effect is also present with a high SERP bias in a reduced way.

5 Discussion

In this study, we investigated the effects of stance labels and different levels of bias metrics on opinionated users' clicking diversity and search behaviors. We conducted a between-subject study on three debated topics (i.e., *atheism*, *intellectual property rights*, and *school uniforms*) where participants take part in an open-ended and low-stakes scenario (i.e. inform themselves on a specific topic by conducting a web search). For **RQ1**, we found that participants exposed to biased search results proportionally consume more attitude-opposing content (i.e., less diverse search results). Furthermore, a post hoc analysis considering users who clicked on search results reveals that they consumed more diverse content in the presence of stance labels than plain search results (in line with Wu et al. [40]). For **RQ2**, we did not find evidence of search behavior differences among our SERP bias and display conditions. However, biased search results highlighted a trend of fewer interactions within the search page (e.g., fewer clicks and less time spent searching). Furthermore, a posthoc analysis shows that users who abandoned the web search without clicking any results issued more queries, visited more pages, and spent less time searching.

We can identify two main reasons why we did not find any significant differences in search behavior in our study. We simulated a realistic scenario and did not put any limit on the search time session and the number of interactions in general. Therefore, users could naturally explore the web search without any particular restrictions concerning search behavior. However, dealing with opinionated users considering no task restrictions and exposing them to biased search results led to a web search abandonment [1] for 19% of participants. Second, the potential backfire effect that triggered users to abandon the web search may hint at a typical conduct present in System 1 conceptual thinking [4], where people's biases affect the way they perceive and process new information, especially if the information is counter-intuitive, conflicting or induce uncertainty [38].

5.1 Implications

Firstly, our findings suggest that it is crucial to examine the behaviors of users who abandon their search. Our post-hoc analysis indicated that 19% of participants abandoned the web search without interacting with any search result while issuing other search behaviors. These findings may suggest a backfire effect [37], leading users to reject attitude-opposing content and quickly end their search. Second, users who interacted with search results and were exposed to stance labels consumed more diverse content, especially with biased search results. These results align with previous research on bias mitigation using informative labels [14,32,40], making them a valuable tool for increasing content diversity consumption. These results bring the need to inspect additional factors on strongly opinionated users, like their characteristics (e.g., intellectual humility [31]), their perception of search results such as relevance [23] or credibility [29], and expose them to different types of tasks (high-stakes [11] or purposeful [41]).

5.2 Limitations and Future Work

Our study has the following limitations. First, we used a ternary stance detection model to classify search results. So, users will not realize the nuances of the search results they are seeing, which are on a seven-point Likert scale. Furthermore, we did not inform users that the AI predictions on search results stances were always correct. Hence, users were unaware of the correctness of the AI-detected stances, and there is a chance they may have thought the labels were potentially wrong. Future work could investigate how to predict more diversified viewpoint representations and the effects of presenting wrong AI predictions for the detected search results stances. Second, we decided to use the Shannon index to measure clicking diversity since we expected fewer than six clicks for each user, as suggested by previous work on users' behavior [29,32,35]. Further studies are needed to explore other methodologies to measure users' clicking diversity based on search results viewpoints (e.g., fairness metrics [34,43]). Third, we performed a Grid Search to find reasonable search result pages for low and high bias metrics [9], based on the AI correctly predicted search results and our user study conditions. Future research should investigate other methodologies to compute more fine-grained boundaries on different bias metrics and datasets.

6 Conclusion

In this paper, we studied the effects of stance labels and bias metrics on users' clicking diversity and search behavior. We found that opinionated users increased their consumption of attitude-opposing content when exposed to biased search results, with a trend towards fewer interactions within the search page. Furthermore, 19% of users who did not interact with search results spent less time searching while issuing more queries and visiting more pages. Instead, participants who issued clicks significantly increased their content diversity when exposed to stance labels, particularly when the search results were biased. As highlighted in our exploratory and qualitative analyses, a potential motivation for our results may lie in the open-ended task formulation, and the need to inspect additional participants' traits. Follow-up studies need to explore whether different ranking bias metrics and stance label representations influence content diversity consumption, along with constructing different types of open and closed tasks to better investigate the search behaviors of opinionated users.

References

1. Abualsaud, M.: The effect of queries and search result quality on the rate of query abandonment in interactive information retrieval. In: Proceedings of the 2020 Conference on Human Information Interaction and Retrieval, CHIIR 2020, pp. 523–526. Association for Computing Machinery, New York (2020). https://doi.org/10.1145/3343413.3377951
2. Ajjour, Y., Alshomary, M., Wachsmuth, H., Stein, B.: Modeling frames in argumentation. In: Inui, K., Jiang, J., Ng, V., Wan, X. (eds.) Proceedings of the 2019 Conference on Empirical Methods in Natural Language Processing and the 9th International Joint Conference on Natural Language Processing (EMNLP-IJCNLP), pp. 2922–2932. Association for Computational Linguistics, Hong Kong (2019). https://doi.org/10.18653/v1/D19-1290. https://aclanthology.org/D19-1290
3. Allam, A., Schulz, P.J., Nakamoto, K.: The impact of search engine selection and sorting criteria on vaccination beliefs and attitudes: two experiments manipulating google output. J. Med. Internet Res. **16**(4), e100 (2014). https://doi.org/10.2196/jmir.2642, http://www.jmir.org/2014/4/e100/

4. Azzopardi, L.: Cognitive biases in search: a review and reflection of cognitive biases in information retrieval, pp. 27–37 (2021). https://doi.org/10.1145/3406522.3446023
5. Chen, S., Khashabi, D., Yin, W., Callison-Burch, C., Roth, D.: Seeing things from a different angle: discovering diverse perspectives about claims. arXiv:1906.03538 [cs] (2019)
6. Draws, T., Inel, O., Tintarev, N., Baden, C., Timmermans, B.: Comprehensive viewpoint representations for a deeper understanding of user interactions with debated topics. In: ACM SIGIR Conference on Human Information Interaction and Retrieval, pp. 135–145. ACM, Regensburg (2022). https://doi.org/10.1145/3498366.3505812. https://dl.acm.org/doi/10.1145/3498366.3505812
7. Draws, T., Liu, J., Tintarev, N.: Helping users discover perspectives: enhancing opinion mining with joint topic models. In: 2020 International Conference on Data Mining Workshops (ICDMW), pp. 23–30. IEEE, Sorrento (2020). https://doi.org/10.1109/ICDMW51313.2020.00013. https://ieeexplore.ieee.org/document/9346407/
8. Draws, T., et al.: Explainable cross-topic stance detection for search results. In: CHIIR 2023: ACM SIGIR Conference on Human Information Interaction and Retrieval (2023)
9. Draws, T., et al.: Viewpoint diversity in search results. In: Kamps, J., et al. (eds.) ECIR 2023. LNCS, vol. 13980, pp. 279–297. Springer, Heidelberg (2023). https://doi.org/10.1007/978-3-031-28244-7_18
10. Draws, T., Tintarev, N., Gadiraju, U.: Assessing viewpoint diversity in search results using ranking fairness metrics. SIGKDD Explor. Newsl. **23**(1), 50–58 (2021). https://doi.org/10.1145/3468507.3468515
11. Draws, T., Tintarev, N., Gadiraju, U., Bozzon, A., Timmermans, B.: This is not what we ordered: exploring why biased search result rankings affect user attitudes on debated topics. In: Proceedings of the 44th International ACM SIGIR Conference on Research and Development in Information Retrieval, SIGIR 2021, pp. 295–305. Association for Computing Machinery, New York (2021). https://doi.org/10.1145/3404835.3462851. https://dl.acm.org/doi/10.1145/3404835.3462851
12. Dumani, L., Neumann, P.J., Schenkel, R.: A framework for argument retrieval. In: Jose, J.M., Yilmaz, E., Magalhães, J., Castells, P., Ferro, N., Silva, M.J., Martins, F. (eds.) ECIR 2020. LNCS, vol. 12035, pp. 431–445. Springer, Cham (2020). https://doi.org/10.1007/978-3-030-45439-5_29
13. Epstein, R., Robertson, R.E.: The search engine manipulation effect (SEME) and its possible impact on the outcomes of elections. Proc. Natl. Acad. Sci. **112**(33), E4512–E4521 (2015). https://doi.org/10.1073/pnas.1419828112. http://www.pnas.org/lookup/doi/10.1073/pnas.1419828112
14. Epstein, R., Robertson, R.E., Lazer, D., Wilson, C.: Suppressing the search engine manipulation effect (SEME). In: Proceedings of the ACM on Human-Computer Interaction, vol. 1(CSCW), pp. 1–22 (2017). https://doi.org/10.1145/3134677. https://dl.acm.org/doi/10.1145/3134677
15. Faul, F., Erdfelder, E., Buchner, A., Lang, A.G.: Statistical power analyses using g* power 3.1: tests for correlation and regression analyses. Behav. Res. Methods **41**(4), 1149–1160 (2009)
16. Gao, R., Shah, C.: Toward creating a fairer ranking in search engine results. Inf. Process. Manag. **57**(1), 102138 (2020). https://doi.org/10.1016/j.ipm.2019.102138. https://linkinghub.elsevier.com/retrieve/pii/S0306457319304121

17. Gezici, G., Lipani, A., Saygin, Y., Yilmaz, E.: Evaluation metrics for measuring bias in search engine results. Inf. Retr. J. **24**(2), 85–113 (2021). https://doi.org/10.1007/s10791-020-09386-w. http://link.springer.com/10.1007/s10791-020-09386-w
18. Jürgens, P., Stark, B.: Mapping exposure diversity: the divergent effects of algorithmic curation on news consumption. J. Commun. **72**(3), 322–344 (2022). https://doi.org/10.1093/joc/jqac009
19. Küçük, D., Can, F.: Stance detection: a survey. ACM Comput. Surv. **53**(1), 1–37 (2021). https://doi.org/10.1145/3369026. https://dl.acm.org/doi/10.1145/3369026
20. Loecherbach, F., Welbers, K., Moeller, J., Trilling, D., Van Atteveldt, W.: Is this a click towards diversity? explaining when and why news users make diverse choices. In: Proceedings of the 13th ACM Web Science Conference 2021, WebSci 2021, pp. 282–290. Association for Computing Machinery, New York (2021). https://doi.org/10.1145/3447535.3462506
21. Lucien, H., et al.: Benefits of diverse news recommendations for democracy: a user study. Dig. Journalism **10**(10), 1710–1730 (2022). https://doi.org/10.1080/21670811.2021.2021804
22. Mattis, N.M., Masur, P.K., Moeller, J., van Atteveldt, W.: Nudging towards diversity: a theoretical framework for facilitating diverse news consumption through recommender design (2021). https://doi.org/10.31235/osf.io/wvxf5. https://osf.io/preprints/socarxiv/wvxf5
23. Maxwell, D., Azzopardi, L., Moshfeghi, Y.: The impact of result diversification on search behaviour and performance. Inf. Retr. J. **22** (2019). https://doi.org/10.1007/s10791-019-09353-0
24. Mcdonald, G., Macdonald, C., Ounis, I.: Search results diversification for effective fair ranking in academic search. Inf. Retr. J. **25**, 1–26 (2022). https://doi.org/10.1007/s10791-021-09399-z
25. Michiels, L., Leysen, J., Smets, A., Goethals, B.: What are filter bubbles really? a review of the conceptual and empirical work. In: Adjunct Proceedings of the 30th ACM Conference on User Modeling, Adaptation and Personalization, UMAP 2022 Adjunct, pp. 274–279. Association for Computing Machinery, New York (2022). https://doi.org/10.1145/3511047.3538028
26. Nickerson, R.S.: Confirmation bias: a ubiquitous phenomenon in many guises. Rev. Gener. Psychol. **2**(2), 175–220 (1998). https://doi.org/10.1037/1089-2680.2.2.175
27. Pathiyan Cherumanal, S., Spina, D., Scholer, F., Croft, W.B.: Evaluating fairness in argument retrieval. In: Proceedings of the 30th ACM International Conference on Information & Knowledge Management, CIKM 2021, pp. 3363–3367. Association for Computing Machinery, New York (2021). https://doi.org/10.1145/3459637.3482099
28. Pogacar, F.A., Ghenai, A., Smucker, M.D., Clarke, C.L.: The positive and negative influence of search results on people's decisions about the efficacy of medical treatments. In: Proceedings of the ACM SIGIR International Conference on Theory of Information Retrieval, ICTIR 2017, pp. 209–216. Association for Computing Machinery, New York (2017). https://doi.org/10.1145/3121050.3121074
29. Pothirattanachaikul, S., Yamamoto, T., Yamamoto, Y., Yoshikawa, M.: Analyzing the effects of document's opinion and credibility on search behaviors and belief dynamics. In: Proceedings of the 28th ACM International Conference on Information and Knowledge Management, CIKM 2019, pp. 1653–1662. Association for Computing Machinery, New York (2019). https://doi.org/10.1145/3357384.3357886

30. Puschmann, C.: Beyond the bubble: assessing the diversity of political search results. Dig. Journalism **7**(6), 824–843 (2019). https://doi.org/10.1080/21670811.2018.1539626
31. Rieger, A., Bredius, F., Tintarev, N., Pera, M.: Searching for the whole truth: harnessing the power of intellectual humility to boost better search on debated topics. In: CHI 2023 - Extended Abstracts of the 2023 CHI Conference on Human Factors in Computing Systems. Conference on Human Factors in Computing Systems - Proceedings, 23–28 April 2023. Association for Computing Machinery (ACM), United States (2023). https://doi.org/10.1145/3544549.3585693
32. Rieger, A., Draws, T., Tintarev, N., Theune, M.: This item might reinforce your opinion: obfuscation and labeling of search results to mitigate confirmation bias. In: Proceedings of the 32nd ACM Conference on Hypertext and Social Media, HT 2021, pp. 189–199. Association for Computing Machinery, New York (2021). https://doi.org/10.1145/3465336.3475101
33. Sanh, V., Debut, L., Chaumond, J., Wolf, T.: Distilbert, a distilled version of bert: smaller, faster, cheaper and lighter. arXiv preprint arXiv:1910.01108 (2019)
34. Sapiezynski, P., Zeng, W., E Robertson, R., Mislove, A., Wilson, C.: Quantifying the impact of user attentionon fair group representation in ranked lists. In: Companion Proceedings of The 2019 World Wide Web Conference, WWW 2019, pp. 553–562. Association for Computing Machinery, New York (2019). https://doi.org/10.1145/3308560.3317595
35. Suzuki, M., Yamamoto, Y.: Analysis of relationship between confirmation bias and web search behavior. In: Proceedings of the 22nd International Conference on Information Integration and Web-Based Applications & Services, iiWAS 2020, pp. 184–191. Association for Computing Machinery, New York USA (2021). https://doi.org/10.1145/3428757.3429086
36. Suzuki, M., Yamamoto, Y.: Characterizing the influence of confirmation bias on web search behavior. Front. Psychol. **12** (2021). https://api.semanticscholar.org/CorpusID:244897002
37. Swire-Thompson, B., DeGutis, J., Lazer, D.: Searching for the backfire effect: measurement and design considerations. J. Appl. Res. Memory Cogn. **9**(3), 286–299 (2020). https://doi.org/10.1016/j.jarmac.2020.06.006. https://www.sciencedirect.com/science/article/pii/S2211368120300516
38. Tversky, A., Kahneman, D.: Judgment under uncertainty: heuristics and biases. Sci. **185**(4157), 1124–1131 (1974). https://doi.org/10.1126/science.185.4157.1124. https://www.science.org/doi/abs/10.1126/science.185.4157.1124
39. Vrijenhoek, S., Bénédict, G., Gutierrez Granada, M., Odijk, D., De Rijke, M.: Radio - rank-aware divergence metrics to measure normative diversity in news recommendations. In: Proceedings of the 16th ACM Conference on Recommender Systems, RecSys 2022, pp. 208–219. Association for Computing Machinery, New York (2022). https://doi.org/10.1145/3523227.3546780
40. Wu, Z., Draws, T., Cau, F., Barile, F., Rieger, A., Tintarev, N.: Explaining search result stances to opinionated people. In: Longo, L. (ed.) xAI 2023, vol. 1902, pp. 573–596. Springer, Cham (2023). https://doi.org/10.1007/978-3-031-44067-0_29
41. Xu, L., Zhuang, M., Gadiraju, U.: How do user opinions influence their interaction with web search results? In: Proceedings of the 29th ACM Conference on User Modeling, Adaptation and Personalization, UMAP 2021, pp. 240–244. Association for Computing Machinery, New York (2021). https://doi.org/10.1145/3450613.3456824

42. Yom-Tov, E., Dumais, S., Guo, Q.: Promoting civil discourse through search engine diversity. Social Sci. Comput. Rev. **32**(2), 145–154 (2014). https://doi.org/10.1177/0894439313506838. http://journals.sagepub.com/doi/10.1177/0894439313506838
43. Zehlike, M., Bonchi, F., Castillo, C., Hajian, S., Megahed, M., Baeza-Yates, R.: Fa*ir: a fair top-k ranking algorithm. In: Proceedings of the 2017 ACM on Conference on Information and Knowledge Management, CIKM 2017, pp. 1569–1578. Association for Computing Machinery, New York (2017). https://doi.org/10.1145/3132847.3132938

Recommendation Fairness in eParticipation: Listening to Minority, Vulnerable and NIMBY Citizens

Marina Alonso-Cortés, Iván Cantador(✉), and Alejandro Bellogín

Escuela Politécnica Superior, Universidad Autónoma de Madrid, 28049 Madrid, Spain
marina.alonso-cortes@estudiante.uam.es,
{ivan.cantador,alejandro.bellogin}@uam.es

Abstract. E-participation refers to the use of digital technologies and online platforms to engage citizens and other stakeholders in democratic and government decision-making processes. Recent research work has explored the application of recommender systems to e-participation, focusing on the development of algorithmic solutions to be effective in terms of personalized content retrieval accuracy, but ignoring underlying societal issues, such as biases, fairness, privacy and transparency. Motivated by this research gap, on a public e-participatory budgeting dataset, we measure and analyze recommendation fairness metrics oriented to several minority, vulnerable and NIMBY (Not In My Back Yard) groups of citizens. Our empirical results show that there is a strong popularity bias (especially for the minority groups) due to how content is presented and accessed in a reference e-participation platform; and that hybrid algorithms exploiting user geolocation information in a collaborative filtering fashion are good candidates to satisfy the proposed fairness conceptualization for the above underrepresented citizen collectives.

Keywords: Recommender system · Citizen participation · Fairness · Bias · Minority group · Vulnerable group · NIMBY

1 Introduction

Citizen participation (a.k.a. public participation) refers to the active involvement of individuals in government decision-making processes that affect their lives and communities [20]. It entails the citizens' engagement in various forms, such as taking part in consultations, attending public meetings, and collaborating with governments, organizations or other stakeholders to solve citizenry problems and to shape public decisions. It thus has the potential to strengthen democracy by creating more inclusive, transparent and informed problem-solving, empowering citizens to have a more active role in public initiatives and policies, and increasing their trust in public institutions [8].

Citizen participation is progressively being conducted on the internet, through the so-called electronic participation (or *e-participation*) platforms [19].

N. Goharian et al. (Eds.): ECIR 2024, LNCS 14611, pp. 420–436, 2024.
https://doi.org/10.1007/978-3-031-56066-8_31

Participedia[1] is an open-source, collaborative project aimed to document, promote and analyze public participation and democratic innovations around the world. It serves as a global resource with information on developed citizen participation initiatives, allowing individuals and organizations to share their experiences, case studies, and best practices. As of October 2023, with more than 2,400 cases recorded, it shows that 30% (111 out of 362) of the considered participatory method types are online. Representative examples of these methods are e-deliberation, e-voting, and e-participatory budgeting.

By harnessing the power of ICT, e-participation can overcome physical barriers to participation and amplify the voices of individuals who may be marginalized or excluded from the government decision-making [35]. It thus can offer greater accessibility, participation convenience (flexibility) and efficiency, and broader reach and inclusivity. Nonetheless, as stated by the OECD [24], e-participation has difficulties, such as the digital divide (i.e., required digital literacy) and technological barriers (e.g., access to the internet). Moreover, it has been shown to entail certain limitations, such as unequal (biased) citizen representation, and non-in-depth deliberation and polarization [5].

These problems are originated or intensified by the overload of information and the lack of personalization in the used platforms [4], among other causes. For e-participatory processes in large cities or at regional or national levels, the number and size of citizen proposals, discussions and debates may be overwhelming for a user. However, in general, current platforms do not filter or rank the content provided to the user according to her profile and preferences. It is for this situation that researchers have begun to explore the use of recommender systems in e-participation [28].

Among other applications, in the research literature, recommender systems for online citizen participation have been proposed for presenting political candidates with similar ideological positions [12,30], and suggesting relevant citizen proposals based on personal preferences [4,6]. Previous work has mainly focused on the development of algorithmic solutions to be effective in terms of personalized content retrieval accuracy. These solutions, by contrast, have not been studied considering underlying societal issues, such as biases, fairness, privacy and transparency.

Motivated by this research gap, in this work, we aim to investigate whether traditional recommendation algorithms applied on a representative e-participation platform entail bias and unfairness effects for certain citizens. Hence, our study targets the following two research questions:

- **RQ1**. How can recommendation fairness be conceptualized and formalized in e-participation taking societal issues into account?
- **RQ2**. How do standard recommendation algorithms behave on considered social fairness dimensions in a real e-participation case?

To address these questions, we evaluate diverse recommendation algorithms according to both accuracy and fairness metrics, on a public dataset of the *Decide Madrid*[2] e-participatory budgeting platform, and for several general,

[1] https://participedia.net.
[2] https://decide.madrid.es.

underrepresented citizen collectives. More specifically, in the context of making personalized recommendations of citizen proposals publicly available and discussed online in Decide Madrid, we carefully identify a number of minority, vulnerable and NIMBY[3] (Not In My Back Yard) groups of city residents, and assign to them the proposals that deal with their concerns, needs and problems. Then, we propose a novel social fairness conceptualization, and measure and analyze associated fairness metrics based on *Generalized Cross Entropy* [10], with which, for the above groups, we study existing biases of recommendations generated by content-based, collaborative filtering and hybrid algorithms. We release our data and code as open access[4].

2 Related Work

2.1 Recommender Systems for Online Citizen Participation

In the context of cities, recommender systems have been proposed for a large variety of problems and tasks. In [28], the authors survey the state of the art, characterizing and categorizing existing recommendation approaches in terms of several city dimensions, namely *smart economy*, *environment*, *mobility*, *governance*, *living*, and *people* dimensions.

In all these dimensions, we can find purposes and goals for recommender systems where citizens are active participants, such as fostering healthy lifestyle [7] and promoting cultural city heritage [2] in *smart living*, and optimizing parking space usage [34] and supporting evacuation management [18] in *smart mobility*.

For the *smart governance* dimension –aimed to increase efficiency in municipal management, and promote citizen participation and inclusion–, as presented in [9], recommender systems can be further categorized according to whether they are applied to government-to-citizen (G2C), government-to-business (G2B), or government-to-government (G2G) services.

According to [9], G2C recommenders are predominant in the research literature, and mainly focus on providing citizens with personalized government e-notifications and e-services [1]; and keeping the government informed about the citizens' problems, concerns and opinions expressed in social media, and e-consultation and e-participation platforms [30].

Besides, recommender systems for online citizen participation have been proposed for assisting voters by presenting political candidates with similar ideological positions [12,30], retrieving comments from individuals who hold similar and dissimilar opinions [22], providing local news relevant to citizen discussions [17], supporting processes of public participation in urban planning [21], finding relevant citizen proposals based on personal preferences [4,6], and assisting citizens in tagging personal e-petitions [33].

[3] NIMBY phenomenon: residents' opposition to certain development projects, facilities or infrastructures that they believe could have a negative impact on their immediate surroundings.

[4] https://github.com/malonsocortes/fairness-eparticipation-recsys.

As [4,6], in this paper, we consider the task of providing personalized suggestions of citizen proposals existing in an e-participation platform. However, differently to previous work, we evaluate recommendation algorithms by analyzing to what extent they underrepresent proposals about minority and vulnerable collectives, and are biased towards proposals that affect citizen majorities. This approach could be explored in other recommendation tasks for online citizen participation, such as those mentioned above.

2.2 Fairness in Recommender Systems

The evaluation of recommender systems has been mainly conducted from the point of view of the accuracy and effectiveness of the generated recommendations. In recent years, nonetheless, there has been growing interest and concern for issues related to fairness [11,32], since recommenders may have algorithmic biases or may be influenced by biases existing in input data –understanding bias as a systematic deviation of the results that benefits some users or items against others.

The notion of fairness in the recommender systems field is acknowledged to be multi-faceted, entailing several factors, such as the target stakeholders, the type of benefit expected from recommendations, and the context of the recommendation application, among others [11]. Besides, traditional fairness definitions have been associated with a homogeneous distribution of benefits among the different groups of users or items involved [13]. For this, it is usually necessary to define protected groups (e.g., female or black candidates in job recommendations) or sensitive attributes, either from users or items (e.g., the user's race or gender, or a job's leadership requirements).

Recent work, by contrast, challenges this idea, and shows that, depending on the application context, definitions of fairness related to non-homogeneous distributions might be needed [10], in part because not all the stakeholders understand fairness under the same perspective. For instance, a streaming service company with free and paid user accounts could consider it fair that the error of a not relevant recommendation would be smaller for a free account than for a paid account.

This consideration was modeled in [10] through a probabilistic framework based on an adaptation of the Generalized Cross Entropy (GCE) metric, which is used to quantify the difference between two probability distributions. Specifically, in the context of recommendation fairness, one probability distribution would be estimated according to the output provided by the recommendation algorithm, whereas the other one would be the ideal or target distribution, either uniform (as the standard *equality* fairness perspective) or non-uniform. Moreover, depending on the probability space, the metric could be used to assess user-, item- or even context-oriented fairness.

The flexibility granted by GCE makes it suitable for our work, as we shall test it with different definitions of fairness according to item attributes. In particular, being a novel contribution to the field, we will use GCE to measure recommendation fairness based on sensitive item distributions, going beyond item popularity, as done in [10].

Nonetheless, while our study advances the understanding of fairness in recommender systems, it does not address their potential rôle in promoting polarization, confirmation bias, and echo chamber effects. These phenomena undermine (political) deliberation and diversity of thought, as highlighted in Pariser's work on filter bubbles [25] and Nguyen et al.'s work on echo chambers [23]. Future research should focus on identifying and mitigating these effects, ensuring exposure to diverse viewpoints in recommender systems.

3 Case Study

3.1 Decide Madrid: A Participatory Budgeting e-Platform

As a specific citizen participation method, participatory budgeting is a democratic process of deliberation and decision-making in which residents choose how to spend part of a municipal budget. In this process, participants raise awareness of problems and issues related to their city within a wide range of topics –e.g., urban planning and housing, environment, education, health, transport and security–, and propose, debate and support public investment in solutions and initiatives for such problems and issues.

Participatory budgeting was originated in 1988 in Porto Alegre, Brazil [29], and since then it has gained popularity spreading to over 7,000 cities around the world[5], especially after the adoption of ICT and the digitization of the process through the so-called e-participatory budgeting. In this context, ad hoc e-participation platforms have entailed increasing transparency and saving time for citizens [26].

Decide Madrid is a representative example of this type of platforms, and has supported the annual participatory budgeting campaigns of Madrid, Spain, since 2016. It is a website that allows the city residents to propose, discuss and vote for proposals which the City Council commits to implement if they satisfy certain feasibility and citizen support requirements. The tool is built upon the CONSUL framework[6], which is accessible as open source, and, as of October 2013, has been used by at least 135 institutions of 35 countries and 90 million citizens around the world. Its architecture and user interface are analogous to those of other popular platforms, such as the Stanford Participatory Budgeting[7] and EU Open Budgets[8] platforms, and are based on traditional *web forums* with tree structures of conversation threads (i.e., nested comments), which are commonly used by other e-participation tools.

In this work, we focus our attention on the citizen proposals posted in Decide Madrid, and the comments that the proposals received from registered users. Due to the large number of proposals (around four thousand per year) in the platform, the need for personalized proposal recommendations is justified. In fact, there already exist scientific papers on the topic [4,6]. However, to the best of our knowledge, no previous research has addressed the potential bias and unfairness effects that recommender systems for e-participation may have.

[5] https://www.participatorybudgeting.org/about-pb/#what-is-pb.
[6] https://consuldemocracy.org.
[7] https://pbstanford.org.
[8] https://openbudgets.eu.

Table 1. Statistics of the Decide Madrid dataset.

Campaign	Users (U)	Proposals (P)	Comments (C)	Sparsity	C/U	C/P
C1	8,009	9,677	34,149	99.96%	4.26	3.52
C2	2,374	3,805	7,190	99.94%	3.02	1.88
C3	1,275	2,814	4,075	99.91%	3.19	1.44
C4	1,210	2,746	4,829	99.89%	3.99	1.75
Total	**12,868**	**19,042**	**50,243**			

3.2 The Decide Madrid Dataset

Among hundreds of data collections, the open data portal[9] of the Madrid City Council gathers and makes publicly available citizen-generated content of Decide Madrid; specifically, its citizen proposals with their metadata and comments. From this open data collection, in [4,5], the authors built a dataset for the proposals made in four participatory budgeting campaigns (2015–2019), which we extended as explained in Sect. 4, and used for the work presented herein. In the dataset, the comments that users made on the proposals are considered as feedback of interest, and are assumed as implicit, unary ratings. Besides, processing the textual metadata in Spanish of the proposals (i.e., titles, tags, summaries and descriptions), the authors assigned categories, topics and locations (districts) to the proposals. Table 1 shows some statistics of the dataset for the covered campaigns. The rating sparsity levels are extremely high (99.9%), being greater than those of popular datasets in the recommender systems research field.

4 Proposed Recommendation Fairness for e-Participation

4.1 Fairness Conceptualization

As mentioned in Sect. 2.2, to address **RQ1**, we propose using the Generalized Cross Entropy (GCE) metric for measuring recommendation fairness. This metric allows considering distinct conceptualizations of fairness, and is based on the difference between the distribution of recommendations generated by an algorithm and an ideal or target distribution (*perspective*) of recommendations with respect to a certain user/item/context variable (*attribute*).

For e-participation, we propose to instantiate and analyze the GCE metric by taking as item **attribute** of interest the belonging of a citizen proposal to a minority, vulnerable or NIMBY citizen group, potentially discriminated by a recommendation algorithm. From now on, we will refer to this attribute as *group*. Its possible values correspond to three broad categories: `Minority` (for minority and vulnerable citizen groups), `NIMBY` (for NIMBY groups), and `Other` (for other groups, presumably non-underrepresented). Additionally, we measure several versions of the GCE metric from the following **perspectives**: uniform (p_u),

[9] https://datos.madrid.es.

Table 2. Possible values of the proposed fairness attribute and perspective.

Attribute	`Minority`
	`NIMBY`
	`Other`
Perspective	Uniform (p_u)
	Test proportion (p_t)
	Biased towards discriminated groups: minority (p_m), NIMBY (p_n), both (p_{m+n})

test proportion (p_t), and biased towards discriminated groups, namely minority/vulnerable (p_m), NIMBY (p_n), and both minority/vulnerable and NIMBY (p_{m+n}). Table 2 gathers all the above attribute and perspective values.

For establishing the minority, vulnerable and NIMBY groups to study, we proceeded in a two-fold process, counting on the collaboration of a political science professor expert in citizen participation. First, we generated a list of potential groups from a compilation of technical reports and scientific papers (e.g., [14,27]). Next, we checked if the Decide Madrid dataset had citizen proposals belonging to any of the identified groups. For such purpose, using the Apache Lucene library[10], we created a search index storing all the citizen proposals of the dataset. The index allowed us to launch regular expression-based queries against the proposals' titles and summaries[11].

In an iterative fashion, we collaboratively defined a vocabulary of terms (i.e., keywords and regular expressions) for each group. For instance, the 'People with disabilities' minority group was represented with terms such as `disabilit*`, `handicap*`, `impairment*`, `accessibility`, `reduced mobility`, and `architectural barrier*`, where the asterisk `*` stands for none or several word letters. Each term was carefully selected so that it did not generate ambiguity in finding its corresponding group. We provide the groups and their vocabularies in our public repository.

Then, in the index, for each term, we searched for proposals whose title or summary had the term. Boosting in the queries the term matches occurred on the titles, we created aggregated scores for every retrieved proposal and group, by counting the frequencies of a group's terms in the proposal's title and summary. Finally, each proposal was assigned to its matched group with the highest score. Table 3 shows the minority/vulnerable and NIMBY groups, respectively listed within the broad categories `Minority` and `NIMBY`. The groups are sorted by decreasing number of proposals in the dataset, showing which citizen collectives are less represented in Decide Madrid.

We note that regardless of its category, any of the sets of proposals corresponds to an unrepresented collective of citizens which is not limited to Madrid, but could exist in any (large) city of the world. We thus believe that the built

[10] https://lucene.apache.org.

[11] We discarded using the proposals' descriptions since they entailed information noise and ambiguities on the proposals' main topics.

Table 3. Considered minority/vulnerable and NIMBY groups with their respective numbers of citizen proposals in the dataset.

Minority and vulnerable collectives		NIMBY issues	
People with mental disorders	9	Shantytowns	6
Evicted and rehoused	10	Ethnicities and xenophobia	11
LGTBIQ+ collective	11	Squatting	12
Poor and people in social exclusion	14	Gambling and betting houses	14
People with dependency or special needs	16	Landfills, incinerators and crematoriums	28
Parents [family conciliation]	18	Antennas and electrical towers	34
Retirees (pensioners)	33	Drugs [on the streets]	35
Women [equality and gender violence]	33	Prostitution	44
Immigrants and refugees	33	Urban planning [at city level]	69
Cyclists and motorcyclists [no bike lanes issues]	57	Alcohol [on the streets]	81
Homeless (vagrants and destitute people)	62	Terraces	86
Unemployed	65	Burials	107
Young people (youth)	75	Resident parking	121
Elderly (Third Age)	142	Noise [on the streets]	173
People with disabilities	262	Feces and urine [on the streets]	207
Children (childhood)	709	Garbage [no recycling issues]	235
	1,549		**1,263**

group list and vocabularies, as well as the sets of citizen proposals, may be of interest for the research community.

4.2 Fairness Formalization

In [10], the GCE metric is defined as follows:

$$GCE_{\beta}(A, R; p_f) = \frac{1}{\beta(1-\beta)}\left[\sum_{a \in A} p_f^{\beta}(a) \cdot p_R^{(1-\beta)}(a) - 1\right] \quad (1)$$

where A is the attribute space upon which probability distributions are defined, R is the recommendation algorithm whose fairness is assessed, and p_f is the ideal or target fairness distribution, against which GCE will compare the estimated p_R distribution from R–in particular, if $p_R = p_f$ then $GCE = 0$, i.e., R is considered a perfectly fair model. By definition, GCE outputs negative values, so the closer they are to 0, the closer the two distributions are and, hence, p_R might be considered fairer with respect to p_f. The choice of β is critical and entails a particular divergence metric depending on its value. As in [10], we shall use a value of $\beta = 2$, which corresponds to Pearson's χ^2, since the metric becomes more robust to outliers.

To estimate p_R, we first obtain the mapping att(i) of each recommended item i to the **attribute space** A. As specified in the previous section, A allows encoding whether a citizen proposal (item) refers to a minority, NIMBY, or other group. Hence, $|A| = 3$. Based on this, we estimate the value p_R of each element $a \in A$ by considering how often the items of the group appear in a given recommendation list:

Table 4. Percentage of test citizen proposals that belong to each group by campaign.

Campaign	`Minority`	`NIMBY`	`Other`
C1	6.7%	10.3%	83.0%
C2	6.2%	7.9%	85.9%
C3	4.4%	7.0%	88.6%
C4	7.7%	7.3%	85.0%

$$p_R(a) = \frac{1}{Z} \sum_{i \in \mathcal{I}:\mathtt{att}(i)=a} rg_R(i) \quad (2)$$

$$rg_R(i) = \sum_{u \in \mathcal{U}} \phi(i, Rec_u^K) \cdot \text{gain}(u, i, r) \quad (3)$$

where $Z = \sum_{i \in \mathcal{I}} rg_R(i)$ is the normalization factor so that $\sum p_R(a_j) = 1$, Rec_u^K is the set of top-K items recommended by model R to user u, and $\phi(i, Rec_u^K) = 1$ if $i \in Rec_u^K$ and 0 otherwise. The function $\text{gain}(u, i, r)$ encodes the recommendation gain of item i for user u in position $r = \text{rank}(i, Rec_u^K)$. In this work, we define the gain as the normalized Discounted Cumulative Gain (nDCG) metric. We refer the reader to [10] for further details and possible configurations of GCE.

Afterwards, to compute the metric, we have to specify the *fairness perspective* to assess. By doing this, we define what is considered a fair recommendation. As explained before, in GCE, this is equivalent to setting a target distribution probability p_f, with the constraint that such distribution needs to live in the same attribute space A as p_R. We shall consider three possibilities, as summarized in Table 2.

By considering a **uniform perspective**, we would assume that fairness is equivalent to *equality*; that is, a fair recommendation algorithm is such that it equally suggests items of the three categories (`Minority`, `NIMBY`, `Other`): $p_u = \left[\frac{1}{3}, \frac{1}{3}, \frac{1}{3}\right]$. While this perspective puts some stress on the minority, vulnerable and NIMBY groups (since they do not frequently appear in the dataset), it still assumes a compromise between the groups must be met. However, in the long term, it may continue reinforcing a sufficiently large number of recommendations from the `Other` category, as they are the most popular ones. To better understand this rationale, Table 4 shows the number of proposals of each category in the test sets of the four campaigns. It is clear that in the dataset, the `Other` category is the largest one, followed by the `NIMBY` and `Minority` categories, in most of the campaigns.

Under a different perspective, lying at the other extreme of the spectrum, we define fairness as the situation where recommendations match the observed items in the system – i.e., based on *exposure* [3]. This conceptualization assumes that a recommender imposes the user a citizen proposal she would not have seen by herself. Hence, it aims to maintain as much as possible the status of previous user preferences, even if they are biased towards non-underrepresented proposals. This estimation could be measured based on the entire dataset, only the training set, or only the test set. We decided towards the latter. Hence, as shown in Table 2, we refer to it as the **test proportion perspective**. In particular, according to the data split of our experiments (Table 4), the ideal distribution according to this perspective for C1 would be $p_t = [0.067, 0.103, 0.830]$.

Finally, we consider a perspective to specifically account for biases or *discriminations* on the considered sensitive citizen categories (and consequently, on their underlying groups), by increasing the weight of each category. We thus would be imposing (and measuring the extent of) positive discrimination or affirmative actions for the groups. Based on this rationale, we define three ideal distributions: a first one biased only towards the `Minority` category (groups) ($p_m = [0.8, 0.1, 0.1]$), following a **minority perspective**; a second one biased towards the `NIMBY` category ($p_n = [0.1, 0.8, 0.1]$), following a **NIMBY perspective**; and a third one where all the weight is shared across the two sensitive categories ($p_{m+n} = [0.45, 0.45, 0.1]$), following an **underrepresented perspective**.

5 Experiments

5.1 Data Processing

On the dataset described in Sect. 3.2, we removed those proposals without location, category and topic, as this metadata is needed by some of the evaluated recommendation algorithms. We also removed the users with no interactions (i.e., comments) in the dataset, avoiding cold-start cases.

As mentioned in Sect. 3.2, the dataset had citizen-generated content of the four e-participatory budgeting campaigns dating from 2015 to 2019. Some citizen proposals had comments from several campaigns. To avoid this situation, we discarded those comments that did not belong to their proposal's campaign. Consequently, at the campaign level, we then repeated the removal of possible users with no comments.

With all the above, we built four datasets, each of them associated to an isolated campaign. Due to lack of space, in the subsequent sections, we only report and analyze average empirical results from experiments on the four campaigns. Nonetheless, we did not find significant differences in the results across campaigns.

5.2 Recommendation Algorithms

We experimented with five families of recommendation algorithms. In all cases, as done in [4,6], we considered the comment a user makes on a citizen proposal as a (unary) rating, indicating implicit feedback of interest, regardless the comment was positive or negative, i.e., in favor or against the proposal or a previous comment.

Specifically, we evaluated algorithms that generate recommendations randomly (*rand*) and based on popularity according to the number of users commenting a proposal (pop_u) and the number of comments of a proposal (pop_c); collaborative filtering algorithms, both heuristic based on items (*ib*) and users

Table 5. Average ranking-based accuracy values achieved by the recommendation algorithms evaluated on the four campaigns of the dataset. Best values per column in darker colors.

	Precision	Recall	MAP	nDCG	MRR	F1
rand	0.001	0.014	0.002	0.004	0.002	0.001
pop_u	0.006	0.217	0.062	0.096	0.067	0.011
pop_c	0.005	0.198	0.044	0.078	0.051	0.010
ib	0.002	0.041	0.013	0.020	0.017	0.003
ub	0.003	0.063	0.018	0.030	0.027	0.005
mf	0.003	0.120	0.052	0.068	0.057	0.006
bpr	0.003	0.101	0.017	0.035	0.021	0.005
cb_{cat}	0.001	0.023	0.003	0.008	0.005	0.002
cb_{top}	0.002	0.033	0.006	0.013	0.011	0.003
cb_{loc}	0.001	0.033	0.004	0.010	0.006	0.003
$cbib_{cat}$	0.001	0.020	0.004	0.008	0.006	0.002
$cbib_{top}$	0.002	0.046	0.010	0.020	0.015	0.004
$cbib_{loc}$	0.002	0.038	0.010	0.017	0.013	0.003
$cbub_{cat}$	0.002	0.066	0.015	0.028	0.022	0.004
$cbub_{top}$	0.002	0.066	0.026	0.037	0.032	0.004
$cbub_{loc}$	0.003	0.070	0.013	0.027	0.019	0.004

(*ub*), and model-based via matrix factorization (*mf*) and Bayesian personalized ranking (*bpr*); content-based algorithms exploiting user and item profiles with category (cb_{cat}), topic (cb_{top}), or location (cb_{loc}) information; and hybrid algorithms using either the *ib* or *ub* heuristic with content-based similarities, i.e., $cbib_{cat}$, $cbib_{top}$, $cbib_{loc}$. $cbub_{cat}$, $cbub_{top}$, and $cbub_{loc}$.

For some of these algorithms, we used the implementations given in the *Implicit* library[12]. Those algorithms that are not included in the library were implemented on top of it, and their source code was made publicly available.

5.3 Accuracy of Recommendation Algorithms

To evaluate the accuracy of the recommenders, we computed the following ranking-based metrics: Precision, Recall, MAP, nDCG, MRR, and F1, which are described in [15]. As mentioned before, we report and analyze the average of these metrics over the four campaigns of the dataset. To ensure stable results, the algorithms were evaluated through a 5-fold cross-validation process, and tuned with respect to nDCG@100.

Table 5 shows the achieved accuracy values at cutoff 50. From it, we first highlight that the popularity algorithms pop_u and pop_c reached the best results for all ranking metrics. This evidences a strong popularity bias existing in the Decide Madrid platform, which likely also appears in (many) other similar e-participation tools, and certainly in other domains and applications [13]. A possible explanation for this bias is the way the content is presented and accessed

[12] https://github.com/benfred/implicit.

Table 6. Average GCE values (the closer to 0, the better) achieved by the recommendation algorithms on the four campaigns of the dataset. Each column refers to a fairness perspective. nDCG values are included for completeness. Best values per column in darker colors.

	nDCG	p_u	p_t	p_m	p_n	p_{m+n}
rand	0.004	–1.264	–0.007	–3.806	–5.151	–2.604
pop_u	0.096	–1.932	–0.031	–10.243	–2.605	–3.817
pop_c	0.078	–1.005	–0.011	–4.716	–2.702	–2.125
ib	0.020	–1.034	–0.004	–3.974	–3.626	–2.181
ub	0.030	–0.954	–0.005	–3.888	–3.237	–2.034
mf	0.068	–2.071	–0.018	–7.292	–6.393	–4.077
bpr	0.035	–1.270	–0.028	–4.482	–4.496	–2.611
cb_{cat}	0.008	–1.600	–0.007	–6.388	–4.543	–3.219
cb_{top}	0.013	–1.757	–0.010	–6.399	–5.450	–3.505
cb_{loc}	0.010	–1.157	–0.006	–3.549	–4.774	–2.407
$cbib_{cat}$	0.008	–1.633	–0.008	–6.617	–4.509	–3.280
$cbib_{top}$	0.020	–1.864	–0.012	–6.605	–5.873	–3.701
$cbib_{loc}$	0.017	–1.225	–0.020	–2.989	–5.726	–2.529
$cbub_{cat}$	0.028	–1.352	–0.007	–6.228	–3.242	–2.764
$cbub_{top}$	0.037	–1.319	–0.007	–6.035	–3.237	–2.702
$cbub_{loc}$	0.027	1.387	–0.018	–6.780	–2.882	–2.824

in Decide Madrid. When a proposal is voted in the platform, it gets higher relevance for being shown at the top positions of the platform's interface, which consequently increases the probability of being accessed, voted and commented (i.e., rated in the database). It is also interesting to note that pop_u, which is based on the number of users who have commented the proposal, is more accurate than pop_c, which considers popularity as the number of comments a proposal has. The more users have commented (rated) a proposal, the more likely the proposal is in the test, since it has a greater number of ratings from distinct users.

The next best performing algorithms are the *mf*, *bpr* and *ub* collaborative filtering recommenders, followed by the user-based hybrid recommenders. This reinforces the idea that the preferences of users in Decide Madrid can be related to each other for personalized recommendation purposes.

5.4 Fairness Impact of Recommendation Algorithms

Addressing **RQ2**, we report (Table 6) and analyze the GCE values achieved by the evaluated recommendation algorithms according to the fairness perspectives presented in Sect. 4. Recall that *fairness as equality* is represented by the **uniform perspective** (p_u). In this case, *ub* and pop_c achieved the best (highest, closer to 0) results, which means that their recommendations are uniform for the considered fairness attribute values: minority/vulnerable, NIMBY or any other citizen collective. On the other extreme, we observe that pop_u and *mf* as the most biased ones.

According to the **test perspective**, that is, how close the generated recommendations are to the item distribution observed in test, the *ib* and *ub* heuristic collaborative filtering algorithms, those based on content (in particular, cb_{loc}), and the hybrid algorithms achieve the best results. This means that these algorithms are good approaches to recommend items of each citizen collective in the same proportion as they interest the users, i.e., as they actually appear in the test set. However, considering a tradeoff between fairness and accuracy, *ib* performs worse, since its nDCG value is lower. In this context, the success of exploiting location information in a collaborative filtering fashion could indicate that location is a good indicator of fair preferences, reflecting that users, in addition to popular proposals, tend to explore and comment on proposals about their surroundings (districts or neighborhoods).

Finally, regarding the **perspective biased towards discriminated groups**, the algorithms exploiting location information –in particular, $cbib_{loc}$ for p_m and $cbub_{loc}$ for p_n– and the heuristic collaborative filtering algorithms stand out. This could be related to the idea that users in the same surrounding (district or neighborhood) tend to be interested in the same proposals, and in this way, proposals from the minority and NIMBY groups that affect a certain environment are relevant and consequently should be recommended to users in that environment. It is worth noting the negative bias that the pop_u algorithm has on the `Minority` category is accentuated, moving far away from the idea of fairness defined by p_m. This is probably linked to the fact that the minority proposals are not popular on the platform. For the `NYMBY` category, popularity algorithms achieve better results, which could be explained by the fact that, by their nature, the NIMBY proposals are more controversial, and thus tend to have more comments and more users commenting on them [5].

6 Conclusions

The integration of recommender systems in e-participation platforms has been envisioned as a solution to reduce the information overload problem of the platforms. However, for such purpose, it is essential to evaluate the generated recommendations not only in terms of personalized content accuracy, but also according to social fairness dimensions, with the aim of avoiding or mitigating biases that could exist or be amplified by how information is presented and accessed in the platforms.

The work reported in this paper represents a seminal research in that direction. In particular, we have proposed a conceptualization and metrics of recommendation fairness oriented to minority, vulnerable and NIMBY groups of citizens, and have experimented with measuring such metrics for heterogeneous recommender systems on a real e-participation dataset. Our empirical results have been revealing. First, they have confirmed initial suspicions about the fact that there is a strong popularity bias on the platform data that affects the recommendation algorithms. Second, they have shown that those recommendations generated by exploiting the users' geolocation information in a collaborative filtering fashion are less affected by such bias for the target underrepresented citizen collectives.

Nonetheless, we are aware of limitations in our study. Most importantly, we should conduct more exhaustive experiments to further corroborate and generalize our findings and conclusions. In this sense, we plan to thoroughly test further hyperparameter settings and recommendation algorithms, and use additional datasets, such as those published in [6] from the e-participatory budgets of New York City, Miami, and Cambridge, which would allow us to explore citizen-generated content in English and distinct types of participatory budgeting processes. Moreover, we note we have focused on the identification and measuring of unfair recommendations. Ad hoc, fairness-aware recommendation algorithms and mitigation techniques should be investigated. We believe diversification [31] could be an effective approach for such purpose.

A second priority line of future work is extending the analysis of e-participation recommendation fairness to additional attributes of users, items and contexts. On the one hand, we plan to run our experiments for particular groups of citizens (rather than considering all together in the broad `Minority` and `NIMBY` categories), aiming to better understand the nature of the identified biases. On the other hand, looking at the behavior of the algorithms according to geolocation information, we could analyze the fairness with respect to different areas of a city, and study whether biases correlate with certain city or citizen participation features. In fact, similarly to [5], achieved results could be contrasted with demographic, socioeconomic and ideological data of the city's districts and neighborhoods.

Another issue that has arisen in our work is the sensitivity of the GCE metric to the differences in the proportion of test ratings for the different groups. In this sense, a possible line of future research would be incorporating into the approximation of the target recommendation distributions a normalization factor based on the number of ratings of each group. The objective would be to define a metric that is more robust to these differences in testing, and therefore more versatile when applied to other cases.

Regardless of these issues, we must highlight the relevance of the targeted citizen groups and proposals, which address pressing concerns within the community: cases of poverty and hunger, deficiencies in educational and healthcare systems, social inequalities and environmental, urban planning, and economic growth problems. Notably, these concerns are related to the Sustainable Development Goals (SDGs) established by the European Union [16]. Therefore, we believe that our work can be an inspiration for future redesigns of e-participation platforms, in such a way that, through novel, fairness-aware methods of information access, retrieval and recommendation, they would allow policymakers to be more aware of existing (city) problems and act (at local level) more effectively on some of the SDGs.

Acknowledgements. This work was supported by Grant PID2019-108965GB-I00 of the Spanish Ministry of Science and Innovation, and Grant PID2022-139131NB-I00 funded by MCIN/AEI/10.13039/501100011033 and by "ERDF A way of making Europe."

References

1. Al-Hassan, M., Lu, H., Lu, J.: A semantic enhanced hybrid recommendation approach: a case study of e-government tourism service recommendation system. Decis. Supp. Syst. **72**, 97–109 (2015)
2. Barile, F., et al.: ICT solutions for the OR.C.HE.STRA project: from personalized selection to enhanced fruition of cultural heritage data. In: Proceedings of the 10th International Conference on Signal-Image Technology and Internet-Based Systems, pp. 501–507. IEEE (2014)
3. Boratto, L., Fenu, G., Marras, M.: Interplay between upsampling and regularization for provider fairness in recommender systems. User Model. User-Adap. Inter. **31**(3), 421–455 (2021)
4. Cantador, I., Bellogín, A., Cortés-Cediel, M.E., Gil, O.: Personalized recommendations in e-participation: offline experiments for the 'Decide Madrid' platform. In: Proceedings of the International Workshop on Recommender Systems for Citizens, pp. 1–6. ACM (2017)
5. Cantador, I., Cortés-Cediel, M.E., Fernández, M.: Exploiting open data to analyze discussion and controversy in online citizen participation. Inf. Process. Manag. **57**(5), 102301 (2020)
6. Cantador, I., Cortés-Cediel, M.E., Fernández, M., Alani, H.: What's going on in my city? recommender systems and electronic participatory budgeting. In: Proceedings of the 12th ACM Conference on Recommender Systems, pp. 219–223. ACM (2018)
7. Casino, F., Batista, E., Patsakis, C., Solanas, A.: Context-aware recommender for smart health. In: Proceedings of the 1st International Smart Cities Conference, pp. 1–2. IEEE (2015)
8. Church, J., Saunders, D., Wanke, M., Pong, R., Spooner, C., Dorgan, M.: Citizen participation in health decision-making: past experience and future prospects. J. Public Health Policy **23**, 12–32 (2002)
9. Cortés-Cediel, M.E., Cantador, I., Gil, O.: Recommender systems for e-governance in smart cities: state of the art and research opportunities. In: Proceedings of the International Workshop on Recommender Systems for Citizens, pp. 1–6. ACM (2017)
10. Deldjoo, Y., Anelli, V.W., Zamani, H., Bellogín, A., Di Noia, T.: A flexible framework for evaluating user and item fairness in recommender systems. User Model. User-Adap. Inter. **31**, 457–511 (2021)
11. Deldjoo, Y., Jannach, D., Bellogín, A., Difonzo, A., Zanzonelli, D.: Fairness in recommender systems: research landscape and future directions. User Model. User-Adap. Inter. (2023)
12. Dyczkowski, K., Stachowiak, A.: A recommender system with uncertainty on the example of political elections. In: Greco, S., Bouchon-Meunier, B., Coletti, G., Fedrizzi, M., Matarazzo, B., Yager, R.R. (eds.) IPMU 2012. CCIS, vol. 298, pp. 441–449. Springer, Heidelberg (2012). https://doi.org/10.1007/978-3-642-31715-6_47

13. Ekstrand, M.D., et al.: All the cool kids, how do they fit in?: popularity and demographic biases in recommender evaluation and effectiveness. In: Proceedings of the 1st Conference on Fairness, Accountability and Transparency, vol. 81, pp. 172–186. PMLR (2018)
14. Eranti, V.: Re-visiting NIMBY: from conflicting interests to conflicting valuations. Sociol. Rev. **65**(2), 285–301 (2017)
15. Gunawardana, A., Shani, G.: Evaluating recommender systems. In: Ricci, F., Rokach, L., Shapira, B. (eds.) Recommender Systems Handbook, pp. 265–308. Springer, Boston, MA (2015). https://doi.org/10.1007/978-1-4899-7637-6_8
16. Hák, T., Janoušková, S., Moldan, B.: Sustainable Development Goals: a need for relevant indicators. Ecol. Ind. **60**, 565–573 (2016)
17. Kavanaugh, A., Ahuja, A., Pérez-Quiñones, M., Tedesco, J., Madondo, K.: Encouraging civic participation through local news aggregation. In: Proceedings of the 14th International Conference on Digital Government Research, pp. 172–179. ACM (2013)
18. Lujak, M., Billhardt, H., Dunkel, J., Fernández, A., Hermoso, R., Ossowski, S.: A distributed architecture for real-time evacuation guidance in large smart buildings. Comput. Sci. Inf. Syst. **14**(1), 257–282 (2017)
19. Macintosh, A.: Characterizing e-participation in policy-making. In: Proceedings of the 37th Hawaii International Conference on System Sciences, p. 10. IEEE (2004)
20. Macintosh, A.: E-democracy and e-participation research in Europe. In: Chen, H., et al. (eds.) Digital Government: e-Government Research, Case Studies, and Implementation, vol. 17, pp. 85–102. Springer, Heidelberg (2008). https://doi.org/10.1007/978-0-387-71611-4_5
21. Marsal-Llacuna, M.-L., de la Rosa-Esteva, J.-L.: The representation for all model: an agent-based collaborative method for more meaningful citizen participation in urban planning. In: Murgante, B., Misra, S., Carlini, M., Torre, C.M., Nguyen, H.-Q., Taniar, D., Apduhan, B.O., Gervasi, O. (eds.) ICCSA 2013. LNCS, vol. 7973, pp. 324–339. Springer, Heidelberg (2013). https://doi.org/10.1007/978-3-642-39646-5_24
22. Nelimarkka, M., et al.: Comparing three online civic engagement platforms using the "spectrum of public participation" framework. In: Oxford Internet, Policy & Politics Conference, pp. 1–22 (2014)
23. Nguyen, T.T., Hui, P.M., Harper, F.M., Terveen, L., Konstan, J.A.: Exploring the filter bubble: the effect of using recommender systems on content diversity. In: Proceedings of the 23rd International Conference on World Wide Web, pp. 677–686. ACM (2014)
24. Organisation for Economic Co-operation and Development: OECD guidelines for citizen participation processes. OECD Publishing (2022)
25. Pariser, E.: The Filter Bubble: What the Internet is Hiding From You. Penguin (2011)
26. Peixoto, T.: Beyond theory: e-participatory budgeting and its promises for eParticipation. Eur. J. ePract. **7**(5), 1–9 (2009)
27. Peroni, L., Timmer, A.: Vulnerable groups: the promise of an emerging concept in European Human Rights Convention law. Int. J. Constitut. Law **11**(4), 1056–1085 (2013)
28. Quijano-Sánchez, L., Cantador, I., Cortés-Cediel, M.E., Gil, O.: Recommender systems for smart cities. Inf. Syst. **92**, 101545 (2020)
29. de Sousa Santos, B.: Participatory budgeting in Porto Alegre: toward a redistributive democracy. Polit. Soc. **26**(4), 461–510 (1998)

30. Terán, L., Meier, A.: A fuzzy recommender system for eElections. In: Andersen, K.N., Francesconi, E., Grönlund, Å., van Engers, T.M. (eds.) EGOVIS 2010. LNCS, vol. 6267, pp. 62–76. Springer, Heidelberg (2010). https://doi.org/10.1007/978-3-642-15172-9_6
31. Vargas, S., Castells, P.: Rank and relevance in novelty and diversity metrics for recommender systems. In: Proceedings of the 5th ACM Conference on Recommender Systems, pp. 109–116. ACM (2011)
32. Wang, Y., Ma, W., Zhang, M., Liu, Y., Ma, S.: A survey on the fairness of recommender systems. ACM Trans. Inf. Syst. **41**(3), 1–43 (2023)
33. Yang, Z., Feng, J.: Explainable multi-task convolutional neural network framework for electronic petition tag recommendation. Electron. Commer. Res. Appl. **59**, 101263 (2023)
34. Yavari, A., Jayaraman, P.P., Georgakopoulos, D.: Contextualised service delivery in the internet of things: parking recommender for smart cities. In: Proceedings of the 3rd World Forum on Internet of Things, pp. 454–459. IEEE (2016)
35. Zheng, Y., Schachter, H.L.: Explaining citizens' e-participation use: the role of perceived advantages. Public Organ. Rev. **17**, 409–428 (2017)

Responsible Opinion Formation on Debated Topics in Web Search

Alisa Rieger[1](✉), Tim Draws[2], Nicolas Mattis[3], David Maxwell[4], David Elsweiler[5], Ujwal Gadiraju[1], Dana McKay[6], Alessandro Bozzon[1], and Maria Soledad Pera[1]

[1] Delft University of Technology, Delft, The Netherlands
{a.rieger,u.k.gadiraju,a.bozzon,m.s.pera}@tudelft.nl
[2] OTTO, Hamburg, Germany
tim.draws@otto.de
[3] Vrije Universiteit Amsterdam, Amsterdam, The Netherlands
n.m.mattis@vu.nl
[4] Booking.com, Amsterdam, The Netherlands
maxwelld90@acm.org
[5] University of Regensburg, Regensburg, Germany
david.elsweiler@ur.de
[6] RMIT University Melbourne, Melbourne, Australia

Abstract. Web search has evolved into a platform people rely on for opinion formation on debated topics. Yet, pursuing this search intent can carry serious consequences for individuals and society and involves a high risk of biases. We argue that web search can and should empower users to form opinions responsibly and that the information retrieval community is uniquely positioned to lead interdisciplinary efforts to this end. Building on digital humanism—a perspective focused on shaping technology to align with human values and needs—and through an extensive interdisciplinary literature review, we identify challenges and research opportunities that focus on the searcher, search engine, and their complex interplay. We outline a research agenda that provides a foundation for research efforts toward addressing these challenges.

Keywords: web search · opinion formation · debated topics

1 Introduction

Web search engines provide fast and convenient access to the often overwhelming amount of resources that could potentially satisfy users' information needs [70]. Nevertheless, search engines are not merely neutral tools for retrieving relevant resources; they act as information gatekeepers and, as a result, play a vital part in shaping individual and collective knowledge [34,68,78].

A. Rieger and T. Draws—Contributed equally.

D. Maxwell—The work undertaken by this author is not related to Booking.com's activities.

N. Goharian et al. (Eds.): ECIR 2024, LNCS 14611, pp. 437–465, 2024.
https://doi.org/10.1007/978-3-031-56066-8_32

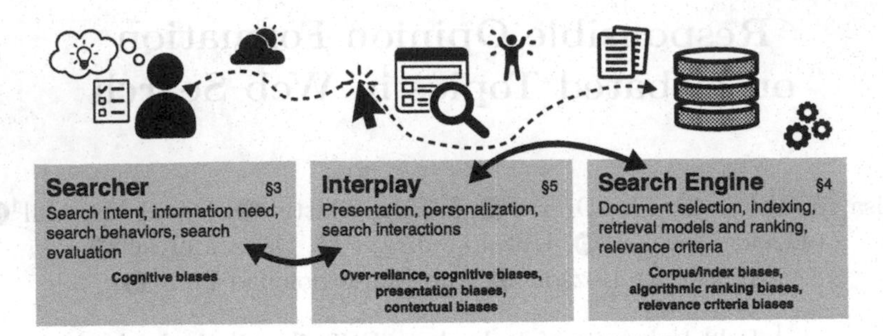

Fig. 1. Search on debated topics. Biases hindering interactions to gain well-rounded knowledge can emerge from the searcher, the search engine, and their interplay. Ultimately, search on debated topics can shape cognitive processes (e.g., attitude change) and concrete actions (e.g., voting in an election).

By providing access to information that can directly or indirectly shape users' views and beliefs, web search assumes an important role in opinion formation [35,52,58,59,124,160,181,208]–developing one's view on a topic to satisfy a personal interest or seeking advice on an issue of personal, business, or societal concern [36]. Opinion formation may involve shallow issues (e.g., outfit choices), but it can also refer to more impactful and even contentious matters: **debated topics**. Debated topics are socio-scientific issues of ongoing discussion that do not convey—at least according to some debate participants or observers—a straightforward solution [176]. They include extremely one-sided matters with clear scientific stances (e.g., whether the Earth is a sphere) and more divisive issues with legitimate arguments on both sides of the spectrum (e.g., whether zoos should exist). Searches on debated topics can impact individual users' opinion formation and subsequent decision-making (e.g., on whether to embrace veganism [59], what financial strategy to employ [208], or whom to vote for [52]) and thus, on aggregate, democratic societies at large.

Conventional search engines fall short of aiding complex, consequential information needs [64,126,181,183], prompting the question *how web search can support information seeking on debated topics*. By that, we do not mean guiding searchers toward a particular view or ideology but instead assisting and empowering them in actively and thoroughly engaging with diverse viewpoints; critically evaluating information to **form opinions responsibly** [102,151]. Although users may intend to expose themselves to diverse viewpoints when searching for debated topics [3,124], responsible opinion formation can be impeded by factors like over-relying on the system to provide accurate and reliable resources [183]. Engaging with information on debated topics is naturally demanding and can trigger emotionally charged behavior, as it has the potential to challenge the searcher's core beliefs and values [81,156]. Thus, search on debated topics inherently requires cognitive effort, particularly to overcome *biases* that can occur during the search process. Such biases may emerge from the user (e.g. cognitive biases) [79,213,217], the search engine (e.g., data, relevance criteria, and algorithmic ranking biases) [28,45,62], or the interaction between them (e.g., presentation, over-reliance, and contextual biases) [11,15,183]. These considerations highlight the complex, mutually evolving interplay of the searcher and the search engine (see Fig. 1), as illustrated in representations of the search process such as the *Information-Seeking and Retrieval Model* [86, Chapter 6].

As search engines are widely used, they can and should be platforms to explore debated topics in all their nuances. The information retrieval (IR) community has dedicated efforts to comprehending the evolving needs of searchers and society and developing technology to support them [142,181]. Given the role search engines play in opinion formation—a search intent they were not explicitly designed for—the importance of advancing the understanding of the associated challenges, as well as the development of system functions that foster responsible opinion formation becomes apparent. Although IR research has already explored and experimented with *fairness* [8,57,224], *diversity* [1,48,177], *argument retrieval* [27,50,150,158,206], and user interface adaptations [37,92,121,219], whether and how web search engines should cater to users' opinion formation and deal with debated topics remains largely unanswered. Resonating with the ideals for future technological development of *digital humanism*, web search should be shaped following individual and societal values and needs instead of letting web search shape individuals and society [209]. To do so, it is essential to recognize opinion formation on debated topics as a distinct search intent, characterized by (1) the heightened risk of searcher and search engine biases and (2) its consequential nature on individuals and society at large, and warranting dedicated research efforts. The IR community is uniquely positioned to spearhead interdisciplinary efforts to advance such socio-technical research endeavors.

In this paper, we delve into the role of web search engines in users' opinion formation, delineating the distinct characteristics of web search on debated topics through an extensive review of interdisciplinary literature. We illuminate the challenges inherent to the searcher (§3), the search engine (§4), and their interplay (§5) and outline a research agenda (§6) encompassing methodological considerations, high-level challenges, and initial research questions towards responsible opinion formation through web search.

2 Digital Humanism and Responsible Opinion Formation

Digital Humanism advocates for reflecting on the relationship between humans and technology. Fostering human-centered design, it prioritizes better lives and societal progress over mere economic growth [209]. Designing technology to embody these ideals is not a linear process as technology and humans co-evolve, mutually shaping one another in an intricately intertwined manner [140,214].

Web search is one of the primary information gateways, impacting searchers' knowledge, choices, and actions [34]. Searchers have cultivated a sense of trust that makes them rely on the system's evaluation and differentiation of resources on their behalf [183]. Yet, search engines are not subject to regulations for content quality and diversity necessary for an informed citizenry, unlike the standards applied for responsible reporting within traditional media outlets [75].

Opaque relevance and ranking criteria are far from value-neutral but function as algorithmic curators that serve a goal, e.g., user satisfaction and profit generation [34,138,210]. Given the profound impact of web search, recent work has called for revisiting relevance criteria and search system design to better align with the needs and values of individuals and democratic societies [32,64,181]. However, it is non-trivial to balance values that might be in tension with each other [34]. These tensions are particularly evident for search on debated topics, where relevance to user needs might not be aligned with relevance to democratic values, necessitating a critical evaluation of value trade-offs.

Forming opinions responsibly involves gathering evidence and critically assessing it [102,151]. In the context of web search, this translates to searchers actively and thoroughly engaging with search results encompassing **diverse viewpoints**. Yet, this is not the norm as *Search Engine Result Pages* (SERPs) often lack viewpoint diversity [45], and searchers tend to primarily interact with information that aligns with their own viewpoints [174,182,203,210].

Viewpoint diversity in people's exposure to information concerning debated topics represents a long-standing research topic in the communication sciences [14,25,54,117,215]. Different democratic notions of viewpoint diversity can be applied depending on the objectives of a system [74]. Which particular notion of viewpoint diversity is appropriate in an opinion formation-related search scenario, however, might depend on both the topic and the user [74,204,205]. For instance, one could argue that viewpoint diversity is vital for unresolved issues but that web search engines should represent topics with a solid scientific basis in a more one-sided fashion. While it may seem obvious that scientifically answerable topics should be presented as such, previous research has shown that exposing strongly opinionated users to nothing but opposing viewpoints can result in a backlash effect; where they become more entrenched in their beliefs [141]. This can increase polarization by leading users to shift their attention away from mainstream and toward more niche information sources [141]. Similarly, increased diversity can also lead to false perceptions of existing evidence, e.g., balancing climate change believers and deniers can create a false image of an open debate that may be worse than an approach that accounts for different weights of evidence [40]. The desirable degree of viewpoint diversity may thus not always be either the minimum or maximum [16] and can depend on the topic and individual user characteristics [122,134].

IR research has largely used binary (e.g., democrat/republican) or ternary taxonomies (e.g., against/neutral/in favor) [60,155,162,221] as *viewpoint representations* for search results. Recent work, however, has shown that such labels unnecessarily reduce complex viewpoints to generic categories, which limits the insight gained in research using them [42]. Researchers have added more nuance to such labels by using ordinal scales [46,172], continuous scales [105,106], multi-categorical perspectives [38], or building on outcomes from communication sciences [13,25] to yield a two-dimensional viewpoint label that includes a nuanced notion of *stance* (e.g., strongly supporting) and *logics of evaluation* (i.e., representing the reasons underlying a stance, e.g., supporting zoos because of their animal conservation efforts) [42]. Despite these advancements, there is a need to analyze existing viewpoint representation frameworks for comprehensibility, practical applicability, and meaningfulness for users and practitioners.

3 The Searcher

The searcher (*information seeker*), turns to a search engine to execute a search intent, motivated by an underlying *information need.* This develops from a perceived problem, a knowledge gap, an internal inconsistency related to their understanding, or some conflict of evidence [19]. Once the searcher enters a *query* into the system, their interaction with the system begins (§5). Such interactions include evaluating the information encountered in search results and can affect searchers' knowledge and attitude towards the search topic [52,99].

Research on how users search the web for debated topics [79], or how they form opinions in non-biased scenarios [61,124] is in its infancy. Progress depends on conducting user studies into behavioral patterns as users search for debated topics (e.g., queries used [3], if they engage with counter-attitudinal viewpoints, or when they stop searching) and searchers' preferences (e.g., whether users prefer diverse or filtered viewpoints [96]). Also crucial are methods to correctly interpret user behavior, e.g., clicks on search results are often seen as a proxy for engagement [46,172] but users may engage with them in a variety of ways that can be just as meaningful for opinion formation [101]. Researchers should investigate how to support users' reflections on their search processes and outcomes (e.g., awareness of their biases and knowledge level) and investigate long-term opinion formation (e.g., changes in search behavior and opinions over time).

Cognitive Biases. To reduce the cognitive demands of processing information on debated topics, searchers frequently (and subconsciously) employ shortcuts, which can introduce cognitive biases [11,61,197]. *Confirmation bias*, searchers' tendency to prioritize information that confirms prior attitudes [136,203,210], can prevent engagement with diverse viewpoints during search on debated topics. This bias has been observed at various stages of the search process, e.g., query formulation [79], and search result selection [145,203,217]. Other studies have noted searchers' inclination to engage with positive (i.e., query-affirming) [213] and mainstream content [61]. Triggered by the search result presentation, other cognitive biases that hinder diligent search behavior can arise (§5). Identifying how to facilitate search in this context requires a thorough understanding of factors affecting searchers' intentions, behavior, vulnerability to biases, and evaluation of the encountered information. It also requires approaches to support and empower searchers for unbiased and diligent search behavior.

Context. The vulnerability to biased search behavior is contingent upon the searcher's context. For instance, when searching purposelessly, as opposed to specifically looking for information on a particular debated topic, searchers' vulnerability to cognitive biases increases [217]. Stressful conditions (e.g., time pressure) may strengthen the influence of cognitive biases [153,180]. This calls for investigating how the searcher's context influences search behavior and the vulnerability to cognitive biases when engaging with debated topics; also how to create search environments that foster unbiased and diligent search behavior and reduce contextual conditions leading to high vulnerability to biases.

User Characteristics. Search behavior, susceptibility to cognitive biases, and reaction to elements of the user interface are affected by *situational* and *stable* user characteristics [198]. Situational factors include attitude strength and certainty [98,201] and involvement with and prior knowledge of the topic [116,133,211]. Stable factors that affect engagement with debated topics include searchers' *need for cognition* (i.e., an individual's tendency to organize their experience meaningfully) [33,152,196], *receptiveness to opposing views* (i.e., willingness to impartially access and evaluate opposing views) [128], and *intellectual humility* (i.e., an individual's tendency to recognize the fallibility of their beliefs and the limits of their knowledge) [30,41,63,104,112,156]. Open research directions include advancing the understanding of how different user characteristics affect search on debated topics throughout the search process, from search intent to search evaluation, and if concepts such as searchers' *moral values* [115,170] play a role. Researchers should also investigate how efforts to support unbiased and diligent search behavior may require adaptation to cater to the diverse needs of searchers with distinct characteristics.

Vulnerable Groups. It is crucial to study and accommodate vulnerable user groups such as children, elderly people, or neurodivergent users in search for opinion formation. These users have certain characteristics (e.g., fewer cognitive resources or low technological literacy) that may make them more vulnerable to viewpoint biases and less likely to enact responsible opinion formation [94,108,118,127]. For instance, children are less likely to judge or explore search results [108] and are more susceptible to opinion formation through misinfomation [118]. Elderly users similarly have increased tendencies toward sharing and interacting with fake news [66,91]. Research is needed to identify who those vulnerable groups are specifically, what particular factors make them vulnerable, and how web search engines can support these users in their opinion formation.

Boosting Searchers' Competencies. Boosting interventions are effective in fostering web literacy skills, such as resilience to misinformation [113,175], detecting micro-targeting [119], and improving privacy behavior [144]. These interventions, which promote individuals' cognitive or motivational competencies [77,103,120], contain a learning component and thus could remain effective even after the intervention. The specific challenges posed by web search on debated topics might require an expansion of traditional web and information literacy constructs [69], for instance by incorporating intellectual virtues [64]. Although boosting interventions that target such virtues have been suggested [171], their effect on search behavior and opinion formation is not fully understood.

4 The Search Engine

Contemporary search engines provide a means of sifting through large volumes of information to find the proverbial *needle in the haystack*. Key to search engines are three inputs: (i) a **document index**, a data structure representing a collection of documents (or corpus, typically a *crawled* [95] collection of web documents for web search engines); (ii) a **retrieval model**, that is responsible for identifying and scoring (and ranking) documents that are deemed *relevant* to what is being searched for, based on a series of *relevance criteria* (e.g., [89,186,220]); and (iii) a **query**, a construct of an *information need* as provided by the searcher, typically formulated as a series of tokens, e.g., `'should zoos exist'`. Search engines—as with other systems—are not immune from biases [132]. Indeed, the design of the retrieval model can raise several areas in which biases can (and do) arise, such as leading to *undue emphasis* on particular perspectives [106].

Corpus/Index Biases. Search results can only list documents that are included in a web search engine's index. With commercial web search engine crawlers indexing huge swathes of the *World Wide Web*, the population of content creators who generate the documents in this collection is unlikely to represent the global human population [62], and follows a highly unequal distribution concerning the number of documents generated per content creator [5,7,200]. Such collections may thus include a *creation bias*, i.e., they do not contain balanced or society-representative viewpoint distributions on all debated topics [146,184]. Moreover, the way in which a retrieval system indexes documents can affect the distribution of available documents. An *indexing bias*—whereby the search engine is programmed to systematically ignore particular documents—may further skew the data that the retrieval system can process [28,154,202].

Algorithmic Ranking Biases. Search engines may (unintentionally) exacerbate viewpoint biases in the indexed corpus through algorithmically-biased relevance criteria [57,147,148]. *Ranking biases* may cause documents that express certain viewpoints to rank higher than others, and therefore receive more attention from searchers (§ 5). This can occur when search result rankings solely focus on relevance criteria that optimize for maximizing searchers' satisfaction [199].

Relevance Criteria Bias. Determining the relevance of a search result is central to search engines. With debated topics, the relevance criteria employed by conventional search engines—which mostly target user satisfaction to maximize profit and efficiency [199]—may prove inadequate. Disregarding relevance to the unbiased knowledge gain of the searcher—as well as relevance to society and public welfare—can impede searchers from gaining a comprehensive understanding of a debated topic and its various arguments [64,68,75,186]. Prior work has found viewpoint biases in highly-ranked search results concerning health information [210,212], politics [161], and other debated topics [45].

Research and practical applications require automatic viewpoint classification methods to evaluate and foster viewpoint diversity. This primarily concerns the development of bias metrics and diversification algorithms.

Viewpoint Detection. Applications for search on debated topics need efficient and reliable methods to assign viewpoint labels to documents, e.g., measuring or mitigating search result viewpoint biases in real-time. Recent research has seen the emergence of *Natural Language Processing* (NLP) tasks like *stance detection* [9,71,129,130,185,207] and *argument mining* [31,109,110,114,137,187], which aim to automatically detect different viewpoint components in text. Other works have used unsupervised topic models [192,194,228] or hybrid approaches (i.e., automatic methods combined with crowdsourcing) [12] to overcome the limitations of supervised stance detection models. However, practitioners will ultimately need fully automatic methods to classify search results into broad viewpoint representations. *Large Language Models* (LLMs) have recently shown promise in this area, still further work is needed. Researchers should build on the existing efforts in stance detection, argument mining, and argument retrieval [2,26] to develop such advanced methods.

Viewpoint Bias Assessment. Assessing viewpoint bias requires metrics that accommodate the chosen ethical notion of viewpoint diversity and viewpoint representation. Current rank-aware viewpoint bias metrics applicable to search results consider categorical stance labels (e.g., against/neutral/in favor) [204, 218], continuous stance labels (e.g., ranging from -strongly opposing to strongly supporting) [106], or multi-dimensional viewpoint labels (i.e., stance and logic of evaluation [45]. Thus far, viewpoint biases in search results are primarily assessed as a deviation from viewpoint balance [45,47,52], deviation from the overall distribution across ranks [57,106], or the presence of scientifically false information [155,212]. Yet, it is unclear what metric may best apply in what scenario, how metrics compare, and what intuitive degrees of viewpoint bias different metric scores suggest. Existing metrics do not distinguish among data, algorithm, or presentation bias, and there is no guideline as to what specific *discount factor* to apply for rank-awareness [178]. There is a need to develop comprehensive viewpoint bias metrics, (simulation) studies to compare metrics, interpretation guidelines (i.e., including metric thresholds where viewpoint biases may become problematic), and best practices for using those metrics.

Viewpoint Diversification. Earlier work has diversified search results for more general user intents [1,82,93,177,226], and even made first steps to manually or automatically diversify viewpoints [45,53]. While some of these works have considered advanced viewpoint labels [45], how to diversify search results for different diversity notions or viewpoint representations, and how to dynamically adapt diversification algorithms to searcher needs (e.g., due to changes in search topic or user context) remains to be determined. Researchers could further explore solutions for data, algorithmic, and presentation biases individually and develop pipelines that increase diversity at each level.

5 The Searcher and Search Engine Interplay

Search engines present the SERP to the searcher, featuring search results that may be personalized, taking into account several contextual factors, such as previous search interactions [100,190]. Searchers interact with the SERP, for instance by querying, scanning the results, and clicking on selected items to access the web page. Substantial challenges associated with searching on debated topics emerge from the intricate interplay of the searcher and the search engine.

Over-Reliance and Cognitive Biases. Searchers rely on search engines and assume that highly-ranked search results are relevant and accurate [61,183] - a notion that may be explained with the perceived quality of top-ranked results (e.g. see work on the related context of news selection [65]), or as a response to information overload. Indeed, prior work shows that when the amount of available information exceeds one's processing capacities, searchers tend to be more selective and prone to cognitive biases [188]. For complex tasks, this reliance may impede searchers from expending the needed cognitive effort, thus turning into over-reliance [183]. Opaque relevance criteria further hinder searchers' ability to assess information completeness [126]. Reliance on the search engine is exemplified in searchers' *position bias* (i.e., users typically tend to pay much more attention to search results at higher ranks [88,149]) as well as the *Search Engine Manipulation Effect* (SEME) [21,52,155], where users tend to change their attitudes following viewpoint biases in search results. So far, little prior work has explored what gives rise to phenomena such as SEME [46]. Effects emerging from the interplay between the searcher and search engine might also be related to additional cognitive biases, such as the *availability bias* (i.e., overestimate the prevalence of information that is easily accessible) [11], or *anchoring bias* (i.e., the top-ranked search result may color the searcher's attitude) [11,139,213]. Such phenomena typically occur without users' awareness [61] and are unlikely what users aim for when they search the web for debated topics. Moreover, as web search results get increasingly augmented or replaced by highly pleasing and personalized answers from artificial intelligence chat systems (e.g., *ChatGPT*) that require exerting even less cognitive effort when searching, over-reliance and cognitive biases among users may become even more prevalent.

Presentation Biases. Search results are typically presented as ranked lists (i.e., split into pages of ten search results each; although other presentation formats have been proposed [90]). Each result is displayed with a *title*, a *snippet* (i.e., a brief excerpt from the document text), and the relevant *URL*. Common web search engines often display additional information such as *entity cards* [29], direct answers [20], or suggestions for alternative queries [125]. These different factors provide ample room for presentation biases in search results [15,17,222]. Viewpoint-related presentation biases could occur due to a more prominent presentation of particular viewpoints, e.g., by more favorable snippets [21,22] or representation in entity cards [121]. Moreover, the impact of presentation biases could be largely hidden as users often engage with search results without clicking on them (e.g., only reading the titles and snippets) [101].

Context. Contextual factors emerging from the searcher-system interplay may aggravate biases [85]. For instance, search result rankings may be affected by users' prior searches, preferences, or location [143,223], viewpoint biases in earlier interactions may lead to biased follow-up search queries [3], and presentation biases may depend on the device that users employ [97].

The biases and artifacts arising from the mutually evolved interplay between searchers and search engines can obstruct fruitful searches that facilitate responsible opinion formation. Thus, there is a need to disentangle and understand this convoluted interplay and design search interfaces that facilitate and motivate thorough engagement with diverse viewpoints.

Exposure and Interaction. The search results users are exposed to (and subsequently interact with) can strongly influence users' opinions [4,21,52,155]. *How* users interact with search results plays an important role here: even when exposed to viewpoint-biased search results on social and political information, search behavior is still characterized by searcher-rooted interaction bias, with searchers prioritizing search results that align with their beliefs [174,182]. While searchers may somewhat defy the impact of exposure effects, they could still lead to more subtle and enduring consequences over time [174]. These observations stress the need for deeper insights into the dynamics of exposure and interaction biases. Considering that viewpoint changes often begin with information encounters on social media [73,124], researchers should moreover explore the relation of exposure and interaction effects across different information settings.

Interfaces. Interface modifications can support unbiased and diligent search behavior, e.g., presenting search results in alternate formats [92], providing information about the search topic or the ranking [121,219], visualizing viewpoints and biases in search results [37,53,216], suggesting alternative queries [157], or highlighting documents with diverse viewpoints [39,221]. Also promising are behavioral interventions to support unbiased search interactions (e.g. warning labels) [53,152,172,173]. Researchers should investigate how different viewpoint representations, notions of viewpoint diversity and additional features, e.g., search result explanations affect searchers [43,166,216]. Interventions that can be customized by the searcher (i.e., *self-nudging* [167]) have worked in the news context [18,24,72] and merit investigation in the realm of web search. As users sparingly utilize customization features and adhere to default configurations [24,191], research is needed to identify user-friendly options and optimize default settings. Increasing search engine *transparency* (e.g., by explaining what factors influenced the ranking or providing meta-information for search results) as a means to raise awareness of system biases and foster appropriate reliance should be investigated. This could boost searchers' technological and information literacy [76,183]. Still, providing meaningful explanations poses several challenges, including decisions regarding the level of detail and presentation [49].

Personalization. Users have diverse characteristics, tendencies, and pre-search opinions [21,46]. This raises the question whether degrees of viewpoint diversity or presentation formats (e.g., stance labels) should be adapted to different searchers [169,172]. Personalization with regards to searchers' opinions, cognitive biases, moral values, and other relevant constructs would require methods to automatically predict these psychometric variables [123]. However, such endeavours would also raise substantial privacy concerns [193]. Whether and how to customize search results and the interface based on factors like user characteristics, past behavior, and the specific topic remains an open question that warrants ethical and research discussion. This may also affect general personalization efforts by web search engines [100,159,190].

6 Research Agenda

The intricate dynamics among the searcher, the search engine, and their interplay (§ 3–5) call for reflecting on research methods and broader research challenges. We outline some of these considerations and challenges, along with research questions to guide efforts on web search on debated topics.

Data Collection and Public Data Sets. Developing and evaluating methods to assign viewpoint labels or foster viewpoint diversity in search results, and user studies on search behavior require high-quality, human-annotated ground truth data sets with search results and viewpoint labels. Creating such data sets is not easy: recent research has shown that different worker characteristics and cognitive biases can reduce the quality of data annotations, especially in subjective tasks such as annotating viewpoints [44,47,51,83]. More work is needed to identify best practices and publish openly available data sets with search results and comprehensive viewpoint labels for different debated topics.

User Studies. Evaluating perceptions of viewpoint representations and viewpoint diversity, understanding factors influencing searchers' behavior, and determining how to support unbiased and diligent search requires qualitative and quantitative studies. Winter and Butler [214] stress the value of ongoing dialogues between the technology developers and users, for responsible technology design. As we delve into issues concerning information access and societal well-being, it is crucial to comprehensively and longitudinally assess design choices and interventions in real-world settings, ensuring that they do mitigate harm rather than inadvertently exacerbating it [56]. Comprehending the impact of various factors on searchers and their behavior needs carefully designed, controlled studies with large sample sizes to grasp subtle differences [111]. Simultaneously, the uncertainty of the complex socio-technical dynamics, normative dimensions, and related risks might necessitate more exploratory research methods [179]. A promising new avenue in this regard that has recently gained traction in the communication sciences may be data donations. While they present legal, ethical and technical challenges, data donations offer externally valid and highly granular insights by enabling researchers to retroactively analyse authentic search queries (e.g. from donated browser histories) [6,23,80].

Cultural Diversity. Different societies, countries, and cultures have vastly different ways of searching about and discussing debated topics [84,107]. Contemporary academic research is almost exclusively conducted in English, so is previous work related to web search on debated topics. Yet, web searchers across the globe may experience viewpoint biases and their undesired effects. It is therefore essential that future research considers web search on debated topics and all related challenges from a multi-lingual and multi-cultural perspective.

Misinformation. Balancing the dangers of exposing users to search results containing false claims with viewpoint diversity while preserving freedom of speech and avoiding (perceptions of) censorship is a particularly difficult issue that requires further investigation. Researchers and practitioners who work in the search for opinion formation space should be aware that misinformation may be particularly impactful here, and therefore closely monitor and leverage ongoing research efforts on misinformation detection and mitigation [55,189,225,227].

Alternative Search Paradigms. In this paper, we have focused on the traditional and dominant idea of search engines that present results as ranked lists. However, there are several alternative paradigms for which the retrieval process, result presentation, and user behavior diverge. Considering these differences becomes pivotal when designing interfaces that synthesize results from different resources into seemingly relevant and coherent written or spoken text [165,168,195]. Conversational interfaces are relatively more engaging than conventional web interfaces in various contexts [10,67,131,163], including potential in supporting long-term memorability [164]. Notably, the pursuit of improving user engagement and experience can be orthogonal to supporting responsible opinion formation. This dichotomy is perfectly captured by the well-established notions of '*seamless*' versus '*seamful*' design in human-computer interaction

(HCI). While seamless design emphasizes clarity, simplicity, ease of use, and consistency to facilitate interaction with technologies, seamful design emphasizes configurability, user appropriation, and the revelation of complexity, ambiguity, or inconsistency [87]. There are several arguments in favor of creating *seamless* interactions with search systems to satisfy user information needs. However, such design choices may not adequately foster responsible opinion formation. Users may also turn towards LLM-based tools like *ChatGPT* [135], which may provide incomplete, misleading, or even inaccurate information due to model hallucinations. Natural language aids comprehension and offers opportunities to directly provide diverse viewpoints (i.e., serving as a *seamless* mode of interaction). However, Shah and Bender [181] warn that such interactions can hinder users' ability to identify incorrect or biased information and to actively explore different resources to construct a model of the knowledge space, building information literacy (i.e., facets that can be supported through *seamful* design). More research is urgently required to better understand whether and how responsible opinion formation can be supported in the context of such emerging search paradigms.

Malicious Intent. Thus far, we have assumed no malicious intention from any actor, i.e., framing biases and harmful effects as unintended byproducts of web search. Yet, malicious actors may use research findings and practical applications for their purposes, e.g., to steer public opinion or manipulate targeted individuals. This solicits methods to detect and safeguard against such actions. Researchers and practitioners need to discuss this possibility in their work.

Research Questions

The research opportunities and challenges discussed in this paper may appear abundant and intimidating. To provide a more approachable starting point, we propose a set of research questions, which are by no means exhaustive.

Foundations: (i) What obligations should search engines bear concerning individual and societal well-being? (ii) Which values and principles should guide the system design process? (iii) What framework can comprehensively represent viewpoints on SERPs? (iv) Which notions of viewpoint diversity would benefit individuals and society? (v) Should the notion of viewpoint diversity be adjusted depending on the specific topic and searcher?

Searcher: (i) Which patterns of search behavior and searcher characteristics can be linked to knowledge gain and attitude change? (ii) Which traits affect searchers' vulnerability to ranking and cognitive biases? (iii) What user-centered interventions can empower unbiased and diligent search behavior?

Search Engine: (i) How should relevance criteria be adjusted for search on debated topics? (ii) What crowdsourcing, automatic, or hybrid methods can accurately and efficiently detect viewpoints expressed in search results? (iii) Which re-ranking strategies meaningfully increase viewpoint diversity?

Interplay: (i) What factors shape the interplay of search engine-rooted exposure biases and searcher-rooted interaction biases? (ii) What interface-centered interventions can empower unbiased and diligent search behavior? (iii) How can the interface be leveraged to enhance the transparency of relevance criteria to the searcher?

7 Concluding Remarks

Drawing upon perspectives from digital humanism and an extensive body of interdisciplinary literature, we offer an in-depth analysis of the distinguishing characteristics and challenges associated with web search on debated topics. We outline a research agenda toward web search that fosters responsible opinion formation by focusing on the searcher, the search engine, and their complex interplay. While rooted in IR, advancements in this area demand a multi- and interdisciplinary approach with input from various domains, including philosophy, psychology, information science, and the communication sciences. With this paper, we aspire to motivate researchers, practitioners, and policymakers across domains to engage in the collective effort of addressing the pressing socio-technical challenges and creating an enriching, unbiased, and trustworthy web search experience. Ultimately, the pursuit of such endeavors would benefit both individuals and society by promoting democratic values, such as an informed citizenry, opinion diversity, and tolerance for differing viewpoints.

Acknowledgements. This project has received funding from the European Union's Horizon 2020 research and innovation programme under the Marie Skłodowska-Curie grant agreement No 860621.

References

1. Agrawal, R., Gollapudi, S., Halverson, A., Ieong, S.: Diversifying search results. In: Proceedings of the Second ACM International Conference on Web Search and Data Mining - WSDM 2009, p. 5. ACM Press, Barcelona (2009). https://doi.org/10.1145/1498759.1498766. https://portal.acm.org/citation.cfm?doid=1498759.1498766
2. Ajjour, Y., Braslavski, P., Bondarenko, A., Stein, B.: Identifying argumentative questions in web search logs. In: Proceedings of the 45th International ACM SIGIR Conference on Research and Development in Information Retrieval, SIGIR 2022, pp. 2393–2399. Association for Computing Machinery, New York (2022). https://doi.org/10.1145/3477495.3531864

3. Alaofi, M., et al.: Where do queries come from? In: Proceedings of the 45th International ACM SIGIR Conference on Research and Development in Information Retrieval, SIGIR 2022, pp. 2850–2862. Association for Computing Machinery, New York (2022). https://doi.org/10.1145/3477495.3531711
4. Allam, A., Schulz, P.J., Nakamoto, K.: The impact of search engine selection and sorting criteria on vaccination beliefs and attitudes: two experiments manipulating google output. J. Med. Internet Res. **16**(4), e100 (2014). https://doi.org/10.2196/jmir.2642. http://www.jmir.org/2014/4/e100/
5. Antelmi, A., Malandrino, D., Scarano, V.: Characterizing the behavioral evolution of twitter users and the truth behind the 90-9-1 rule. In: Companion Proceedings of The 2019 World Wide Web Conference, WWW 2019, pp. 1035–1038. Association for Computing Machinery, New York (2019). https://doi.org/10.1145/3308560.3316705
6. Araujo, T., et al.: Osd2f: an open-source data donation framework. Comput. Commun. Res. **4**(2), 372–387 (2022)
7. Arthur, C.: What is the 1% rule? (2006). https://www.theguardian.com/technology/2006/jul/20/guardianweeklytechnologysection2
8. Asudeh, A., Jagadish, H.V., Stoyanovich, J., Das, G.: Designing fair ranking schemes. In: Proceedings of the 2019 International Conference on Management of Data, pp. 1259–1276. ACM, Amsterdam (2019). https://doi.org/10.1145/3299869.3300079. https://dl.acm.org/doi/10.1145/3299869.3300079
9. Augenstein, I., Rocktäschel, T., Vlachos, A., Bontcheva, K.: Stance detection with bidirectional conditional encoding. In: Proceedings of the 2016 Conference on Empirical Methods in Natural Language Processing, pp. 876–885. Association for Computing Machinery, New York (2016)
10. Avula, S., Chadwick, G., Arguello, J., Capra, R.: Searchbots: User engagement with chatbots during collaborative search. In: Proceedings of the 2018 Conference on Human Information Interaction & Retrieval, pp. 52–61 (2018)
11. Azzopardi, L.: Cognitive biases in search: a review and reflection of cognitive biases in information retrieval. In: Proceedings of the 2021 Conference on Human Information Interaction and Retrieval, pp. 27–37. ACM, Canberra (2021). https://doi.org/10.1145/3406522.3446023. https://dl.acm.org/doi/10.1145/3406522.3446023
12. Baden, C., Kligler-Vilenchik, N., Yarchi, M.: Hybrid content analysis: toward a strategy for the theory-driven, computer-assisted classification of large text corpora. Commun. Methods Meas. **14**(3), 165–183 (2020). https://doi.org/10.1080/19312458.2020.1803247. https://www.tandfonline.com/doi/full/10.1080/19312458.2020.1803247
13. Baden, C., Springer, N.: Com(ple)menting the news on the financial crisis: The contribution of news users' commentary to the diversity of viewpoints in the public debate. Eur. J. Commun. **29**(5), 529–548 (2014). https://doi.org/10.1177/0267323114538724. http://journals.sagepub.com/doi/10.1177/0267323114538724
14. Baden, C., Springer, N.: Conceptualizing viewpoint diversity in news discourse. Journalism **18**(2), 176–194 (2017). https://doi.org/10.1177/1464884915605028. http://journals.sagepub.com/doi/10.1177/1464884915605028
15. Baeza-Yates, R.: Bias on the web. Commun. ACM **61**(6), 54–61 (2018). https://doi.org/10.1145/3209581. https://dl.acm.org/doi/10.1145/3209581
16. Bail, C.A., et al.: Exposure to opposing views on social media can increase political polarization. Proc. Natl. Acad. Sci. **115**(37), 9216–9221 (2018)

17. Bar-Ilan, J., Keenoy, K., Levene, M., Yaari, E.: Presentation bias is significant in determining user preference for search results-a user study. J. Am. Soc. Inf. Sci. Technol. **60**(1), 135–149 (2009)
18. Beam, M.A.: Automating the news: how personalized news recommender system design choices impact news reception. Commun. Res. **41**(8), 1019–1041 (2014)
19. Belkin, N.J.: Anomalous states of knowledge as a basis for information retrieval. Can. J. Inf. Sci. **5**(1), 133–143 (1980)
20. Bernstein, M.S., Teevan, J., Dumais, S., Liebling, D., Horvitz, E.: Direct answers for search queries in the long tail. In: Proceedings of the SIGCHI Conference on Human Factors in Computing Systems, pp. 237–246 (2012)
21. Bink, M., Schwarz, S., Draws, T., Elsweiler, D.: Investigating the influence of featured snippets on user attitudes. In: ACM SIGIR Conference on Human Information Interaction and Retrieval, CHIIR 2023. ACM, New York (2023). https://doi.org/10.1145/3576840.3578323
22. Bink, M., Zimmerman, S., Elsweiler, D.: Featured snippets and their influence on users' credibility judgements. In: ACM SIGIR Conference on Human Information Interaction and Retrieval, CHIIR 2022, pp. 113–122. Association for Computing Machinery, New York (2022). https://doi.org/10.1145/3498366.3505766
23. Blassnig, S., Mitova, E., Pfiffner, N., Reiss, M.V.: Googling referendum campaigns: analyzing online search patterns regarding swiss direct-democratic votes. Media Commun. **11**(1), 19–30 (2023)
24. Van den Bogaert, L., Geerts, D., Harambam, J.: Putting a human face on the algorithm: co-designing recommender personae to democratize news recommender systems. Digital Journal. 1–21 (2022)
25. Boltanski, L., Thévenot, L.: On Justification: Economies of Worth, vol. 27. Princeton University Press, Princeton (2006)
26. Bondarenko, A., et al.: Overview of touché 2022: argument retrieval. In: Barrón-Cedeño, A., et al. (eds.) Experimental IR Meets Multilinguality, Multimodality, and Interaction, vol. 13390, pp. 311–336. Springer, Cham (2022). https://doi.org/10.1007/978-3-031-13643-6_21
27. Bondarenko, A., et al.: Overview of touché 2021: argument retrieval. In: Candan, K.S., et al. (eds.) Experimental IR Meets Multilinguality, Multimodality, and Interaction, vol. 12880, pp. 450–467. Springer, Cham (2021)
28. Van den Bosch, A., Bogers, T., De Kunder, M.: Estimating search engine index size variability: a 9-year longitudinal study. Scientometrics **107**(2), 839–856 (2016)
29. Bota, H., Zhou, K., Jose, J.M.: Playing your cards right: the effect of entity cards on search behaviour and workload. In: Proceedings of the 2016 ACM on Conference on Human Information Interaction and Retrieval, CHIIR 2016, pp. 131–140. Association for Computing Machinery, New York (2016). https://doi.org/10.1145/2854946.2854967
30. Bowes, S.M., Costello, T.H., Lee, C., McElroy-Heltzel, S., Davis, D.E., Lilienfeld, S.O.: Stepping outside the echo chamber: is intellectual humility associated with less political myside bias? Pers. Soc. Psychol. Bull. **48**, 150–164 (2022). https://doi.org/10.1177/0146167221997619
31. Budzynska, K., Reed, C.: Advances in argument mining. In: Proceedings of the 57th Annual Meeting of the Association for Computational Linguistics: Tutorial Abstracts, pp. 39–42. Association for Computational Linguistics, Florence (2019). https://doi.org/10.18653/v1/P19-4008. https://www.aclweb.org/anthology/P19-4008
32. Burke, R.: Personalization, fairness, and post-userism. In: Perspectives on Digital Humanism, p. 145 (2022)

33. Cacioppo, J.T., Petty, R.E., Morris, K.J.: Effects of need for cognition on message evaluation, recall, and persuasion. J. Pers. Soc. Psychol. **45**, 805–818 (1983). https://doi.org/10.1037/0022-3514.45.4.805
34. Canca, C.: Did you find it on the internet? ethical complexities of search engine rankings. In: Perspectives on Digital Humanism, p. 135 (2022)
35. Carroll, N..: In search we trust: exploring how search engines are shaping society. Int. J. Knowl. Soc. Res. **5**(1), 12–27 (2014). https://doi.org/10.4018/ijksr.2014010102. https://services.igi-global.com/resolvedoi/resolve.aspx?doi=10.4018/ijksr.2014010102
36. Chacoma, A., Zanette, D.H.: Opinion formation by social influence: from experiments to modeling. PLoS ONE **10**(10), e0140406 (2015)
37. Chamberlain, J., Kruschwitz, U., Hoeber, O.: Scalable visualisation of sentiment and stance. In: Calzolari, N., et al. (eds.) Proceedings of the Eleventh International Conference on Language Resources and Evaluation (LREC 2018). European Language Resources Association (ELRA), Miyazaki (2018). https://aclanthology.org/L18-1660
38. Chen, S., Khashabi, D., Yin, W., Callison-Burch, C., Roth, D.: Seeing Things from a Different Angle: Discovering Diverse Perspectives about Claims. arXiv:1906.03538 [cs] (2019)
39. Chen, T., Yin, H., Ye, G., Huang, Z., Wang, Y., Wang, M.: Try this instead: personalized and interpretable substitute recommendation. In: Proceedings of the 43rd International ACM SIGIR Conference on Research and Development in Information Retrieval, pp. 891–900 (2020)
40. Cushion, S., Thomas, R.: From quantitative precision to qualitative judgements: professional perspectives about the impartiality of television news during the 2015 UK general election. Journalism **20**(3), 392–409 (2019)
41. Deffler, S.A., Leary, M.R., Hoyle, R.H.: Knowing what you know: intellectual humility and judgments of recognition memory. Pers. Individ. Differ. **96**, 255–259 (2016). https://doi.org/10.1016/j.paid.2016.03.016
42. Draws, T., Inel, O., Tintarev, N., Baden, C., Timmermans, B.: Comprehensive Viewpoint Representations for a Deeper Understanding of User Interactions With Debated Topics. In: Proceedings of the 2022 ACM SIGIR Conference on Human Information Interaction and Retrieval, CHIIR 2022, p. 11. ACM, New York (2022). https://doi.org/10.1145/3498366.3505812. https://drive.google.com/file/d/1cMUzKX9QkAGfTAM8WaDKRK7y23auzNn5/view?usp=sharing
43. Draws, T., et al.: Explainable cross-topic stance detection for search results. In: ACM SIGIR Conference on Human Information Interaction and Retrieval, CHIIR 2023. ACM, New York (2023). https://doi.org/10.1145/3576840.3578296
44. Draws, T., Rieger, A., Inel, O., Gadiraju, U., Tintarev, N.: A checklist to combat cognitive biases in crowdsourcing. In: Proceedings on the Ninth AAAI Conference on Human Computation and Crowdsourcing, HCOMP 2021. AAAI (2021)
45. Draws, T., et al.: Viewpoint diversity in search results. In: Kamps, J., et al. (eds.) Advances in Information Retrieval. Lecture Notes in Computer Science, vol. 13980, pp. 279–297. Springer, Cham (2023)
46. Draws, T., Tintarev, N., Gadiraju, U., Bozzon, A., Timmermans, B.: This is not what we ordered: exploring why biased search result rankings affect user attitudes on debated topics. In: Proceedings of the 44th International ACM SIGIR Conference on Research and Development in Information Retrieval, pp. 295–305. ACM, Virtual Event Canada (2021). https://doi.org/10.1145/3404835.3462851. https://dl.acm.org/doi/10.1145/3404835.3462851

47. Draws, T.A.: Understanding Viewpoint Biases in Web Search Results. Phd thesis, Delft University of Technology, Delft, Netherlands (2023). https://doi.org/10.4233/uuid:1b177026-6af7-48f3-ba04-ab7109db3c36
48. Drosou, M., Pitoura, E.: Search result diversification. SIGMOD Rec. **39**(1), 7 (2010)
49. van Drunen, M.Z., Helberger, N., Bastian, M.: Know your algorithm: what media organizations need to explain to their users about news personalization. Int. Data Privacy Law **9**(4), 220–235 (2019)
50. Dumani, L., Neumann, P.J., Schenkel, R.: A framework for argument retrieval. In: Jose, J.M., et al. (eds.) ECIR 2020. LNCS, vol. 12035, pp. 431–445. Springer, Cham (2020). https://doi.org/10.1007/978-3-030-45439-5_29
51. Eickhoff, C.: Cognitive biases in crowdsourcing. In: Proceedings of the Eleventh ACM International Conference on Web Search and Data Mining, pp. 162–170. ACM, Marina Del Rey (2018). https://doi.org/10.1145/3159652.3159654. https://dl.acm.org/doi/10.1145/3159652.3159654
52. Epstein, R., Robertson, R.E.: The search engine manipulation effect (SEME) and its possible impact on the outcomes of elections. Proc. Natl. Acad. Sci. **112**(33), E4512–E4521 (2015). https://doi.org/10.1073/pnas.1419828112. http://www.pnas.org/lookup/doi/10.1073/pnas.1419828112
53. Epstein, R., Robertson, R.E., Lazer, D., Wilson, C.: Suppressing the search engine manipulation effect (SEME). In: Proceedings of the ACM on Human-Computer Interaction, vol. 1(CSCW), 1–22 (2017). https://doi.org/10.1145/3134677. https://dl.acm.org/doi/10.1145/3134677
54. Eskens, S., Helberger, N., Moeller, J.: Challenged by news personalisation: five perspectives on the right to receive information. J. Media Law **9**(2), 259–284 (2017). https://doi.org/10.1080/17577632.2017.1387353. https://www.tandfonline.com/doi/full/10.1080/17577632.2017.1387353
55. Figueira, Á., Oliveira, L.: The current state of fake news: challenges and opportunities. Procedia Comput. Sci. **121**, 817–825 (2017). https://doi.org/10.1016/j.procs.2017.11.106
56. Freiling, I., Krause, N.M., Scheufele, D.A.: Science and ethics of "Curing" misinformation. AMA J. Ethics **25**, 228–237 (2023). https://doi.org/10.1001/amajethics.2023.228
57. Gao, R., Shah, C.: Toward creating a fairer ranking in search engine results. Inf. Process. Manag. **57**(1), 102138 (2020). https://doi.org/10.1016/j.ipm.2019.102138. https://linkinghub.elsevier.com/retrieve/pii/S0306457319304121
58. Gevelber, L.: How Mobile Has Changed How People Get Things Done: New Consumer Behavior Data. Think with Google (2016). https://think.storage.googleapis.com/docs/mobile-search-consumer-behavior-data.pdf
59. Gevelber, L.: It's all about 'me'-how people are taking search personally. Technical report (2018). https://www.thinkwithgoogle.com/marketing-strategies/search/personal-needs-search-trends/
60. Gezici, G., Lipani, A., Saygin, Y., Yilmaz, E.: Evaluation metrics for measuring bias in search engine results. Inf. Retr. J. **24**(2), 85–113 (2021). https://doi.org/10.1007/s10791-020-09386-w. http://link.springer.com/10.1007/s10791-020-09386-w
61. Ghenai, A., Smucker, M.D., Clarke, C.L.: A think-aloud study to understand factors affecting online health search. In: Proceedings of the 2020 Conference on Human Information Interaction and Retrieval, pp. 273–282. ACM, Vancouver (2020). https://doi.org/10.1145/3343413.3377961. https://dl.acm.org/doi/10.1145/3343413.3377961

62. Giunchiglia, F., Kleanthous, S., Otterbacher, J., Draws, T.: Transparency paths - documenting the diversity of user perceptions. In: Adjunct Proceedings of the 29th ACM Conference on User Modeling, Adaptation and Personalization, pp. 415–420. ACM, Utrecht (2021). https://doi.org/10.1145/3450614.3463292. https://dl.acm.org/doi/10.1145/3450614.3463292
63. Gorichanaz, T.: Relating information seeking and use to intellectual humility. J. Am. Soc. Inf. Sci. **73**, 643–654 (2022). https://doi.org/10.1002/asi.24567
64. Gorichanaz, T.: Virtuous search: a framework for intellectual virtue in online search. J. Assoc. Inf. Sci. Technol. (2023). https://doi.org/10.1002/asi.24832
65. Groot Kormelink, T., Costera Meijer, I.: A user perspective on time spent: temporal experiences of everyday news use. Journal. Stud. **21**(2), 271–286 (2020)
66. Guess, A., Nagler, J., Tucker, J.: Less than you think: prevalence and predictors of fake news dissemination on facebook. Sci. Adv. **5**(1), eaau4586 (2019)
67. Gupta, A., Basu, D., Ghantasala, R., Qiu, S., Gadiraju, U.: To trust or not to trust: how a conversational interface affects trust in a decision support system. In: Proceedings of the ACM Web Conference 2022, pp. 3531–3540 (2022)
68. Haider, J., Sundin, O.: Invisible Search and Online Search Engines: The Ubiquity of Search in Everyday Life. Taylor & Francis (2019). https://doi.org/10.4324/9780429448546
69. Haider, J., Sundin, O.: Information literacy challenges in digital culture: conflicting engagements of trust and doubt. Inf. Commun. Soc. **25**, 1176–1191 (2022). https://doi.org/10.1080/1369118X.2020.1851389
70. Halavais, A.: Search Engine Society. John Wiley & Sons, Hoboken (2017)
71. Hanselowski, A., et al.: A retrospective analysis of the fake news challenge stance detection task. arXiv preprint arXiv:1806.05180 (2018)
72. Harambam, J., Bountouridis, D., Makhortykh, M., Van Hoboken, J.: Designing for the better by taking users into account: a qualitative evaluation of user control mechanisms in (news) recommender systems. In: Proceedings of the 13th ACM Conference on Recommender Systems, pp. 69–77 (2019)
73. Hassoun, A., Beacock, I., Consolvo, S., Goldberg, B., Kelley, P.G., Russell, D.M.: Practicing information sensibility: how gen Z engages with online information. In: Proceedings of the 2023 CHI Conference on Human Factors in Computing Systems, CHI 2023, pp. 1–17. Association for Computing Machinery, New York (2023). https://doi.org/10.1145/3544548.3581328
74. Helberger, N.: On the democratic role of news recommenders. Dig. Journal. **7**(8), 993–1012 (2019). https://doi.org/10.1080/21670811.2019.1623700. https://www.tandfonline.com/doi/full/10.1080/21670811.2019.1623700
75. Helberger, N., Kleinen-von Königslöw, K., van der Noll, R.: Regulating the new information intermediaries as gatekeepers of information diversity. Info **17**, 50–71 (2015). https://doi.org/10.1108/info-05-2015-0034
76. Hermann, E.: Artificial intelligence and mass personalization of communication content-an ethical and literacy perspective. New Media Soc. **24**(5), 1258–1277 (2022)
77. Hertwig, R., Grüne-Yanoff, T.: Nudging and boosting: steering or empowering good decisions. Perspect. Psychol. Sci. **12**, 973–986 (2017). https://doi.org/10.1177/1745691617702496
78. Hinman, L.M.: Searching Ethics: The Role of Search Engines in the Construction and Distribution of Knowledge. Springer, Heidelberg (2008). https://doi.org/10.1007/978-3-540-75829-7

79. van Hoof, M., Meppelink, C.S., Moeller, J., Trilling, D.: Searching differently? how political attitudes impact search queries about political issues. New Media Soc. 14614448221104405 (2022)
80. van Hoof, M., Trilling, D., Meppelink, C., Moeller, J., Loecherbach, F.: Googling politics? the computational identification of political and news-related searches from web browser histories (2023)
81. Howe, L.C., Krosnick, J.A.: Attitude Strength. Ann. Rev. Psychol. **68**, 327–351 (2017). https://doi.org/10.1146/annurev-psych-122414-033600
82. Hu, S., Dou, Z., Wang, X., Sakai, T., Wen, J.R.: Search result diversification based on hierarchical intents. In: Proceedings of the 24th ACM International on Conference on Information and Knowledge Management, pp. 63–72. ACM, Melbourne (2015). https://doi.org/10.1145/2806416.2806455. https://dl.acm.org/doi/10.1145/2806416.2806455
83. Hube, C., Fetahu, B., Gadiraju, U.: Understanding and mitigating worker biases in the crowdsourced collection of subjective judgments. In: Proceedings of the 2019 CHI Conference on Human Factors in Computing Systems, pp. 1–12. ACM, Glasgow (2019). https://doi.org/10.1145/3290605.3300637. https://dl.acm.org/doi/10.1145/3290605.3300637
84. Hwa-Froelich, D.A., Vigil, D.C.: Three aspects of cultural influence on communication: a literature review. Commun. Disord. Q. **25**(3), 107–118 (2004)
85. Ingwersen, P., Järvelin, K.: Information retrieval in context: Irix. SIGIR Forum **39**(2), 31–39 (2005). https://doi.org/10.1145/1113343.1113351
86. Ingwersen, P., Järvelin, K.: The Turn: Integration of Information Seeking and Retrieval in Context. Springer, Heidelberg (2005). https://doi.org/10.1007/1-4020-3851-8
87. Inman, S., Ribes, D.: "beautiful seams" strategic revelations and concealments. In: Proceedings of the 2019 CHI Conference on Human Factors in Computing Systems, pp. 1–14 (2019)
88. Joachims, T., Granka, L., Pan, B., Hembrooke, H., Gay, G.: Accurately interpreting clickthrough data as implicit feedback. ACM SIGIR Forum **51**(1), 8 (2016)
89. Joachims, T., Swaminathan, A., Schnabel, T.: Unbiased learning-to-rank with biased feedback. In: Proceedings of the Tenth ACM International Conference on Web Search and Data Mining, pp. 781–789. ACM, Cambridge (2017). https://doi.org/10.1145/3018661.3018699. https://dl.acm.org/doi/10.1145/3018661.3018699
90. Joho, H., Jose, J.M.: A comparative study of the effectiveness of search result presentation on the web. In: Lalmas, M., MacFarlane, A., Rüger, S., Tombros, A., Tsikrika, T., Yavlinsky, A. (eds.) ECIR 2006. LNCS, vol. 3936, pp. 302–313. Springer, Heidelberg (2006). https://doi.org/10.1007/11735106_27
91. Jones-Jang, S.M., Mortensen, T., Liu, J.: Does media literacy help identification of fake news? information literacy helps, but other literacies don't. Am. Behav. Sci. **65**(2), 371–388 (2021)
92. Kammerer, Y., Gerjets, P.: How search engine users evaluate and select web search results: the impact of the search engine interface on credibility assessments. In: Web search engine research. Emerald Group Publishing Limited (2012)
93. Kaya, M., Bridge, D.: Subprofile-aware diversification of recommendations. User Model. User-Adapt. Interact. **29**(3), 661–700 (2019). https://doi.org/10.1007/s11257-019-09235-6. http://link.springer.com/10.1007/s11257-019-09235-6
94. Kennedy, A.M., Jones, K., Williams, J.: Children as vulnerable consumers in online environments. J. Consum. Aff. **53**(4), 1478–1506 (2019)
95. Khder, M.A.: Web scraping or web crawling: state of art, techniques, approaches and application. Int. J. Adv. Soft Comput. Appl. **13**(3) (2021)

96. Kim, D.H., Pasek, J.: Explaining the diversity deficit: value-trait consistency in news exposure and democratic citizenship. Commun. Res. **47**(1), 29–54 (2020)
97. Kim, J., Thomas, P., Sankaranarayana, R., Gedeon, T., Yoon, H.J.: Understanding eye movements on mobile devices for better presentation of search results. J. Am. Soc. Inf. Sci. **67**(11), 2607–2619 (2016)
98. Knobloch-Westerwick, S., Meng, J.: Looking the other way: selective exposure to attitude-consistent and counterattitudinal political information. Commun. Res. **36**, 426–448 (2009). https://doi.org/10.1177/0093650209333030
99. Knobloch-Westerwick, S., Johnson, B.K., Westerwick, A.: Confirmation bias in online searches: impacts of selective exposure before an election on political attitude strength and shifts. J. Comput.-Mediat. Commun. **20**(2), 171–187 (2015). https://doi.org/10.1111/jcc4.12105. https://academic.oup.com/jcmc/article/20/2/171-187/4067554
100. Koene, A., et al.: Ethics of personalized information filtering. In: Tiropanis, T., Vakali, A., Sartori, L., Burnap, P. (eds.) INSCI 2015. LNCS, vol. 9089, pp. 123–132. Springer, Cham (2015). https://doi.org/10.1007/978-3-319-18609-2_10
101. Kormelink, T.G., Meijer, I.C.: What clicks actually mean: exploring digital news user practices. Journalism **19**(5), 668–683 (2018)
102. Kornblith, H.: Justified belief and epistemically responsible action. Phil. Rev. **92**, 33–48 (1983). https://doi.org/10.2307/2184520
103. Kozyreva, A., Lewandowsky, S., Hertwig, R.: Citizens versus the internet: confronting digital challenges with cognitive tools. Psychol. Sci. Public Interest **21**, 103–156 (2020). https://doi.org/10.1177/1529100620946707
104. Krumrei-Mancuso, E.J., Haggard, M.C., LaBouff, J.P., Rowatt, W.C.: Links between intellectual humility and acquiring knowledge. J. Posit. Psychol. **15**, 155–170 (2020). https://doi.org/10.1080/17439760.2019.1579359
105. Kulshrestha, J., et al.: Quantifying search bias: investigating sources of bias for political searches in social media. In: Proceedings of the 2017 ACM Conference on Computer Supported Cooperative Work and Social Computing, pp. 417–432. ACM, Portland (2017). https://doi.org/10.1145/2998181.2998321. https://dl.acm.org/doi/10.1145/2998181.2998321
106. Kulshrestha, J., et al.: Search bias quantification: investigating political bias in social media and web search. Inf. Retr. J. **22**(1–2), 188–227 (2019). https://doi.org/10.1007/s10791-018-9341-2. http://link.springer.com/10.1007/s10791-018-9341-2
107. Kwak, H., An, J., Salminen, J., Jung, S.G., Jansen, B.J.: What we read, what we search: Media attention and public attention among 193 countries. In: Proceedings of the 2018 World Wide Web Conference, WWW 2018, pp. 893–902. International World Wide Web Conferences Steering Committee, Republic and Canton of Geneva, CHE (2018). https://doi.org/10.1145/3178876.3186137
108. Landoni, M., Aliannejadi, M., Huibers, T., Murgia, E., Pera, M.S.: Have a clue! the effect of visual cues on children's search behavior in the classroom. In: ACM SIGIR Conference on Human Information Interaction and Retrieval, pp. 310–314 (2022)
109. Lawrence, J., Reed, C.: Combining argument mining techniques. In: Proceedings of the 2nd Workshop on Argumentation Mining, pp. 127–136. Association for Computational Linguistics, Denver (2015). https://doi.org/10.3115/v1/W15-0516. http://aclweb.org/anthology/W15-0516
110. Lawrence, J., Reed, C.: Argument mining: a survey. Comput. Linguist. **45**(4), 765–818 (2020)

111. Lazar, J., Feng, J.H., Hochheiser, H.: Research Methods in Human-Computer Interaction. Morgan Kaufmann, Burlington (2017)
112. Leary, M.R., et al.: Cognitive and interpersonal features of intellectual humility. Pers. Soc. Psychol. Bull. **43**, 793–813 (2017). https://doi.org/10.1177/0146167217697695
113. Lewandowsky, S., van der Linden, S.: Countering misinformation and fake news through inoculation and prebunking. Eur. Rev. Soc. Psychol. **32**, 348–384 (2021). https://doi.org/10.1080/10463283.2021.1876983
114. Lippi, M., Torroni, P.: Context-independent claim detection for argument mining. In: Twenty-Fourth International Joint Conference on Artificial Intelligence (2015)
115. Liscio, E.: Axies: identifying and evaluating context-specific values. In: Proceedings of the 20th International Conference on Autonomous Agents and Multiagent Systems, AAMAS 2021, p. 10 (2021)
116. Liu, J., Zhang, X.: The role of domain knowledge in document selection from search results. J. Am. Soc. Inf. Sci. **70**, 1236–1247 (2019). https://doi.org/10.1002/asi.24199
117. Loecherbach, F., Moeller, J., Trilling, D., van Atteveldt, W.: The unified framework of media diversity: a systematic literature review. Dig. Journal. **8**(5), 605–642 (2020). https://doi.org/10.1080/21670811.2020.1764374. https://www.tandfonline.com/doi/full/10.1080/21670811.2020.1764374
118. Loos, E., Ivan, L., Leu, D.: "save the pacific northwest tree octopus": a hoax revisited. or: How vulnerable are school children to fake news? Inf. Learn. Sci. (2018)
119. Lorenz-Spreen, P., Geers, M., Pachur, T., Hertwig, R., Lewandowsky, S., Herzog, S.M.: Boosting people's ability to detect microtargeted advertising. Sci. Rep. **11**(1), 1–9 (2021)
120. Lorenz-Spreen, P., Lewandowsky, S., Sunstein, C.R., Hertwig, R.: How behavioural sciences can promote truth, autonomy and democratic discourse online. Nat. Hum. Behav. **4**, 1102–1109 (2020). https://doi.org/10.1038/s41562-020-0889-7
121. Ludolph, R., Allam, A., Schulz, P.J.: Manipulating google's knowledge graph box to counter biased information processing during an online search on vaccination: application of a technological debiasing strategy. J. Med. Internet Res. **18**(6), e137 (2016). https://doi.org/10.2196/jmir.5430. http://www.jmir.org/2016/6/e137/
122. Mattis, N., Masur, P., Möller, J., van Atteveldt, W.: Nudging towards news diversity: a theoretical framework for facilitating diverse news consumption through recommender design. New Media Soc. 14614448221104413 (2022)
123. McDuff, D., Thomas, P., Craswell, N., Rowan, K., Czerwinski, M.: Do affective cues validate behavioural metrics for search? In: Proceedings of the 44th International ACM SIGIR Conference on Research and Development in Information Retrieval, pp. 1544–1553 (2021)
124. McKay, D., et al.: We are the change that we seek: information interactions during a change of viewpoint. In: Proceedings of the 2020 Conference on Human Information Interaction and Retrieval, pp. 173–182 (2020)
125. Meij, E., Bron, M., Hollink, L., Huurnink, B., de Rijke, M.: Learning semantic query suggestions. In: Bernstein, A., et al. (eds.) ISWC 2009. LNCS, vol. 5823, pp. 424–440. Springer, Heidelberg (2009). https://doi.org/10.1007/978-3-642-04930-9_27
126. Miller, B., Record, I.: Justified belief in the digital age: on the epistemic implications of secret internet technologies. Episteme **10**, 117–134 (2013). https://doi.org/10.1017/epi.2013.11

127. Milton, A., Pera, M.S.: Into the unknown: exploration of search engines' responses to users with depression and anxiety. ACM Trans. Web (2021)
128. Minson, J.A., Chen, F.S., Tinsley, C.H.: Why won't you listen to me? measuring receptiveness to opposing views. Manag. Sci. (2019). https://doi.org/10.1287/mnsc.2019.3362
129. Mohammad, S.M., Kiritchenko, S., Sobhani, P., Zhu, X., Cherry, C.: Semeval-2016 task 6: detecting stance in tweets. In: Proceedings of the International Workshop on Semantic Evaluation, SemEval 2016, San Diego, California (2016)
130. Mohammad, S.M., Sobhani, P., Kiritchenko, S.: Stance and sentiment in tweets. Spec. Sect. ACM Trans. Internet Technol. Argument. Social Media **17**(3) (2017)
131. Moore, R.J., Arar, R.: Conversational UX Design: A Practitioner's Guide to the Natural Conversation Framework. Morgan & Claypool, San Rafael (2019)
132. Mowshowitz, A., Kawaguchi, A.: Assessing bias in search engines. Inf. Process. Manag. **38**(1), 141–156 (2002)
133. Mummolo, J.: News from the other side: how topic relevance limits the prevalence of partisan selective exposure. J. Polit. **78**(3), 763–773 (2016)
134. Munson, S.A., Resnick, P.: Presenting diverse political opinions: how and how much. In: Proceedings of the SIGCHI Conference on Human Factors in Computing Systems, pp. 1457–1466 (2010)
135. N.D: Introducing chatgpt (2023). https://openai.com/blog/chatgpt/
136. Nickerson, R.S.: Confirmation bias: a ubiquitous phenomenon in many guises. Rev. Gen. Psychol. **2**(2), 175–220 (1998)
137. Niculae, V., Park, J., Cardie, C.: Argument Mining with Structured SVMs and RNNs. arXiv:1704.06869 [cs] (2017)
138. Noble, S.U.: Algorithms of Oppression. New York University Press, New York (2018)
139. Novin, A., Meyers, E.: Making sense of conflicting science information: exploring bias in the search engine result page. In: Proceedings of the 2017 Conference on Conference Human Information Interaction and Retrieval, pp. 175–184. ACM, Oslo (2017). https://doi.org/10.1145/3020165.3020185. https://dl.acm.org/doi/10.1145/3020165.3020185
140. Nowotny, H.: Digital humanism: navigating the tensions ahead. In: Perspectives on Digital Humanism, p. 317 (2022)
141. Nyhan, B., Reifler, J.: When corrections fail: the persistence of political misperceptions. Polit. Behav. **32**(2), 303–330 (2010)
142. Oliveira, B., Teixeira Lopes, C.: The evolution of web search user interfaces - an archaeological analysis of google search engine result pages. In: Proceedings of the 2023 Conference on Human Information Interaction and Retrieval, CHIIR 2023, pp. 55–68. Association for Computing Machinery, New York (2023). https://doi.org/10.1145/3576840.3578320
143. Olteanu, A., Castillo, C., Diaz, F., Kıcıman, E.: Social data: biases, methodological pitfalls, and ethical boundaries. Front. Big Data **2**, 13 (2019)
144. Ortloff, A.M., Zimmerman, S., Elsweiler, D., Henze, N.: The effect of nudges and boosts on browsing privacy in a naturalistic environment. In: Proceedings of the 2021 Conference on Human Information Interaction and Retrieval, CHIIR 2021, pp. 63–73. Association for Computing Machinery, New York (2021). https://doi.org/10.1145/3406522.3446014
145. Oswald, M.E., Grosjean, S.: Confirmation bias. Cogn. Illus. Handb. Fallacies Biases Thinking Judge. Memory **79**, 83 (2004)

146. Otterbacher, J.: Addressing social bias in information retrieval. In: Bellot, P., Trabelsi, C., Mothe, J., Murtagh, F., Nie, J.Y., Soulier, L., SanJuan, E., Cappellato, L., Ferro, N. (eds.) CLEF 2018. LNCS, vol. 11018, pp. 121–127. Springer, Cham (2018). https://doi.org/10.1007/978-3-319-98932-7_11
147. Otterbacher, J., Bates, J., Clough, P.: Competent men and warm women: gender stereotypes and backlash in image search results. In: Proceedings of the 2017 CHI Conference on Human Factors in Computing Systems, pp. 6620–6631. ACM, Denver (2017). https://doi.org/10.1145/3025453.3025727. https://dl.acm.org/doi/10.1145/3025453.3025727
148. Otterbacher, J., Checco, A., Demartini, G., Clough, P.: Investigating user perception of gender bias in image search: the role of sexism. In: The 41st International ACM SIGIR Conference on Research & Development in Information Retrieval, pp. 933–936. ACM, Ann Arbor (2018). https://doi.org/10.1145/3209978.3210094. https://dl.acm.org/doi/10.1145/3209978.3210094
149. Pan, B., Hembrooke, H., Joachims, T., Lorigo, L., Gay, G., Granka, L.. In google we trust: users' decisions on rank, position, and relevance. J. Comput.-Mediat. Commun. **12**(3), 801–823 (2007). https://doi.org/10.1111/j.1083-6101.2007.00351.x. https://academic.oup.com/jcmc/article/12/3/801-823/4582975
150. Pathiyan Cherumanal, S., Spina, D., Scholer, F., Croft, W.B.: Evaluating fairness in argument retrieval. In: Proceedings of the 30th ACM International Conference on Information & Knowledge Management, pp. 3363–3367 (2021)
151. Peels, R.: Responsible Belief: A Theory in Ethics and Epistemology. Oxford University Press, Oxford (2016)
152. Pennycook, G., Rand, D.G.: Lazy, not biased: susceptibility to partisan fake news is better explained by lack of reasoning than by motivated reasoning. Cognition **188**, 39–50 (2019). https://doi.org/10.1016/j.cognition.2018.06.011. https://linkinghub.elsevier.com/retrieve/pii/S001002771830163X
153. Phillips-Wren, G., Jefferson, T., McKniff, S.: Cognitive bias and decision aid use under stressful conditions. J. Decis. Syst. **28**(2), 162–184 (2019)
154. Pirkola, A.: The effectiveness of web search engines to index new sites from different countries. Inf. Res. Int. Electron. J. **14**(2) (2009)
155. Pogacar, F.A., Ghenai, A., Smucker, M.D., Clarke, C.L.: The positive and negative influence of search results on people's decisions about the efficacy of medical treatments. In: Proceedings of the ACM SIGIR International Conference on Theory of Information Retrieval, pp. 209–216. ACM, Amsterdam (2017). https://doi.org/10.1145/3121050.3121074. https://dl.acm.org/doi/10.1145/3121050.3121074
156. Porter, T., Elnakouri, A., Meyers, E.A., Shibayama, T., Jayawickreme, E., Grossmann, I.: Predictors and consequences of intellectual humility. Nat. Rev. Psychol. **1**, 524–536 (2022). https://doi.org/10.1038/s44159-022-00081-9
157. Pothirattanachaikul, S., Yamamoto, T., Yamamoto, Y., Yoshikawa, M.: Analyzing the effects of "People also ask" on search behaviors and beliefs. In: Proceedings of the 31st ACM Conference on Hypertext and Social Media, HT 2020, pp. 101–110. Association for Computing Machinery, New York (2020). https://doi.org/10.1145/3372923.3404786
158. Potthast, M., et al.: Argument search: assessing argument relevance. In: Proceedings of the 42nd International ACM SIGIR Conference on Research and Development in Information Retrieval, pp. 1117–1120 (2019)
159. Prem, E.: Our digital mirror. In: Perspectives on Digital Humanism, p. 89 (2022)
160. Purcell, K., Rainie, L., Brenner, J.: Search engine use 2012 (2012)

161. Puschmann, C.: Beyond the bubble: assessing the diversity of political search results. Dig. Journal. **7**(6), 824–843 (2019). https://doi.org/10.1080/21670811.2018.1539626. https://www.tandfonline.com/doi/full/10.1080/21670811.2018.1539626
162. Qiu, M., Jiang, J.: A latent variable model for viewpoint discovery from threaded forum posts. In: Proceedings of the Conference of the North American Chapter of the Association for Computational Linguistics-Human Language Technologies, pp. 1031–1040. Association for Computational Linguistics (2013). https://ink.library.smu.edu.sg/sis_research/1890/
163. Qiu, S., Gadiraju, U., Bozzon, A.: Improving worker engagement through conversational microtask crowdsourcing. In: Proceedings of the 2020 CHI Conference on Human Factors in Computing Systems, pp. 1–12 (2020)
164. Qiu, S., Gadiraju, U., Bozzon, A.: Towards memorable information retrieval. In: Proceedings of the 2020 ACM SIGIR on International Conference on Theory of Information Retrieval, pp. 69–76 (2020)
165. Radlinski, F., Craswell, N.: A theoretical framework for conversational search. In: Proceedings of the 2017 Conference on Conference Human Information Interaction and Retrieval, CHIIR 2017, pp. 117–126. Association for Computing Machinery, New York (2017). https://doi.org/10.1145/3020165.3020183
166. Ramos, J., Eickhoff, C.: Search result explanations improve efficiency and trust. In: Proceedings of the 43rd International ACM SIGIR Conference on Research and Development in Information Retrieval. pp. 1597–1600 (2020)
167. Reijula, S., Hertwig, R.: Self-nudging and the citizen choice architect. Behav. Public Policy **6**, 119–149 (2022). https://doi.org/10.1017/bpp.2020.5
168. Ren, P., Chen, Z., Ren, Z., Kanoulas, E., Monz, C., De Rijke, M.: Conversations with search engines: SERP-based conversational response generation. ACM Trans. Inf. Syst. **39**, 47:1–47:29 (2021). https://doi.org/10.1145/3432726
169. Reuver, M., et al.: Are we human, or are we users? the role of natural language processing in human-centric news recommenders that nudge users to diverse content. In: Proceedings of the 1st Workshop on NLP for Positive Impact, pp. 47–59 (2021)
170. Rezapour, R., Dinh, L., Diesner, J.: Incorporating the measurement of moral foundations theory into analyzing stances on controversial topics. In: Proceedings of the 32st ACM Conference on Hypertext and Social Media, pp. 177–188. ACM, Virtual Event USA (2021). https://doi.org/10.1145/3465336.3475112. https://dl.acm.org/doi/10.1145/3465336.3475112
171. Rieger, A., Bredius, F., Tintarev, N., Pera, M.S.: Searching for the whole truth: harnessing the power of intellectual humility to boost better search on debated topics. In: Extended Abstracts of the 2023 CHI Conference on Human Factors in Computing Systems, xCHI EA 2023, pp. 117–126. Association for Computing Machinery, New York (2023). https://doi.org/10.1145/3544549.3585693
172. Rieger, A., Draws, T., Theune, M., Tintarev, N.: This item might reinforce your opinion: Obfuscation and labeling of search results to mitigate confirmation bias. In: Proceedings of the 32nd ACM Conference on Hypertext and Social Media, pp. 189–199 (2021)
173. Rieger, A., Draws, T., Theune, M., Tintarev, N.: Nudges to mitigate confirmation bias during web search on debated topics: support vs. manipulation. ACM Trans. Web (2023). https://doi.org/10.1145/3635034
174. Robertson, R.E., Green, J., Ruck, D.J., Ognyanova, K., Wilson, C., Lazer, D.: Users choose to engage with more partisan news than they are exposed to on Google Search. Nature 1–7 (2023). https://doi.org/10.1038/s41586-023-06078-5

175. Roozenbeek, J., van der Linden, S.: Fake news game confers psychological resistance against online misinformation. Palgrave Commun. **5**, 1–10 (2019). https://doi.org/10.1057/s41599-019-0279-9
176. Salmerón, L., Kammerer, Y., García-Carrión, P.: Searching the web for conflicting topics: page and user factors. Comput. Hum. Behav. **29**(6), 2161–2171 (2013)
177. Santos, R.L.T., Macdonald, C., Ounis, I.: Search result diversification. Found. Trends^{TM} Inf. Retr. **9**(1), 1–90 (2015). https://doi.org/10.1561/1500000040. http://www.nowpublishers.com/article/Details/INR-040
178. Sapiezynski, P., Zeng, W., E Robertson, R., Mislove, A., Wilson, C.: Quantifying the impact of user attentionon fair group representation in ranked lists. In: Companion Proceedings of The 2019 World Wide Web Conference, pp. 553–562. ACM, San Francisco (2019). https://doi.org/10.1145/3308560.3317595. https://dl.acm.org/doi/10.1145/3308560.3317595
179. Schiaffonati, V., et al.: Explorative experiments and digital humanism: Adding an epistemic dimension to the ethical debate. In: Perspectives on Digital Humanism, p. 77 (2022)
180. Schmitt, J.B., Debbelt, C.A., Schneider, F.M.: Too much information? predictors of information overload in the context of online news exposure. Inf. Commun. Soc. **21**(8), 1151–1167 (2018). https://doi.org/10.1080/1369118X.2017.1305427. https://www.tandfonline.com/doi/full/10.1080/1369118X.2017.1305427
181. Shah, C., Bender, E.M.: Situating search. In: ACM SIGIR Conference on Human Information Interaction and Retrieval, CHIIR 2022, pp. 221–232. Association for Computing Machinery, New York (2022). https://doi.org/10.1145/3498366.3505816
182. Slechten, L., Courtois, C., Coenen, L., Zaman, B.: Adapting the selective exposure perspective to algorithmically governed platforms: the case of google search. Commun. Res. **49**, 1039–1065 (2022). https://doi.org/10.1177/00936502211012154
183. Smith, C.L., Rieh, S.Y.: Knowledge-context in search systems: toward information-literate actions. In: Proceedings of the 2019 Conference on Human Information Interaction and Retrieval, CHIIR 2019, pp. 55–62. Association for Computing Machinery, New York (2019). https://doi.org/10.1145/3295750.3298940
184. Springer, A., Garcia-Gathright, J., Cramer, H.: Assessing and addressing algorithmic bias-but before we get there... In: AAAI Spring Symposia (2018)
185. Sun, Q., Wang, Z., Zhu, Q., Zhou, G.: Stance detection with hierarchical attention network. In: Proceedings of the 27th International Conference on Computational Linguistics, pp. 2399–2409 (2018)
186. Sundin, O., Lewandowski, D., Haider, J.: Whose relevance? web search engines as multisided relevance machines. J. Am. Soc. Inf. Sci. **73**, 637–642 (2022). https://doi.org/10.1002/asi.24570
187. Swanson, R., Ecker, B., Walker, M.: Argument mining: extracting arguments from online dialogue. In: Proceedings of the 16th Annual Meeting of the Special Interest Group on Discourse and Dialogue, pp. 217–226. Association for Computational Linguistics, Prague (2015). https://doi.org/10.18653/v1/W15-4631. http://aclweb.org/anthology/W15-4631
188. Swar, B., Hameed, T., Reychav, I.: Information overload, psychological ill-being, and behavioral intention to continue online healthcare information search. Comput. Hum. Behav. **70**, 416–425 (2017)
189. Swire-Thompson, B., Lazer, D.: Public health and online misinformation: challenges and recommendations. Ann. Rev. Public Health **41**, 433–451 (2020). https://doi.org/10.1146/annurev-publhealth-040119-094127

190. Tavani, H.: Search engines and ethics. In: Zalta, E.N. (ed.) The Stanford Encyclopedia of Philosophy. Metaphysics Research Lab, Stanford University, fall 2020 edn. (2020). https://plato.stanford.edu/archives/fall2020/entries/ethics-search/
191. Thaler, R.H., Sunstein, C.R.: Nudge: The Final Edition. Yale University Press, New Haven (2021)
192. Thonet, T., Cabanac, G., Boughanem, M., Pinel-Sauvagnat, K.: users are known by the company they keep: topic models for viewpoint discovery in social networks. In: Proceedings of the 2017 ACM on Conference on Information and Knowledge Management, pp. 87–96. ACM, Singapore (2017). https://doi.org/10.1145/3132847.3132897. https://dl.acm.org/doi/10.1145/3132847.3132897
193. Toch, E., Wang, Y., Cranor, L.F.: Personalization and privacy: a survey of privacy risks and remedies in personalization-based systems. User Model. User-Adap. Inter. **22**, 203–220 (2012). https://doi.org/10.1007/s11257-011-9110-z
194. Trabelsi, A., Zaiane, O.R.: Finding arguing expressions of divergent viewpoints in online debates. In: Proceedings of the 5th Workshop on Language Analysis for Social Media (LASM), pp. 35–43. Association for Computational Linguistics, Gothenburg (2014). https://doi.org/10.3115/v1/W14-1305. http://aclweb.org/anthology/W14-1305
195. Trippas, J.R., Spina, D., Thomas, P., Sanderson, M., Joho, H., Cavedon, L.: Towards a model for spoken conversational search. Inf. Process. Manag. **57**, 102162 (2020). https://doi.org/10.1016/j.ipm.2019.102162
196. Tsfati, Y., Cappella, J.N.: Why do people watch news they do not trust? the need for cognition as a moderator in the association between news media skepticism and exposure. Media Psychol. **7**, 251–271 (2005)
197. Tversky, A., Kahneman, D.: Judgment under uncertainty: heuristics and biases: biases in judgments reveal some heuristics of thinking under uncertainty. Science **185**(4157), 1124–1131 (1974)
198. Valkenburg, P.M., Peter, J.: The differential susceptibility to media effects model. J. Commun. **63**(2), 221–243 (2013)
199. Van Couvering, E.: Is relevance relevant? market, science, and war: discourses of search engine quality. J. Comput.-Mediat. Commun. **12**, 866–887 (2007). https://doi.org/10.1111/j.1083-6101.2007.00354.x
200. Van Mierlo, T.: The 1% rule in four digital health social networks: an observational study. J. Med. Internet Res. **16**(2), e2966 (2014)
201. van Strien, J.L.H., Kammerer, Y., Brand-Gruwel, S., Boshuizen, H.P.A.: How attitude strength biases information processing and evaluation on the web. Comput. Hum. Behav. **60**, 245–252 (2016). https://doi.org/10.1016/j.chb.2016.02.057
202. Vaughan, L., Thelwall, M.: Search engine coverage bias: evidence and possible causes. Inf. Process. Manag. **40**(4), 693–707 (2004)
203. Vedejová, D., Čavojová, V.: Confirmation bias in information search, interpretation, and memory recall: evidence from reasoning about four controversial topics. Think. Reason. **28**(1), 1–28 (2022)
204. Vrijenhoek, S., Bénédict, G., Gutierrez Granada, M., Odijk, D., De Rijke, M.: Radio-rank-aware divergence metrics to measure normative diversity in news recommendations. In: Proceedings of the 16th ACM Conference on Recommender Systems, pp. 208–219 (2022)
205. Vrijenhoek, S., Kaya, M., Metoui, N., Möller, J., Odijk, D., Helberger, N.: Recommenders with a mission: assessing diversity in newsrecommendations. arXiv:2012.10185 [cs] (2020)

206. Wachsmuth, H., Syed, S., Stein, B.: Retrieval of the best counterargument without prior topic knowledge. In: Proceedings of the 56th Annual Meeting of the Association for Computational Linguistics, vol. 1: Long Papers, pp. 241–251 (2018)
207. Wang, R., Zhou, D., Jiang, M., Si, J., Yang, Y.: A survey on opinion mining: from stance to product aspect. IEEE Access **7**, 41101–41124 (2019). https://doi.org/10.1109/ACCESS.2019.2906754. https://ieeexplore.ieee.org/document/8672602/
208. Weber, I., Jaimes, A.: Who uses web search for what: and how. In: Proceedings of the Fourth ACM International Conference on Web Search and Data Mining, pp. 15–24 (2011)
209. Werthner, H., Prem, E., Lee, E.A., Ghezzi, C. (eds.): Perspectives on Digital Humanism. Springer, Heidelberg (2022). https://doi.org/10.1007/978-3-030-86144-5
210. White, R.: Beliefs and biases in web search. In: Proceedings of the 36th International ACM SIGIR Conference on Research and Development in Information Retrieval, pp. 3–12. ACM, Dublin (2013). https://doi.org/10.1145/2484028.2484053. https://dl.acm.org/doi/10.1145/2484028.2484053
211. White, R.W., Dumais, S.T., Teevan, J.: Characterizing the influence of domain expertise on web search behavior. In: Proceedings of the Second ACM International Conference on Web Search and Data Mining, WSDM 2009, pp. 132–141. Association for Computing Machinery, New York (2009). https://doi.org/10.1145/1498759.1498819
212. White, R.W., Hassan, A.: Content bias in online health search. ACM Trans. Web **8**(4), 1–33 (2014). https://doi.org/10.1145/2663355. https://dl.acm.org/doi/10.1145/2663355
213. White, R.W., Horvitz, E.: Belief dynamics and biases in web search. ACM Trans. Inf. Syst. **33**(4), 1–46 (2015). https://doi.org/10.1145/2746229. https://dl.acm.org/doi/10.1145/2746229
214. Winter, S.J., Butler, B.S.: Responsible technology design: conversations for success. In: Perspectives on Digital Humanism, pp. 271–275 (2022)
215. Wolfgang, J.D., Vos, T.P., Kelling, K., Shin, S.: Political journalism and democracy: how journalists reflect political viewpoint diversity in their reporting. Journal. Stud. **22**(10), 1339–1357 (2021)
216. Wu, Z., Draws, T., Cau, F., Barile, F., Rieger, A., Tintarev, N.: Explaining search result stances to opinionated people. In: Longo, L. (ed.) xAI 2023. Communications in Computer and Information Science, vol. 1902, pp. 573–596. Springer, Cham (2023). https://doi.org/10.1007/978-3-031-44067-0_29
217. Xu, L., Zhuang, M., Gadiraju, U.: How do user opinions influence their interaction with web search results? In: Proceedings of the 29th ACM Conference on User Modeling, Adaptation and Personalization, pp. 240–244 (2021)
218. Yang, K., Stoyanovich, J.: Measuring fairness in ranked outputs. In: Proceedings of the 29th International Conference on Scientific and Statistical Database Management, pp. 1–6. ACM, Chicago (2017). https://doi.org/10.1145/3085504.3085526. https://dl.acm.org/doi/10.1145/3085504.3085526
219. Yang, K., Stoyanovich, J., Asudeh, A., Howe, B., Jagadish, H., Miklau, G.: A nutritional label for rankings. In: Proceedings of the 2018 International Conference on Management of Data, pp. 1773–1776. ACM, Houston (2018). https://doi.org/10.1145/3183713.3193568. https://dl.acm.org/doi/10.1145/3183713.3193568
220. Yin, D., et al.: Ranking relevance in yahoo search. In: Proceedings of the 22nd ACM SIGKDD International Conference on Knowledge Discovery and Data Mining, KDD 2016, pp. 323–332. Association for Computing Machinery, New York (2016). https://doi.org/10.1145/2939672.2939677

221. Yom-Tov, E., Dumais, S., Guo, Q.: Promoting civil discourse through search engine diversity. Social Sci. Comput. Rev. **32**(2), 145–154 (2014). https://doi.org/10.1177/0894439313506838. http://journals.sagepub.com/doi/10.1177/0894439313506838
222. Yue, Y., Patel, R., Roehrig, H.: Beyond position bias: examining result attractiveness as a source of presentation bias in clickthrough data. In: Proceedings of the 19th International Conference on World Wide Web, pp. 1011–1018 (2010)
223. Zamani, H., Bendersky, M., Wang, X., Zhang, M.: Situational context for ranking in personal search. In: Proceedings of the 26th International Conference on World Wide Web, WWW 2017, pp. 1531–1540. International World Wide Web Conferences Steering Committee, Republic and Canton of Geneva, CHE (2017). https://doi.org/10.1145/3038912.3052648
224. Zehlike, M., Yang, K., Stoyanovich, J.: Fairness in ranking, part i: score-based ranking. ACM Comput. Surv. **55**(6), 1–36 (2022)
225. Zhang, D., Vakili Tahami, A., Abualsaud, M., Smucker, M.D.: Learning trustworthy web sources to derive correct answers and reduce health misinformation in search. In: Proceedings of the 45th International ACM SIGIR Conference on Research and Development in Information Retrieval, SIGIR 2022, pp. 2099–2104. Association for Computing Machinery, New York (2022). https://doi.org/10.1145/3477495.3531812
226. Zheng, K., Wang, H., Qi, Z., Li, J., Gao, H.: A survey of query result diversification. Knowl. Inf. Syst. **51**(1), 1–36 (2017)
227. Zhou, X., Zafarani, R.: A survey of fake news: fundamental theories, detection methods, and opportunities. ACM Comput. Surv. **53**, 109:1–109:40 (2020). https://doi.org/10.1145/3395046
228. Zhu, L., He, Y., Zhou, D.: Hierarchical viewpoint discovery from tweets using Bayesian modelling. Expert Syst. Appl. **116**, 430–438 (2019). https://doi.org/10.1016/j.eswa.2018.09.028. https://linkinghub.elsevier.com/retrieve/pii/S0957417418306055

The Impact of Differential Privacy on Recommendation Accuracy and Popularity Bias

Peter Müllner[1(✉)], Elisabeth Lex[2], Markus Schedl[3,4], and Dominik Kowald[1,2]

[1] Know-Center GmbH, Graz, Austria
{pmuellner,dkowald}@know-center.at
[2] Graz University of Technology, Graz, Austria
elisabeth.lex@tugraz.at
[3] Johannes Kepler University Linz, Linz, Austria
markus.schedl@jku.at
[4] Linz Institute of Technology, Linz, Austria

Abstract. Collaborative filtering-based recommender systems leverage vast amounts of behavioral user data, which poses severe privacy risks. Thus, often random noise is added to the data to ensure Differential Privacy (DP). However, to date, it is not well understood in which ways this impacts personalized recommendations. In this work, we study how DP affects recommendation accuracy and popularity bias when applied to the training data of state-of-the-art recommendation models. Our findings are three-fold: First, we observe that nearly all users' recommendations change when DP is applied. Second, recommendation accuracy drops substantially while recommended item popularity experiences a sharp increase, suggesting that popularity bias worsens. Finally, we find that DP exacerbates popularity bias more severely for users who prefer unpopular items than for users who prefer popular items.

Keywords: Recommender Systems · Collaborative Filtering · Differential Privacy · Accuracy · Popularity Bias

1 Introduction

Modern collaborative filtering-based recommender systems aim to generate personalized recommendations that cater to the specific preferences of each individual user. Such recommender systems need to provide recommendations of high accuracy and must ensure that the recommendations do not exhibit popularity bias, i.e., overestimate the relevance of popular items. For this, vast amounts of user data need to be processed, which exposes the users to many severe privacy risks [7,10,49,57], e.g., the disclosure of rating data [10] or the inference of sensitive user attributes [21,52]. Thus, besides recommendation accuracy and popularity bias, user privacy is another important aspect of recommender systems research. Hence, it is critical to leverage privacy-preserving techniques such as *Differential Privacy (DP)* [14] to devise privacy-aware recommender systems.

N. Goharian et al. (Eds.): ECIR 2024, LNCS 14611, pp. 466–482, 2024.
https://doi.org/10.1007/978-3-031-56066-8_33

Many mechanisms utilized to establish DP include the injection of random noise into the users' interaction data, which typically decreases the overall recommendation accuracy [8,59]. For recommender systems, some widely used mechanisms are the Gaussian or Laplacian Input Perturbation [14,19], Plausible Deniability [16,39], or the 1-Bit mechanism [11,13]. In particular, the 1-Bit mechanism is a natural match to the binary feedback data prevalent in modern recommender systems. This mechanism randomly substitutes parts of the positive feedback data with negative or missing feedback data, and then, this modified data is used to train the recommendation model. Specifically, the amount of positive feedback data that is randomly substituted depends on the privacy budget ϵ, i.e., a hyperparameter that controls how much random noise is incorporated into the recommendation process and what level of DP is achieved.

However, how DP impacts personalized recommendations is not well understood. Specifically, it remains unclear whether DP impacts the recommendations of all users, or just some users, and research on the connection between the ϵ value and the drop in recommendation accuracy is scarce. Also, how DP and the ϵ value impact the item popularity distribution in the respective recommendation lists and thus, popularity bias, is an open research topic. To shed light on these issues, in this work, we address the following three research questions:

- *How many users are impacted by DP? (RQ1)*
- *How does the privacy budget ϵ influence the accuracy drop? (RQ2)*
- *In which ways does DP impact popularity bias? (RQ3)*
  a. *How does DP impact the popularity distribution of the recommendations?*
  b. *How does DP impact popularity bias for different user groups?*

Accordingly, we perform experiments with a neural matrix-factorization model (i.e., *ENMF* [12]), a graph convolution network model (i.e., *LightGCN* [25]), and a variational autoencoder model (i.e., *MultVAE* [33]), and use datasets from the movie (i.e., *MovieLens 1M* [23]), music streaming (i.e., *LastFM User Groups* [29]), and online retail domain (i.e., *Amazon Grocery & Gourmet* [43]). Plus, we test various ϵ values to cater for different levels of privacy.

Our results show that nearly all users are impacted by DP, i.e., their recommendations are different from those generated without DP. Plus, this difference increases when ϵ becomes smaller (*RQ1*). With respect to recommendation accuracy, we find that DP leads to a substantial drop, which is most severe for small ϵ values. This highlights the trade-off between recommendation accuracy and privacy (*RQ2*). Similarly, we present strong evidence that DP increases popularity bias, in particular, when ϵ becomes smaller. This underlines an important trade-off between popularity bias and privacy. Moreover, we identify a "the poor get poorer" effect: DP increases popularity bias, especially for users that are already prone to strong popularity bias without DP, i.e., users that prefer unpopular items (cf. the unfairness of popularity bias [2,29]).

Overall, this work extends existing research on the trade-off between recommendation accuracy and privacy, and contributes novel insights on the connection between DP and popularity bias.

2 Related Work

Several previous works [7,10,49,57] identified many critical privacy risks for users in collaborative filtering-based recommender systems. For example, through the recommendations, the recommender system could leak user data to malicious parties [10,24,53], or an adversary could infer sensitive attributes of the user, e.g., gender [21,52,58]. To address these privacy risks, privacy-enhancing techniques, such as *Federated Learning* [34,36], *Homomorphic Encryption* [22,26], or *Differential Privacy (DP)* [14,35,39] need to be incorporated into the recommender system. However, Homomorphic Encryption has high computational complexity, and Federated Learning can still leak sensitive user information [42,45].

Therefore, in the past years, DP has emerged as a prominent choice in the recommender systems research community. However, one important shortcoming of DP is its negative impact on recommendation accuracy: DP typically leads to a substantial accuracy drop, since it incorporates random noise into the recommendation process [8,20]. Several works address this trade-off between recommendation accuracy and privacy by applying DP in different ways [20,41], e.g., by applying DP only to parts of the dataset [39,54], or by carefully tuning the degree of noise [59]. In detail, Zhu et al. [59] monitor how strong the item-to-item similarities would change if a piece of user data was not present. This way, they can better estimate what minimal level of random noise is necessary to ensure DP, and increase recommendation accuracy over comparable approaches. In a recommender system, there are typically a few users that are willing to openly share their data and many users that are less inclined to share their data. Xin and Jaakkola [54] exploit this, and protect only a subset of users with DP, which facilitates recommendation accuracy. Müllner et al. [39] attach to this, and modify the recommendation process of user-based KNN to minimize the number of users to which DP needs to be applied.

Besides privacy, another critical problem of recommender systems is popularity bias, i.e., the recommender system overestimates the relevance of popular items and therefore, popular items are overrepresented in the recommendations [2]. This can be regarded as disadvantaged, or "unfair" treatment of users that prefer unpopular items, since the recommendations do not match these users as well as users that prefer popular items. In theory, DP and fairness are closely connected to each other [15,56], since for both, a user's data needs to be hidden from the recommender system, either to preserve privacy, or to prohibit discrimination based on, e.g., age or gender. In practice, correlations in the dataset can still reveal age or gender and, therefore, lead to unfairness [6,17]. In this vein, several works [4,17,55] investigate the trade-off between fairness and privacy. For example, Sun et al. [50] use user data that is protected with DP to rectify the recommendations to increase fairness. Similarly, also Yang et al. [55] use post-processing to optimize for fairness with respect to recommendation accuracy. They observe that regarding recommendation accuracy, DP can lead to more unfairness; however, they do not address DP's impact on popularity bias.

Despite few existing works, how DP impacts personalized recommendations remains an understudied problem and many research gaps exist. For example, whether the recommendations of all users are impacted, or how beyond-accuracy objectives, such as reducing popularity bias or increasing diversity, are impacted. Thus, our work attaches to existing work with respect to studying DP's impact on recommendation accuracy, and in addition, we provide novel insights to DP's impact on the trade-off between privacy and popularity bias.

3 Method

In this section, we explain how DP is applied to the user data and then, we present multiple evaluation metrics to quantify DP's impact on recommendation accuracy and popularity bias. Also, we describe the datasets used in this study, and provide all preprocessing steps. Finally, we detail the experimental setup including the hyperparameters, recommendation models, and our precise evaluation protocol. We also provide our source-code to foster reproducibiltity.

3.1 Differential Privacy for Implicit Feedback

To ensure DP, we use the DP-mechanism from Ding et al. [13], which is a natural match to the binary implicit feedback data prevalent in today's recommender systems [11]. With this mechanism, for positive feedbacks $\mathcal{D}^+$ and negative or missing feedbacks $\mathcal{D}^-$ between users and items, the probability that the feedback $f_{u,i}$ between user u and item i is present in the DP dataset $\mathcal{D}^+_{\mathcal{DP}}$ is:

$$Pr[f_{u,i} \in \mathcal{D}^+_{DP}] = \begin{cases} \frac{e^\epsilon}{e^\epsilon+1} & \text{if } f_{u,i} \in \mathcal{D}^+ \\ \frac{1}{e^\epsilon+1} & \text{if } f_{u,i} \in \mathcal{D}^- \end{cases} \tag{1}$$

where ϵ is the privacy budget [14] (i.e., it quantifies how much privacy loss is tolerated; the higher, the less noise is added). In addition to positive feedback data, also negative or missing feedback data can be randomly added to $\mathcal{D}^+_{DP}$. However, the recommendation model is unable to identify these feedbacks and assumes that all feedbacks in $\mathcal{D}^+_{DP}$ are positive. By applying this mechanism to the training data of the recommendation model, the recommendations shall not leak information about the data that has been used in the recommendation process. For computational efficiency, we follow Chen et al. [11] and randomly sample one negative feedback for each positive feedback.

3.2 Evaluation Metrics

To identify users that are impacted by DP, we compute the Jaccard distance between a user u's recommendation list $\mathcal{R}(u)$ generated without DP and u's recommendation list $\mathcal{R}_{DP}(u)$ generated with DP applied to the training data. For the recommendation lists, we use the common cut-off of $n = 10$ items. Formally, the set of users impacted by DP (i.e., $U_{impacted}$) is given by:

$$U_{impacted} = \left\{ u \in U : 1 - \frac{|\mathcal{R}(u) \cap \mathcal{R}_{DP}(u)|}{|\mathcal{R}(u) \cup \mathcal{R}_{DP}(u)|} > 0 \right\} \tag{2}$$

where U is the set of all users. This means that we consider a user u as impacted, if DP leads to at least one different item in u's recommendation list (cf. [18]).

Overall, we quantify to what degree DP impacts recommendation accuracy and popularity bias of a user u's recommendations via measuring the relative change of an evaluation metric μ, when DP is applied (cf. [31,40]):

$$\text{Relative change } \Delta\mu(u) = \frac{\mu_{DP}(u) - \mu(u)}{\mu(u)} \tag{3}$$

$$\text{Average relative change } \Delta\mu = \frac{1}{|U_{impacted}|} \sum_{u \in U_{impacted}} \Delta\mu(u) \tag{4}$$

where $\mu_{DP}(u)$ is the value of the metric for user u when DP is applied and $\mu(u)$ is the value of the metric without applying DP. Furthermore, $\Delta\mu$ denotes the average change over all impacted users.

Recommendation Accuracy. To study the impact of DP on recommendation accuracy, we compute the ranking-agnostic *Recall* [44] metric. In this work, we use ranking-agnostic metrics since they fit to the way in which we identify impacted users, i.e., whether any item in the recommendation list changes due to DP, disregarding the ordering of the items within the recommendation list. We do not additionally include *Precision*, since $\Delta Recall = \Delta Precision$[1].

Popularity Bias. We evaluate DP's impact on popularity bias via measuring the *Average Recommendation Popularity (ARP)* [28], i.e., the average fraction of users that interacted with a recommended item:

$$ARP(u) = \frac{1}{|\mathcal{R}(u)|} \sum_{i \in \mathcal{R}(u)} \phi(i) \tag{5}$$

where $\mathcal{R}(u)$ is the recommendation list of user u, and item i's popularity $\phi(i) = |U_i|/|U|$ is the number of users that interacted with i, i.e., $|U_i|$, divided by the number of all users $|U|$. Several works suggest [2,29] that users that prefer unpopular items experience more popularity bias than users that prefer popular items. Thus, we use *Popularity Lift (PopLift)* [3] to quantify popularity bias for distinct user groups. Specifically, this metric indicates whether the *ARP* matches the average item popularity $\Gamma(\cdot)$ in the average user's profile of user group G:

$$PopLift(G) = \frac{\sum_{u \in G} ARP(u) - \sum_{u \in G} \Gamma(u)}{\sum_{u \in G} \Gamma(u)} \tag{6}$$

We inspect two user groups: users that prefer items of low popularity, i.e., U_{low}, and users that prefer items with high popularity, i.e., U_{high}. We follow Abdollahpouri et al. [2] and correspondingly define U_{low} as the set of the 20% of users with

[1] The number of recommended relevant items is divided by the number of all relevant items (i.e., *Recall*), or by the length of the recommendation list (i.e., *Precision*). When DP is applied, $\Delta Recall$ and $\Delta Precision$ only depend on how the number of recommended relevant items changes and therefore, the relative change is the same.

Table 1. Descriptive statistics of the three datasets. *Users* is the number of users, *Items* is the number of items, *Interactions* is the amount of interactions in the dataset, i.e., positive feedback, *Profile Size* is the average number of interactions per user, and *Density* is the inverse sparsity of the dataset in percent.

Dataset	Users	Items	Interactions	Profile Size	Density
MovieLens 1M	6,038	3,533	575,281	95.28	2.70%
LastFM User Groups	2,999	78,799	348,437	116.18	0.15%
Amazon Grocery & Gourmet	3,222	6,839	72,176	22.40	0.33%

the lowest fraction of popular items in their profile, and U_{high} as the set of the 20% of users whose profiles contain the highest fraction of popular items. The set of popular items is given by the 20% of items with the highest item popularity scores $\phi(i)$. In addition, we test whether there exists a *Disparate Impact* [37] of DP on U_{low} and U_{high}. Therefore, we measure the *Gap* [38], i.e., the absolute difference between the *PopLift* values of the two user groups:

$$Gap = |PopLift(U_{low}) - PopLift(U_{high})| \tag{7}$$

3.3 Datasets

For our experiments, we use three datasets, i.e., *MovieLens 1M* [23], *LastFM User Groups* [29], and *Amazon Grocery & Gourmet* [43] (see Table 1). *MovieLens 1M* and *Amazon Grocery & Gourmet* comprise rating scores in the range of 1 to 5, whereas *LastFM User Groups* comprises listening events between users and music artists [32,48]. For consistency and comparability, we follow [47] and sum the listening events per artist, followed by scaling the resulting scores to the range of 1 to 5. For *MovieLens 1M* and *LastFM User Groups*, we perform 20-core user pruning, followed by discarding scores below the respective mean value, i.e., 3.58 for *MovieLens 1M* and 1.13 for *LastFM User Groups*. We follow common practice [51], and regard all scores below this threshold, as well as missing scores, as negative feedback. For *Amazon Grocery & Gourmet*, we additionally perform 5-core item pruning before filtering the scores according to a threshold of 4.24.

3.4 Evaluation Protocol

We split each user's data into 60% training data used for model training, 20% validation data used for hyperparameter tuning, and 20% test data used for evaluation. After hyperparameter tuning (see Sect. 3.5), to research the impact of DP on personalized recommendations, we add DP to the training data (see Eq. 1) and retrain the recommendation models to calculate the evaluation metrics (see Sect. 3.2). Specifically, we retrain each model with five different random seeds and average the evaluation metrics to cope for random fluctuations in the training process. We repeat this procedure for multiple privacy budget values, i.e., $\epsilon \in$

Table 2. Model hyperparameters used in our experiments (learning rate α, dropout probability ρ, embedding dimensionality d, negative weight ω, L_2 regularization factor λ, number of propagation layers n, number of hidden units h).

Model	*MovieLens 1M*	*LastFM User Groups*	*Amazon Grocery & Gourmet*
ENMF	$\alpha = 0.01, \rho = 0.1, d = 32, \omega = 0.25$	$\alpha = 0.001, \rho = 0.25, d = 128, \omega = 0.25$	$\alpha = 0.001, \rho = 0.25, d = 64, \omega = 0.25$
LightGCN	$\alpha = 0.0001, d = 128, n = 1, \lambda = 0.0001$	$\alpha = 0.001, d = 128, n = 4, \lambda = 0.01$	$\alpha = 0.001, d = 128, n = 2, \lambda = 0.001$
MultVAE	$\alpha = 0.01, \rho = 0.5, d = 64, h = 800$	$\alpha = 0.001, \rho = 0.5, d = 128, h = 600$	$\alpha = 0.0001, \rho = 0.5, d = 128, h = 600$

$\{5, 4, 3, 2, 1, 0.1, 0.01\}$. To foster the reproducibility of our research, we publish our source code[2].

3.5 Recommendation Models and Parameter Settings

To cover different kinds of recommender systems, we experiment with a neural matrix-factorization model, i.e., *ENMF* [12], a graph convolution network model, i.e., *LightGCN* [25], and a variational autoencoder model, i.e., *MultVAE* [33].

- *ENMF* [12] is an efficient neural matrix-factorization model that does not leverage negative sampling. Instead, a negative weighting scheme is used, which benefits training efficiency and recommendation accuracy.
- *LightGCN* [25] is a lightweight graph convolution network, which, in contrast to more complex approaches, only uses neighborhood aggregation and does not include feature transformations or nonlinear activations.
- *MultVAE* [33] is a variational autoencoder that generates recommendations based on a multinomial likelihood. This way, it aims to mimic the generative process of implicit feedback data as prevalent in recommender systems.

For model training, we use Adam [27] to optimize the models for 5,000 epochs with a batch size of 4,096, and we employ an early stopping threshold of 50. We perform grid search for every model-dataset pair and determine the hyperparameters of the model with the highest *Recall* on the validation data (see Table 2). Note that hyperparameter tuning is performed on the original training data without DP. *LightGCN* requires negative samples and therefore, we sample one negative feedback for each positive feedback uniformly at random. After a careful inspection, we find that with the given hyperparameters, *LightGCN* cannot produce personalized recommendations for *Amazon Grocery & Gourmet*. To solve this, we manually adapt the learning rate to 0.001 and the batch size to 1,024. In all experiments, the top 10 ranked items are recommended to each user.

4 Results and Discussion

In this section, we present our results with respect to the three research questions. First, we measure for how many users the recommendations differ when DP is

[2] https://github.com/pmuellner/ImpactOfDP.

Table 3. Absolute values of the evaluation metrics when no DP is used. This serves as baseline for our results in the remainder of this paper, which measure the relative change of the evaluation metrics when DP is applied. For calculating the metrics, we use all impacted users.

Model	*MovieLens 1M*			*LastFM User Groups*			*Amazon Grocery & Gourmet*		
	Recall ↑	*ARP* ↓	*PopLift* ↓	*Recall* ↑	*ARP* ↓	*PopLift* ↓	*Recall* ↑	*ARP* ↓	*PopLift* ↓
ENMF	0.1697	0.2172	0.7084	0.0971	0.0836	1.8816	0.0932	0.0180	0.5143
LightGCN	0.1669	0.1958	0.5405	0.0925	0.0800	1.7585	0.0836	0.0259	1.1796
MultVAE	0.1694	0.1990	0.5657	0.0835	0.0576	0.9864	0.0643	0.0199	0.6734

applied, and we measure how strong these differences are (*RQ1*). Second, we detail these differences with respect to the relative change of recommendation accuracy (*RQ2*) and popularity bias (*RQ3a*). Plus, we investigate the impact of DP on popularity bias from the perspective of two user groups: users that prefer unpopular items and users that prefer popular items (*RQ3b*). As a baseline, Table 3 includes the absolute values of our evaluation metrics without DP.

4.1 Differences Between Recommendations

First, we approach *RQ1* and quantify how many users are impacted by DP (see Table 4). We find that for all datasets, recommendation models, and ϵ values, DP impacts nearly all users, i.e., different items are recommended than without DP. For these impacted users, the average difference, i.e., the Jaccard distance between the recommendations with and without DP, in most cases, lies above 0.5. Thus, on average, more than every second item in the recommendation list would not have been recommended without DP. Overall, the impact of DP increases when ϵ becomes smaller, i.e., when more noise is added to the training data of the recommendation models. Specifically, for $\epsilon = 0.1$ and across all recommendation models and datasets, more than 99.99% of users are impacted by DP, and the average Jaccard distance lies between 0.8058 and 0.9743.

This gives strong evidence that DP fundamentally impacts the generated recommendations for nearly all users (RQ1).

4.2 Impact on Recommendation Accuracy

Next, we build on our results from *RQ1*, and study how DP's impact on the recommendation lists affects recommendation accuracy (*RQ2*). We find that DP leads to a substantial drop in recommendation accuracy, as measured by *Recall* (see Fig. 1). In contrast to *MovieLens 1M* and *LastFM User Groups*, the recommendation accuracy for *Amazon Grocery & Gourmet* already drops in case $\epsilon = 5$, which is possibly due to DP being applied to the (on average) small user profiles in this dataset (see Table 1). In case of *ENMF* and *MultVAE*

Table 4. *No. Users* is the percentage of users that are impacted by DP and *Avg. J.* is the average Jaccard distance between the recommendations with and without DP. The worst results are given in **bold**. We find that nearly all users are impacted by DP and that the recommendations substantially differ from those generated without DP (*RQ1*).

ϵ	Model	*MovieLens 1M* No. Users ↓	*MovieLens 1M* Avg. J. ↓	*LastFM User Groups* No. Users ↓	*LastFM User Groups* Avg. J. ↓	*Amazon Grocery & Gourmet* No. Users ↓	*Amazon Grocery & Gourmet* Avg. J. ↓
5	*ENMF*	99.41%	0.5118	98.06%	0.4988	99.96%	0.7872
	LightGCN	97.40%	0.4207	99.14%	0.5112	99.94%	0.7382
	MultVAE	99.71%	0.5903	99.68%	0.6983	**100.00%**	0.9204
2	*ENMF*	99.85%	0.5974	99.64%	0.5757	**100.00%**	0.8620
	LightGCN	99.86%	0.6252	99.92%	0.6518	99.99%	0.8132
	MultVAE	99.93%	0.6828	**100.00%**	0.7950	**100.00%**	0.9447
1	*ENMF*	99.99%	0.7006	99.95%	0.6858	**100.00%**	0.9253
	LightGCN	**99.99%**	0.7352	99.99%	0.7464	**100.00%**	0.8775
	MultVAE	**100.00%**	0.7592	**100.00%**	0.8408	**100.00%**	0.9567
0.1	*ENMF*	**100.00%**	**0.8183**	**100.00%**	**0.8058**	**100.00%**	**0.9743**
	LightGCN	**99.99%**	**0.8300**	**100.00%**	**0.8490**	**100.00%**	**0.9360**
	MultVAE	**100.00%**	**0.8447**	**100.00%**	**0.9250**	**100.00%**	**0.9635**

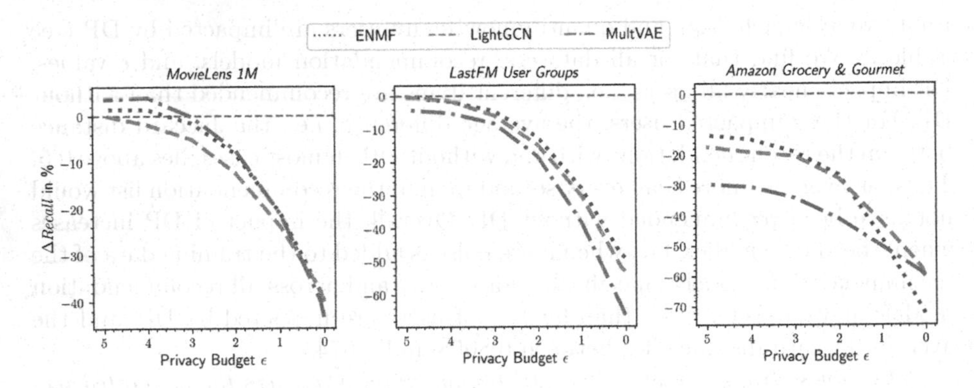

Fig. 1. DP's impact on recommendation accuracy as measured by *ΔRecall*. DP leads to a severe drop in recommendation accuracy. In particular, this drop becomes more serious for small ϵ values that provide a high level of privacy. This corresponds to the well-known accuracy-privacy trade-off (*RQ2*).

on *MovieLens 1M*, the recommendation accuracy increases slightly for large ϵ values, which can be possibly due to the fact that the noise introduced by DP acts as Tikhonov regularization for the model [9]. However, when more noise is added, i.e., $\epsilon < 3$, the recommendation accuracy for these models and dataset drops as well. Overall, the drop in recommendation accuracy gets worse when ϵ becomes smaller. Specifically, for $\epsilon = 0.1$ and across all recommendation models, the recommendation accuracy drops by at least 37.39% (*MovieLens 1M*), 48.00% (*LastFM User Groups*), or 57.10% (*Amazon Grocery & Gourmet*). Since lower ϵ values lead to higher levels of privacy, this corresponds to the well-known trade-off between recommendation accuracy and privacy [8,59].

In summary, DP leads to a substantial drop in recommendation accuracy, and this drop becomes more severe for smaller ϵ values (RQ2).

4.3 Impact on Popularity Bias

In this section, we study how DP impacts popularity bias (*RQ3*). First, in Fig. 2, we monitor how DP impacts the average recommendation popularity (*ARP*) and the popularity lift (*PopLift*). Then, we investigate DP's impact on popularity bias from the perspective of two user groups: users that prefer unpopular items and users that prefer popular items.

Impact on Recommendation Popularity. We find that DP leads to a substantial increase with respect to *ARP* (see Fig. 2a). Specifically, the increase in *ARP* gets worse, when ϵ becomes smaller. For example, for $\epsilon = 0.1$ and across all recommendation models, *ARP* increases by at least 19.75% (*MovieLens 1M*), 47.00% (*LastFM User Groups*), or 132.85% (*Amazon Grocery & Gourmet*). We investigate these differences in more detail, and find that the increase is especially high for datasets, for which the baseline value without DP is small (see Table 3). This means that without DP, also items of low popularity are recommended, which are typically hard to recommend (cf. the item cold-start problem [46]). With the noise introduced by DP, these items are even harder to recommend, which increases the *ARP* value. Thus, more popular items are recommended as ϵ becomes smaller, which indicates a trade-off between privacy and popularity bias. In adddition to *ARP*, we also use *PopLift* to quantify popularity bias, since it relates *ARP* to a user's preference for popular items (see Fig. 2b). As in case of *ARP*, also *PopLift* increases when the ϵ value becomes smaller, i.e., the popularity of the recommended items increasingly mismatches the item popularity distribution in the users' profiles. Specifically, for $\epsilon = 0.1$ and across all recommendation models, *PopLift* increases by at least 36.16% (*MovieLens 1M*), 28.49% (*LastFM User Groups*), or 128.38% (*Amazon Grocery & Gourmet*). This means that as ϵ becomes smaller, there is an increasing mismatch between the recommendation popularity and the item popularity distribution of the users.

Therefore, DP makes the recommendations more biased towards popular items, which strongly overestimates the users' preferences for popular items. This underlines the important trade-off between privacy and popularity bias (RQ3a).

Disparate Impact on User Groups. Building upon our finding that DP makes popularity bias worse, we finally investigate whether the strength of this

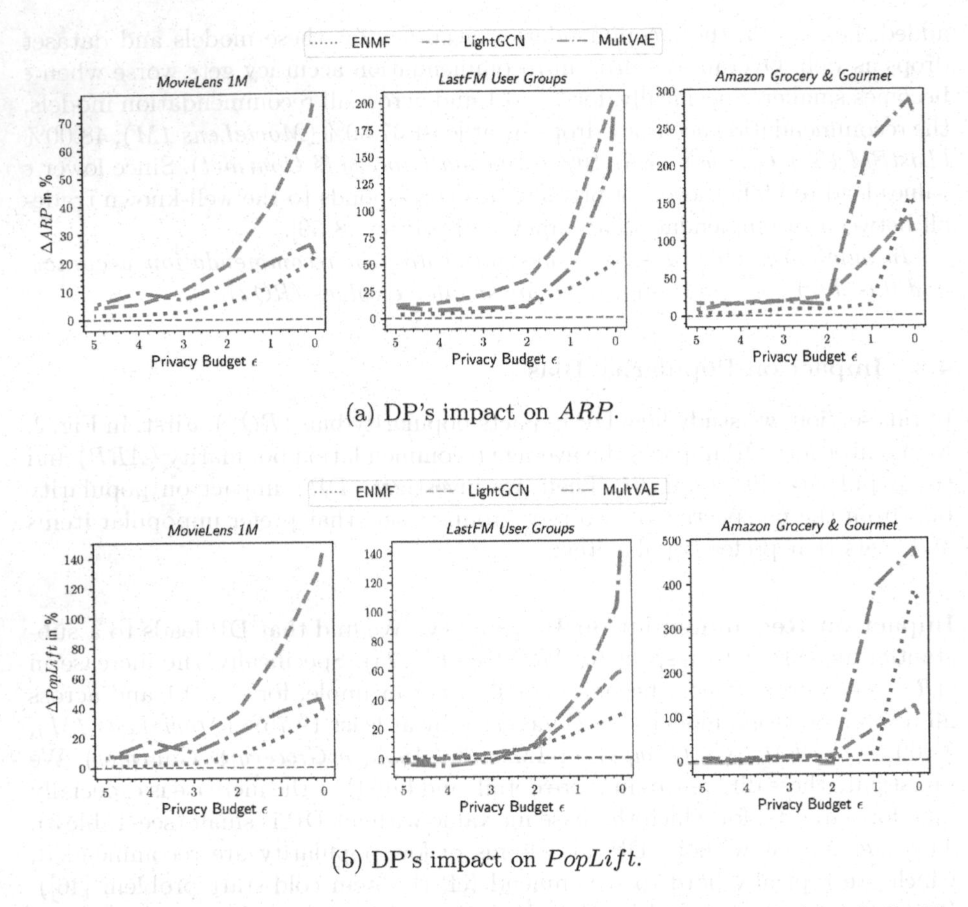

Fig. 2. DP's impact on popularity bias as measured by ΔARP and $\Delta PopLift$. We find that DP increases *ARP*, which becomes more severe the smaller the ϵ value is (see Fig. 2a). Plus, the recommendation popularity mismatches the item popularity distribution in the user profiles (see Fig. 2b). Overall, this gives strong evidence that DP makes popularity bias worse (*RQ3*).

effect differs between users that prefer popular items (i.e., U_{high}) and users that prefer unpopular items (i.e., U_{low}). For both user groups, *PopLift* increases for small ϵ values (see Table 5). Similarly, also the *Gap* between both user groups' *PopLift* values grows when ϵ becomes smaller, which suggests that there exists a disparate impact of DP (cf. [55]). We investigate *Gap* and *PopLift* in more detail and find that in general, *PopLift* increases more severely for U_{low} than for U_{high}. This can be regarded as a "poor get poorer" effect, since these disadvantaged users already experience strong popularity bias without DP. However, in case of *MultVAE* and *LastFM User Groups*, *PopLift* is higher for U_{high} than for

Table 5. The absolute *PopLift* values for users that prefer unpopular items (U_{low}) and users that prefer popular items (U_{high}), and the *Gap*, i.e., the absolute difference, between both. The worst results are given in **bold**. Popularity bias increases for both user groups with decreasing ϵ, but as *Gap* suggests, popularity bias increases especially for users that prefer unpopular, niche items (*RQ3b*).

		MovieLens 1M		*LastFM User Groups*		*Amazon Grocery & Gourmet*	
		PopLift ↓	*Gap* ↓	*PopLift* ↓	*Gap* ↓	*PopLift* ↓	*Gap* ↓
ϵ	Method	U_{low}/U_{high}		U_{low}/U_{high}		U_{low}/U_{high}	
No DP	*ENMF*	1.0923/0.4800	0.6124	4.1028/1.1578	2.9450	1.4079/0.0845	1.3235
	LightGCN	0.4225/0.5296	0.1072	2.7848/1.3273	1.4576	1.8253/0.7109	1.1144
	MultVAE	0.6247/0.4901	0.1347	0.7441/0.9092	0.1651	1.3910/0.2259	1.1650
5	*ENMF*	1.0940/0.4903	0.6037	4.0972/1.1629	2.9343	1.4043/0.0896	1.3147
	LightGCN	0.4625/0.5566	0.0940	2.7952/1.2790	1.5162	1.7539/0.6766	1.0773
	MultVAE	0.6538/0.5227	0.1311	0.7244/0.8928	0.1685	1.4372/0.2783	1.1589
2	*ENMF*	1.2088/0.5147	0.6941	4.5492/1.2334	3.3158	1.4977/0.1380	1.3597
	LightGCN	0.7447/0.6206	0.1241	3.1516/1.2894	1.8623	2.0951/0.7736	1.3215
	MultVAE	0.8409/0.5814	0.2595	0.1894/1.0524	0.8629	1.4175/0.2014	1.2161
1	*ENMF*	1.3658/0.5612	0.8046	5.2311/1.3309	3.9001	1.5723/0.1517	1.4206
	LightGCN	1.1633/0.7265	0.4368	3.8118/1.5267	2.2851	3.1031/1.2728	1.8303
	MultVAE	1.0044/0.6233	0.3811	0.3395/1.3139	**0.9744**	6.0433/2.0317	4.0117
0.1	*ENMF*	**1.5276/0.6445**	**0.8831**	**5.7217/1.4448**	**4.2769**	**4.9652/1.2375**	**3.7277**
	LightGCN	**1.7767/0.8415**	**0.9352**	**5.7233/1.6460**	**4.0773**	**4.5163/1.5600**	**2.9563**
	MultVAE	**1.1595/0.6370**	**0.5225**	**1.0760/1.7873**	0.7113	**7.5308/2.3216**	**5.2092**

U_{low}. It is known that for some datasets[3] *MultVAE* is able to recommend many unpopular items from the long-tail [5]. This results in lower *ARP* values than in case of the other datasets, i.e., 0.0184 for U_{low} and 0.0933 for U_{high} (without DP), which especially benefits U_{low}. Therefore, this helps to maintain a low *PopLift* value for U_{low}, and may explain why in this specific case, *PopLift* is lower for U_{low} than for U_{high}.

Overall, DP makes popularity bias worse for both user groups, but most severely for users that prefer unpopular items (RQ3b).

4.4 Discussion

Overall, DP impacts nearly all users (*RQ1*) and leads to reduced recommendation accuracy (*RQ2*) and increased popularity bias (*RQ3*). Plus, especially users that prefer unpopular items experience a sharp increase in popularity bias.

However, the impact of DP strongly depends on the level of privacy that shall be ensured, i.e., the ϵ value. This suggests that carefully choosing ϵ is essential in balancing the trade-off between privacy, accuracy, and popularity bias (*RQ2*, *RQ3*). Moreover, DP's impact on popularity bias is especially severe for users

[3] No clear pattern across datasets can be observed [5] and thus, this behavior of *MultVAE* needs to be researched in the future.

that prefer unpopular items (*RQ3b*). Thus, the trade-off between privacy, recommendation accuracy, and popularity bias can differ between groups of user, which underlines that group-specific mitigation strategies may be required. We hope that our results can inform research in this direction.

5 Conclusion and Future Work

In this work, we investigated in which ways Differential Privacy (DP) impacts personalized recommendations. In experiments with three datasets and three recommendation algorithms, we added DP to the training data of state-of-the-art recommendation models, and found that nearly all users' recommendations change when DP is applied. Also, for higher levels of privacy, recommendation accuracy drops substantially while popularity bias increases. In addition, we detail DP's impact on popularity bias and observe a "poor get poorer" effect: DP exacerbates popularity bias more severely for users who already experience strong popularity bias without DP, i.e., users who prefer unpopular items. Overall, our work further researches the trade-off between recommendation accuracy and privacy and, in addition, provides novel insights on the important trade-off between popularity bias and privacy.

Future Work. In the future, we plan to research how users that are especially disadvantaged by DP, i.e., users that prefer unpopular items, can reach a satisfactory trade-off between recommendation accuracy, popularity bias, and privacy. Specifically, we aim to test whether popularity bias mitigation strategies can help to prohibit the exacerbation of popularity bias for disadvantaged user groups. One limitation of this work is that we investigated the impact of DP only on the users, but not on other stakeholders of the recommender system. Thus, we plan to investigate the impact of DP also on providers and creators of items [1]. Additionally, we aim to evaluate DP-based recommendations also in more privacy-sensitive domains such as job recommendations [30].

Acknowledgments. This research is funded by the "DDAI" COMET Module within the COMET - Competence Centers for Excellent Technologies Programme, funded by the Austrian Federal Ministry for Transport, Innovation and Technology (bmvit), the Austrian Federal Ministry for Digital and Economic Affairs (bmdw), the Austrian Research Promotion Agency (FFG), the province of Styria (SFG) and partners from industry and academia. The COMET Programme is managed by FFG. Moreover, this research received support by the Austrian Science Fund (FWF): DFH-23 and P36413; and by the State of Upper Austria and the Federal Ministry of Education, Science, and Research, through grants LIT-2020-9-SEE-113 and LIT-2021-10-YOU-215. For open access purposes, the author has applied a CC BY public copyright license to any author accepted manuscript version arising from this submission.

References

1. Abdollahpouri, H., et al.: Multistakeholder recommendation: survey and research directions. User Model. User-Adap. Inter. **30**, 127–158 (2020)
2. Abdollahpouri, H., Mansoury, M., Burke, R., Mobasher, B.: The unfairness of popularity bias in recommendation. In: Workshop on Recommendation in Multi-stakeholder Environments (RMSE), in Conjunction With the 13th ACM Conference on Recommender Systems (RecSys) (2019)
3. Abdollahpouri, H., Mansoury, M., Burke, R., Mobasher, B.: The connection between popularity bias, calibration, and fairness in recommendation. In: Proceedings of the 14th ACM Conference on Recommender Systems (RecSys), pp. 726–731 (2020)
4. Agarwal, S.: Trade-offs between fairness, interpretability, and privacy in machine learning. Master's thesis, University of Waterloo (2020)
5. Anelli, V.W., Bellogín, A., Di Noia, T., Jannach, D., Pomo, C.: Top-n recommendation algorithms: a quest for the state-of-the-art. In: Proceedings of the 30th ACM Conference on User Modeling, Adaptation and Personalization (UMAP), pp. 121–131 (2022)
6. Bagdasaryan, E., Poursaeed, O., Shmatikov, V.: Differential privacy has disparate impact on model accuracy. In: Proceedings of the 33rd International Conference on Neural Information Processing Systems (NeurIPS), pp. 15479–15488 (2019)
7. Beigi, G., Liu, H.: A survey on privacy in social media: identification, mitigation, and applications. ACM Trans. Data Sci. (TDS) **1**(1), 1–38 (2020)
8. Berkovsky, S., Kuflik, T., Ricci, F.: The impact of data obfuscation on the accuracy of collaborative filtering. Expert Syst. Appl. **39**(5), 5033–5042 (2012)
9. Bishop, C.M.: Training with noise is equivalent to tikhonov regularization. Neural Comput. **7**(1), 108–116 (1995)
10. Calandrino, J.A., Kilzer, A., Narayanan, A., Felten, E.W., Shmatikov, V.: "you might also like:" privacy risks of collaborative filtering. In: 2011 IEEE Symposium on Security and Privacy (S&P), pp. 231–246 (2011)
11. Chen, C., Zhou, J., Wu, B., Fang, W., Wang, L., Qi, Y., Zheng, X.: Practical privacy preserving poi recommendation. ACM Trans. Intell. Syst. Technol. (TIST) **11**(5), 1–20 (2020)
12. Chen, C., Zhang, M., Zhang, Y., Liu, Y., Ma, S.: Efficient neural matrix factorization without sampling for recommendation. ACM Trans. Inf. Syst. (TOIS) **38**(2), 1–28 (2020)
13. Ding, B., Kulkarni, J., Yekhanin, S.: Collecting telemetry data privately. In: Proceedings of the 31st International Conference on Neural Information Processing Systems (NeurIPS), pp. 3574–3583 (2017)
14. Dwork, C.: Differential privacy: a survey of results. In: International conference on Theory and Applications of Models of Computation (TAMC), pp. 1–19 (2008)
15. Dwork, C., Hardt, M., Pitassi, T., Reingold, O., Zemel, R.: Fairness through awareness. In: Proceedings of the 3rd Innovations in Theoretical Computer Science Conference (ITCS), pp. 214–226 (2012)
16. Dwork, C., Roth, A., et al.: The algorithmic foundations of differential privacy. Now Publishers, Inc. (2014)
17. Ekstrand, M.D., Joshaghani, R., Mehrpouyan, H.: Privacy for all: ensuring fair and equitable privacy protections. In: Proceedings of ACM Conference on Fairness, Accountability, and Transparency (FAccT), pp. 35–47 (2018)

18. Eskandanian, F., Sonboli, N., Mobasher, B.: Power of the few: analyzing the impact of influential users in collaborative recommender systems. In: Proceedings of the 27th ACM Conference on User Modeling, Adaptation and Personalization, pp. 225–233 (2019)
19. Friedman, A., Berkovsky, S., Kaafar, M.A.: A differential privacy framework for matrix factorization recommender systems. User Model. User-Adapt. Interact. (UMUAI) **26**(5), 425–458 (2016)
20. Friedman, A., Knijnenburg, B.P., Vanhecke, K., Martens, L., Berkovsky, S.: Privacy aspects of recommender systems. In: Ricci, F., Rokach, L., Shapira, B. (eds.) Recommender Systems Handbook, pp. 649–688. Springer, Boston, MA (2015). https://doi.org/10.1007/978-1-4899-7637-6_19
21. Ganhör, C., Penz, D., Rekabsaz, N., Lesota, O., Schedl, M.: Unlearning protected user attributes in recommendations with adversarial training. In: Proceedings of the 45th International ACM SIGIR Conference on Research and Development in Information Retrieval (SIGIR), pp. 2142–2147. Springer, Heidelberg (2022)
22. Gentry, C.: A fully homomorphic encryption scheme. Ph.D. thesis, Stanford university (2009)
23. Harper, F.M., Konstan, J.A.: The movielens datasets: history and context. ACM Trans. Interact. Intell. Syst. (TiiS) **5**(4), 1–19 (2015)
24. Hashemi, H., et al.: Data leakage via access patterns of sparse features in deep learning-based recommendation systems. Workshop on Trustworthy and Socially Responsible Machine Learning (TSRML), in Conjunction with the 36th Conference on Neural Information Processing Systems (NeurIPS) (2022)
25. He, X., Deng, K., Wang, X., Li, Y., Zhang, Y., Wang, M.: Lightgcn: simplifying and powering graph convolution network for recommendation. In: Proceedings of the 43rd International ACM SIGIR Conference on Research and Development in Information Retrieval (SIGIR), pp. 639–648. Springer, Heidelberg (2020)
26. Kim, S., Kim, J., Koo, D., Kim, Y., Yoon, H., Shin, J.: Efficient privacy-preserving matrix factorization via fully homomorphic encryption. In: Proceedings of the 11th ACM on Asia Conference on Computer and Communications Security (ASIACCS), pp. 617–628 (2016)
27. Kingma, D.P., Ba, J.: Adam: a method for stochastic optimization. In: Proceedings of 3rd International Conference on Learning Representations (ICLR) (2015)
28. Klimashevskaia, A., Elahi, M., Jannach, D., Trattner, C., Skjærven, L.: Mitigating popularity bias in recommendation: potential and limits of calibration approaches. In: Advances in Information Retrieval: Workshop on Algorithmic Bias in Search and Recommendation (BIAS) in conjunction with the 42nd European Conference on IR Research (ECIR), pp. 82–90. Springer, Heidelberg (2022). https://doi.org/10.1007/978-3-031-09316-6_8
29. Kowald, D., Schedl, M., Lex, E.: The unfairness of popularity bias in music recommendation: a reproducibility study. In: Jose, J.M., et al. (eds.) ECIR 2020. LNCS, vol. 12036, pp. 35–42. Springer, Cham (2020). https://doi.org/10.1007/978-3-030-45442-5_5
30. Lacic, E., Reiter-Haas, M., Kowald, D., Reddy Dareddy, M., Cho, J., Lex, E.: Using autoencoders for session-based job recommendations. User Model. User-Adap. Inter. **30**, 617–658 (2020)
31. Lesota, O., et al.: Analyzing item popularity bias of music recommender systems: are different genders equally affected? In: Proceedings of the 15th ACM Conference on Recommender Systems (RecSys), pp. 601–606 (2021)
32. Lex, E., Kowald, D., Schedl, M.: Modeling popularity and temporal drift of music genre preferences. Trans. Int. Soc. Music Inf. Retr. **3**(1) (2020)

33. Liang, D., Krishnan, R.G., Hoffman, M.D., Jebara, T.: Variational autoencoders for collaborative filtering. In: Proceedings of the World Wide Web Conference (TheWebConf), pp. 689–698 (2018)
34. Lin, Y., et al.: Meta matrix factorization for federated rating predictions. In: Proceedings of the 43rd International ACM SIGIR Conference on Research and Development in Information Retrieval (SIGIR), pp. 981–990. Springer, Heidelberg (2020)
35. Long, J., Chen, T., Nguyen, Q.V.H., Yin, H.: Decentralized collaborative learning framework for next poi recommendation. ACM Trans. Inf. Syst. **41**(3) (2023). https://doi.org/10.1145/3555374
36. McMahan, B., Moore, E., Ramage, D., Hampson, S., Arcas, B.A.: Communication-efficient learning of deep networks from decentralized data. In: Proceedings of the 20th International Conference on Artificial Intelligence and Statistics (AISTATS), pp. 1273–1282 (2017)
37. Mehrabi, N., Morstatter, F., Saxena, N., Lerman, K., Galstyan, A.: A survey on bias and fairness in machine learning. ACM Comput. Surv. (CSUR) **54**(6), 1–35 (2021)
38. Melchiorre, A.B., Rekabsaz, N., Parada-Cabaleiro, E., Brandl, S., Lesota, O., Schedl, M.: Investigating gender fairness of recommendation algorithms in the music domain. Inf. Process. Manag. (IP&P) **58**(5), 102666 (2021)
39. Müllner, P., Lex, E., Schedl, M., Kowald, D.: Reuseknn: neighborhood reuse for differentially-private knn-based recommendations. ACM Trans. Intell. Syst. Technol. (2023). https://doi.org/10.1145/3608481
40. Muellner, P., Kowald, D., Lex, E.: Robustness of meta matrix factorization against strict privacy constraints. In: Hiemstra, D., Moens, M.-F., Mothe, J., Perego, R., Potthast, M., Sebastiani, F. (eds.) ECIR 2021. LNCS, vol. 12657, pp. 107–119. Springer, Cham (2021). https://doi.org/10.1007/978-3-030-72240-1_8
41. Müllner, P., Lex, E., Schedl, M., Kowald, D.: Differential privacy in collaborative filtering recommender systems: a review. Front. Big Data **6** (2023). https://doi.org/10.3389/fdata.2023.1249997
42. Nasr, M., Shokri, R., Houmansadr, A.: Comprehensive privacy analysis of deep learning: Passive and active white-box inference attacks against centralized and federated learning. In: Proceedings of the IEEE Symposium on Security and Privacy (S&P), pp. 739–753 (2019)
43. Ni, J., Li, J., McAuley, J.: Justifying recommendations using distantly-labeled reviews and fine-grained aspects. In: Proceedings of the Conference on Empirical Methods in Natural Language Processing and the 9th International Joint Conference on Natural Language Processing (EMNLP-IJCNLP), pp. 188–197 (2019)
44. Parra, D., Sahebi, S.: Recommender systems: sources of knowledge and evaluation metrics. In: Advanced Techniques in Web Intelligence-2: Web User Browsing Behaviour and Preference Analysis, pp. 149–175. Springer, Heidelberg (2013). https://doi.org/10.1007/978-3-642-33326-2_7
45. Ren, H., Deng, J., Xie, X.: GRNN: generative regression neural network-a data leakage attack for federated learning. ACM Trans. Intell. Syst. Technol. (TIST) **13**(4), 1–24 (2022)
46. Saveski, M., Mantrach, A.: Item cold-start recommendations: learning local collective embeddings. In: Proceedings of the 8th ACM Conference on Recommender systems (RecSys), pp. 89–96 (2014)
47. Schedl, M., Bauer, C.: Distance-and rank-based music mainstreaminess measurement. In: Adjunct Publication of the 25th Conference on User Modeling, Adaptation and Personalization (UMAP): Workshop on Surprise, Opposition, and Obstruction in Adaptive and Personalized Systems (SOAP), pp. 364–367 (2017)

48. Schedl, M., Bauer, C., Reisinger, W., Kowald, D., Lex, E.: Listener modeling and context-aware music recommendation based on country archetypes. Front. Artif. Intell. **3**, 508725 (2021)
49. Lam, S.K., Frankowski, D., Riedl, J.: Do you trust your recommendations? an exploration of security and privacy issues in recommender systems. In: Müller, G. (ed.) ETRICS 2006. LNCS, vol. 3995, pp. 14–29. Springer, Heidelberg (2006). https://doi.org/10.1007/11766155_2
50. Sun, J.A., Pentyala, S., Cock, M.D., Farnadi, G.: Privacy-preserving fair item ranking. In: Kamps, J., et al. (eds.) ECIR 2023, vol. 13981, pp. 188–203. Springer, Heidelberg (2023). https://doi.org/10.1007/978-3-031-28238-6_13
51. Sun, Z., et al.: Are we evaluating rigorously? benchmarking recommendation for reproducible evaluation and fair comparison. In: Proceedings of the 14th ACM Conference on Recommender Systems (RecSys), pp. 23–32 (2020)
52. Weinsberg, U., Bhagat, S., Ioannidis, S., Taft, N.: Blurme: inferring and obfuscating user gender based on ratings. In: Proceedings of the 6th ACM Conference on Recommender Systems (RecSys), pp. 195–202 (2012)
53. Xin, X., et al.: On the user behavior leakage from recommender system exposure. ACM Trans. Inf. Syst. (TOIS) **41**(3), 1–25 (2023)
54. Xin, Y., Jaakkola, T.: Controlling privacy in recommender systems. In: Proceedings of the 27th International Conference on Neural Information Processing Systems (NeurIPS), pp. 2618–2626. MIT Press, Cambridge (2014)
55. Yang, Z., Ge, Y., Su, C., Wang, D., Zhao, X., Ying, Y.: Fairness-aware differentially private collaborative filtering. In: Companion Proceedings of the ACM Web Conference (TheWebConf), pp. 927–931 (2023)
56. Zemel, R., Wu, Y., Swersky, K., Pitassi, T., Dwork, C.: Learning fair representations. In: International conference on machine learning (ICML), pp. 325–333 (2013)
57. Zhang, M., et al.: Membership inference attacks against recommender systems. In: Proceedings of the ACM SIGSAC Conference on Computer and Communications Security (CCS), pp. 864–879 (2021)
58. Zhang, S., Yin, H.: Comprehensive privacy analysis on federated recommender system against attribute inference attacks. IEEE Trans. Knowl. Data Eng. (TKDE) (2023)
59. Zhu, T., Li, G., Ren, Y., Zhou, W., Xiong, P.: Differential privacy for neighborhood-based collaborative filtering. In: Proceedings of the IEEE/ACM International Conference on Advances in Social Networks Analysis and Mining (ASONAM), pp. 752–759 (2013)

Author Index

N. Goharian et al. (Eds.): ECIR 2024, LNCS 14611, pp. 483–484, 2024.
https://doi.org/10.1007/978-3-031-56066-8

Printed in the United States
by Baker & Taylor Publisher Services